The Insect Ovary

The Insect Ovary

Ultrastructure, previtellogenic growth and evolution

Jürgen Büning
Professor of Zoology
University of Erlangen-Nürnberg
Germany

CHAPMAN & HALL
London · Glasgow · Weinheim · New York · Tokyo · Melbourne · Madras

Published by Chapman & Hall, 2–6 Boundary Row, London SE1 8HN, UK

Chapman & Hall, 2–6 Boundary Row, London SE1 8HN, UK

Blackie Academic & Professional, Wester Cleddens Road, Bishopbriggs, Glasgow G64 2NZ, UK

Chapman & Hall GmbH, Papelallee 3, 69469 Weinheim, Germany

Chapman & Hall Inc., One Penn Plaza, 41st Floor, New York NY 10119, USA

Chapman & Hall Japan, Thomson Publishing Japan, Hirakawacho Nemoto Building, 6F, 1–7–11 Hirakawa-cho, Chiyoda-ku, Tokyo 102, Japan

Chapman & Hall Australia, Thomas Nelson Australia, 102 Dodds Street, South Melbourne, Victoria 3205, Australia

Chapman & Hall India, R. Seshadri, 32 Second Main Road, CIT East, Madras 600 035, India

First edition 1994

Typeset in 10/12 pt Palatino by ROM-Data Corporation, Falmouth, Cornwall

Printed in Great Britain at the University Press, Cambridge

ISBN 0 412 36080 2

A catalogue record for this book is available from the British Library

Library of Congress Catalog Card Number 93-74897

∞ Printed on permanent acid-free text paper, manufactured in accordance with ANSI/NISO Z39.48–1992 and ANSI/NISO Z39.48–1994 (Permanence of Paper).

In memory of Karlheinz Bier who introduced me to the wonder world of developmental biology.

Contents

Preface

This book will give an overview of insect ovaries, showing the diversities and the common traits in egg growth processes. The idea to write this book developed while looking at the flood of information which appeared in the early 1980s on early pattern formation in *Drosophila* embryos. At this time a significant breakthrough was made in studies of this little fly, combining molecular biological methods with classical and molecular genetics. The answers to questions about early pattern formation raised new questions about the architecture of ovaries and the growth of eggs within these ovaries. However, by concentrating only on *Drosophila* it is not possible to form an adequate picture of what is going on in insect ovaries, since the enormous diversity found among insects is not considered sufficiently.

Almost forgotten, but the first to study the architecture of ovaries, was Alexander Brandt writing in 1878 in *Über das Ei und seine Bildungsstaette* (On the egg and its organ of development). More than 100 years later, a series of ten books or more would be required to survey all the serious information we have today on insect oogenesis. Thus, this book is a personal selection and personal view on the theme, and the authors must be excused by all those scientists whose papers could not be included.

The book briefly describes the ectodemes, i.e. the ectodermal parts of insect ovaries and focuses on the germ cells, their differentiation and interconnections to mesodermal tissues of the ovary. However themes such as vitellogenesis and chorionogenesis are cut short in favour of previtellogenic growth and phylogeny of insect ovarial types. The book is written for scientists and students whose interests lie in developmental biology in the widest sense, as well as for those interested in phylogeny.

My special thanks go to Szczepan Bilinski, who not only wrote the section on the entognathan ovary and the notes to the accessory nuclei found in growing insect eggs, but who encourages me to go on with the whole story even in bad times. Thanks to my wife Siegrun for her continuous and never ending patience. Thanks to all my colleagues for helpful discussions and for sending illustrations so promptly. I express my special thanks to my students and to the students of my colleague Szczepan Bilinski; without their work this book would never have been written. Furthermore, I would like to thank the editorial staff at Chapman & Hall, especially those who edited my poor English into a readable form. Thanks to them all.

1

Introduction

1.1 WHY WE ARE INTERESTED IN INSECT REPRODUCTION

One of the fundamental events during lifetime is founding the next generation. To gain optimum adaptability to changes in the environment within the next generation, sexual reproduction is found in all higher animal populations. When life goes through sexual reproduction a very serious and dangerous phase has to be passed especially for metazoans, since life will be reduced to single, haploid cells, the gametes, which are the basis of genetic exchange. The 'coming together' of sexual cells, fertilization and the fusion of gametes forming the zygote, as well as the differentiation into the new individual can not be understood without knowing how sexual cells are built and what potentials they have. This book will support this understanding and will present some aspects of the ultrastructure, genetics and physiology of oocytes during growth as well as during the early events of ovarian development which determine the different ovariole types we will find among insects.

Why are we interested in insects? Many reasons could be stated including their short generation time and their high reproductive rates which make genetic analysis possible. Another reason, a very practical one, is the relatively low costs of insect cultures. Furthermore, it is of general interest to elucidate reproduction strategies in a group to which two-thirds of all living species belong. Last, but not least, a variety of insect species are pests of cultured plants and these cereals, fruits and vegetables are the basic food of mankind. Thus, detailed knowledge of insect reproduction is needed for biological pest control, in order to maintain our environment for future generations.

1.2 A SHORT HISTORICAL REVIEW

Our interrelation with insects is conflicting. On the one hand we admire insects, for instance beautiful coloured butterflies or some, mostly tropical, bugs and beetles. On the other hand, most people have an irrational fear of insects, possibly on account of their strange appearance and behaviour which does not fit into our human or mammalian pattern. This ambivalence is very old and is shown in historical times, for instance in the symbol of the Scarabaeus, a dung beetle which was adored in old Egypt as a God for recreation or regeneration. People thought the dung pill would change directly into life by some irrational forces of the beetle, fussily rolling the dung pill. A more negatively attributed report from the same historical background is that of the seven plagues described in the Bible. One of the plagues was giant swarms of grasshoppers, a crop pest, still feared today. The rise of this pest is grounded in its enormous fecundity, coupled with a short generation cycle. These most admirable, but to men harmful, properties of grasshoppers and of many other insects, can produce, under certain environmental conditions, these fearful plagues of past and present.

Scientific reports on insect life cycles and the anatomy of males and females is grounded in the seventeenth century. It was the famous M. Malphigi, who first illustrated the insect ovary as long tubes in 1669, analysing the inner organs of the silk moth, *Bombyx mori* (Figure 1.1).

In the eighteenth century we remember the work of J. Swammerdam; he described the ovaries of bees, wasps and lice in his famous book *Bybel der Natuure* in 1737. He was the first to discover the end-standing ligament structure of the insect ovary. The first great breakthrough in elucidating the morphology of insect ovaries is coupled strongly with the rise of light microscopy and its methodical applications during the last century. During the first decades all inner organs were described using freshly made whole mount preparations. The objects were prepared on slides in a drop of tap water, fresh albumen or insect haemolymph to provide desiccation. A coverslip on top allowed better optical brightness. The microscope itself was rather primitive, compared with light microscopes today. Diffuse daylight was used to penetrate object and lenses and only the bright-field method was known.

Of the hundreds of articles written on insect oogenesis in the past century we want to stress only a few, which all share the germ plasm–germ line theory formulated by Weismann in 1885 and 1892.

In 1825, Purkinje published his observations on hens' eggs and he described the 'germinal vesicles' for the first time. Some years later, in 1827, von Baer showed that the germinal vesicles found in mammals, including man, are the common entities which are equivalent to the germinal vesicles of hens' eggs. This generalization was extended to insect eggs predominantly

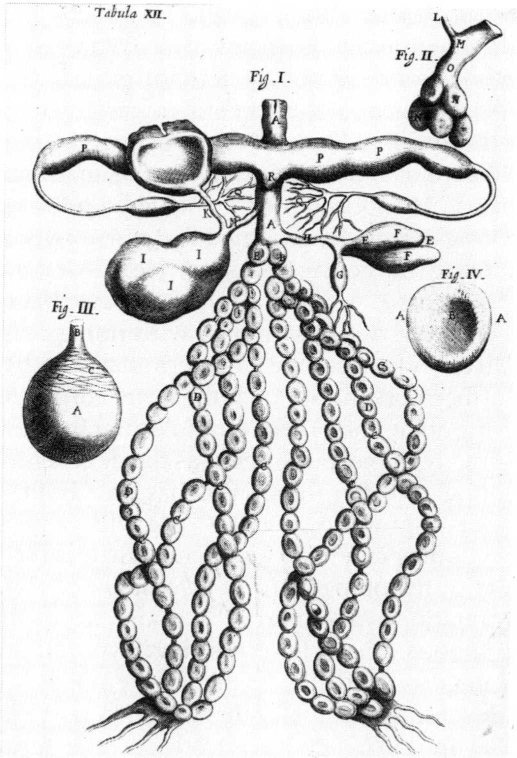

Figure 1.1 The ovary of *Bombyx mori*, prepared, analysed and depicted by Morio Malpighi in 1669 (his Figure 1). Common oviduct (A); accessory glands (Q,P); bursa copulatrix (I); ductus bursa (M); ductus vulva (K); ductus receptaculi (H); receptaculi seminis (E,F); receptacular gland (G); lateral oviduct (B); eggs in pedicel region (D); follicular stalk (C). Four ovarioles on either side are shown fusing partially. This aspect is typical for lepidopteran ovaries, since ovarioles are connected tightly in anterior regions by the outer ovary envelope. As is typical for this species, in older adults, lateral oviducts and pedicel regions of ovarioles are continuously filled with chorionated, ripe eggs, while younger stages of egg development like vitellogenesis and previtellogenesis have already ceased.

by the work of R. Wagner in 1836. Some scientists in those days had the correct interpretation of an oocyte nucleus, others felt the germinal vesicle was the true oocyte cell.

At this time, Müller (1825) described the ovary of phasmids, rediscovering the terminal filament of ovarioles. He found a tight coupling of these structures with the insect heart. However, he misinterpreted the terminal filaments as blood vessels and thereby opened one of the first scientific debates on insect ovary structures. He proposed that all elements housing the ovariole may come into the ovariole tube via the terminal filament: a theory by which oocytes and follicle cells had a common origin and developed one from another inside the ovariole. We must remember that in those days the cell theory had not yet been introduced (Schleiden, 1838; Schwann, 1839). A more correct interpretation was given by Stein (1847), who showed that the terminal filament does not open into the insect heart. He was the first to give us a detailed comparative anatomical analysis of the coleopteran ovary. He noticed the monolayered follicular epithelium, enclosing the growing oocyte. He assumed that nurse cells might have the same roots as the oocyte (the germinal vesicle). He showed that coleopteran ovaries fall into two classes, to one of which all adephage coleopterans belong, in which each oocyte is assembled with nurse cells. In the other class, embracing all other beetles, one large terminal chamber exists, housing nurse cells. Furthermore, he adopted the term 'corpus luteum' for the coloured tissue rising at the end of each ovariole after the first eggs are laid. He interpreted correctly the corpus luteum as the degenerating remainder of the follicular epithelium after the chorion has been built.

In 1866, Metschnikoff proposed that the 'pole cells' of the insect embryo are the precursor cells of oocytes and nurse cells. He was the first to try and give an entire story of the ontogenetic development of gonads and eggs. His correct interpretation was ignored by many scientists, since Claus (1864) and other famous scientists of the day favoured another theory, based on the interpretation of Müller. In this theory, oocytes and nurse cells were thought to develop from small cells, which in part, give rise also to the follicular cells. In 1867, Leydig introduced the term 'germ cells' (Keimzellen) for those cells which have an opaque nucleus and are found in the terminal chamber. He showed that these cells are the founder cells of oocytes and nurse cells. Nevertheless, he followed Claus and his theory in later publications.

The debate was solved after some new techniques became well established: tissue fixation, paraffin embedding, and the microtome technique. All these developments were made in the second half of the nineteenth century.

In 1885, Balbiani published his investigations on the reproductive organs in the midge *Chironomus*. As Metschnikoff, 20 years before, he found pole cells as the progenitors of germ cells. Korschelt (1886) assumed common ancestors of nurse cells and oocytes in the fly ovary. However, he still followed the theory of Claus. In 1891 and more precisely in 1895 Heymons published his research on insect embryos and gonads. He could clearly show that germ cells develop very early and do not share in the differentiation processes, leading to the three germ layers ectoderm, endoderm and mesoderm. Furthermore, he showed that the somatic cells (i.e. follicle cells, cells of the terminal filament and the cells of the outer epithelial sheath) are all descendants of the mesoderm. These results were a clear confirmation of Metschnikoff's proposals and forced the theory of Weismann (1885, 1892) who claimed the continuity of germ cells through generation cycles. In 1901, Giardina published his famous work on the origin of oocytes and nurse cells of the water beetle, *Dytiscus*. He showed clearly the rise of oocytes and nurse cells out of a common germ cell by differential mitoses. In addition, he described an unusual germ cell nuclear body which appears only in the oocyte and in the directly preceding germ cell line and not in those cells which differentiate to nurse cells. He believed that this germ line body would cause the differentiation into oocytes and nurse cells, a hypothesis which was not correct, since we know today other insects in which these bodies occur in nurse cells [for instance in tipulids (Bayreuther, 1952, 1956), or in the polytrophic flea *Hystrichopsylla* (Büning and Sohst, 1989; Büning, 1990)]. Furthermore, in most insects housing meroistic ovaries, a prominent nuclear body is missing. The nature of this body was discovered some decades later and will be discussed in more detail in Section 3.3.

The less we know, the more hypotheses will exist, and even those which are shown to be inadequate will be traded for a long time. Thus, the hypothesis of Claus (1864) and others was defeated by Paulcke (1900) when he investigated the polytrophic meroistic ovary of bees, and even Deegener kept the discussion open when he reviewed information on insect ovaries in 1928.

At the end of our historical review we should remember the main points which led to the well-accepted description of three morphological and physiological types in the insect ovary. One of the first who described insect ovaries comparatively was Dufour, for instance in his work on anatomy and physiology of Hemiptera (1831), or Orthoptera, Hymenoptera and Neuroptera (1841). Stein's comparative work in 1847 on beetle ovaries has been mentioned before. In 1858, Huxley described the ovary of aphids. He showed for the first time a nutritive cord, connecting the nurse cells with the oocyte. In the same year (1858) Lubbock gave a remarkable review of the knowledge of his time about the insect ovary. He found nutritive cords in some other hemipterans, too. He provided evidence that the terminal chamber of hemipteran ovarioles may be equivalent to the terminal chamber found in polyphage coleopterans. In 1878, Brandt published a monograph on the insect ovary, which summarized all the knowledge on insect ovaries up to that time. He introduced some years before (1874) the term 'panoistic' for those ovarioles in which only oocytes will develop, and the term 'meroistic' for those ovarioles in which nurse cells and oocytes differentiate.

In 1903, Gross published his work on the insect ovary. He introduced two additional terms which we use today. Those meroistic ovaries in which nurse cells and oocytes alternate, he termed 'polytrophic'. Those in which nurse cells are kept in a single, end-standing terminal chamber and to which oocytes are connected by nutritive cords, he named 'telotrophic'. Three insect groups, the Hemiptera, the Coleoptera–Polyphaga, and the Sialidae he stated correctly to be telotrophic and he assumed a different origin for all three types. From about 1890 up to the beginning of World War I, most of the histology on insect ovaries was elucidated by means of light microscopy. Some of these papers will be discussed in more detail in the appropriate sections.

During the following four decades the interest in histological problems was not as intense as before. Other branches of biology, such as classical genetics or physiology became predominant and many of the old findings were not passed on to the next scientific generations. Two classic and widespread works on insect histology, summarizing and selecting the old findings, influenced succeeding generations of scientists. These are Hermann Weber's book *Lehrbuch der Entomologie* from 1933 and Richard E. Snodgrass's book *Insect morphology* published in 1935.

New dimensions were opened with the development of the electron microscope and the necessary applications for biological materials, such as new fixation procedures, plastic embedding and ultrathin slicing. All these developments became practicable in the late 1950s and early 1960s. In parallel, some light microscope techniques became widespread as did phase contrast microscopy combined with semi-thin techniques, and interference phase contrast microscopy. Most of this book is founded on the scientific work since then.

Today, we are again in the midst of a historical turning point. New techniques will open new dimensions in light and electron microscopy. Computer-enhanced video and laser techniques will show us new aspects of cell dynamics and cell structures. Gene structures have become more and more clear using DNA sequence analysis; their products can be analysed by genetic engineering. Monoclonal antibodies allow us to follow the life cycle of defined proteins at the light and electron microscopical level.

At the beginning of the twentieth century biological science began to spread into more and more specialized disciplines. The time is now coming when this specialized knowledge will fuse again in interdisciplinary research.

2

The ovary of Entognatha

Szczepan Bilinski, Jagiellonian University of Krakow, Poland

2.1 INTRODUCTION

The Entognatha are currently classified as a sister group of the Insecta s. str. (for further discussion of hexapodan phylogeny see section 2.5). Traditionally, the entognathans are subdivided into three well-characterized orders: the proturans (Protura), collembolans or springtails (Collembola) and diplurans (Diplura). The structure of the female gonad as well as the course of oogenesis in representatives of all these orders have been thoroughly investigated (Krzysztofowicz, 1975, 1977; Bilinski, 1976, 1977, 1983a,b; Palevody, 1976). These studies have shown that among general similarities between the ovaries of Entognatha and Insecta s. str. some fundamental differences occur. They concern such important processes as the germ cell cluster formation, yolk accumulation and deposition of egg envelopes (Bilinski, 1976, 1977, 1983a,b). Since entognathan oogenesis has not been broadly reviewed till now, the aim of this section is to summarize the present knowledge of this process.

2.2 GROSS ARCHITECTURE AND ACCESSORY STRUCTURES

Two well-defined categories of female gonads can be discerned among Entognatha. In the Protura, Collembola and Diplura: Campodeidae the paired ovaries are sac shaped and not divided into discrete ovarioles (Palevody, 1971; Bareth, 1972; Jura, 1975; Krzysztofowicz, 1977; Asaba and Ando, 1978) (Figures 2.1–2.3). They are localized ventro-laterally in the abdomen of proturans and collembolans, or laterally along the alimentary tract in the abdomen of campodeids.

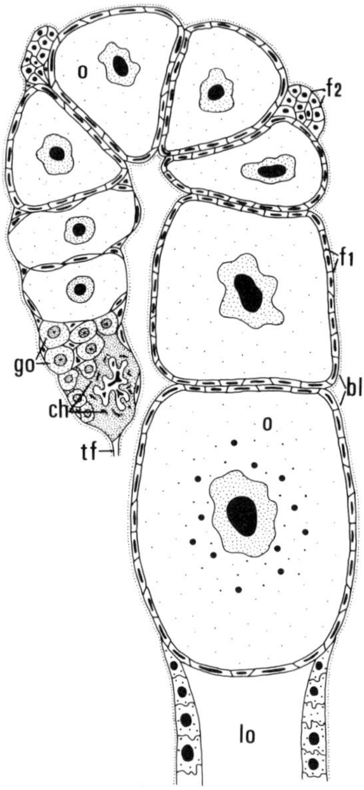

Figure 2.1 Diagram of a premature ovary of the proturan *Acerentomon* sp. Basement lamina (bl); apical chamber (ch); follicular cells of the first generation (f1); follicular cells of the second generation (f2); germarial oocytes (go); lateral oviduct (lo); oocyte (o); terminal filament (tf); nuclei of somatic cells are shown in black.

In all investigated species the anterior tip of the sac is equipped with a poorly developed terminal filament (Figures 2.1–2.3). The posterior end is connected to a lateral oviduct. Both lateral oviducts join forming a median, common oviduct that opens to a bursa copulatrix (campodeids), or directly to the exterior (proturans and collembolans). According to Matsuda (1976) a simple spermatheca connected to the common oviduct or bursa copulatrix occurs in the campodeids and collembolans.

The ovarial sacs are covered by a thin layer of basement lamina only. In contrast to Insecta s. str. the outer epithelial sheath is absent even in premature or larval stages (Bilinski, 1976, 1983a; Klag, unpublished). In *Campodea*, the ovaries and oviducts are entwined with loosely arranged, isolated tracheae, which might be regarded as remnants of a degenerating or reduced outer epithelial sheath.

Two regions, a germarium and a vitellarium, can be distinguished within each ovarial sac. The germaria are located laterally as in collembolans (Figure 2.2) or apically as in proturans and campodeids (Figures 2.1, 2.3, 2.4a).

Within vitellaria individual oocytes (proturans), or egg chambers (campodeids and collembolans) are ordered linearly. In the last two groups all oocytes occurring in a given specimen develop synchronously (Palevody, 1976; Krzysztofowicz, 1977; Bilinski, 1979), whereas the proturan ovaries contain oocytes in consecutive stages of oogenesis (Jura, 1975). The ovaries of Protura deserve particular attention. In this group the tip of the ovary (i.e. the terminal filament, the germarium and the anterior part of the vitellarium) is bent caudo-ventrally and lies directly attached to the older oocytes (Figure 2.1). Within the germarial region, besides young gonial cells, a characteristic apical chamber occurs (Figure 2. 1). For further discussion

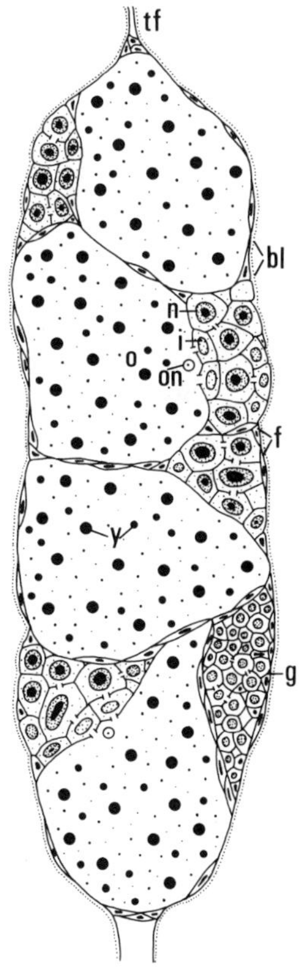

Figure 2.2 Diagram of an ovary of the collembolan *Tetrodontophora bielanensis* in the vitellogenic stage. Basement lamina (bl); follicular cells (f); germarium (g); intermediate cells (i); nurse cells (n); oocyte (o); oocyte nucleus (on); terminal filament (tf); yolk spheres (y); nuclei of somatic cells are shown in black.

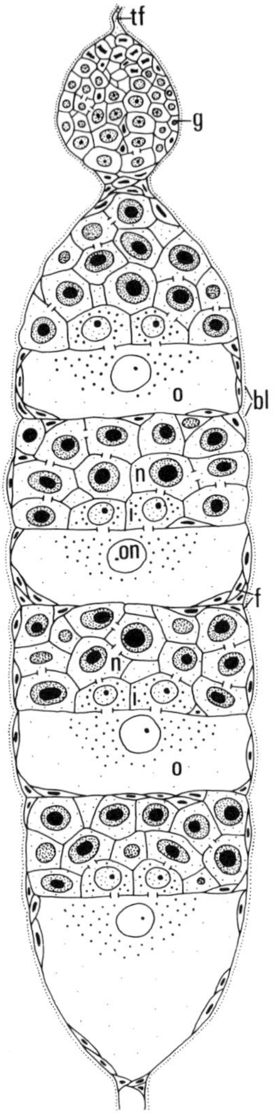

Figure 2.3 Diagram of an ovary of the dipluran *Campodea* sp. in the previtellogenic stage. Basement lamina (bl); follicular cells (f); germarium (g); intermediate cells (i); nurse cells (n); oocyte (o); oocyte nucleus (on); terminal filament (tf); yolk spheres are shown by dots in oocytes and intermediate cells; nuclei of somatic cells are shown in black.

Figure 2.4 (a) The apical part of the ovary of *Campodea*. Note the germarium (g) and constriction separating it from the vitellarium (v). Phase contrast × 150. (b) The panoistic ovariole of *Catajapyx aquilonaris*. Fat body (fb); lateral oviduct (lo); terminal filament (arrow). Phase contrast × 100. (c) Follicular cells (fc) surrounding the vitellogenic oocyte (o) of *Campodea*. Fat body cells are close to follicles. × 1200. (d) Follicular cells of *Campodea*. Vitellogenesis. Basement lamina (bl), Golgi complex (Gc), oocyte (o), secretory vacuole (v) × 20 200. (e) Follicular cells (fc) of *Tetrodontophora bielanensis*. Previtellogenesis. Oocyte (o), cytoplasmic projections (*) × 25 000.

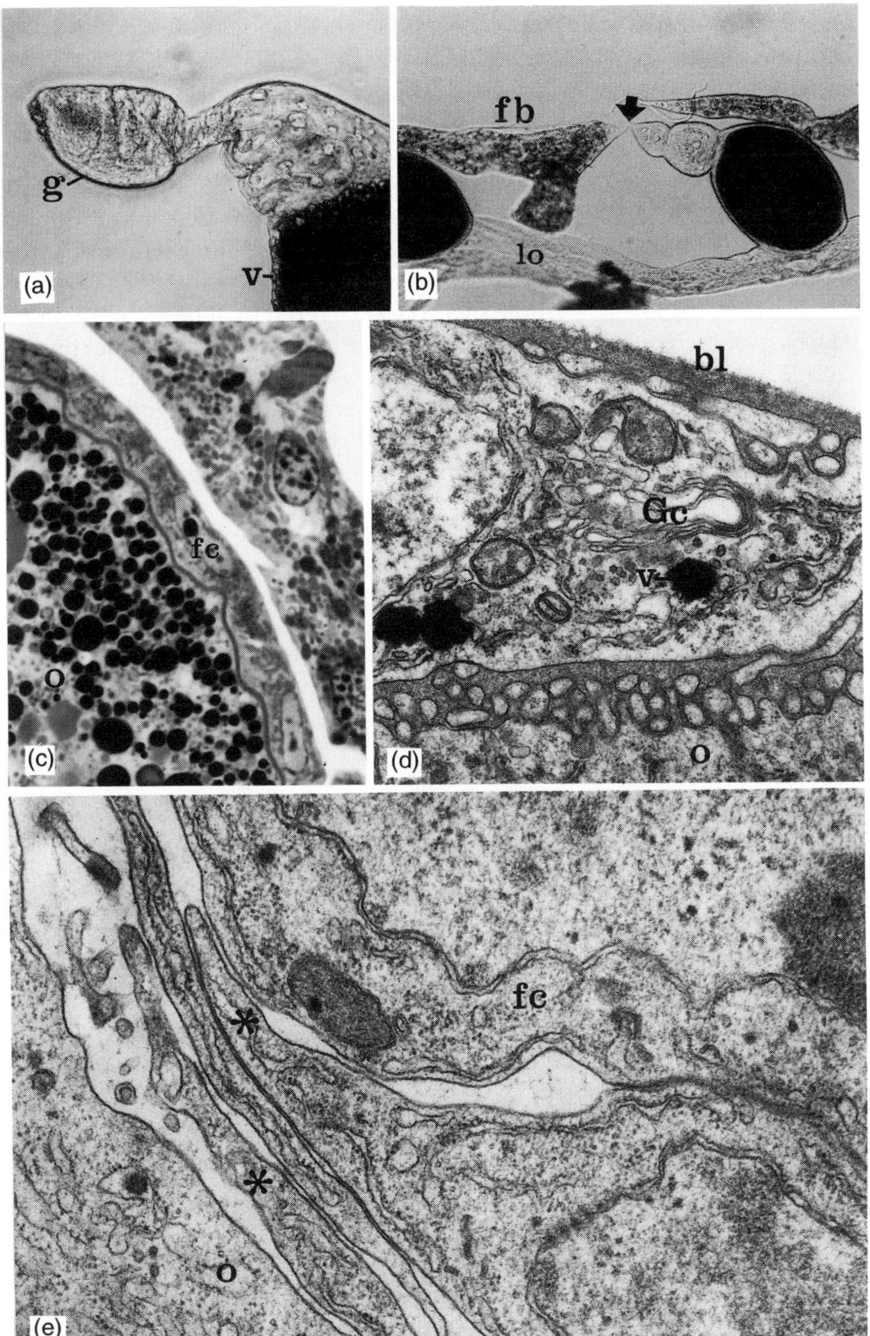
g
v
(a)
fb
lo
(b)
fc
o
(c)
bl
Gc
v
o
(d)
fc
*
*
o
(e)

of the structure and possible significance of this chamber see sections 2.4.1 and 2.5. An entirely different type of ovary has been described in the Diplura: Japygidae (Grassi, 1887; Bilinski and Szklarzewicz, 1992). Here, each of the two ovaries is composed of seven segmentally (metamerically) arranged panoistic ovarioles. The two lateral oviducts bearing the ovarioles extend through the abdomen and merge posteriorly, forming a common oviduct (Figure 2.5a). The latter, according to Matsuda (1976), opens to a simple bursa copulatrix. Each individual ovariole is differentiated into an apical terminal filament, a germarium comprising young gonial cells and a posterior vitellarium, which contains three to four linearly arranged oocytes (Figures 2.4b, 2.5b). The whole ovariole is covered by a basement lamina. The base of each ovariole is connected directly to the lateral oviduct (Figure 2.4b, 2.5b); a pedicel (ovariole stalk) is absent. The terminal filaments of ovarioles comprising the individual ovary are not joined together. Each of them is attached separately to the closest fat body lobe (Figure 2.4b, 2.5b).

The female reproductive systems of all investigated entognathans are

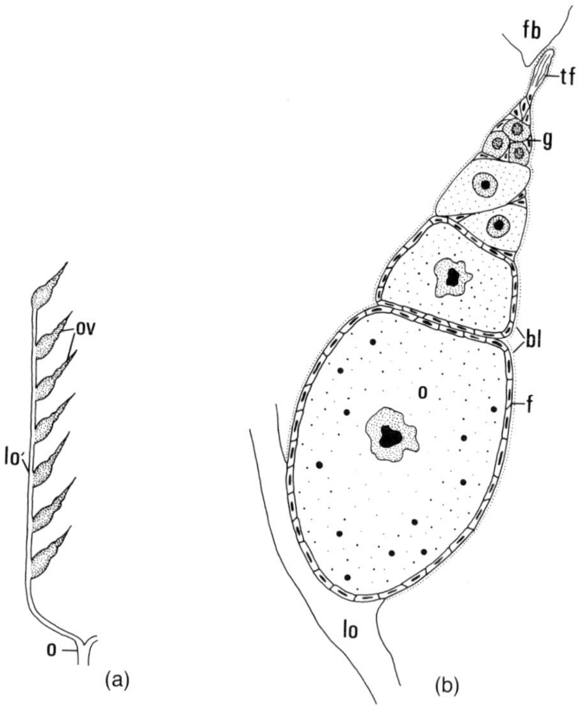

Figure 2.5 (a) Diagram of the ovary of *Catajapyx aquilonaris*. Lateral oviducts (lo); common oviduct (o); ovariole (ov). (b) Diagram of an ovariole of *C. aquilonarius*. Basement lamina (bl); fat body (fb); follicular epithelium (f); germarium (g); lateral oviduct (lo); oocyte (o); terminal filament (tf); nuclei of somatic cells are shown in black.

devoid of accessory glands (Matsuda, 1976). Evolutionary significant characters of the ovaries in major entognathan groups are summarized in Table 2.1.

2.3 SOMATIC CELLS

2.3.1 Terminal filament

The terminal filaments of the Entognatha have never been described with the use of the electron microscope and in consequence, their structure remains practically obscure. In Collembola, Protura and Diplura: Campodeidae they are thin and short, 'almost invisible with the light microscope' (Asaba and Ando, 1978). The terminal filaments of ovarioles in the japygids are comparatively thicker and more complex. In *Catajapyx aquilonaris* (Bilinski and Szklarzewicz, 1992) each filament consists of a few somatic cells. The apical part of these cells bears numerous projections that are predominantly orientated parallel to the long axis of the ovariole. The cells of the terminal filament as well as their projections are surrounded by

Table 2.1 Important characters of ovaries (ovarioles) in major entognathan groups

			'Diplura'	
	Protura	Collembola	Campodeidae	Japygidae
Category of the ovary	Sac-like	Sac-like[a]	Sac-like[a]	Metameric, composed of discrete ovarioles[b]
Type	Secondary panoistic	Polytrophic meroistic[a]	Polytrophic meroistic[a]	Panoistic[b]
Localization of the germarium	Apical (the tip of ovary bent caudo-ventrally)	Lateral	Apical	Apical
Germ cell clusters	?	Chain-like with intermediate cells[a]	Chain-like with intermediate cells[a]	Absent
Oocyte nucleus	Active in rRNA synthesis	Inactive[a]	Inactive[a]	Active in rRNA synthesis
Ooplasm of previtellogenic oocytes	Devoid of membranous organelles	Filled with numerous rough ER elements and Golgi complexes[a]	Filled with numerous rough ER elements and Golgi complexes[a]	?

[a]Synapomorphic characters shared by the Collembola and Campodeidae.
[b]Plesiomorphic hexapod conditions.

a basement lamina. A transverse septum separating the terminal filament from the germarium is absent.

2.3.2 Prefollicular and follicular cells

According to authors using traditional, histological methods, the ovaries of the Entognatha are devoid of a follicular epithelium, but instead possess single parietal cells ('cellules parietales': Palevody, 1971; 1976; Bareth, 1972; 'ovarian sheath cells': Matsuzaki, 1973). More recent ultrastructural studies, however, have indicated that at least in some developmental stages the entognathan oocytes are covered by the continuous layer of flat somatic cells (Bilinski, 1977; 1979; Klag, 1978). Their location beneath the basement lamina as well as their involvement in the formation of the egg envelopes clearly show that they constitute a simple but 'true' follicular epithelium (Bilinski and Klag, 1977; Bilinski, 1983b). This suggestion has been additionally supported by the discovery of gap junctions between the oocyte and enveloping cells (heterocellular gap junctions) in the ovary of *Acerentomon* (Bilinski and Klag, 1982) and *Campodea* (Bilinski, 1987).

As in insects proper, the follicular epithelium of the entognathans develops from somatic 'prefollicular' cells. In *Campodea* small, spindle-shaped prefollicular cells are localized in a constriction that separates the germarium from the vitellarium (Figures 2.5 and 2.6f). In contrast, the prefollicular cells of proturans and collembolans are scattered throughout the germaria between dividing and/or developing germ cells (Klag, 1978).

In the polytrophic ovaries of the Collembola and Campodeidae the oocyte nurse cell complexes (see section 2.4.1) are invested with the prefollicular cells within the germarium or while migrating from the germarium to the vitellarium. Initially, these cells do not form the continuous epithelium but lie on the surface of the germ cells singly or in small groups consisting of 2–4 cells. In this stage of morphogenesis, the follicular cells are markedly flattened and equipped with long cytoplasmic projections by which individual cells may communicate (Figure 2.4e). The cytoplasm of these cells contains numerous free ribosomes, mitochondria and single elements of rough endoplasmic reticulum (ER). During growth

Figure 2.6 (a) Follicular cells of the first generation in *Acerentomon*. Desmosome (d); Golgi complex (Gc); perioocytic space (po); primary sheath (arrows) × 11 400. (b) The heterocellular gap junction (arrows) between the oocyte and follicular cell (fc) in *Campodea* × 22 700. (c) The lateral oviduct (lo) of *Acerentomon*. Cross-section, vacuoles (v) × 1000. (d) The apical part of the germarium of *Campodea*. Note the metaphasal plates × 1100. (e) The unbranched chain of cystocytes in the germarium of *Campodea* × 1000. (f) Differentiating germ cell complexes in the basal part of the germarium of *Campodea*. Prefollicular cells (pf); presumptive nurse cells (nc); oocyte (o) × 1000.

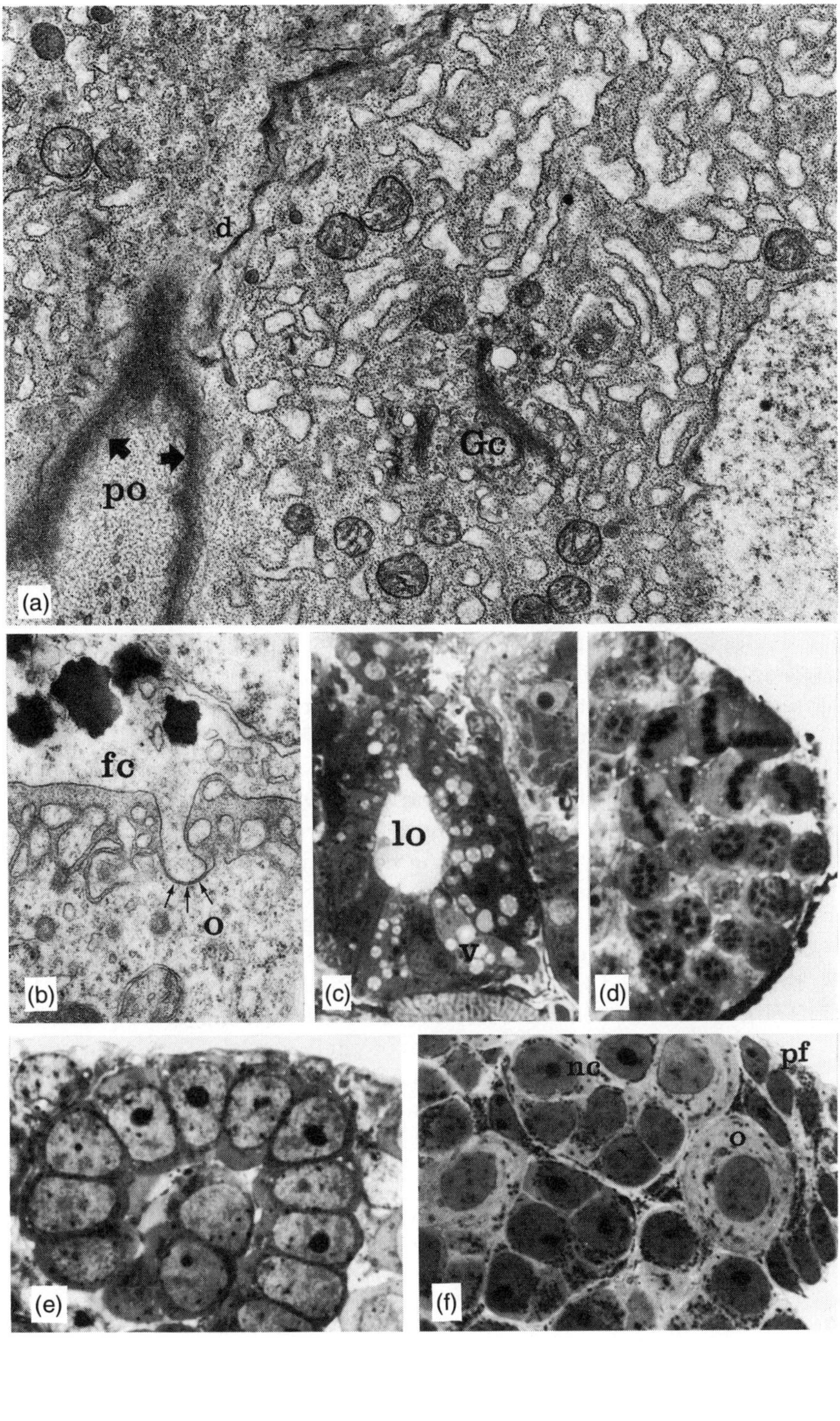

(a)
d
po
Gc
(b)
fc
o
(c)
lo
v
(d)
(e)
(f)
nc
pf
o

stages, follicular cells multiply by mitotic divisions and enlarge markedly. As a result, in the final phase of vitellogenesis, the continuous follicular epithelium is formed around each oocyte (Figure 2.2). Nurse cells have been absorbed meanwhile (see below). Concurrently, in the cytoplasm of the follicular cells numerous elements of rough ER and Golgi complexes are accumulated (Figure 2.4d). These organelles are subsequently involved in the synthesis and secretion of the precursors of the egg envelopes (Bilinski, 1983b)(see section 2.4.5).

An entirely different morphogenesis of the follicular epithelium has been described in the proturan *Acerentomon*. Here, two morphologically and functionally distinct generations of follicular cells operate in succession (Bilinski and Klag, 1977). The first generation, termed 'the follicular cells proper', form a continuous epithelium on the oocyte surface as early as the onset of previtellogenesis (Klag, 1978). Although these cells are extremely flat, they contain numerous dilated elements of rough ER, active Golgi complexes, free ribosomes and mitochondria (Figure 2.6a). The adjacent cells are linked by simple desmosomes. Beneath the epithelium, on the surface of the oocyte, a thin granular layer termed 'a primary sheath' is deposited (Figure 2.6a). It has been suggested, on the basis of ultrastructural observations, that the precursors of this layer are synthesized by the described 'proper' epithelium (Bilinski, 1977). Toward the end of vitellogenesis both the sheath and the epithelium degenerate (Bilinski and Klag, 1977). The follicular cells of the second generation are initially clustered at one pole of the oocyte where they form a well-defined aggregation. After yolk formation and degeneration of the follicular cells proper, individual cells leave this aggregation and spread over the surface of the oocyte covering it completely (Bilinski and Klag, 1977). Since the follicular cells of the second generation are involved in chorion formation they have been referred to as 'chorion producing cells'. These specialized cells contain numerous organelles (Figure 2.7c) connected with their secretory activity and are joined by well-developed septate junctions.

As has been mentioned above, the follicular cells of some entognathans remain connected with the oocyte via heterocellular gap junctions (Figure 2.6b). In *Acerentomon* and *Campodea* (Bilinski and Klag, 1982; Bilinski, 1987) these junctions develop at the onset of vitellogenesis. They are assumed to play a role in the coordination of the joint differentiation of the oocyte and the surrounding follicular cells and/or in the metabolic cooperation between the coupled cells (Bilinski, 1987).

Figure 2.7 (a) Previtellogenic oocyte of *Acerentomon*. Nucleus (n); nucleolus (nu); mitochondria (m) × 11 800. (b) Scanning electron micrograph of a proturan egg. On left side outer envelope (o) was removed and smooth surface of inner envelope (i) is seen × 700. (c) Follicular cells (fc) of the second generation in *Acerentomon*. Envelope (e); Golgi complex (Gc); oocyte (o); secretory vacuole (v) × 13 500.

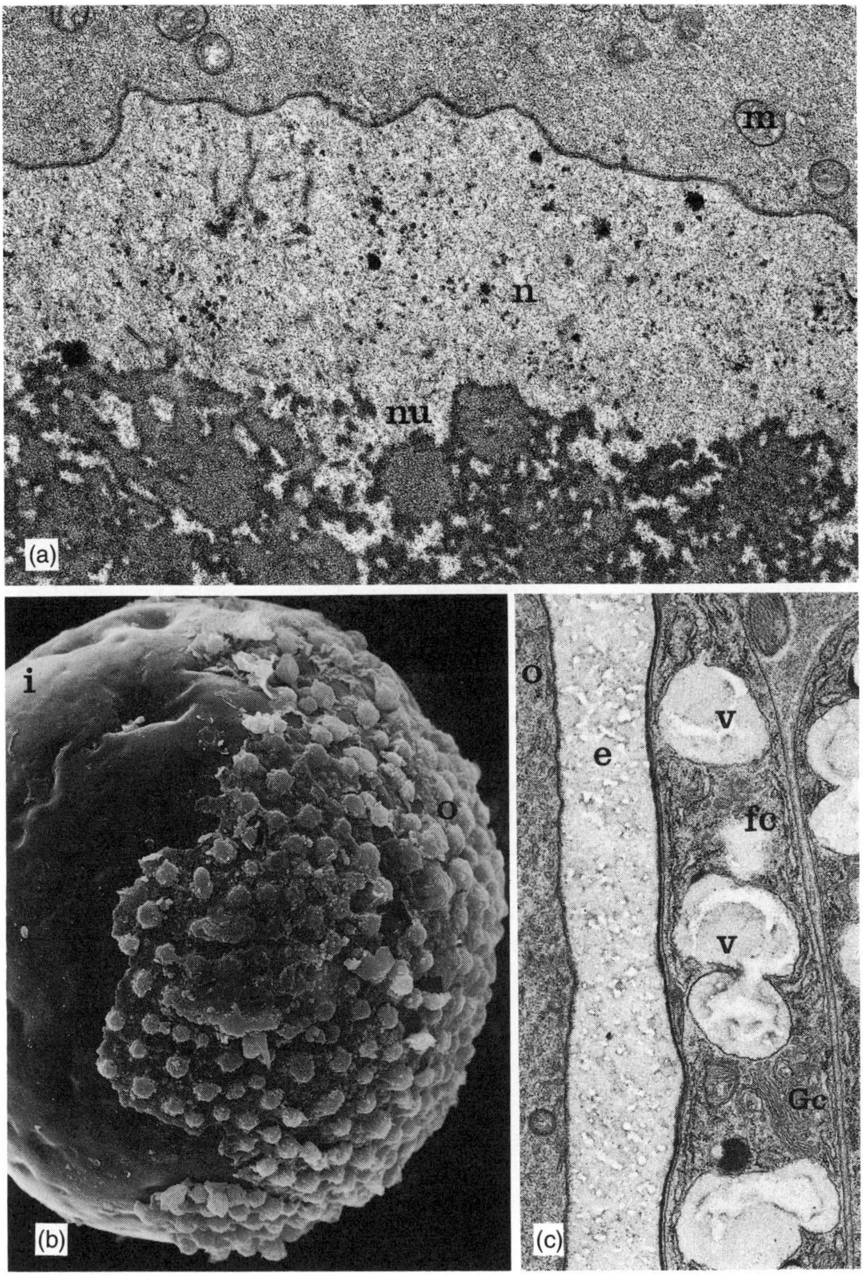
m
n
nu
(a)
i
o
(b)
o
e
v
fc
v
Gc
(c)

2.3.3 Oviduct

In the Entognatha the morphology and ultrastructure of oviducts has been investigated only in representatives of the proturans (Jura, 1975; Jaglarz, unpublished). The lateral oviducts in this group are short and relatively wide (Figure 2.1) (Jura, 1975). Their walls are composed of a single layer of large, roughly cuboidal, epithelial cells, resting on a thin basal lamina (Figure 2.6c). The lateral membranes of the adjacent cells are closely apposed and linked by extensive septate junctions (Jaglarz, unpublished). The membranes delimiting the oviduct lumen are equipped with short, irregularly distributed microvilli. The cytoplasm of the epithelial cells is tightly filled with numerous dilated elements of rough ER, well-developed Golgi complexes and prominent secretory vacuoles (Figure 2.6c). The latter are often observed in direct contact with the adluminal plasma membrane. These observations suggest that the lateral oviducts of the proturans synthesize macromolecules that are secreted later to the oviduct lumen. In the seventh abdominal segment, the two lateral oviducts fuse, forming a medially located, common oviduct. The latter is composed of markedly flattened epithelial cells that are lined by a thin cuticular intima. Individual cells are firmly joined by a complex system of narrow interdigitating cytoplasmic processes. Since the presence (or absence) of the intima indicates the origin of a particular group of cells it might be supposed that in the proturans the lateral oviducts are of mesodermal origin, whereas the common oviduct is ectodermal (Jaglarz, unpublished).

In contrast to the majority of the entognathans, the oviducts of the japygids are long and extend throughout the whole abdomen (Figure 2.5a). Preliminary ultrastructural observations have shown that the walls of the lateral oviducts in *Catajapyx aquilonaris* are composed of flattened epithelial cells that are joined by numerous alternating processes. Within the cytoplasm of these cells, elements of rough ER, mitochondria and single Golgi complexes occur (Szklarzewicz, in preparation).

2.4 GERM CELLS

2.4.1 Germarium and the formation of germ cell clusters

Among the entognathans, polytrophic meroistic ovaries occur in the Collembola (Krzysztofowicz, 1971, 1977; Palevody, 1971, 1976; Matsuzaki, 1973) and Diplura: Campodeidae (Bareth, 1972; Bilinski, 1979). In both groups the germaria are relatively large and contain dividing and/or differentiating germ cell clusters (clones). As in insects proper, here too, each cluster arises as a result of several consecutive mitotic divisions of an initial cystoblast (Palevody, 1976; Bilinski, 1983a). In *Campodea* the first mitotic cycles are synchronous (Figure 2.6d); thereafter synchrony of the divisions is lost and in consequence the final number of sibling cells (cystocytes) per clone is approximately 36 (Bilinski, 1983a). The germ cell clusters in *Campodea*, in contrast to those of most Insecta, are non-branched and chain-like (Figure 2.6e). The sibling cells comprising a given chain are morphologically identical and remain connected via simple intercellular bridges (see section 2.4.3). Obviously, each cystocyte, excluding the first and the last, possesses only two bridges. Fusomes and consequently polyfusomes have never been observed within intercellular bridges of *Campodea*.

The formation of germ cell clusters in the collembolans is not well documented. However, it has been shown that in *Folsomia*, *Isotomurus* (Palevody, 1973, 1976), *Tomocerus* (Matsuzaki, 1973), *Tetrodontophora* (Klag, unpublished), *Allacma* and *Arrhopalites* (Jablonska *et al.*, 1993) clones are also linear and nonbranched. The number of the cystocytes per sibling group is variable and amounts to 8–9 (*Folsomia*), 16 (*Isotomurus*), 8–12 (*Tomocerus*) and more than 20 (*Tetrodontophora*, *Allacma* and *Arrhopalites*). The intercellular bridges connecting the germ cells in these species are, as in *Campodea*, simple and devoid of fusomes (Palevody, 1976).

Although fusomes, their engagement in cluster formation and possible function in oocyte–nurse cell determination will be discussed extensively in sections 3.3.4, 3.4 and 3.5.2, a brief summary will be given here, to make differences between both groups more clear.

Most germ cell clusters of Insecta are branched and most of them contain, at least after breakdown of the mitotic spindle, a gelatinous mass, called the fusome. Adjacent fusomes, formed after subsequent divisions, fuse to a polyfusome that extends through all intercellular bridges. Each new spindle is orientated in this way so that one pole touches the polyfusome. This ensures that only one cell of each dividing pair receives all previously formed bridges, the other cell none (Storto and King, 1989). The two innermost cells, which have the greatest number of intercellular bridges, differentiate as pro-oocytes: all others are transformed to nurse cells. It has been concluded that the polyfusome might be responsible for branching of clusters and for oocyte determination (King *et al.*, 1982; Storto and King, 1989). Chapter 3 will show, however, that oocyte–nurse cell determination in true insects is still not fully understood.

It might be postulated that the main difference between the polytrophy of the entognathans and the insects proper lies in the formation of linear (1) versus branched clusters and that this difference is related to the lack of fusomes (2) in the former group. The first character cannot be valued as strong, since recently linear clusters in Ephemeroptera were found (Gottanka, 1992; Gottanka and Büning, 1990, 1993; sections 3.4 and 3.5.3). Furthermore, these exceptional clusters contain polyfusomes during the

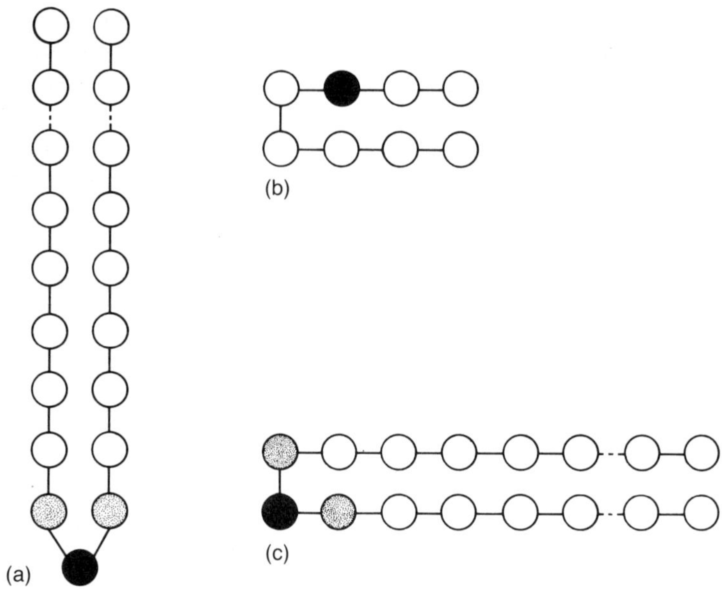

Figure 2.8 Arrangement of sibling cells within germ cell clusters of Entognatha: (a) *Campodea*, (b) *Folsomia*, (c) *Tetrodontophora bielanensis*. Nurse cells (open circles); oocytes (solid circles); intermediate cells (dotted circles). (Figure b modified from Palevody, 1976.)

mitotic division phase. Thus, the second character seems to be of higher value. On the other hand, in some groups of insects proper (e.g. heteropterans, aphids) polyfusomes have been lost or reduced.

According to Palevody (1973), in springtails (Collembola) all cystocytes of a cluster enter meiotic prophase and form synaptonemal complexes during pachynema. The distinction between the oocyte and presumptive nurse cells is therefore possible as late as the post pachytene stages. In all collembolans and campodeids studied so far the oocyte differentiates from the cell placed more or less centrally within the chain (Figures 2.6f and 2.8).

Figure 2.9 (a) The apical chamber (ch) of *Acerentomon* in contact with older previtellogenic oocytes (o). Follicular cells (fc); nuage material (*); extensions of somatic cells (arrows) × 13 000. (b) The previtellogenic egg chamber of *Campodea*. Intermediate cell (ic); nurse cells (nc); oocyte (o) × 1000. (c) Incorporation of ^{3}H-uridine into nurse cells of *Tetrodontophora bielanensis* (courtesy of A. Krzysztofowicz, Jagiellonian University, Kraków) × 800. (d) The dense basophilic granule in the cytoplasm of the intermediate cell of *Campodea* × 26 000. (e) Tangential section through the intercellular bridge connecting gonial cells of *Tetrodontophora bielanensis*. Note the dense network and the elements of rough ER (arrows). × 25 500. (Courtesy of J. Klag, Jagiellonian University, Kraków.)

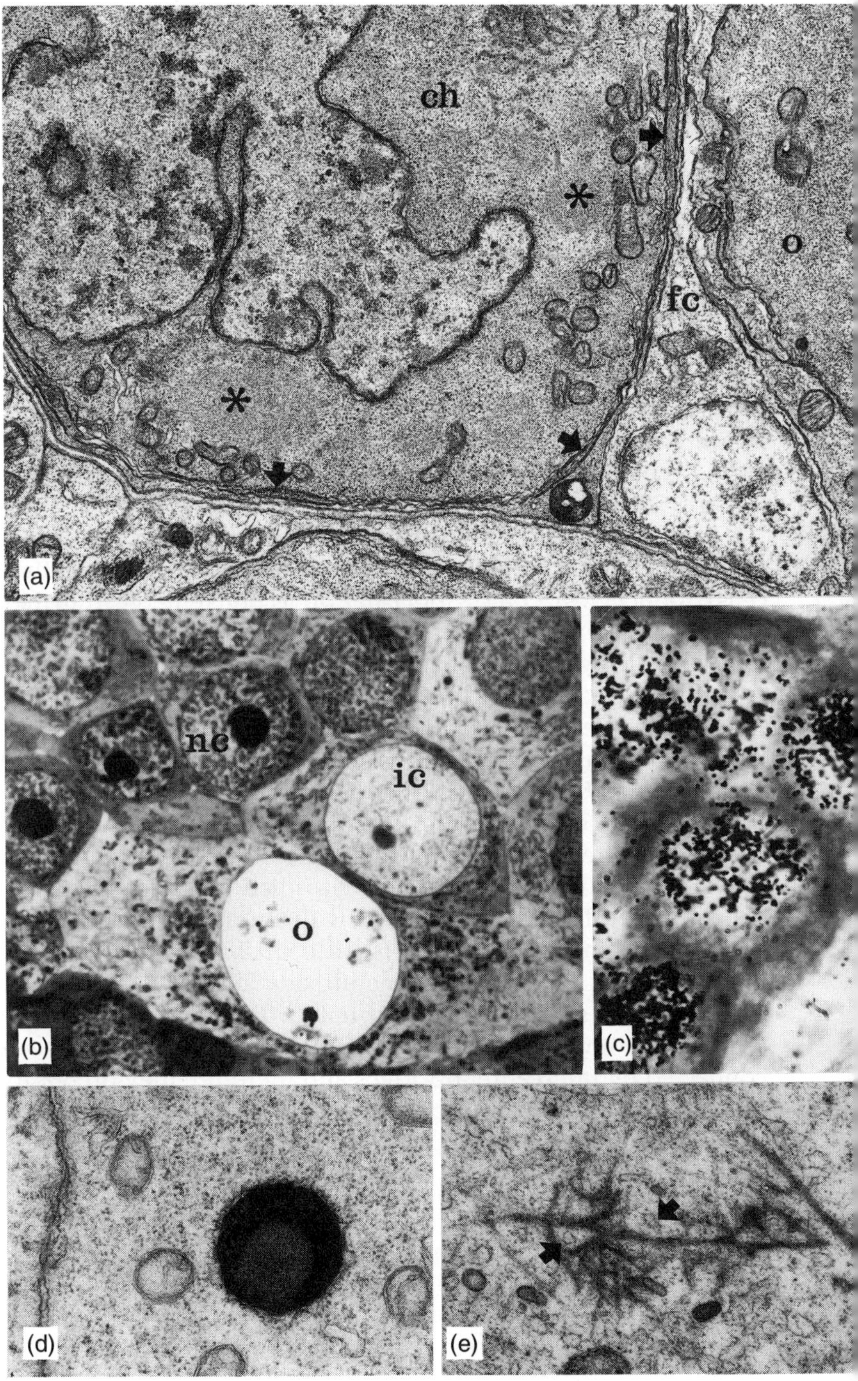
ch
*
*
o
fc
(a)
nc
ic
o
(b)
(c)
(d)
(e)

Other cells become the presumptive nurse cells. In collembolans (*Tetrodontophora, Allacma* and *Arrhopalites*) and in campodeids (*Campodea*) the presumptive nurse cells diversify additionally into two categories. The two cells adjoining the oocyte and connected with it via intercellular bridges develop into specific intermediate cells while others become the nurse cells proper (Bilinski, 1983a; Bilinski and Tylek, 1987; Jablonska *et al.*, 1993). Differentiated germ cell clones are invested with the prefollicular cells (see section 2.3.2) and migrate to the vitellarium. A functional unit composed of one germ cell clone and surrounded by the somatic follicular cells constitutes an egg chamber.

The germaria of proturans consists of two well-defined zones (Figure 2.1). The apical zone comprises a large cytoplasmic area termed an apical chamber while the basal one contains small oocytes (Klag and Bilinski, 1984). In the nuclei of the germarial oocytes synaptonemal complexes occur indicating that these cells have already entered the meiotic prophase (Klag, 1978). The apical chamber possesses a single, highly branched nucleus (Figures 2.1 and 2.9a).

The cytoplasm of the chamber is abundant in mitochondria, free ribosomes and large aggregations of nuage material and resembles that of nurse cells (trophocytes). Long cytoplasmic projections emerge from the chamber and penetrate into the spaces between the germarial oocytes. These projections are filled with ribosomes and parallel orientated microtubules (Klag and Bilinski, 1984). The whole chamber is firmly enveloped by the somatic cells, or their very thin extensions, and the huge cell inside never contacts the basement lamina that surrounds the whole ovary (Figure 2.9a). All the above data led Klag and Bilinski to suggest that the chamber cell is of germ line origin. Recently, Stys and Bilinski (1990) proposed that the chamber represents a 'retained trace of the original meroism'. Consequently, the ovaries of the Protura have been classified as secondary panoistic (= neopanoistic). This idea agrees well with the current genealogical hypotheses on entognathan phylogeny (see section 2.5).

The germaria of larval and mature ovarioles of the japygid *Catajapyx aquilonaris* are small and contain only a limited number of germ cells (Figures 2.4b and 2.5b) (Bilinski and Szklarzewicz, 1992). These cells are not connected by intercellular bridges and therefore they do not form clusters. Hence the ovaries of japygids might be regarded as primarily panoistic.

Figure 2.10 (a) Nurse cells of *Campodea*. Nuage material (nm); nucleolus (nu); nucleus (n) × 14 500. (b) Two closely apposed intercellular bridges of *Campodea*. Note regular striation of tangentially sectioned rim, cross-sectioned rims (*) × 38 000.

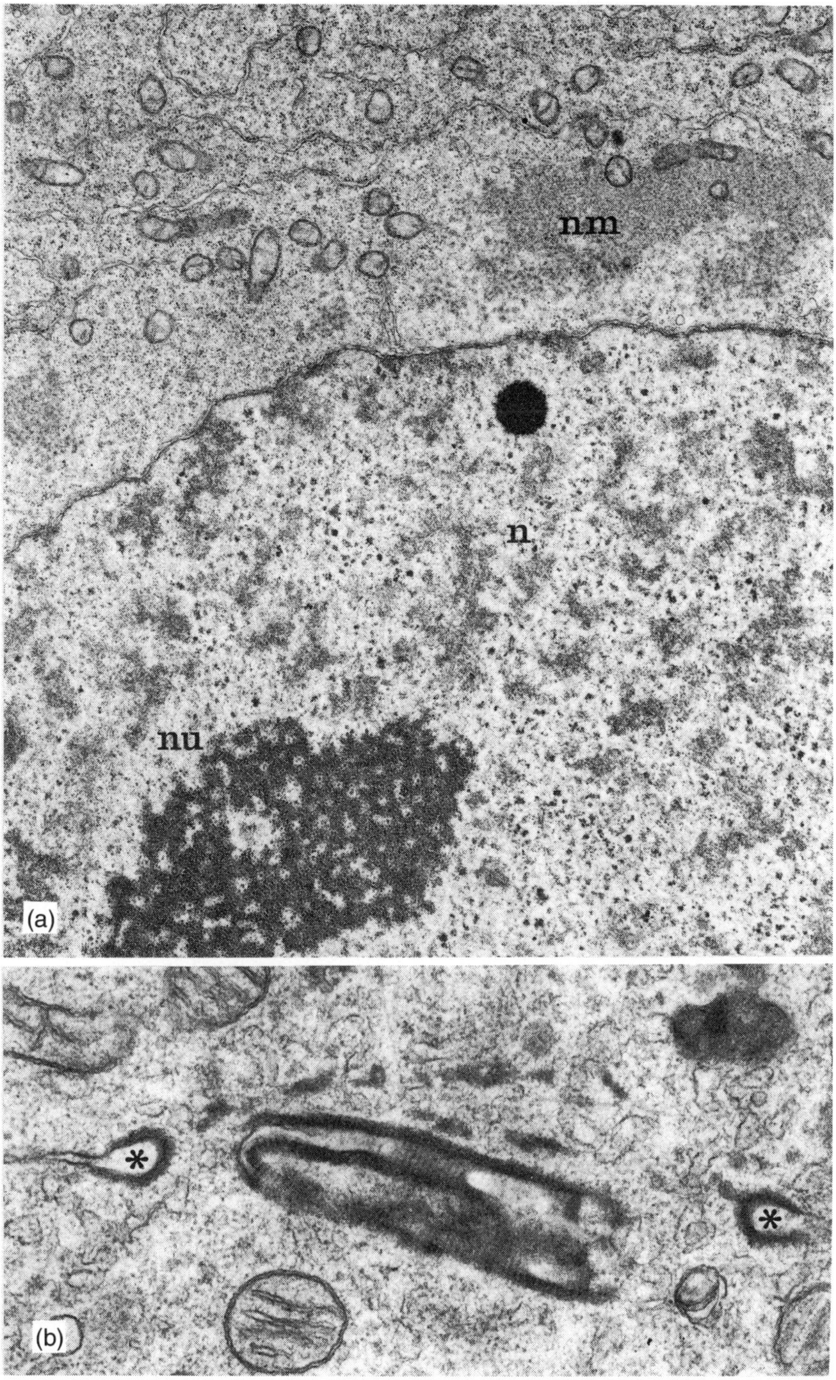
nm
n
nu
(a)
(b)

2.4.2 Ultrastructure and differentiation of nurse cells

The ovary of *Campodea* will be used to illustrate various aspects of the differentiation and diversification of the entognathan nurse cells. Presumptive nurse cells in this species are comparatively small and isodiametric (Figure 2.6f). Their nuclei are spherical and contain single nucleoli. In the karyoplasm chromatin aggregations and numerous electron-opaque granules are observed. Within the cytoplasm, cisternae of rough ER, Golgi complexes, mitochondria, small accumulations of nuage material and single symbiotic bacteroids occur. As a result of the diversification of the presumptive nurse cells, the typical nurse cells and intermediate cells arise. The former are larger and apparently polyploidal. Their nuclei possess large, well-developed often lobate nucleoli (Figures 2.9b and 2.10a).

Small clumps of heterochromatin and numerous clusters of electron-opaque granules occur in the karyoplasm (Figure 2.10a). In the perinuclear cytoplasm large accumulations of nuage material closely associated with mitochondria are present. The remaining cytoplasm contains elements of rough ER, Golgi complexes, mitochondria and an abundance of free ribosomes (Figure 2.10a). The above observations indicate that in the nurse cell nuclei of the entognathans, as in those of the Insecta, an intensive synthesis of ribosomal RNA (rRNA) takes place. The resulting ribosomes are subsequently transferred via the intercellular bridges to the oocyte (Bilinski, 1979).

The intermediate cells are attached to the anterior pole of the oocyte (Figure 2.9b). They contain large, but comparatively 'transparent' nuclei in which small spherical nucleoli occur (Figure 2.9b). The level of polyploidy of these cells is lower than that of the typical nurse cells (Bilinski and Tylek, 1987). These facts show that nuclei of the intermediate cells play only a minor role in rRNA synthesis (Bilinski and Tylek, 1987). On the other hand, the intermediate cells maintain some features in common with the oocyte, and most significantly participate in the synthesis and formation of lipid droplets and dense basophilic granules (Figure 2.9d) (see section 2.4.4). The occurrence of the intermediate cells in the egg chambers of *Campodea* led to the suggestion that at least in this species the factor responsible for oocyte determination has a gradient character and that the ultimate fate of a given cell within a clone might be correlated with the received concentration of a hypothetical 'determinant' (Bilinski, 1983a).

The intermediate cells of *Tetrodontophora bielanenesis*, originally termed 'blocking cells', are structurally and functionally much more specialized (Krzysztofowicz, 1971, 1975). Here, these cells form a characteristic, flattened cap that covers a low elevation of the ooplasm containing the oocyte nucleus (Figure 2.2). The cytoplasm of the intermediate cells of *Tetrodontophora* possesses exceptionally numerous mitochondria, elements of rough ER and single, randomly distributed microtubules. The latter are more frequent in the vicinity of the intercellular bridges (Krzysztofowicz, 1975). Autoradiographic and histochemical investigations have shown that the nuclei of these cells are inactive in rRNA synthesis (Krzysztofowicz, 1971, 1975). It has been proposed that the intermediate cells of *Tetrodontophora* are involved in the transport of macromolecules and organelles from the nurse cells proper to the oocyte.

The typical nurse cells of *Tetrodontophora* and other collembolans contain large, spherical or slightly irregular, polyploidal nuclei that possess well-developed nucleoli (Matsuzaki, 1973; Sikora, 1975; Palevody, 1976; Jablonska, 1993). This is in line with autoradiographic and histochemical experiments showing that these nuclei are very active in rRNA synthesis (Figure 2.9c) (Krzysztofowicz, 1971). The nuclear envelope is perforated by numerous pores. It has been estimated that in the nurse cells of *Tetrodontophora* the pore density exceeds $60/\mu m^2$ (Sikora, 1975). The cytoplasm of the nurse cells of the collembolans is rich in free ribosomes, mitochondria and accumulations of nuage material, but Golgi complexes and elements of ER are rare. Beside these organelles, stacks of annulate lamellae, lipid droplets and microorganisms are observed in the nurse cell cytoplasm of different species (Bilinski, 1974; Palevody, 1976). In *Allacma* and *Arrhopalites* the nurse cells of the distal part of the nurse cell chain contain much more nuage material associated with their nuclei than the typical nurse cells described above (Jablonska *et al.*, 1993).

The nurse cells of the entognathans are active till the final phase of yolk accumulation. Later, they are resorbed or engulfed by the oocyte that is subsequently surrounded by the continuous follicular epithelium (Palevody, 1971; Rosciszewska, 1976).

2.4.3 Intercellular bridges connecting germ cells

Recent electron microscopical studies have revealed that the intercellular bridges connecting germ cells in the Collembola and Campodeidae are nearly identical (Palevody, 1976; Bilinski, 1983a; Kisiel, 1987). Their rims are slightly dilated and covered with dense filamentous material, which on tangential sections shows a regular striation (Figure 2.10b). Plasma membranes adjacent to the bridge are parallel and often joined by simple desmosomes. As has been discussed earlier, in the intercellular bridges of the entognathans the fusomes are not present. Instead of them, young germarial bridges are filled with an irregular network composed of a dense substance. This network is clearly discernible only in sections tangential to the plane of the bridge (Figure 2.9e). Individual microtubules and elongated cisternae of endoplasmic reticulum run across the empty spaces of the network (Figure 2.9e). In the later stages of oogenesis the network disappears enabling communication between interconnected cells. Since, within the intercellular bridges, mitochondria, free ribosomes and elements of the rough ER are present it might be suggested that these organelles are transferred from the nurse cells to the oocyte.

2.4.4 Previtellogenic growth of oocytes

In the entognathans, the smallest previtellogenic oocytes are largely devoid of membranous structures and contain primarily only free ribosomes and mitochondria (Bilinski, 1975; Klag, 1978). In this group the formation and accumulation of cytomembranes take place during the first phase of the previtellogenic growth and are always connected with the activity (proliferation, blebbing) of the nuclear envelope. The only detailed description of such a process in the entognathans comes from studies on young meiotic oocytes of *Tetrodontophora bielanensis* (Bilinski, 1975). Here, small vesicles, derived by localized blebbing of the outer nuclear membrane, fuse to generate short cisternae. These in turn assemble, giving rise to Golgi complexes (dictyosomes). Eventually each complex is encircled by an endoplasmic reticulum lamella and drifts off into the cytoplasm. In the oocytes of *Tetrodontophora* and *Campodea* the nuclear envelope is also involved in the formation of rough ER. This is evidenced by the frequent continuities between the outer membrane of the nuclear envelope and rough ER elements. As a result of the above processes, the ooplasm of both species becomes populated with numerous membranous organelles (rough ER, Golgi complexes) that subsequently participate in the autosynthesis of yolk precursors. Detailed discussion of the 'mixed' type of vitellogenesis in entognathans is given elsewhere (Bilinski, 1979; Bitsch and Palevody, 1980; Szklarzewicz, 1989). The participation of the nuclear envelope in the generation of the membranes in the oocytes is discussed in detail by Kessel *et al.* (1986).

In the last phase of previtellogenic growth, in the ooplasm of the collembolans and diplurans, lipid droplets and dense basophilic granules are accumulated. The latter are formed within dilated elements of rough ER without the participation of Golgi complexes (Figure 2.9d). After completion of vitellogenesis, these granules migrate to the peripheral ooplasm and form a distinct cortical layer (Figure 2.4c). The fate and function of the basophilic granules are not clear. However, their staining properties, unusual location within the oocyte as well as the mode of their origin strongly suggest that they do not represent yolk spheres (Bilinski and Tylek, 1987).

In addition to the organelles described, previtellogenic oocytes of the entognathans contain mitochondria, stacks of annulate lamellae, accessory nuclei, dense lamellar structures and crystalline inclusions (Palevody, 1972; Bilinski, 1974). The oocyte surface is devoid of microvilli during previtellogenic growth (Figure 2.4e). These structures develop in the last phase of previtellogenesis, just before the onset of yolk acccumulation (Figure 2.6b).

The structure and function of the oocyte nucleus (germinal vesicle) are entirely different in the panoistic (Protura and Japygidae) and meroistic polytrophic (Collembola and Campodeidae) ovaries. In the latter, the oocyte nuclei are generally small, 'transparent' and inactive in RNA synthesis (Krzysztofowicz, 1971). During previtellogenesis they are predominantly localized close to the intercellular bridges connecting the oocyte with the nurse cells (or intermediate cells) (Figures 2.2, 2.3 and 2.9b). In contrast, germinal vesicles of the proturans (Figure 2.7a) are large and centrally located. They contain huge but single nucleoli, suggesting that the amplification of genes of rRNA (rDNA) does not take place here. Ultrastructural investigations have shown that the nucleoli in the oocytes of *Acerentomon* are very active (Figure 2.7a). The resulting rRNA is transported via nuclear pores to the cytoplasm which in consequence is densely and uniformly filled with abundant free ribosomes (Figure 2.7a). Additional to the highly active nucleolus, intranuclear annulate lamellae are present in the germinal vesicle of *Acerentomon* (Bilinski and Klag, 1985).

2.4.5 Eggs and egg coverings

The eggs of entognathans are roughly spherical and devoid of any regional specializations like micropyles, aeropyles and respiratory appendages (compare with Insecta s. str., section 3.2.5).

Usually, they are covered with two simple and structureless envelopes. Since homologies of the entognathan egg coverings with those of true insects are not clear, the widely used terms 'chorion' and 'vitelline membrane' will be replaced in this section by more general ones: the outer and inner envelope.

In proturan *Eosentomon* the inner envelope is homogenous and smooth, while the outer is equipped with numerous, regularly spaced hemispherical projections (Figure 2.7b). Similarly, the outer egg envelope of some collembolans is studded with granules (*Heteromurus*) or mushroom-like structures (*Hypogastrura*) (Larink and Bilinski, 1989). It has been suggested that all these surface structures serve a respiratory function, holding on the egg a gas film, 'the respiratory plastron' (Bernard, 1979; Larink and Bilinski, 1989). In contrast to other entognathans, campodeids deposit their eggs in groups. Each group (batch) is attached to the substrate by a supporting stalk (Bilinski and Larink, 1989). The eggs of *Campodea* are covered by one comparatively thick and rigid envelope.

Ultrastructural investigations on the formation of egg envelopes have shown that in entognathans this process may follow various strategies in different groups (species). In the collembolan *Tetrodontophora* the first (and definitively outer) envelope is formed by the follicular epithelium, whereas the inner one is secreted by the deposited egg cell (Krzysztofowicz and Kisiel, 1989). Similarly, in *Acerentomon* and *Campodea* the first egg envelope is produced by the follicular cells (Bilinski and Klag, 1977; Bilinski, 1983a,b). In the latter genus this envelope is additionally reinforced by the secretion synthesized by the oviduct cells (Asaba and Ando, 1978). Finally, in some collembolans (*Tomocerus, Folsomia*) the oocyte is engaged in the formation of the inner envelope (Matsuzaki, 1973; Palevody, 1976).

2.5 EVOLUTION OF ENTOGNATHAN OVARIES

It is generally accepted that the common ancestors of the hexapods, as well as those of the insects, were characterized by panoistic ovaries (King and Büning, 1985; Stys and Bilinski, 1990). Among the entognathans, two groups, the Collembola and the Campodeidae, possess ovaries of the more advanced, meroistic polytrophic type. As has been shown earlier, the ovaries of these two groups show essential similarities (summarized in Table 2.1). Moreover, ultrastructural analysis of the germ cell cluster formation has also indicated that this process in the collembolans and campodeids retains fundamentally important similarities such as: the linear arrangement of germ cells within clusters, the lack of fusomes and polyfusomes, the peculiar structure of intercellular bridges. In this light, the convergent origin of the polytrophy in the collembolans and campodeids is rather improbable. Consequently, it might be postulated that among the entognathans polytrophic ovaries evolved only once.

The major taxa of the Hexapoda are shown in two cladograms (Figure 2.11): the current orthodox hypothesis maintained by Hennig (1981), Kristensen (1981) and many other authors (Figure 2.11a), and the new alternative proposed by Stys and Bilinski (1990) (Figure 2.11b). Cladogram b is based on the idea that the 'Diplura' represent a paraphyletic group and should be split into two equally ranked taxa; the Japygina and Campodeina, and that the former group is more closely related to the Insecta s. str. than the latter. This genealogical hypothesis has been discussed in depth by Stys and Bilinski (1990) and Bilinski (1993) and is strongly supported by the analysis of several morphological characters of recent and extinct 'diplurans', like the new Carboniferous family Testajapygidae (Kukalova-Peck, 1987). Analysis of presented cladograms indicates that in the entognathans the transition from the ancestral panoism to the more evolved 'linear polytrophy' could take place in the common ancestor of the Parainsecta (the Protura and Collembola) and the Diplura (cladogram a asterisk) or in the common ancestor of the Parainsecta and Campodeina (cladogram b asterisk). Since the ovaries of the japygids are primarily panoistic (see above), cladogram b seems to adopt better to the known distribution of the ovary types in the entognathan groups. It should be stressed finally, that in the light of both genealogical hypotheses the ovaries of the Protura should be regarded as secondary panoistic ('neopanoistic'). This is in line with the discovery of the apical chamber in the germarial zone of *Acerentomon* (Klag and Bilinski, 1984; see section 2.4.1). It might be added here, that similar reversals to the ancestral mode of panoism have also been

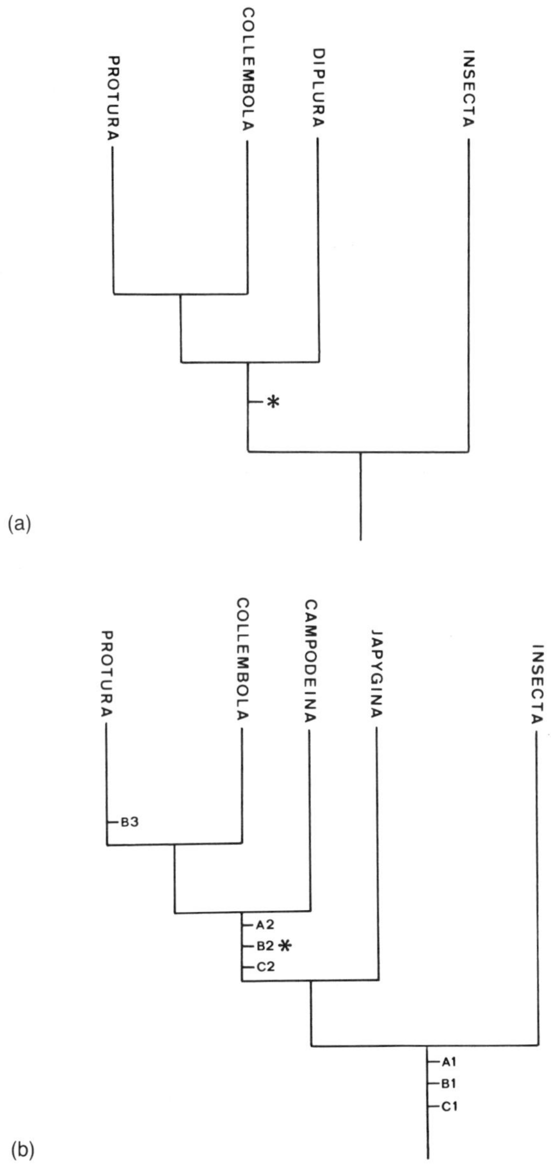

Figure 2.11 Cladogram of the major taxa of Hexapoda. Current orthodox hypothesis (a), new alternative proposed by Stys and Bilinski (1990) (b). Asterisks show where the transition from panoism to 'linear polytrophy' could take place. Selected characters of the ovaries (ovarioles) are marked in cladogram (b) (Insecta s. str. are not considered). Explanation of characters: A. Ovary: (1) divided into metamerically arranged ovarioles; (2) sac shaped; B. Type of ovary: (1) panoistic; (2) polytrophic-meroistic; (3) secondary panoistic; C. Germ cell clusters: (1) absent; (2) linear.

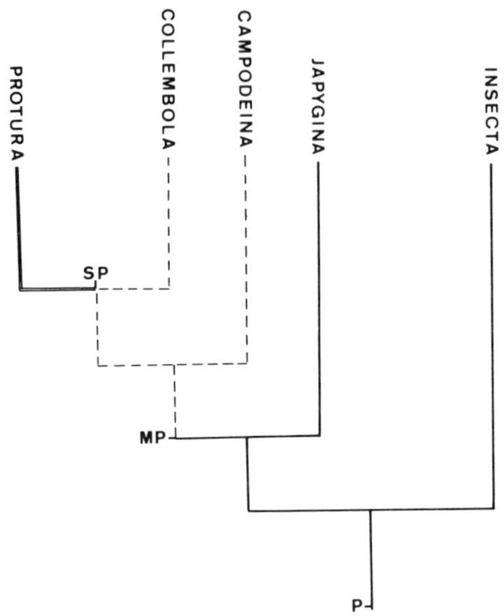

Figure 2.12 Proposed evolution of the ovary types in the entognathans. Panoism (P); polytrophy (MP); secondary panoism (neopanoism) (SP).

suggested among the Insecta s. str. (King and Büning, 1985; Pritsch and Büning, 1989; Gottanka and Büning, 1990; Stys and Bilinski, 1990; Büning, 1993; section 3.6). Figure 2.12 summarizes the proposed evolution of ovary types in the entognathans.

3

The ovary of Ectognatha, the Insecta s. str.

This chapter will review articles dealing with structure, physiology, some genetic aspects and phylogeny of insect ovaries. Our main interest lies in the ontogenetic development of ovaries and oocytes, i.e. the formation of mesodermal tissues into ovaries and the behaviour of germ cells during these ontogenetic processes. Therefore, the sections on accessory organs and the physiology of vitellogenesis are kept short and the reader is referred to recent reviews dealing with these topics (accessory organs: Davey, 1985; Gillott, 1988; Kaulenas, 1991; Wyatt, 1991; vitellogenesis: Kunkel and Nordin, 1985; Postlethwait and Giorgi, 1985; Wyatt, 1991). Further references are given in the appropriate sections.

Two other aspects of oogenesis are not included: the influence and control of morphogenetic processes via hormones and the influence of parasitism and symbiosis on insect ovaries. Both topics are large and should be reviewed separately (for hormone and nervous control see among others: Raabe, 1986; Wyatt, 1991; for symbiosis and parasitism see: Buchner, 1965; Schwemmler, 1979).

3.1 OVARIAN ACCESSORY ORGANS OF ECTODERMAL ORIGIN

The insect ovary develops from mesodermal tissue, which shelters germ cells and their derivatives, oocytes and eventually nurse cells. Each ovary opens into a lateral oviduct, which is also of mesodermal origin, at least in part. These structures and their functioning will be described in the forthcoming sections. A brief survey of the ectodermal organs will be given here, all of which mediate between the mesodermal ovary and the environment. Thus, they function as organs necessary for successful egg storage, insemination and egg laying. All these structures are tube-like, derived from median invaginations of the last abdominal segments.

3.1.1 Anatomy

The description of the gross morphology follows the arguments of Snodgrass (1935), Weber and Weidner (1974) and Martoja (1977) (Figure 3.1), although we now know that some of their conclusions concerning ontogeny and phylogeny on diverse tracts and appendices are questionable. Primarily, the mesodermal lateral oviducts open at the end of the seventh abdominal

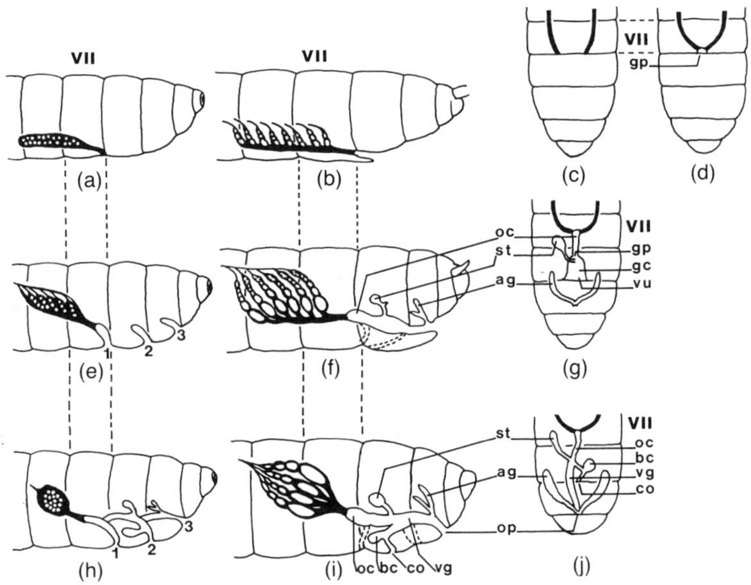

Figure 3.1 Synopsis of the ectodermal parts of the insect ovary, ordered from simple (a–c: Ephemeroptera; d: Dermaptera), to more complex arrangements (e–g: 'orthopteroid' species s.l.; h–j: Lepidoptera). Mesodermal lateral oviducts, ovarioles and anlagen (a, e, h) are shown black; ectodermal pouches and derivatives are blank. Accessory gland (ag); bursa copulatrix (bc); copulatory opening (co); genital chamber (gc); gonopore (gp); oviductus communis (oc); oviporus (op); spermatheca (st); vagina (vg); vulva (vu). Primary median pouches of seventh (VII), eighth, and ninth abdominal segments are signed 1, 2 and 3. Degenerating openings are stippled. Mesodermal ovarian envelopes are omitted. (a, b, e, f, h, i are after Weber and Weidner, 1974; and c, d, g, j are after Snodgrass, 1935.)

segment as seen in the Ephemeroptera (Palmén, 1884; Gillott, 1988). In all other insect orders the two lateral oviducts narrow and fuse with a single tube-like structure, the common median oviduct (Heymons, 1891, 1895; Snodgrass, 1933, 1935). The median oviduct is derived from a pouch, which invaginated from the distal rim of the seventh abdominal segment. The opening discharges the eggs directly into the environment and serves as a gonopore. Such a simple arrangement is known from dermapteran females (Snodgrass, 1935); in other insects the gonopore opens into a pouch of the eighth abdominal segment called the genital chamber. This pouch serves as the female copulatory organ and its opening is named the vulva. In many insects the genital chamber is tube-like, elongated and opens at the end of the ninth segment. In all these cases, the genital chamber is called the vagina and its opening is the oviporus.

The vagina has two fundamental functions: it is the pathway of ripe eggs and is the female copulatory organ. In many insects, copulation forces some adaptations from the female and the male copulatory organs as females have an extra vaginal cavity, named the bursa copulatrix (Figure 3.1). In some butterflies (Lepidoptera), cicadas (Homoptera, Cicadiformes), snake-flies (Raphididae) and scorpion-flies (Panorpidae) both functions are separated by two tubes and openings which can be derived from the primordial median outpocketings of the last abdominal sterna (Figure 3.1) (Weber and Weidner, 1974).

Several additional ectodermal pouches or tubes can insert directly into the vagina or nearby. One of these is the spermatheca, a single, median pouch inserting near the beginning of the vagina. After copulation, insect sperm, mostly packed into spermatophores, are stored and nourished in this organ until they can inseminate the ripe eggs via one of the small openings of the egg shell (chorion), the micropyle(s). This may happen after some hours or in extreme cases, as in honey-bee queens, after some months or years.

In some insects, like *Rhodnius prolixus*, the function of the ectodermal pouch, which is homologous to the spermatheca of other insects, has swapped roles with the collaterial (accessory) glands (see below). The spermatheca is now called the colleterial or cement gland (Davey and Webster, 1967; Huebner, 1980; Lococo and Huebner, 1980a).

In most insects, the ninth abdominal segment houses two, mainly symmetrical, additional ectodermal tubes (ectodemes; Davey, 1985) which can fuse in a median ventral position. These tubes are the accessory (collateral) glands in which various substances are produced, which cover the freshly layed eggs for different purposes.

Mucoproteins are produced by the accessory glands to fix eggs to the substrate, as shown for *Drosophila* (Riley and Forgash, 1967). This function is thought to be primary; however there is no real proof (Gillott, 1988). In many insects which lay their eggs into or on to the water, the secretion swells on contact with water and serves as a protective additional egg shell (Gillott, 1988).

Other functions of the accessory glands may be the production of lubricants which help sperm to move into the spermathecal duct (Callahan and Cascio, 1963; Wilkes, 1965), or which soften the spermatophore wall (Gerber *et al.*, 1971).

In *Musca domestica* hydrolytic enzymes are produced which serve in the fertilization process. The enzymes covering the micropylar region of the eggs disintegrate the micropylar cap substance and help to destroy the acrosome membrane of the sperm, a process necessary for sperm penetration (Leopold and Degrugillier, 1973; Degrugillier and Leopold, 1976; Degrugillier, 1977; Leopold *et al.*, 1978; Leopold, 1981). Recent results clearly demonstrate that both processes are tightly coupled: the gland secretion helps to dissolve the cap substances of the micropylar apparatus and only this mixture is able to dissolve the acrosome membrane (Degrugillier, 1985).

Cockroach females transport their eggs in an ootheca which is synthesized from products of the accessory (colleterial) glands. The left colleterial gland produces the protein (oothecin) of which the ootheca will be formed, together with diphenoloxidase, and phenolic O-β-glucoside. The right gland produces only β-glucosidase. When all substances mix in the vagina where the ootheca is formed, the β-glucosidase starts a cascade of chemical reactions by which phenol is liberated first. It is oxidized by diphenoloxidase to an O-chinone which, in a final reaction, will tan the protein which will form the hard ootheca. Tanning begins automatically when the different products from the left and right colleterial glands are mixed in the common duct (Mercer and Brunet, 1959; Stay and Roth, 1962; Hackman, 1974; Stay and Coop, 1974; Gillott, 1988).

Another variation in the secretion products of accessory glands occurs in some dipterans (damselflies, Glossinea) in which not only the single egg is retained in the uterus, which is a transformed vagina, but the hatched larva is fed via secretions of the milk (collaterial) glands by the mother during the whole larval growth phase. The secretion consists of peptides, free amino acids, phospholipids, cholesterol and triglycerides (Cmelik *et al.*, 1969). We cannot go into the details of this organ and its implications here (for lipids see: Langley and Pimley, 1974; Langley, 1977; Langley *et al.*, 1981; for RNA and protein synthesis see: Tobe and Davey, 1974; Tobe *et al.*, 1973; Tobe and Langley, 1978; Davey, 1985; summarized recently in Gillott, 1988). According to the intensive production of milk during larval growth, milk glands are composed of numerous distal tubes, emerging by branchings of four proximal tubes, which insert themselves in two collecting ducts, ensheathed by a common muscular tube (Tobe *et al.*, 1973).

In Hymenoptera an additional gland exists (Dufour's gland or calyx gland) which opens dorsally into the vagina. *Leptopilina heterotoma* is a parasitoid wasp which deposits her eggs in the haemocoel of *Drosophila* larvae. Normally, all foreign substances are encapsulated by special haemocytes (lamellocytes) of the host. When depositing its eggs, *Leptopilina* injects substances into the host haemolymph from a colleterial gland (Dufour's gland) which selectively destroys the lamellocytes but retains the longevity and health of the host (Ritzki and Ritzki, 1984). These results confirm the earlier studies of Osman (1978) who reported identical findings from another braconid wasp, *Pimpla turionellae*, infecting the larva of a moth, *Galleria melonella*. A third wasp, *Venturia canescens*, which infects larvae of *Ephestia kuehniella*, was investigated by Feddersen *et al.* (1986). Here too, a substance is produced by the calyx gland, covering and protecting the eggs for encapsulation by the host haemocytes. The virus-like particles accumulate first in the nuclei of the gland cells, are transported through the cytoplasm and are exocytosed. When these particles cover the eggs, they are protected against encapsulation by the host haemocytes. Eggs depleted of particles will be affected (Schmidt and Schuchmann-Feddersen, 1989). In *Biosteres longicaudus* (Hymenoptera: Braconidae) the virus-like particles are produced by accessory gland tubules and transported into the Dufour's gland where they accumulate before they are injected into the host (*Anastrepha suspensa*, Hymenoptera: Braconidae) (Lawrence and Akin, 1990). The immune response of the host is not well understood; however, a protein called haemolin which belongs to the immunoglobulin superfamily, might be involved (Sun *et al.*, 1990).

Other functions of the extra tubes of the vagina which are not directly involved in egg or sperm storage, egg laying or fertility are known to occur in some hymenopterans, such as sting bees, wasps, bumble bees or hornets, in which these tubes are transformed into poison glands, producing mixtures of toxins (Kindås-Mügge *et al.*, 1974; Jentsch, 1978).

The insemination of eggs is under the control of the female. Ripe eggs can be stored in the lateral or in the common oviducts. When passing into the vagina, a small portion of sperm is released actively by the female spermatheca via its proximal part which is ensheathed with muscles and serves as a sperm pump. Copulation, egg storage, egg movement, insemination and egg laying are all under nervous and/or humoral control, stimulated by environmental signals (for more details of this interesting and wide field see: Adiyodi and Adiyodi, 1975; Tobe and Langley, 1978; Sugawara, 1981; Davey, 1985; Raabe, 1986; Gillott, 1988; Grillot and Raabe, 1989; Kaulenas, 1991).

3.1.2 Ultrastructure

According to the preceding section we can expect a typical ultrastructure for ectodermal glands, since sperm must be nourished during their stay in the receptaculum seminis and, corresponding to accessory glands, adhesive glues and/or other secretions have to be produced. The secretions of the ectodermal glands have to overcome a barrier, the cuticle, which is produced by the ectoderm itself during an earlier ontogenetic period. In 1974, Noirot and Quennedey reviewed the work on epidermal tissues and glands and presented three types of gland cells, according to their mode of overcoming the cuticular barrier.

The first type does not differ very much from normal epidermal cells (Figure 3.2). It is covered by the cuticle and the secretion passes through the cuticle directly. In cells of the second type, the secreting cell is enclosed by two epidermal cells and the secretion crosses these cells before diffusing through the cuticle (Quennedey, 1971). The third and most common type is developed by the cooperation of four cells, of which one or two can be reduced totally during the secreting phase. The ontogenetic development will be described for the tergal glands of *Blatella* (Noirot and Quennedey, 1974; Sreng and Quennedey, 1976) and by the colleterial gland of *Rhodnius prolixus* (Lococco and Huebner, 1980b; Figure 3.3a).

By two subsequent divisions in which cytokinesis remains incomplete for a longer period, four cells arise. One cell becomes the innermost cell, sheathed by the second, and both by the third, and all three by the fourth cell, like the leaves in a bulbous tuber. In the beginning, all cells contact the cuticle. The innermost cell develops a cilium which grows. This cell and the directly accompanying cell are pushed down creating a duct in the third cell. Additional growth continues the duct into the first cell from which the so-called end apparatus originates. The duct of the third cell becomes the ductulus. The cilium degenerates while the end apparatus is growing. The ductules and end apparatus are invaginations of their cells and are covered consequently by a thin epicuticle. The innermost cell and the fourth cell will degenerate in many glands. The second cell becomes the real secretory cell. The secretion will pass through the cell membrane by exocytosis and enter the end apparatus (Figure 3.4) from which it reaches the lumen via the ductulus.

This third type of secretory cell is found in many epidermal tissues and among these in spermathecae and accessory glands (including collateral, colleterial, Dufour's, and poison glands, irrespective their doubtful phylogenetic relations, Gillott, 1988)(Clements and Potter, 1967; Conti *et al.*, 1972; Suzzoni, 1972; Barbier, 1975; Berry and Johnson, 1975; Happ and Happ, 1975; Dallai, 1975; Johnson and Berry, 1977; Huebner, 1980; Lococo and

Figure 3.2 (a) Electron micrograph of type 1 gland cell, which consists of only one cell, as found in spermatheca of *Rhodnius prolixus* after starvation. Inset gives an overview of epithelial regression. Lumen (L); cell apex (AI); lateral interdigitations (I). (b) General structure of distal epithelial cell type in the spermatheca gland of *Rhodnius*. BL = basal lamina. Bars represent 2 μm. (Huebner, 1980; with permission of E. Huebner and Alan R. Liss, Inc.)

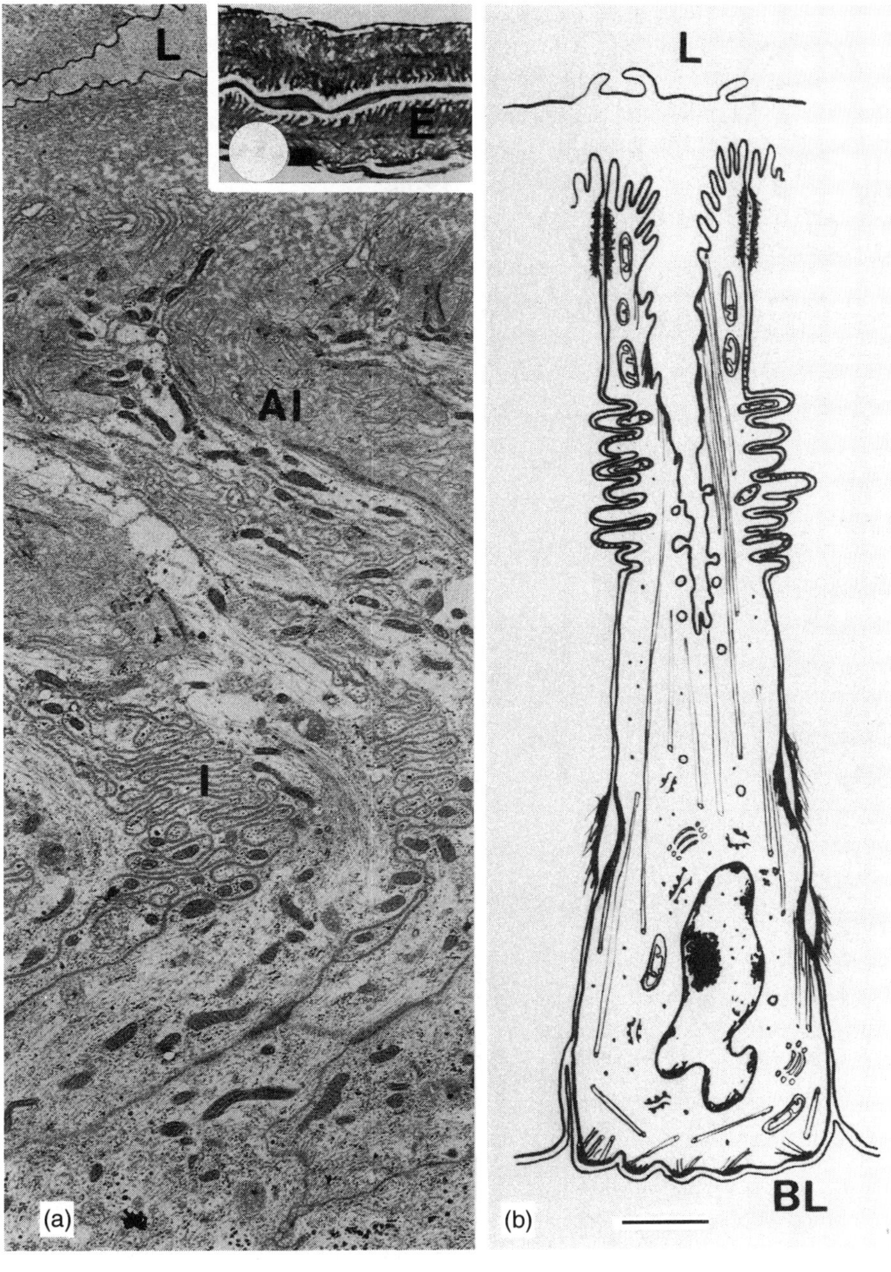
L
E
AI
I
(a)
L
BL
(b)

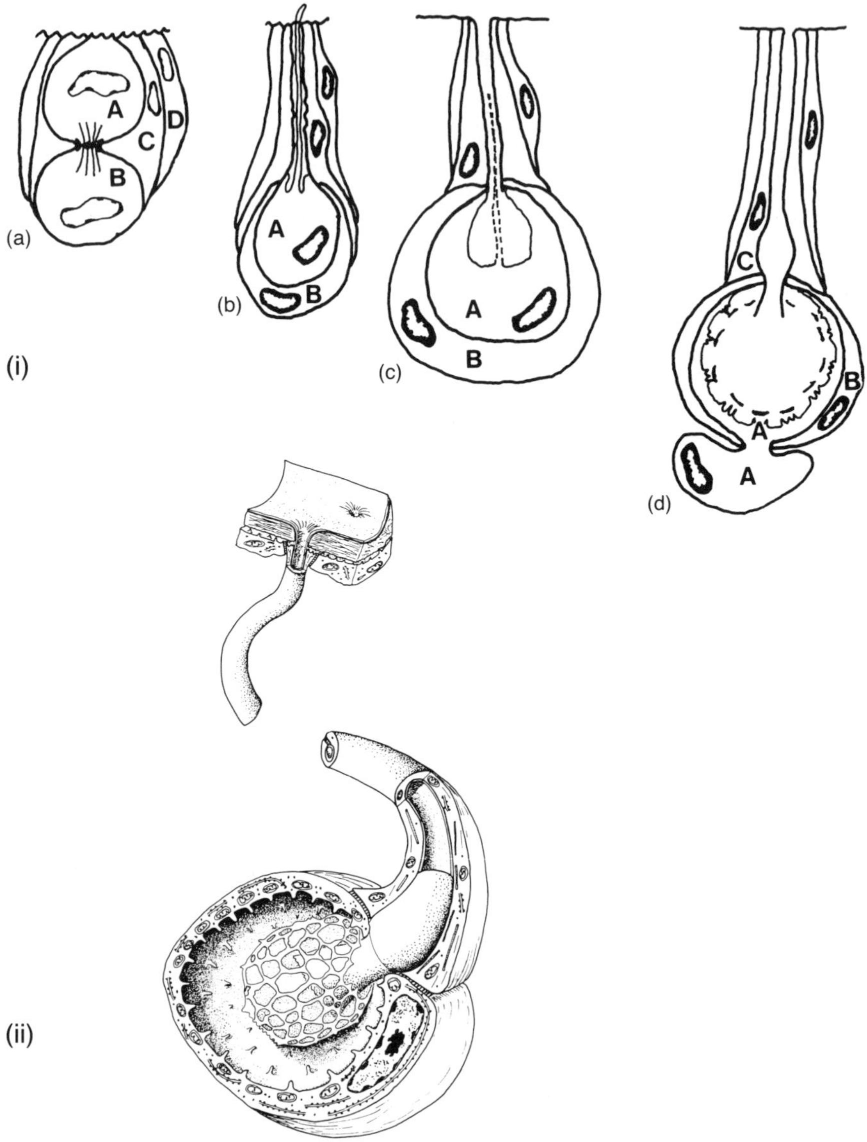

Figure 3.3 (i) Scheme of ontogenetic development of type III epidermal glands combined after Sreng and Quennedey (1976) and Lococo and Huebner (1980b). At least two inner cells (A and B) are sister cells which build the end apparatus (cell A) and the surrounding cell (B). Cell (C) builds the ductule and is encompassed by cell (D). In most type III epidermal glands, the end apparatus is built by transformation of cilium (b, c), which degenerates later on. For further discussion see text. (ii) Drawing of cement gland of *Rhodnius prolixus*, type III epidermal gland, in which no cilium is built during development of the end apparatus. (With permission of E. Huebner and Longman Group Ltd.)

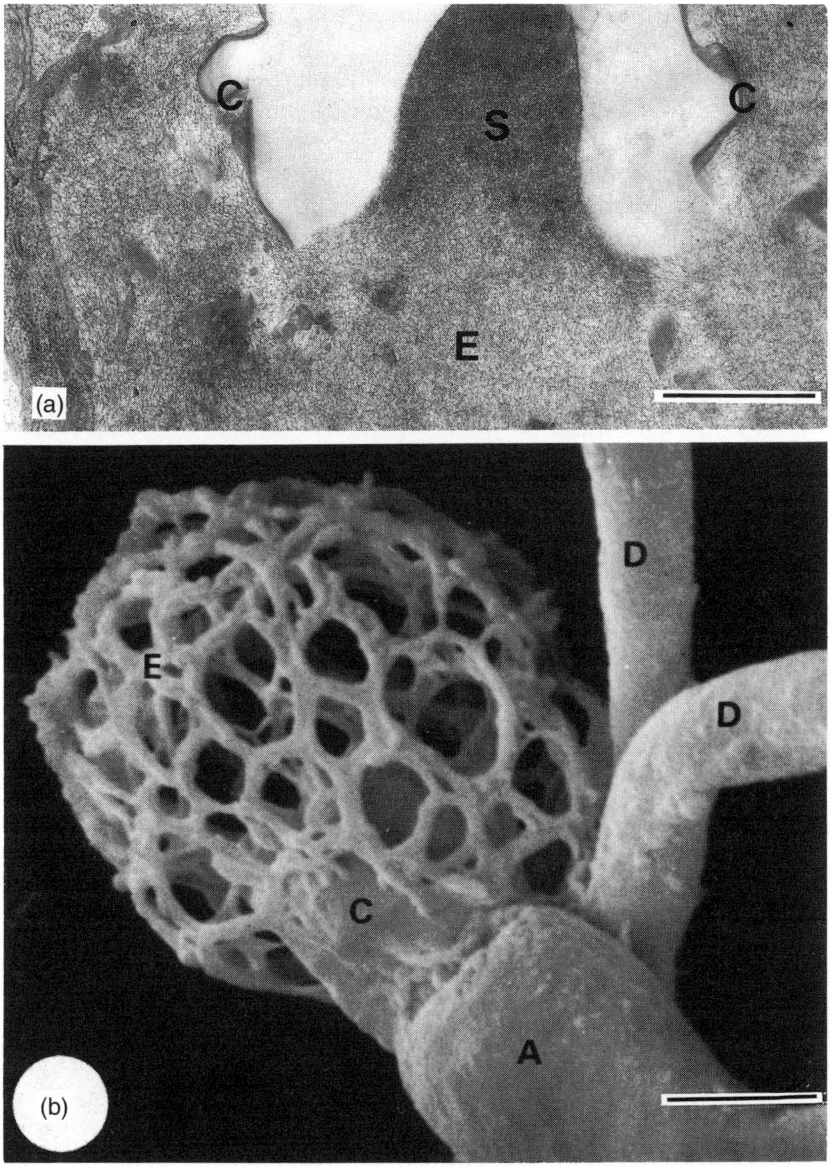

Figure 3.4 (a) Longitudinal section through end apparatus (E), collar region (C), and non-ciliar secretion area (S) during ontogenetic development of female accessory gland in *Rhodnius prolixus*. (b) End apparatus (E) of cement gland secretory cells, type III, as found in *R. prolixus*. Funnel-like collar (C), enlarged ampulla (A), ductule (D); bars represent 1 µm. (Lococco and Huebner, 1980b; with permission of E. Huebner and Longman Group Ltd.)

Huebner, 1980a,b; Al-Khalifa, 1981; Bitsch, 1981; Pal and Ghosh, 1982; Ramadan *et al.*, 1986a,b; Soltani-Mazouni and Bordereau, 1987; Blass and Ruthmann, 1989). Not all of these papers are based on ontogenetic developmental investigations. However, in all glands described, a ductule will be embraced by a cell other than the cell creating the end apparatus, and ductules and end apparatus are of the same architecture. Variations occur concerning the number of entangled cells, as for instance in the colleterial (cement) gland of *Rhodnius prolixus* (Hemiptera) where the fourth cell apparently does not embrace the other three during the formation of the secretory unit, and no transient ciliary structure was detected (Figures 3.3b, 3.4) (Lococco and Huebner, 1980b).

As shown, the third type of gland cell seems to be the most common among female ectodemes. The second type has not yet been reported, but the first type is found for instance in the spermathecae of *Rhodnius* (Figure 3.2) where the cells differ between proximal and distal regions (Huebner, 1980). The proximal duct is ensheathed with well-developed muscles and serves as a sperm pump. According to the proposed function of spermathecae, the apical cell membranes are deeply folded, suggesting an intense communication between 'in-doors' and 'out-doors' at least for smaller molecules, to ensure an optimal environment for sperm survival (Huebner, 1980).

3.2 MESODERMAL TISSUES OF THE OVARY

Ovaries are internal organs which shelter germ cells during their development from their dormant stages as oogonia to fully differentiated oocytes. All somatic tissues of the ovary are derived from the mesodermal sheath. We will first describe the differentiation of ovarioles, which are the final organs of oocyte development in insects. The following sections will describe the differentiated tissues of the ovary of adults.

3.2.1 Ontogenesis of the ovary

The insect ovary is subject to changes in its anatomy, morphology and ultrastructure during the whole life of the animal. We will focus on the period in which the ovary is built, a period which will start after embryogenesis and will come to a preliminary end when oocyte growth begins. The general architecture of an insect ovary does not depend on germ cell interaction with the somatic tissue as is shown by irradiation experiments. Jura (1964a,b) has shown that cleavage nuclei reaching the posterior pole of *Drosophila virilis* embryos will not develop into pole cells (germ cells) after preceding UV irradiation of pole plasm. These posterior cells failed to invade the normal developing ovaries. Similar irradiation experiments were made by Günther (1971), which gave the same results as ligation experiments, both carried out on the long and slender eggs of *Pimpla turionella* (Hymenoptera): during normal development pole cells are built, but they are absent after the experimental procedures. Nevertheless, normal gonads develop, the inner mesodermal cell mass of which is not replaced by immigrating pole cells. Richard-Mercier (1977) did similar experiments in *Leptinotarsa decemlineata* (Coleoptera: Polyphaga). Cauter-

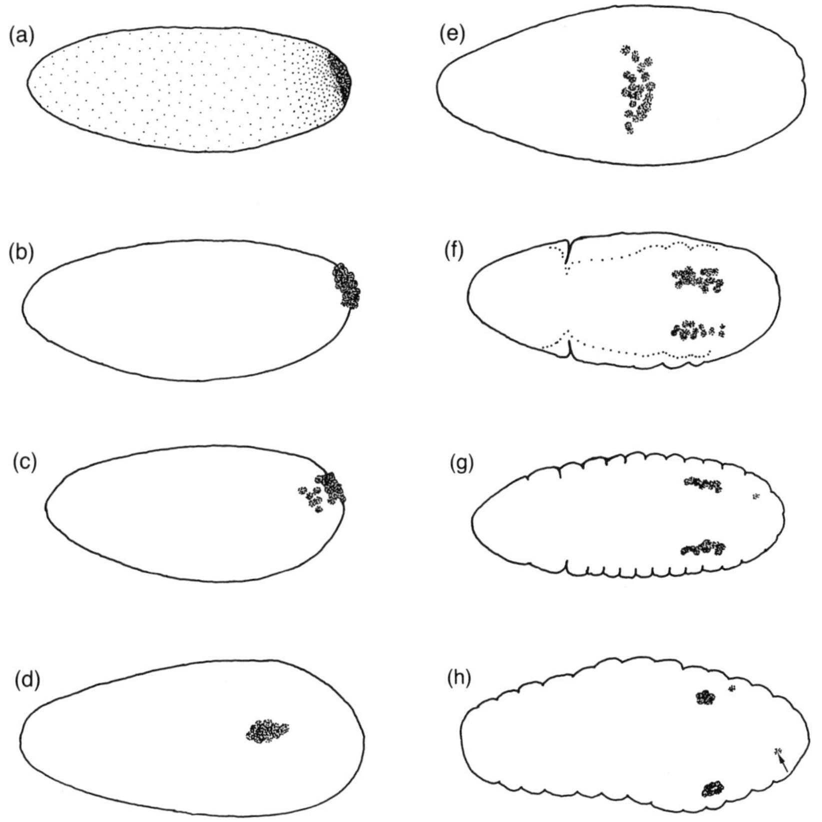

Figure 3.5 Documentation of early movements of germ cells in *Drosophila* wild type embryos. Germ cells are visualized by staining with anti-*vasa* antiserum. (a) Cleavage embryo: note that *vasa* protein is found in the whole embryo but a steep gradient is formed, rising at 20% egg length and ends with its maximum at the posterior pole (0%). (b) Cellular blastoderm, only pole cells are stained. (c) Early gastrulation, pole cells begin to immigrate. (d) Germ band elongation (dorsal view as in all following figures), pole cells shift as one block, associated to posterior midgut primordium. (e) Pole cells begin to move apart from midgut primordium. (f–h) Pole cells assemble at both sides of the alimentary canal in only one spherical gonad anlage each. Arrow in (h) points to a germ cell lost during migration phase. (Drawings after colour pictures of Lasko and Ashburner, 1990.)

ization of pole cells did not prevent development of gonads in either sex. However, interactions between germ cells and somatic cells of the ovary play an important role during oocyte differentiation (sections 3.3.1 and 3.3.2).

After early dislocation of pole cells into two main groups, insect germ cells are housed in two mesodermal tubes, placed on each side of the alimentary canal (Figure 3.5; for further discussion of pole cell migration

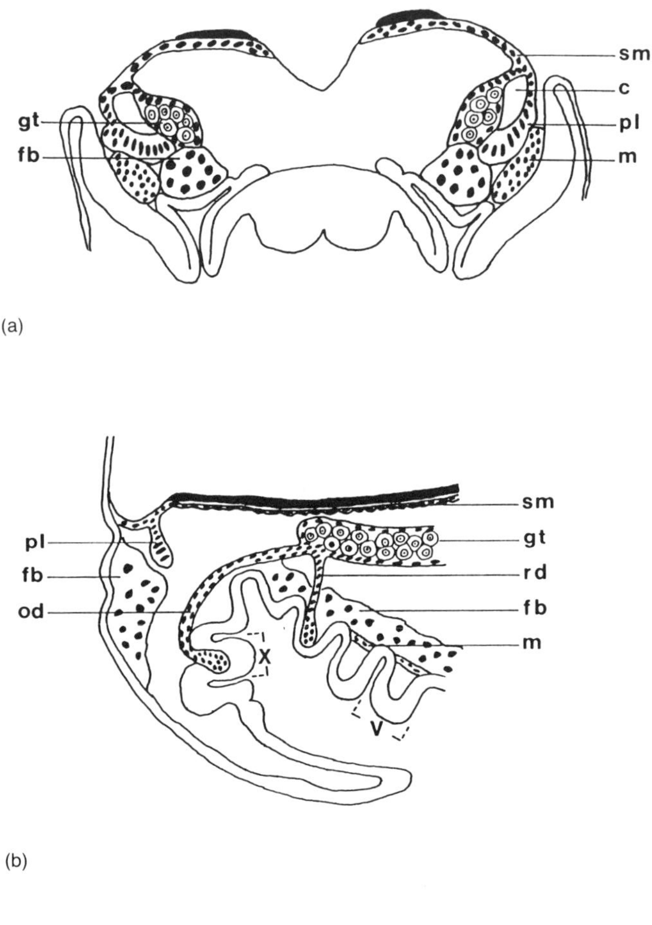

Figure 3.6 Cross-section (a) and longitudinal (sagittal) section (b) through abdomen of *Forficula* sp. embryo, just before katatrepsis (a), and shortly after (b). Female gonads have just been formed as gonadal tubes (gt), enclosing germ cells. Other mesodermal tissues shown by black nuclei only: splanchnic mesoderm (sm); pericardial cells (pl); coelom (c); muscles (m); fat body (fb); reduced gonadal duct (rd); oviduct (od); segments (5, 10) are indicated by roman letters; entoderm is black; ectoderm and its derivatives are shown blank. (After Heymons, 1895.)

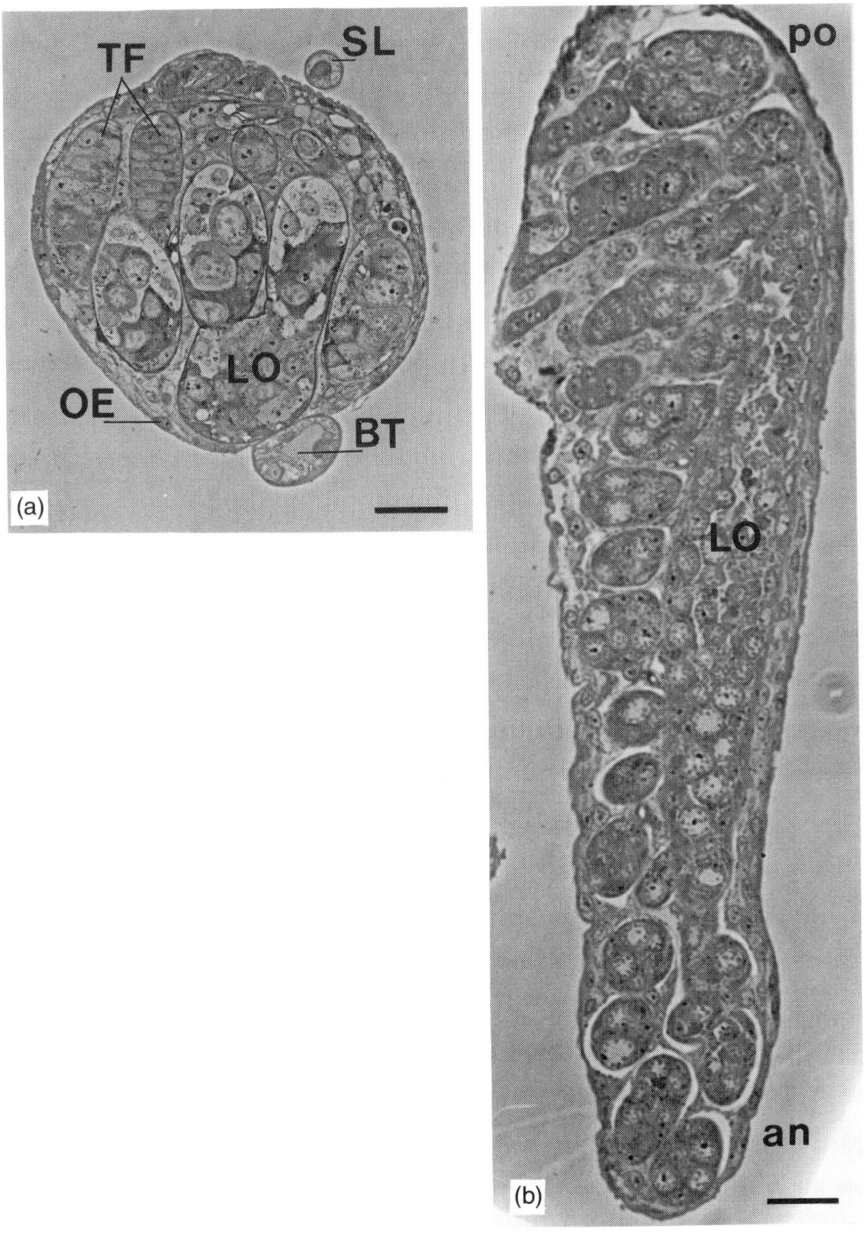

Figure 3.7 (a) Cross-section and longitudinal section (b) through gonadal tubes of *Sialis* larva, showing immigration of germ cells from lateral oviduct (LO) into ovariole anlagen. Basal tube (BT) and suspensory ligament (SL), which connects all terminal filaments (TF), are both ensheathed with thickened basal laminae. Outer envelope (OE) of gonadal tube persists throughout gonadal development. Posterior (caudal) ovariole anlagen (po) precede anterior (cranial) (an) during development. Bars represent 25 μm.

and ovary anlagen see sections 3.3.1 and 3.6.1). Both tubes are in close contact with the fat body. These ovary-anlagen (primordia) are outpocketings of the splanchnic mesoderm (i.e. the mesodermal coelomic wall, next to the alimentary canal) (Figure 3.6) (Heymons, 1891, 1895, 1897; Kahle, 1908; Hirschler, 1909; Seidel, 1924; Weber, 1933; Snodgrass, 1935; Heming, 1979).

The tubes can vary in length from most concentrated ball-like structures, as known from *Drosophila* (King *et al.*, 1968; King, 1970) and other flies like *Chrysomya* (Ullerich, 1981), to elongated tubes as found in many orthopteroid species (Heymons, 1895) and other orders, for example Megaloptera (Figure 3.7) (Büning, 1979c).

Each tube ends in a suspensory ligament which can reach into the thoracic segments where it is anchored next to the heart at the dorsal diaphragm. However, in many insects the suspensory ligament becomes shortened and will attach to the body wall, the fat body, or it will be reduced totally.

The caudal end of each gonadal tube narrows and forms a knob or a small tube, which gives birth to the lateral oviduct. Its basal part will contact and fuse to the ending of the common oviduct, which emanates, as already shown, from the ectoderm (Figure 3.6b). In many species it is the border of the ninth to tenth abdominal segment, which gives birth to the final opening; however, the previous segments contribute to ectodermal parts (see section 3.1; Figure 3.1).

The description of the general pattern of ontogenetic development of insect ovaries follows in the main the description of the *Drosophila* ovary (King, 1970). Deviating developments are described in section 3.5, where special orders are looked at. The phylogenetic significance of differences and a general view of early events is given in section 3.6.1. Within the germ tubes we find differentiation along the tubes (cranio–caudad) and at right angles (i.e. in the dorso-ventral direction) (Figures 3.7 and 3.8).

The inner differentiation of mesodermal parts begins with an accumulation of cells on the dorsal and ventral side of the tube (Figure 3.8a). Within the dorsal plug, some cells differentiate into terminal filament anlagen, contacting ventrally with the germ cells. Between two neighbouring terminal filament anlagen some other somatic cells intervene with the germinal cells, separating the whole mass into smaller, finger-like portions (Figure 3.8b). These somatic cells split into two layers, the innermost will become the inner sheath and, more basally, the prefollicular cells; the outer layer differentiates into the tunica externa. The ventral plug of mesodermal cells differentiates into the lateral oviduct. The lateral cells differentiate into the outer envelope (peritoneal sheath) (Figure 3.8c). In many species, this epithelium disrupts and becomes reticular during growth.

Ovary development starts during late embryogenesis and proceeds during larval (and pupal) stages. However, the precise position in the life

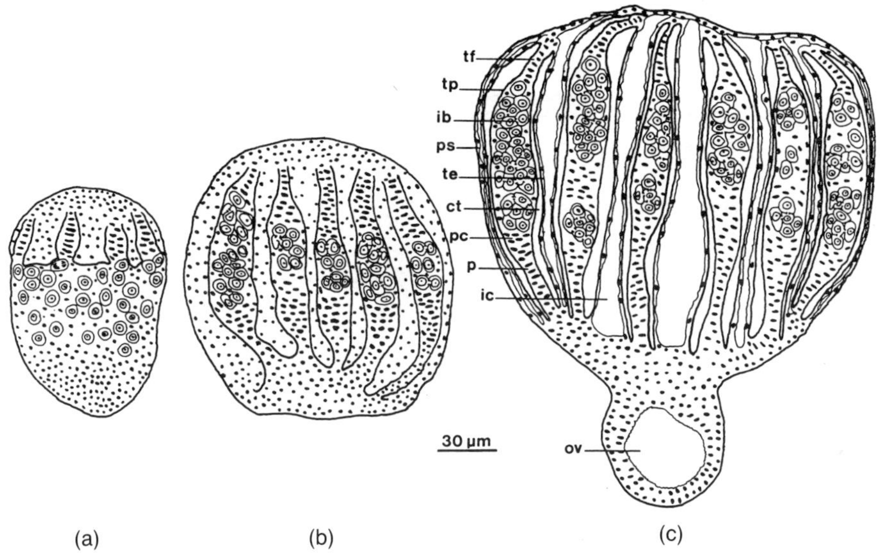

Figure 3.8 Cross-section through female pupal gonads of *Drosophila.* Two hours after puparium formation (a), somatic cells, represented by black dots, are morphologically differentiated as cells of the terminal filament, which become enclosed by the tunica propria (thick line). Germ cells are shown as single cells, one of which, at the base of the terminal filament is in mitosis. Thirty hours after puparium formation, germ cells are enclosed by somatic cells of the ovarioles, just formed (b). The germ cells have already started cystocyte mitoses, indicated by intercellular bridges. Thirty-six hours after puparium formation, all sheaths have been formed and the cavity of the oviduct is built (c). Terminal filament (tf); tunica propria (tp); intercellular bridge between germ cells (ib); peritoneal sheath (outer ovary envelope) (ps); tunica exterrna (te); cleft between tunica externa and tunica propria (ct); prefollicular cells (pc); pedicel (p); interovariolar cleft (ic); oviduct cavity (oc). (Diagrams after King, 1970.)

cycle and its duration is species specific and depends on life time and behaviour of larval and adult stages. For instance, if the imago does not feed, all processes of egg growth (i.e. vitellogenesis and previtellogenesis) are shifted to pupal and larval stages and organogenesis of gonads begins in late embryogenesis and is completed in early larval stages as found in Ephemeroptera or Megaloptera. Short larval phases, as in Diptera, concentrate most developmental phases into the pupal stage. In Figure 3.9 the phases of organogenesis of ovaries and oogenesis are shown as they fit into the life cycles of diverse species.

Ovariole development is not always fully synchronized as noted above, especially in those insects in which comb-like ovaries exist. Metachrony is found for instance in *Sialis* (Megaloptera) (Figure 3.7) (Büning, 1979c),

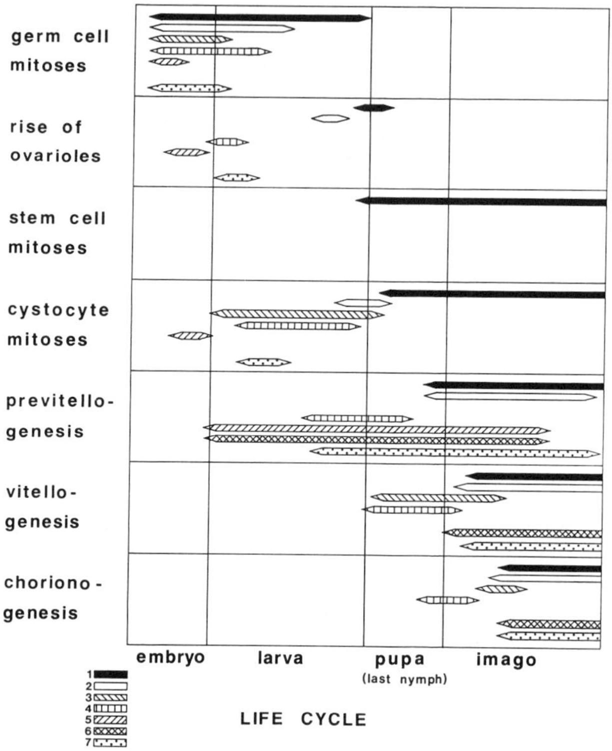

Figure 3.9 Synopsis of oogenetic periods in relation to life cycle stages of diverse species. The bars represent time of duration only. (1) *Drosophila melanogaster* (Diptera) (data from King (1970) and others). This species is the only one known, in which differential stem cell mitoses are going on indefinitely. Most periods are shifted in pupal and adult stages. (2) *Bruchidius obtectus* (Coleoptera); stem cell mitoses are in question and omitted, as in all following examples. Cystocyte mitoses cease early in pupal stages. (3) *Elenchus tenuicornis* (Strepsiptera); in this endoparasitic species somatic tissues are reduced, thus, no ovarioles are formed. Previtellogenesis does not take part. Vitellogenesis is short, embryogenesis up to hatching takes part in the mother. (4) *Sialis flavilatera* (Mecoptera: Sialidae); main phases are shifted into early stages of the life cycle. (5) Virginopare aphids (Hemiptera); extreme shift into embryonic and larval stages. Previtellogenic growth is greatly reduced, but occurs during all larval and early imaginal stages, according to the telescoped generations. Vitellogenesis and chorionogenesis are omitted, embryos are fed by haemolymph resources of the mother and grandmother. (6) Ovipare aphids (Hemiptera); early oogenetic periods are not investigated, vitellogenesis and chorionogenesis are normal. (7) *Gryllus bimaculatus* (Orthoptera); as known from many hemimetabolous insects, early periods of oogenesis are shifted to embryonic and larval stages.

Sciara (Diptera) (Figure 3.10) (Büning, unpublished), *Leptinotarsa* (Coleoptera: Polyphaga) (Richard-Mercier, 1977; Richard-Mercier *et al.*, 1986), *Liposcelis* (Psocoptera) (Goss, 1954) and diverse species of aphids (Blackman, 1978; Büning, 1985). In *Sialis, Liposcelis* and *Sciara* ovarioles of the caudal region develop earlier than apical ovarioles; the opposite occurs in aphids and scale insects. However, in *Liposcelis,* where five ovariole anlagen are built, only one or two from the middle will reach vitellogenic stages. In scale insects, the metachrony is similar, as in aphids, but this pattern is superimposed by some late developing ovariole anlagen, scattered randomly among others (Figure 3.11a) (Emeis, 1915; Huges-Schrader, 1925; Weglarska, 1961, 1962).

Sometimes, ovariole development becomes disturbed and the tunica externa fails to develop between ovarioles, thus, a common tunica propria connects two ovarioles (Ma and Ramaswamy, 1987). Other abnormalities lead to ovarioles with doubled pedicel regions; sometimes, the basal region does not split, giving birth to ovarioles with double heads fusing into a common pedicel. Both abnormalities are found commonly in higher termites like *Odontotermes stercovivorus*, a species in which the number of ovarioles is raised from about 70, when hatching, to more than 3000 at the end of nymphal stage 4 (Figure 3.11b) (Truckenbrodt and Amelung, 1986).

3.2.2 Ovarian sheaths and terminal filament

Each ovary is encompassed by a mesodermal envelope, generating from parts of the splanchnic mesoderm, the gonadal tube. The ultrastructure of these ovarian sheath cells has not been studied during the last decades of ultrastructural research. Thus, we must refer to more preliminary results.

The ovarian envelope of the whole ovary persists in some species as mentioned above. One of these is the ovary of *Sialis* sp., in which the outer sheath of the ovary perpetuates up to the egg-laying period. The sheath is enclosed on both sides by a basement membrane, the structure, function and composition of which are described in section 2.3 (Figure 3.12). Cells and nuclei are discus shaped. Cell to cell contacts are stabilized by desmosomes. During oocyte growth periods and during egg storage, which takes place in this order in the lateral oviducts, the outer membrane expands tremendously and huge holes appear between cells. Parallel ordered microtubules and microfilaments are found. Ectodermal tracheal cells envelop and transmit the sheath in part, going directly to the ovariole envelope; the tunica externa is more reduced in this species than in ovarioles of similar length in other species.

From older literature it is well known that whole ovaries as well as single ovarioles can contract (e.g. Landois, 1869; Curtin and Jones, 1961; King, 1970; Kriger and Davey, 1982; King and Büning, 1985). As described above

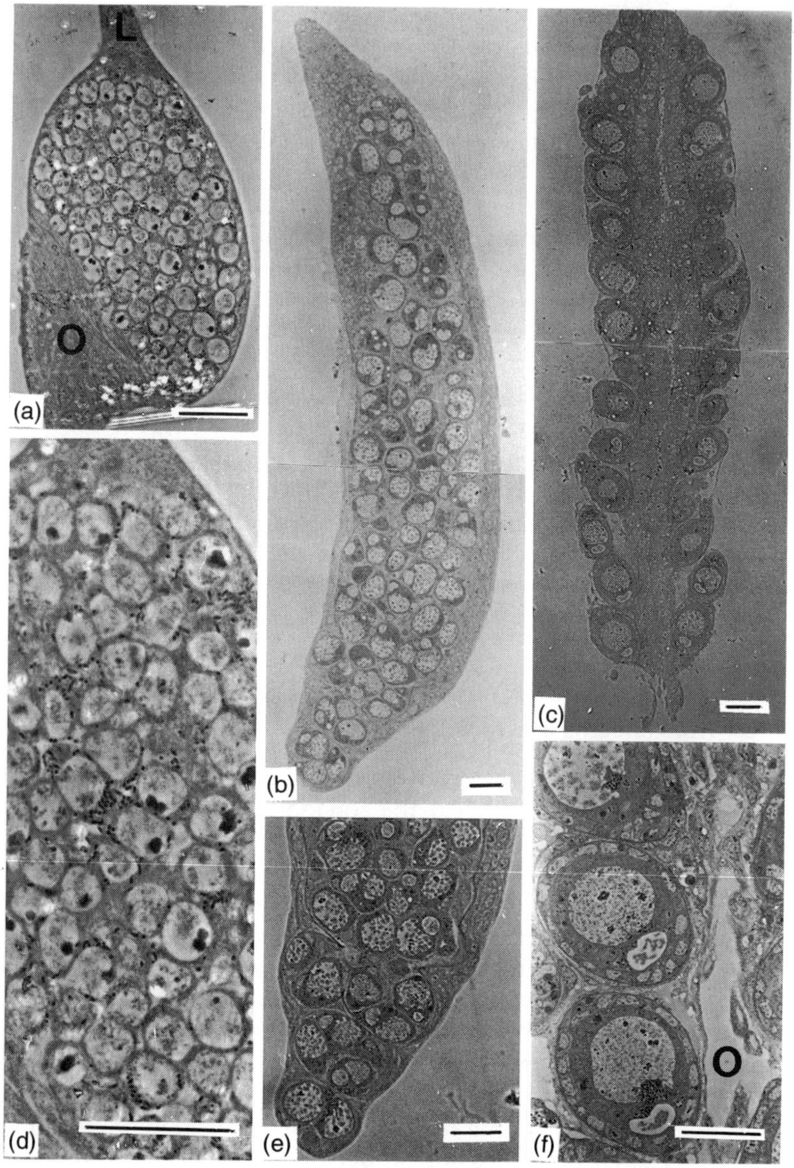

Figure 3.10 Development of ovary in *Sciara coprophila* (Diptera: Sciaridae). (a) Larval ovary at the 'eyespot stage'. Each ovary-anlagen houses about 250 germ cell descendants, enclosed by somatic tissue which has already differentiated an antero-dorsal ligament (L) and a thickened basal part, the future lateral oviduct (O). Enlargement below shows germ cells, equal in size. (b) During pupal stages, each ovary anlagen elongates, and nurse cell–oocyte differentiation becomes apparent. Enlarged nuclei represent nurse cell nuclei during endomitosis, small cells with defined chromosomes are oocytes. Enlargement below shows first stages of ovariole development. Some germ cell clusters become enclosed and separated by somatic interstitial cells and tunica propria. (c) Differentiation of ovarioles is completed in pupae. Enlargement below shows three follicles with one nurse cell and one oocyte each, enclosed by follicular cells. Lateral oviduct (O). Bars represent 20 μm.

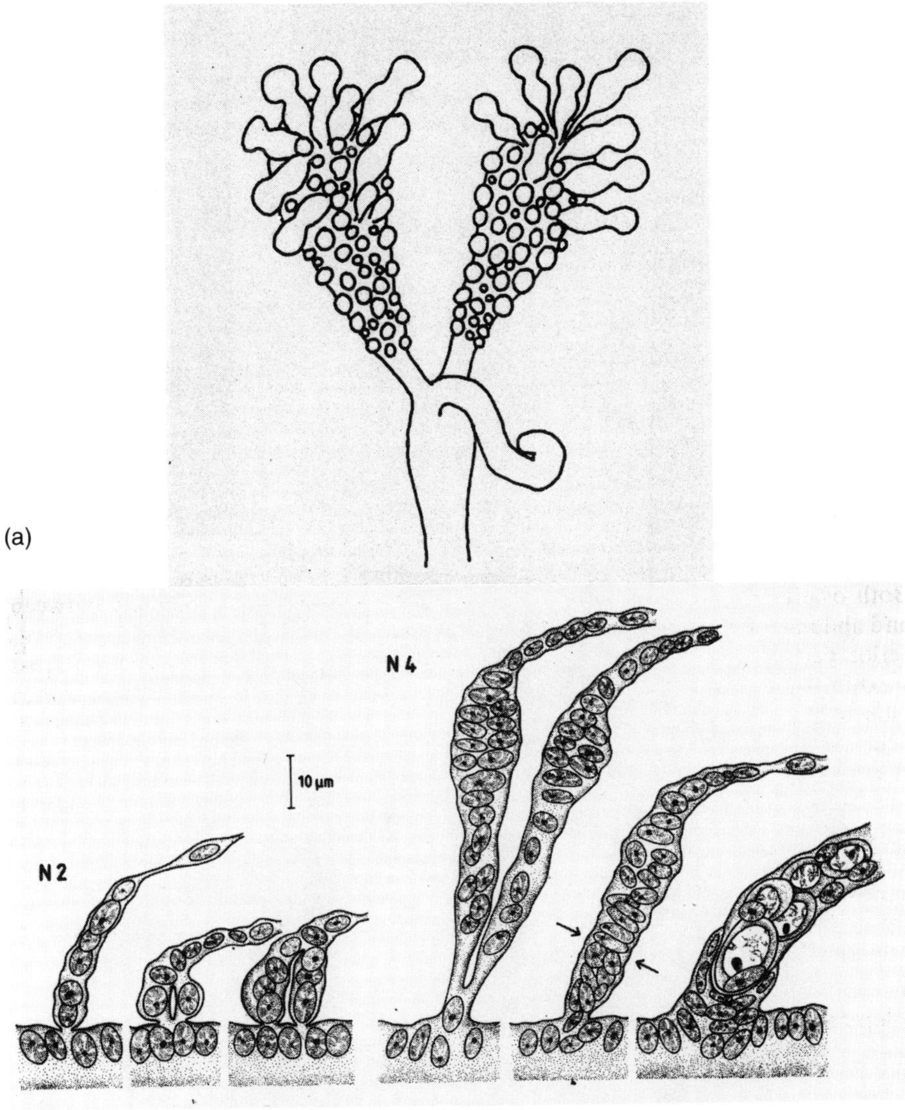

Figure 3.11 (a) Ovary of scale insects. Ovarioles are more advanced in apical regions. However, very young ovariole anlagen are scattered in between (after Weglarska, 1961). (b) Additional ovarioles are built by fission during stages of nymphal development (N2; N4) in a prospective termite queen (*Odontotermes stercorivorus*). Lateral oviduct is indicated at the base. Arrows signify the boundary between ovariole and oviduct. Somatic envelopes are not shown. (From Truckenbrodt and Amelung, 1986; with permission of E. Schweizerbarth'sche Verlagsbuchhandlung, Stuttgart.)

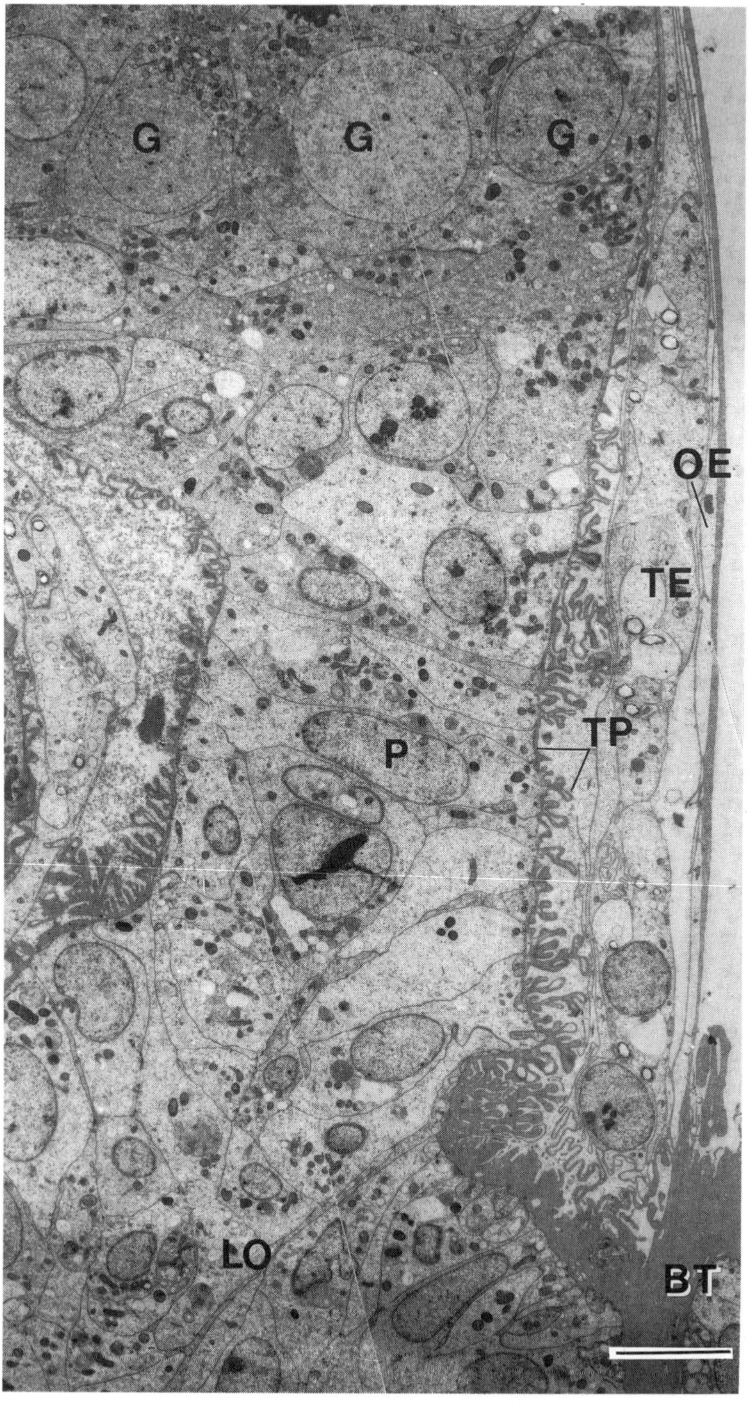
G
G
G
OE
TE
TP
P
LO
BT

for the ovary of *Sialis* (Megaloptera), the ovarian envelope persists in flies, too. Recently, the ovarian muscles of the stable fly *Stomoxys calcitrans* were investigated by Cook and Peterson (1989). They found stellate muscle cells

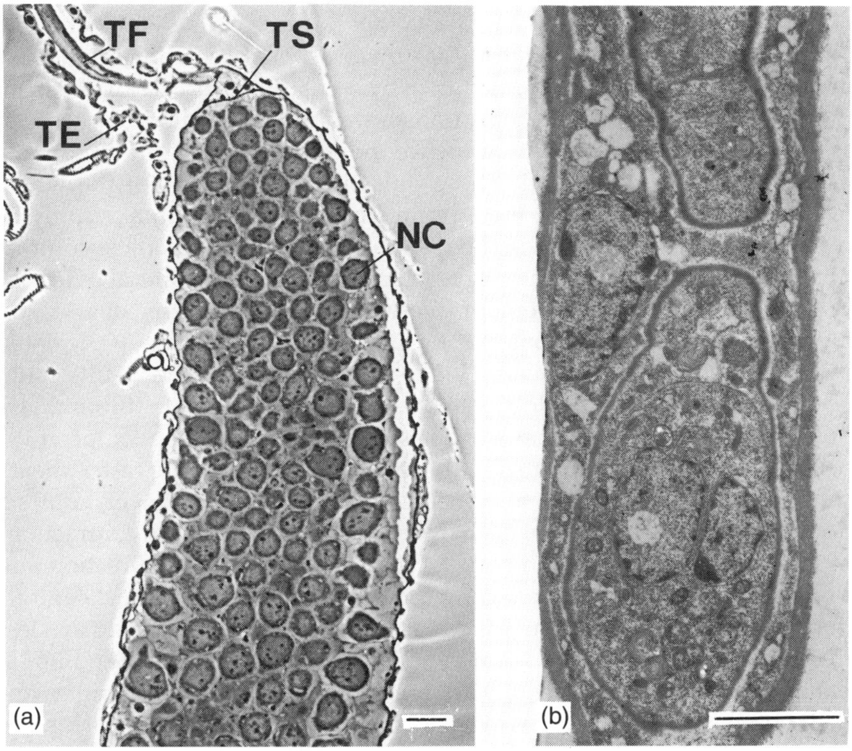

Figure 3.13 (a) Semi-thin section through the apical part of an ovariole of *Coccinella septempunctata* (Coleoptera, Polyphaga: Coccinellidae). Terminal filum (TF) is accompanied by tunica externa (TE). Tranversal septum (TS) splits from basal lamina (BL) of nurse chamber (NC). Bar represents 20 μm. (b) Ultrastructure of terminal filum of *Ctenophthalmus* sp. (Siphonaptera), which is ensheathed by tunica externa. Both tissues are accompanied by basal laminae. Bar represents 2 μm.

Figure 3.12 Lateral oviduct (LO), pedicel region (P), tunica propria (TP) (= basal lamina of the ovariole), tunica externa (TE), and outer envelope (OE) of ovary of *Sialis flavilatera* (Megaloptera: Sialidae). Basal laminae surround each mesodermal tissue, as shown here for tunica externa and outer envelope. Note highly folded appearance of basal laminae in pedicel and lateral oviduct regions, as well as enormous thickness of basal laminae which accompany lateral oviduct and basal tube (BT). Germ cell (G). Bar represents 5 μm. (For a better understanding of the topology of tissues, compare with Figures 3.36 and 3.111.)

which form a network around the whole ovary, overlaid with a net of tracheas. The muscle cells appear striated as are the muscle cells of the ovariole sheath especially in the lower parts which ensheath the pedicel. As in the house fly (Degrugillier and Leopold, 1972), innervation of sheaths was only found in the pedicel region. Pulsating and synchronized contractions of muscle fibres of the outer envelope were recorded with complete cycles of oscillations ranging from 200 μm to 4 s. The oscillation is independent of a direct nerve supply, as shown by oscillating ovaries isolated from the nerve system; however, neurotransmitters and neuromodulators, such as octopamine and proctolin, influence the frequency and amplitude in a manner not yet understood. These pulsating contractions of the whole ovary are interpreted as a local pump to ensure an optimal supply of resources to the ovarioles via the haemolymph. Vertical pedicel contractions promote the extrusion of eggs from the pedicel and lateral oviducts. Whereas the above measurements were made on vitellogenic ovaries, we investigated these pulsations during previtellogenic growth on isolated ovaries of *Drosophila hydei* using video tapes. Generally, the results could be confirmed, but some additional observations were made (Büning, unpublished). Each contraction wave is started in the outer envelope at the pedicel region and terminates in the apex of the ovary. The outer sheath contractions are independent of pedicel contractions. However, contractions of the tunica externa of each ovariole are coupled to the contractions of the pedicel: a contraction wave is induced in the apical direction and slows down to zero when reaching the youngest follicle. Pedicel contractions and their contraction wave along the individual ovariole are less frequent, less constant, and not synchronized among ovarioles. Similar patterns of muscle contractions are found in other types of ovarioles (for their possible role in transport mechanisms see section 3.4). In many insects, the tunica externa is the functioning envelope of ovarioles during the phase of oocyte growth (Figures 3.12, 3.13 and 3.14).

The development of the tunica externa and its accompanying net of trachea and tracheoles normally precedes the growth phases of germ cell descendants. Furthermore, the bigger an animal is, the bigger is the tunica externa and its accompanying tracheal network, a direct consequence of the intense gas and metabolite exchange needed for oocyte growth. One example of these extremes is the aphids, in which ovariole sheaths are highly reduced, especially during virginopare generations; at the other extreme are the queens of social insects such as bees, wasps and ants in which the young ovarioles are wrapped up with highly folded sheaths. The ultrastructure is similar to that described for the outer envelope of the ovary. The tunica externa develops a reticular network of striated muscles (Figure 3.14). Thereby, a developmental gradient can exist from apical to basal regions in each ovariole sheath, as known from flies. Basal parts have a thick musculature which becomes thinner and thinner in the apical

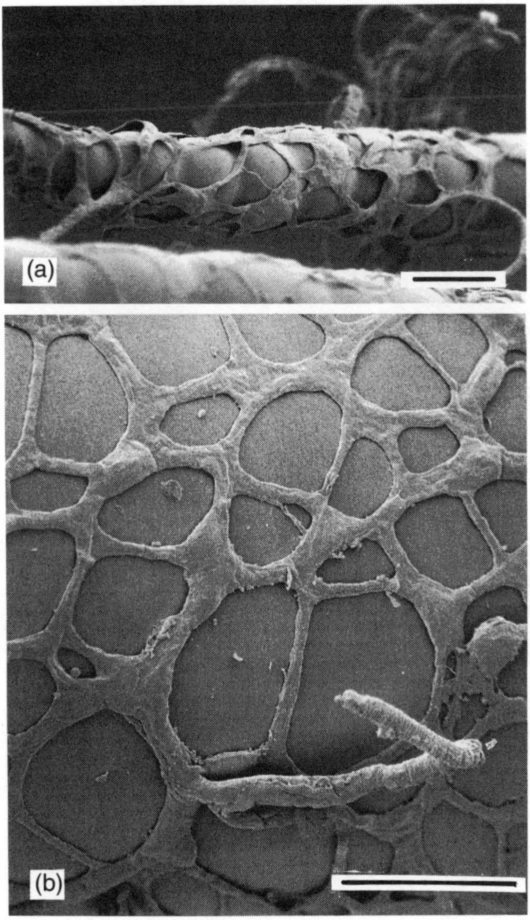

Figure 3.14 SEM of tunica externa of *Locusta migratoria* (Orthoptera). (a) As typical for ovarioles of many species, tunica cells form a network around ovarioles. (b) Detail of the same ovariole at the beginning of vitellogenesis. Bars represent 50 μm. (Courtesy of J. Gottanka, University of Erlangen-Nürnberg, Germany.)

direction. The germarium or the terminal chamber of telotrophic ovarioles is devoid of musculature, but muscle fibres are found in the zone of previtellogenesis.

In stone flies (Plecoptera), ovarioles are devoid of terminal filaments and of tunica externa. The ovarioles develop as outpocketings, a mode also known from coccid ovaries. The ovary sheath is reduced also and embraces only the lateral oviduct as a network of loosely connected cells (Figure 3.15).

The terminal filament is supposed to be the first mesodermal tissue, differentiating in the ovariole anlagen (Figures 3.7 and 3.8). Those cells

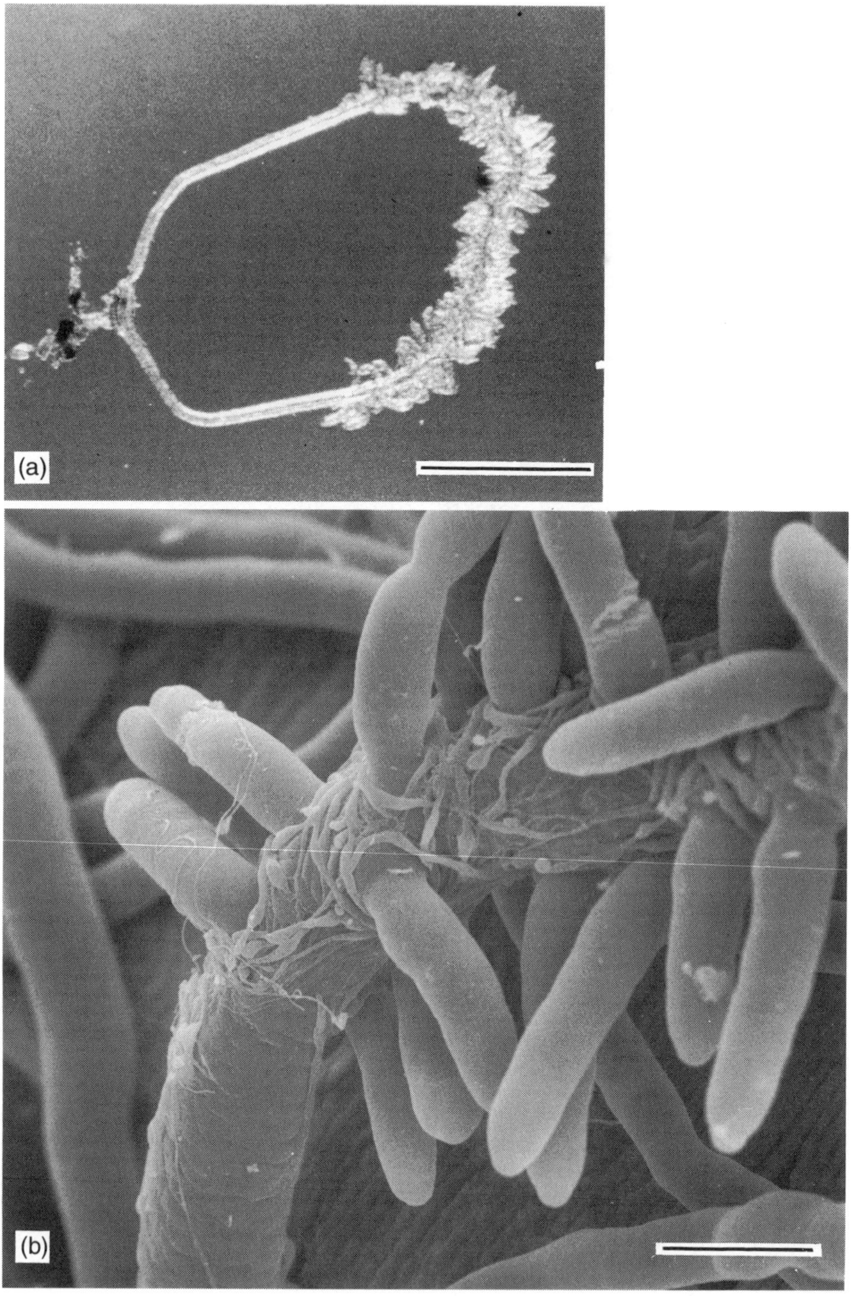
(a)
(b)

situated next to the germ cells are disc-like and orientated perpendicular to the long axis of the ovariole. In many species, such as hemipterans or coleopterans, they are clearly separated by a transverse septum from the cells of the inner sheath (Figure 3.14a). In other species, like *Drosophila* (King, 1970), *Sialis* (Büning, 1979c) and *Parthenothrips* (Pritsch and Büning, 1989), a transverse septum is absent. Its material is continuous with the tunica propria and is discussed in the following section. The disc-shaped appearance of the cells changes with the distance from the germarial region. Cells of the terminal filament are single lined and become more and more elongated. The terminal filament is ensheathed by the tunica externa along its whole length, and the tunica externa of adjacent ovarioles of an ovary can join to a suspensory ligament. This ligament will fuse with the outer envelope of the ovary. Here, muscle cells can be developed (Huebner, 1984a).

3.2.3 Tunica propria and other extracellular matrices

All differentiating mesodermal tissues and blood cells (fibroblast-like haemocytes) secrete extracellular matrices (basal lamina, basement membranes) to all sides which are in direct contact with the body cavity (Francois, 1973, 1978; Wigglesworth, 1973; Ashhurst and Constin, 1974; Ashhurst, 1985; Knibiehler *et al.*, 1987). Therefore, all sheaths of the ovary are embraced with a basal lamina (Figures 3.12, 3.13 and 3.16). The ovariole itself has in its dilating parts, i.e. the basal vitellarium and the pedicel, a very strong basement membrane, the tunica propria, similar in diameter to the basement membrane encompassing the lateral oviduct. The role of follicle cells, which are mesodermal, will be discussed in section 3.2.5.

The extracellular matrix has been studied since evidence accumulated that molecules of this structure play an important role during embryogenesis, especially during organogenesis (Ashhurst, 1985; Bunch and Brower, 1993). In insects, as in vertebrates, decoration of the outer cell membrane surfaces with special proteins, which communicate with the extracellular matrix and/or other outer surfaces of other cells, is supposed to have a leader function for cells and tissues when forming new systems. This was shown first in grasshoppers (Berlot and Goodman, 1984) and later in *Drosophila*, for neuron growth cones, which are guided by special glycoproteins, some

---

Figure 3.15 (a) Overview of freshly prepared ovary of *Nemoura* sp. shown in differential phase contrast. Note that ovary is not separated into two anlagen as found in most insect species, but remains connected over the alimentary canal as shown in SEM picture (b). Bar represents 300 μm. (b) Tunica externa does not encompass ovarioles but forms a loose network around the lateral oviduct. Ovarioles are without terminal filament and without tunica externa. Bar represents 50 μm. (Courtesy of J. Gottanka, University of Erlangen-Nürnberg, Germany.)

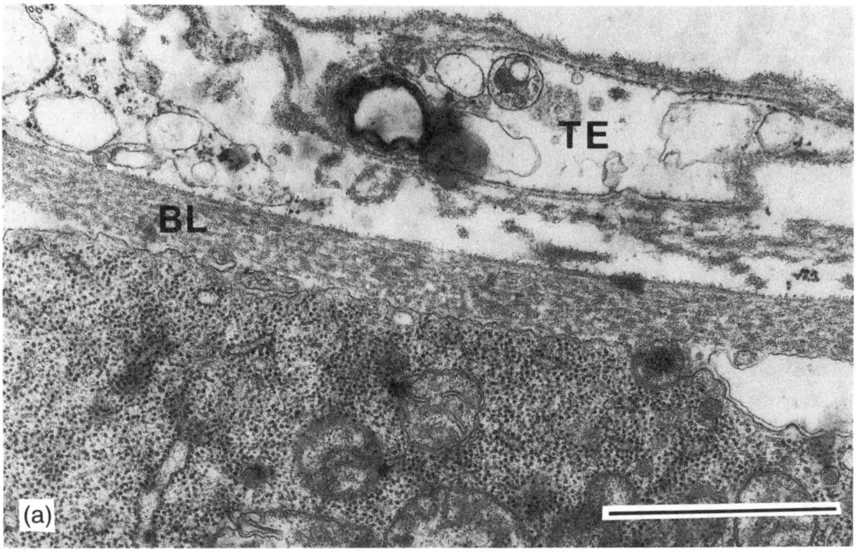

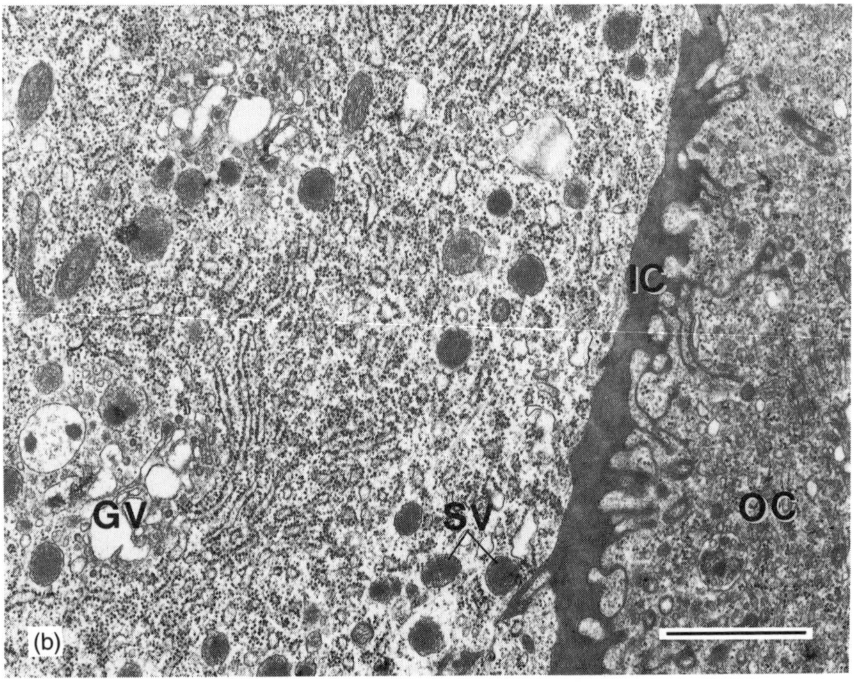

Figure 3.16 (a) Ovary of *Hystrichopsylla talpae* (Siphonaptera). Basal lamina (= tunica propria) (BL) becomes fibrillar during late stages of previtellogenesis. Tracheal cell of tunica externa (TE) is ensheathed by basal lamina at both sides. (b) Ovary of *H. talpae*. Follicle cells have well-developed Golgi vesicles (GV), rough ER and secretion vesicles (SV) filled here with vitelline membrane substances, while endocytosis of vitellogenins into oocyte (OC) via intercellular cleft (IC) is still going on. Bars represent 1 μm.

of which (fasciclin II) are related to vertebrate neural cell adhesion molecules, like N-CAM, a member of the immunoglobulin superfamily; others (fasciclin I and III) are unrelated to previously known molecules (e.g. Zinn *et al.*, 1988; Elkins *et al.*, 1990; Grenningloh *et al.*, 1991). Such guidance by specific proteins communicating with the extracellular matrix might be common during embryonic development, for instance in the migration of germ cells from the pole cell region to their final position in the gonad anlagen. Furthermore, some of these genes (e.g. fasciclin III) not only have influence during nerve cone growth, but are also necessary for the correct ontogenetic development of ovarioles (see section 3.2.6).

The ultrastructure of ovariole basement membranes is not very instructive (Figures 3.12 and 3.13). A distinctive lamina lucida adjacent to follicular membranes does not exist. The basal follicular membrane directly borders the lamina densa, which consists of homogenous material of middle electron density. This aspect changes during vitellogenesis in which rod-like structures occur, mostly orientated parallel to the long axis of the ovariole (Figure 3.16a).

The chemical nature is complex and varies in place and time (Ashhurst, 1985). Out of the collagen family, there is evidence that collagen type I is present. The fibrils, the periodicity of which can be seen in high resolution electron micrographs, are exactly the same as known from vertebrates (Ashhurst and Bailey, 1980). Another collagen, apparently not forming fibrils, is found in the basement membranes of insects. Its gene sequences have been elucidated, showing that this collagen belongs to the type IV family, well known as a major constituent of the basal laminae of vertebrates and nematodes, which indicates its great evolutionary constancy (Blumberg *et al.*, 1987, 1988; Cecchini *et al.* 1987; Knibiehler *et al.*, 1987, 1990; Lunstrum *et al.*, 1988). Other major constituents of basement membranes in insects are laminin (Fessler *et al.*, 1987; Chi and Hui, 1989), proteoglycan (Campbell *et al.*, 1987) and entactin (Blumberg *et al.*, 1988). Recently, laminin has been identified by immunofluorescence of anti-laminin antibodies in basement membranes of *Drosophila* ovarioles (Gutzeit *et al.*, 1991). During vitellogenesis, laminin microfilaments are ordered circumferentially (i.e. at right angles to the longitudinal axis) around each follicle, especially in the anterior half, which encloses the nurse cells. Posterior halves of follicles, which embrace the oocyte and its columnar follicular epithelium, show significantly less fluorescence. The authors discuss this arrangement with the massive transport mechanisms which must occur during these stages of oogenesis in which the nurse cells break down (see section 3.4). Digestion with collagenase clearly shows that basal laminae are necessary to maintain the cell shape and the shape of the whole follicle (Went and Junquera, 1981; Gutzeit and Haas-Assenbaum, 1991).

The physical properties of basal lamina have been investigated by Reddy and Locke (1990), who used colloidal gold particles of defined size to measure the pore width of basal laminae found in lepidopteran larvae. They found that all those tissues which are involved in the import or export mechanisms of haemolymph proteins possess basal laminae which allowed molecules with diameters of 15 nm or less to pass; whereas those tissues which are not engaged in these processes are covered with tighter basal laminae which allowed only particles smaller than 6 nm in diameter to pass. In addition, some tissues have bilayered basal laminae; one layer with wide pores which connects to the haemolymph and an inner one with tighter pores, which borders the cell membrane, as found in Malpigian tubules, heart and muscle.

Ovarian basal laminae were investigated by Giorgi *et al.* (1991), who investigated ovaries of the black blow fly, *Phormia regina*. They used different tracers to study penetration into vitellogenic growing follicles. Peroxidase has a molecular weight of 40 kDa and can pass freely through the basal lamina of living follicles and aldehyde-fixed follicles. The vitellogenin of *Phormia* is composed of four polypeptides with molecular weights of 42–45.5 kDa each (Zou *et al.*, 1988) which pass this barrier. Ferritin cannot pass because it is larger (molecular weight 500 kDa). Cationized ferritin cannot pass either; however, it assembles in regular patches on the surface of the basal lamina, indicating stable anionic sites in the basal lamina, which might be caused by patched concentrations of glycosaminoglycans enriched in negatively charged groups.

The above analyses show that the basal laminae of ovarioles have at least three functions: (1) they serve as a mechanical structure to keep the elongated structure of the ovarioles stable; (2) they serve as physical sieves against the haemolymph; (3) they have some chemical selectivity for macromolecules, and this may play a role in hormone access during previtellogenesis.

Although we have some molecular data, we are still far from understanding the extracellular matrices of insect tissues. The role(s) of the tunica propria urgently needs further elucidation.

3.2.4 Inner sheath, interstitial cells and prefollicular tissue

Inner sheath cells enclose the germ cells on the apical and lateral sides totally. During organogenesis, these cells are more or less cuboidal and mitoses are found. In functioning ovarioles (i.e. during previtellogenesis and vitellogenesis) the tissue becomes more and more flattened and its cells show all the characters of reduced physiological activity: the nuclei are small with a high content of heterochromatic areas connected to the nuclear envelope, the nucleolus is poorly developed and the cytoplasm has few ribosomes, rough and smooth ER, Golgi complexes or mitochondria. This aspect is found in all functioning ovarioles as long as a germarium exists, irrespective of the ovariole type (Büning, 1972, 1978, 1979a,b,c, 1980, 1985;

Büning and Sohst, 1988, 1989; Huebner, 1984a,b; Huebner and Anderson, 1972c; King, 1970; King and Büning, 1985; Pritsch and Büning, 1989;). These morphological data are confirmed by autoradiography using labelled precursors for RNA and protein synthesis, which showed only weak syntheses compared with data from nurse cells, prefollicle and follicle cells (Bier, 1963a,b; 1965a,b; Bier *et al.*, 1967; Mays, 1969, 1972; Büning, 1972; Ullmann, 1973).

In some insects, especially in coleopterans and to a lesser extent in hymenopterans and some other orders, some of these inner sheath cells protrude into the germarium and may intermingle with germ cells. These cells are called interstitial cells. In polyphage coleopterans interstitial cells form a three-dimensional reticular network in the terminal chamber, which itself is the follower of the germarial region (see section 3.5.3) in functioning ovarioles. The ultrastructure is similar to that of the inner sheath cells just described. However, bundles of microfilaments and sometimes of microtubules can occur, indicating their proposed function: the reticular network may stabilize the end-standing trophic chamber (Figures 3.17 and 3.18) (Büning, 1972, 1978, 1979a; King and Büning, 1985).

At the base of the germarial region some cells at the periphery, adjacent to the inner sheath cells, keep their embryonic character. These cells are prefollicular cells, which themselves give rise to the follicular epithelium in functioning ovarioles. The number of cells proliferating at the base of the germarium (terminal chamber in telotrophic ovarioles), is type dependent: in panoistic ovarioles of 'primitive' orders, there are some embryonic cells scattered between arrested oogonia of the germarium (Gottanka and Büning, 1990; Taddei *et al.*, 1992). In polytrophic meroistic ovarioles the architecture has not changed substantially: embryonic prefollicle cells are scattered among rising clusters in the posterior part of the germarium. The quickly dividing cells embrace the recently formed germ cell clusters totally when leaving the germarial region. Thus, the first follicle has been built, which has already gained anterior–posterior polarity (Koch and King, 1966, 1969; King, 1970) (see sections 3.3.4 and 3.5.2 for different orders). In telotrophic meroistic ovarioles the region of quickly dividing prefollicle cells is restricted to the bottom of the terminal (nutritive) chamber and in hemipteran and in polyphage coleopteran ovarioles, some hundreds of embryonic cells can be found, which form a prefollicular tissue (Wick and Bonhang, 1955; Bonhag, 1958; Brunt, 1971; Huebner and Anderson, 1972; Büning, 1972, 1979a; Schreiner, 1977a,b; Huebner, 1984a,b; King and Büning, 1985). However, there are exceptions: the telotrophic ovarioles of Megaloptera and of Ephemeroptera have rather few prefollicle cells and the panoistic ovary of Thysanoptera has a small tissue at the base of the germarium (Büning, 1972, 1979c; Matsuzaki and Ando, 1977; Pritsch and Büning, 1989).

Prefollicle cells are mostly spindle shaped and orientated vertical to the long axis of the ovariole. Many cells are in contact with the tunica propria. In interphase, cells can be polarized, showing lipid droplets and glycogen

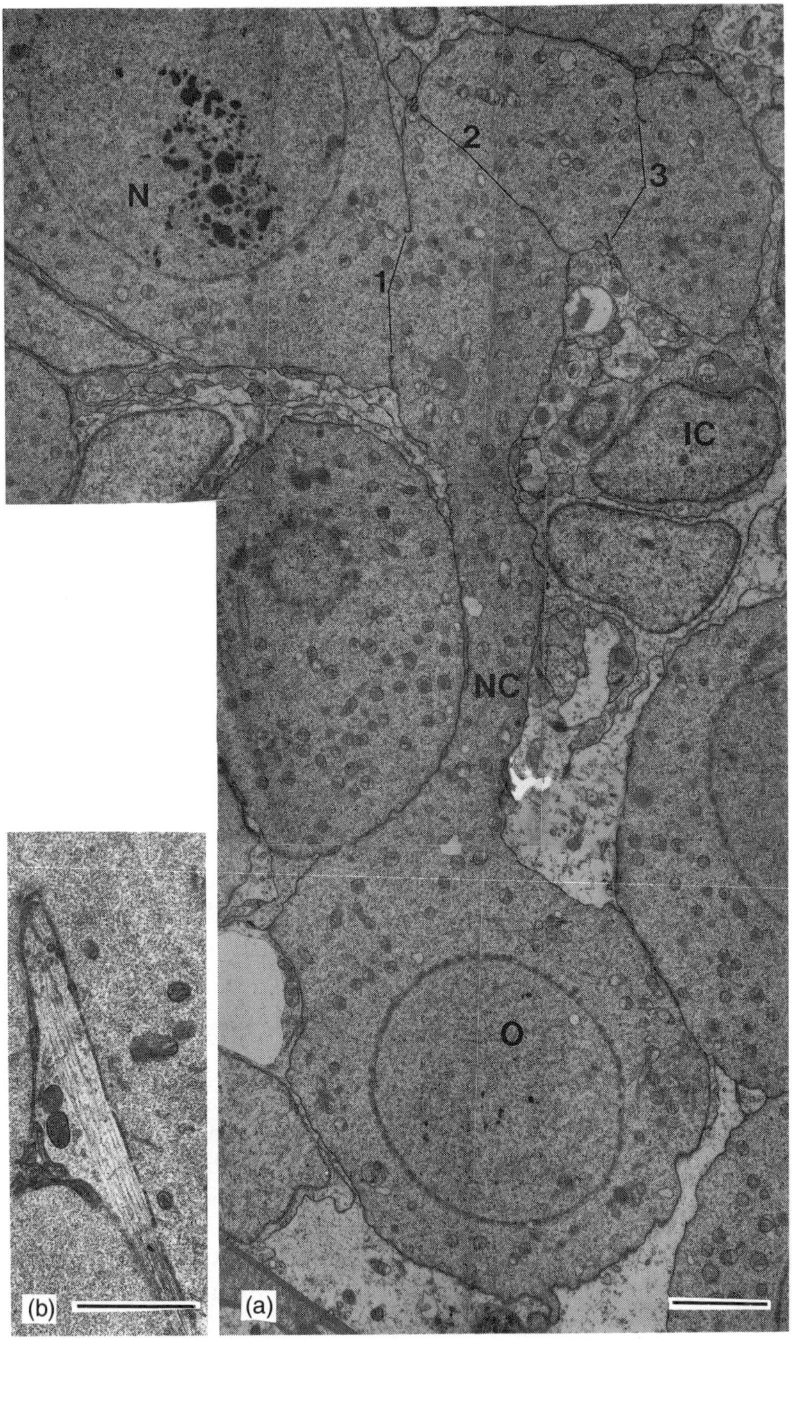
N
1
2
3
IC
NC
O
(b)
(a)

granules in the apical parts, i.e. facing the young oocytes, as found in some hemipterans (Lutz and Huebner, 1980; Schreiner, 1977a,b). The prefollicular tissue is an embryonic tissue which produces cells preferentially in the posterior direction. These cells enclose the resting (i.e. not growing) oogonia. Consequently, a developmental gradient is established in larger prefollicular tissues at least in hemipterans. Thus, Lutz and Huebner (1980) found a more intense RNA synthesis in the upper parts than in the lower parts of the prefollicular tissue when investigating the ovary of *Rhodnius prolixus* by autoradiography.

3.2.5 Follicle cells and their synthesis

Reports on the ultrastructure of follicle cells are numerous, especially those describing the phase of vitellogenesis. We will focus here on earlier events and will briefly describe morphological changes during vitellogenesis. Special aspects of follicle cell activity will be discussed elsewhere, for instance the role of follicle cells in determining the processes of embryonal axes during oocyte growth (section 3.3.2) and the movements of special subsets of follicular cells during oogenesis of *Drosophila* and related species (section 3.5.2).

Generally, follicular cells are of high synthetic activity during their whole life time. This has been demonstrated by RNA and protein autoradiography in flies (Diptera) (Figure 3.19) (King and Burnett, 1959; Bier, 1963a,b, 1965a, 1967a,b); beetles (adephagous Coleoptera: Bier, 1965b; Bier and Ribbert, 1966; Bier *et al.*, 1967), (polyphagous Coleoptera: Büning, 1972; Ullmann, 1973; Ray and Ramamurty, 1979); butterflies (Lepidoptera) (Bier, 1965b; Zalokar, 1965; Pollak and Telfer, 1969); wasps and bees (Hymenoptera) (Bier, 1965b; Engels, 1968, 1973); lacewings (Neuroptera) (Gruzova *et al.*, 1972); gryllids (Orthoptera) (Favard-Séréno and Durand, 1963; Bier, 1970); bugs (Hemiptera) (Mays, 1972); Mecoptera (Ramamurty, 1963); and earwigs (Dermaptera) (Engels, 1969). These autoradiographic data are supported by ultrastructural studies which all showed a pronounced nucleolus in follicular cells during previtellogenesis and vitellogenesis, together with a high content of rough ER and free ribosomes, Golgi complexes and vacuoles (Figures 3.16, 3.20–3.22).

Furthermore, the DNA content of follicle cells is raised during the

---

Figure 3.17 (a) Basal region of trophic chamber of *Asclera coerulea* (Coleoptera: Polyphaga). Somatic interstitial cells (IC) are found in telotrophic ovaries of polyphage Coleoptera. Distribution of three intercellular bridges (1, 2, 3) among nurse cells (N) and resting oocyte (O) at beginning of nutritive cord (NC) indicates branched cluster of germ cells. Bar represents 2 μm. (b) In some species, such as *Melinopterus (Aphodius) prodromus* (Coleoptera: Lamellicornia), bundles of microtubules and microfibrils are found in finger-like projections of interstitial cells, indicating their stabilizing function in the terminal chamber. Bar represents 1 μm.

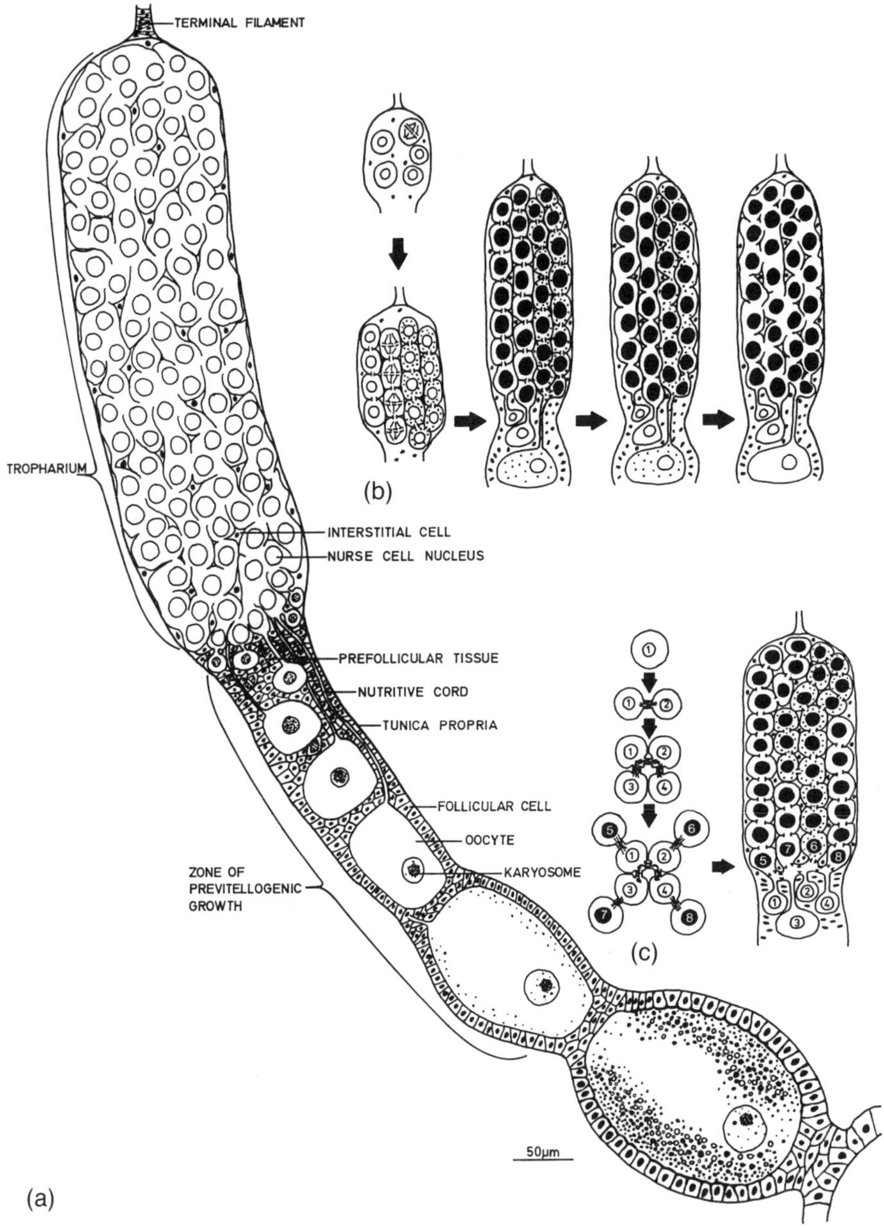
TERMINAL FILAMENT
TROPHARIUM
INTERSTITIAL CELL
NURSE CELL NUCLEUS
PREFOLLICULAR TISSUE
NUTRITIVE CORD
TUNICA PROPRIA
FOLLICULAR CELL
OOCYTE
KARYOSOME
ZONE OF PREVITELLOGENIC GROWTH
50μm
(a)
(b)
(c)

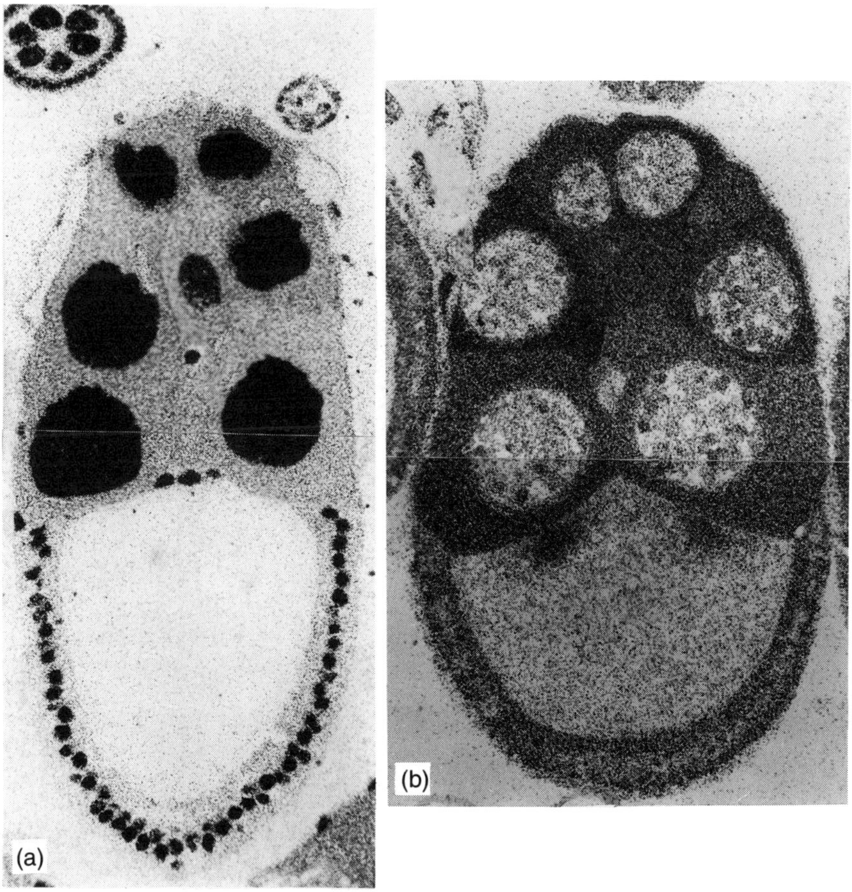

Figure 3.19 Autoradiogram after *in vivo* incubation of ^{3}H-cytidine into follicles of *Musca domestica*. (a) Strong label is apparent over follicle cell and nurse cell nuclei after 1 h of incubation, whereas follicle cell and nurse cell cytoplasm is labelled less strongly. The oocyte cytoplasm is nearly devoid of label. (b) After 5 h of incubation, most label is found in follicle cell and nurse cell cytoplasm, whereas corresponding nuclei stain to a much lesser extent. Note the steep gradient of label in oocyte cytoplasm, emerging from intercellular bridges connecting nurse cells to the oocyte. (Originals from Bier, 1963b.)

Figure 3.18 (a) Scheme of telotrophic meroistic ovary of polyphage Coleoptera. (b) Model of cluster formation, starting with several stem cells, transformed directly to cystoblasts. Linear clusters arise predominantly. The most basally orientated sibling continues as oocyte, all others will be transformed to nurse cells. (a and b after King and Büning, 1985.) (c) Recent investigations prefer another model, starting from one cystoblast, which follows in the beginning the 2^n-rule, forming a branched cluster. Later on, linear subclusters are preferentially built by intercalar divisions. Basally located cells continue as oocytes, all others develop as nurse cells.

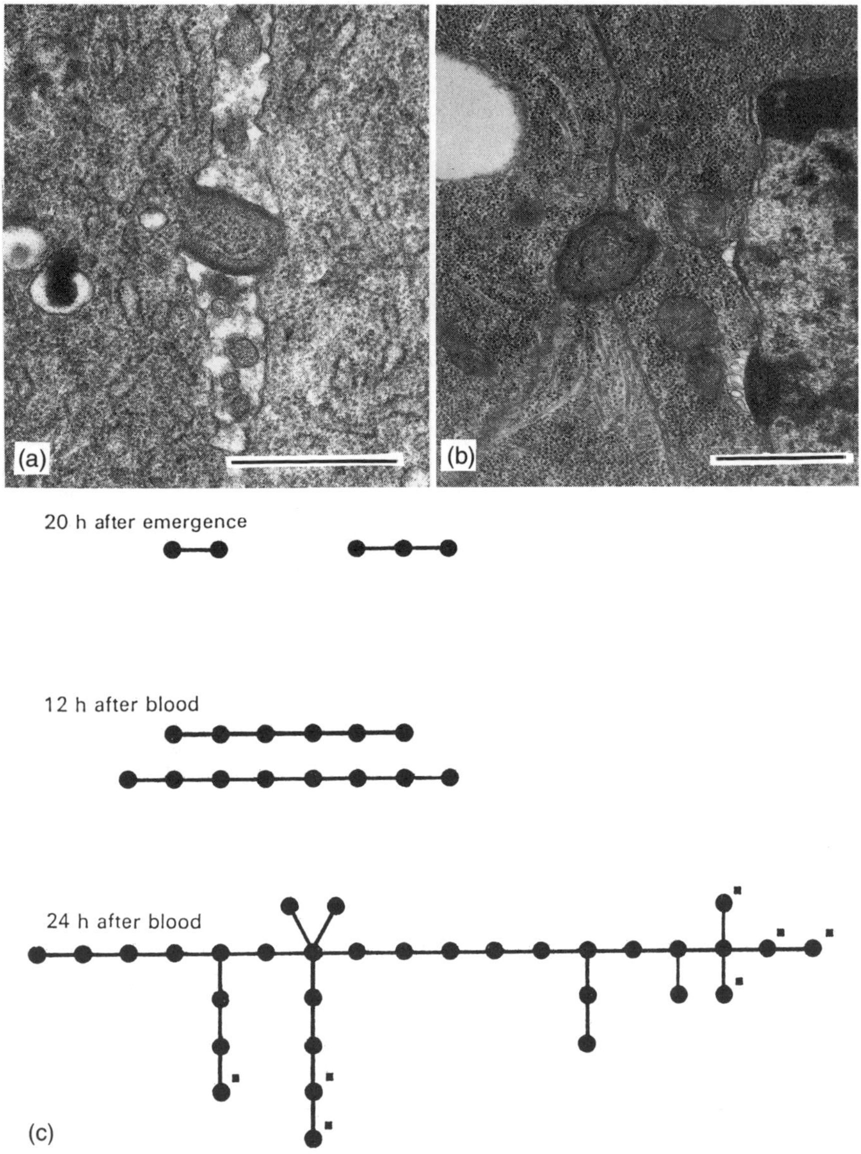

Figure 3.20 (a) Intercellular bridge between follicular cells of *Calliphora erythrocephala* (Diptera). Somatic intercellular bridges are much smaller than those between germ cells. (b) Cross-sectioned intercellular bridge between follicular cells of *Boreus hiemalis* (Mecoptera: Boreidae). Mecoptera (Boreidae) are the third taxon (order) in which intercellular bridges between follicle cells are found. Bars represent 1 μm. (c) Diagram of follicle cell syncytia in *Culex pipiens* (Diptera) during selected periods. Dividing activity stops 24 h after blood meal in follicular tissues. Those cells, which continue outside the series are indicated by squares. (After Fiil, 1978.)

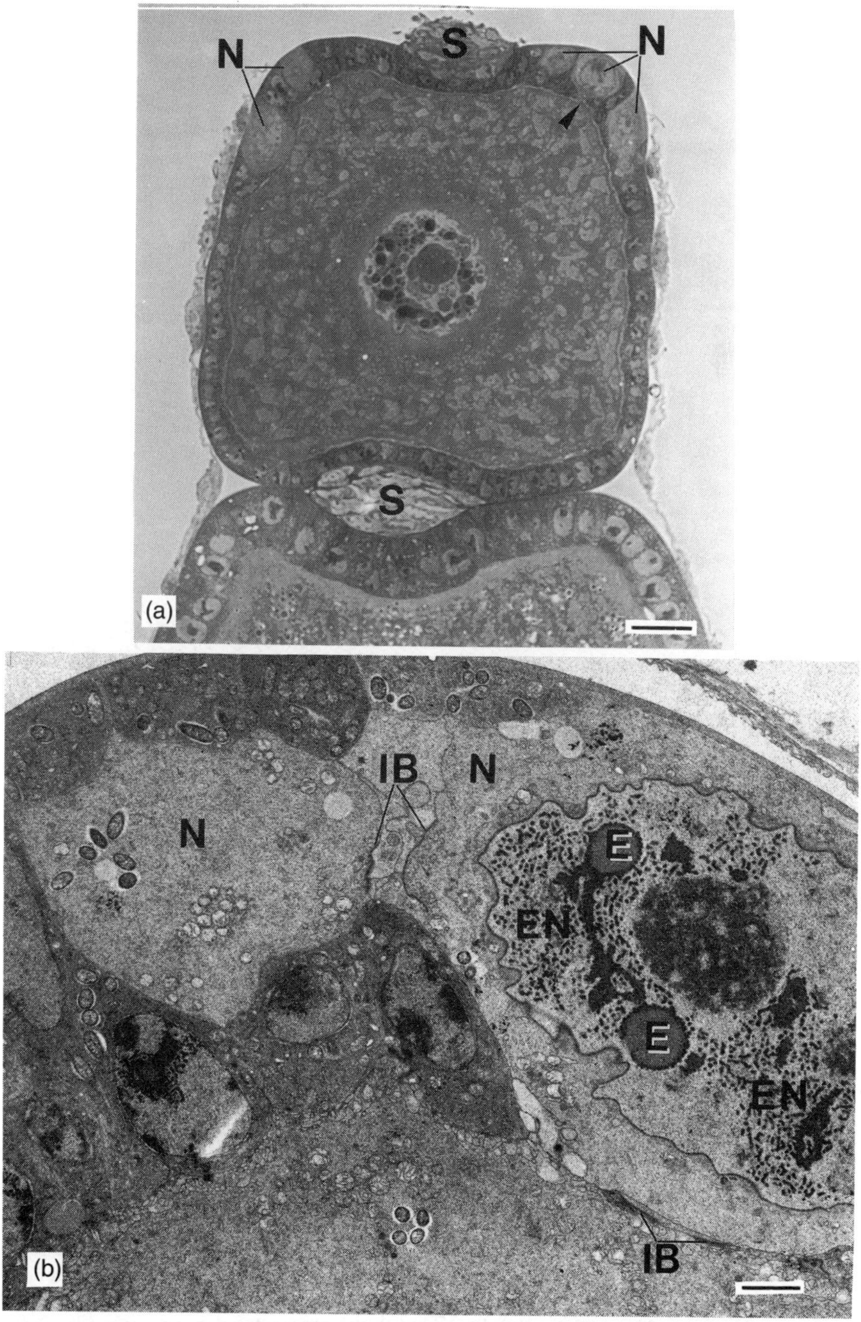

Figure 3.21 Polytrophic meroistic ovary of *Hystrichopsylla talpae*. (a) During late previtellogenesis, nurse cells (N) become integrated into the follicular epithelium of the oocyte. Germinal vesicle is filled with numerous extranuleoli, surrounding a globular endobody. Stalk cells (S); bar represents 20 μm. (b) Ultrastructure of nurse cells. Nuclei show large compartments of extranucleolar material (EN) and endobodies (E); however, nurse cell cytoplasm is of low electron contrast similar to early cytoplasm of oocytes. Rims of intercellular bridges (IB). Bar represents 2 μm.

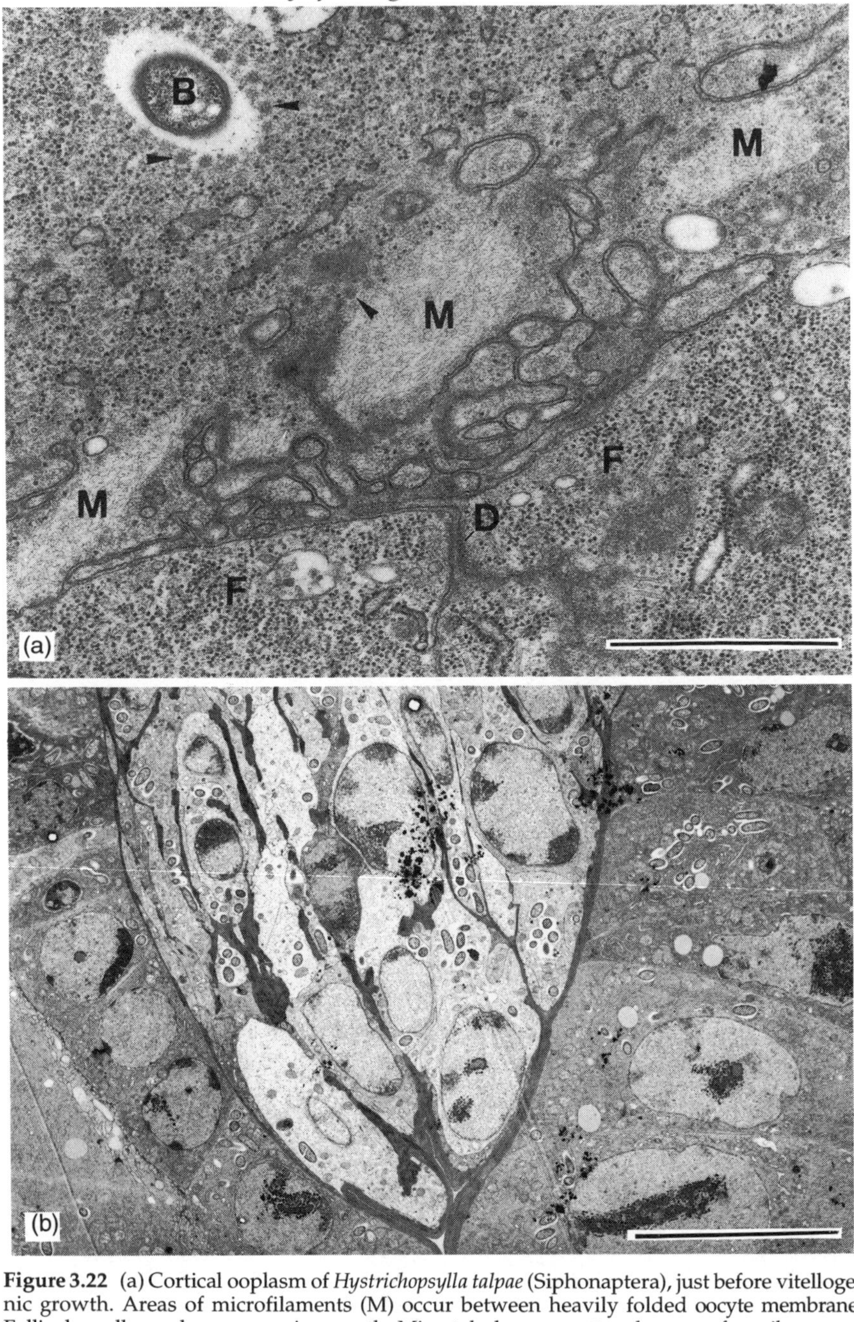

Figure 3.22 (a) Cortical ooplasm of *Hystrichopsylla talpae* (Siphonaptera), just before vitellogenic growth. Areas of microfilaments (M) occur between heavily folded oocyte membrane. Follicular cell membranes remain smooth. Microtubules are scattered among free ribosomes. Ball-like structures of unknown function (arrowheads). Symbiontic bacterium (B); desmosome (D); connecting follicle cells (F). Bar represents 1 μm. (b) Interfollicular stalk cells, compressed between the first vitellogenic growing oocyte and the last previtellogenic growing oocyte. Note the thick basal laminae between cells. Bar represents 10 μm.

growth phases of oocytes. Endomitotic polyploidization was found for instance in *Carausius morosus* (Phasmida) (Pijnacker and Godeke, 1984), *Hyalophora cecropia* (Lepidoptera) (Crippa and Telfer, 1971), *Drosophila melanogaster* (Diptera) (Mahowald *et al.*, 1979), in aphids (Hemiptera) (Orlando and Crema, 1968) and in *Dysdercus intermedius* (Hemiptera) (Dittmann and Maier, 1987), whereby selective underreplication occurs in the beginning at least in *Drosophila* and *Dysdercus*. In *Carausius*, the final polyploid status of 64C is reached after three telophasic restitution cycles and two endoreduplications. Previtellogenic follicle cells have DNA contents of 2C–8C, vitellogenic follicle cells have 16C, whereas vitelline membrane formation is done by cells with 32C, and finally, during chorion formation follicle cells have as their highest value 64C. During this enhancement of DNA cytokinesis is repressed, but, occasionally, nuclei will be divided. Thus, two nuclei can occur which contain the same amount of DNA each, or in which the two nuclei contain different multiples of 2C (Pijnacker and Godeke, 1984). In many insects as in *Lepisma* (Zygentoma), *Aeschna, Gomphus* (Odonata), *Gryllus* (Orthoptera), (Gross, 1903), *Eomenacanthus* (Mallophaga) (Bilinski and Jankowska, 1987), and all bugs (Heteroptera) investigated so far (Gross, 1901, 1903; Huebner and Anderson, 1972; Huebner, 1984a,b; Dittmann and Maier, 1987), the follicle cell nucleus undergoes one amitotic division (Figures 3.23 and 3.24).

As shown for *Dysdercus* (Heteroptera), the amitotic division halves the total DNA content very exactly after selective underreplication (Figure 3.25) (Dittmann and Maier, 1987).

One of these changes in DNA content could be elucidated at the genetic, molecular and ultrastructural level during follicle cell ontogenesis in *Drosophila*. Here, intrachromosomal DNA amplification of chorion genes is found (Figure 3.26) (Spradling and Mahowald, 1980; Osheim and Miller, 1983; De Cicco *et al.*, 1984; Regier and Kafatos, 1985; Osheim *et al.*, 1985, 1988; Spradling, 1987; Spradling and Orr-Weaver, 1987; Heck and Spradling, 1990; recently reviewed by Orr-Weaver, 1991). Cells, which have to divide mitotically, can amplify a special gene only by local duplication processes which are followed by homologous recombinations. Thereby, the linear array of genetic information can be restored and, finally, we will find linear gene duplications, or extrachromosomal rings, bearing amplified genes (Roberts *et al.*, 1983). However, in those somatic cells which have reached their final destination and have stopped dividing, a more simplified mode of gene amplification can occur: special gene origins start replication forks repeatedly, whereby later started replication forks terminate at earlier points. This results in a parallel array of replicated genes (Figures 3.26 and 3.27). In *Drosophila* about 20 chorion genes are known today, most of which are arranged in two clusters on the genome, one at 7F1 on the X chromosome the other at 66D11-15 on the third chromosome (Griffin-Shea *et al.*, 1982; Parks *et al.*, 1986). The X chromosome cluster

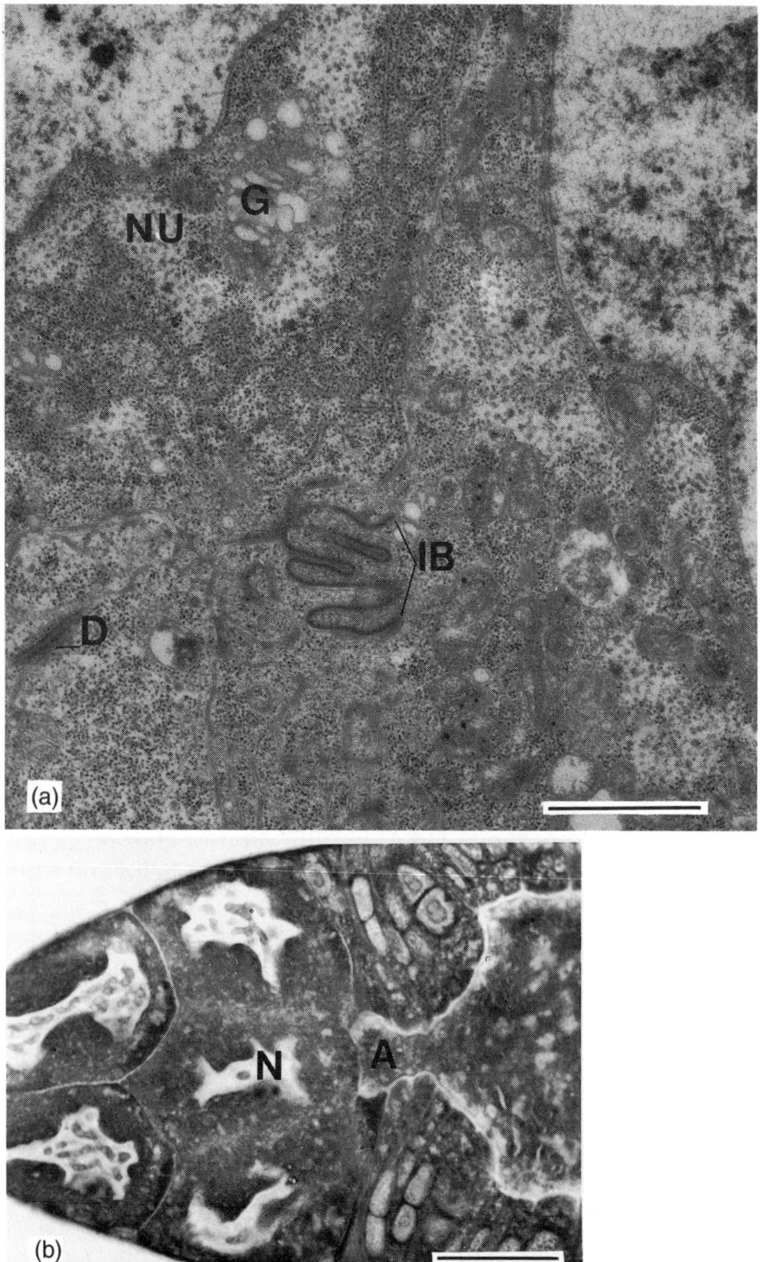

Figure 3.23 (a) Intercellular bridge rims (IB) of bark lice (Psocoptera) can be extremely folded. Note also the areas of nuage material (NU) near to the nurse cell nuclear envelope. Golgi apparatus (G); desmosome (D). (b) Follicle of *Eomenacanthus stramineus* (Mallophaga). Follicular cells show double nuclei, as known from Heteroptera. A nutritive appendix (A) is well developed, which bears the intercellular bridges to directly neighbouring nurse cells (N). Bar represents 20 µm. (From Bilinski and Jankowska, 1987; with permission of S. Bilinski and Gustav Fischer Verlag, Jena.)

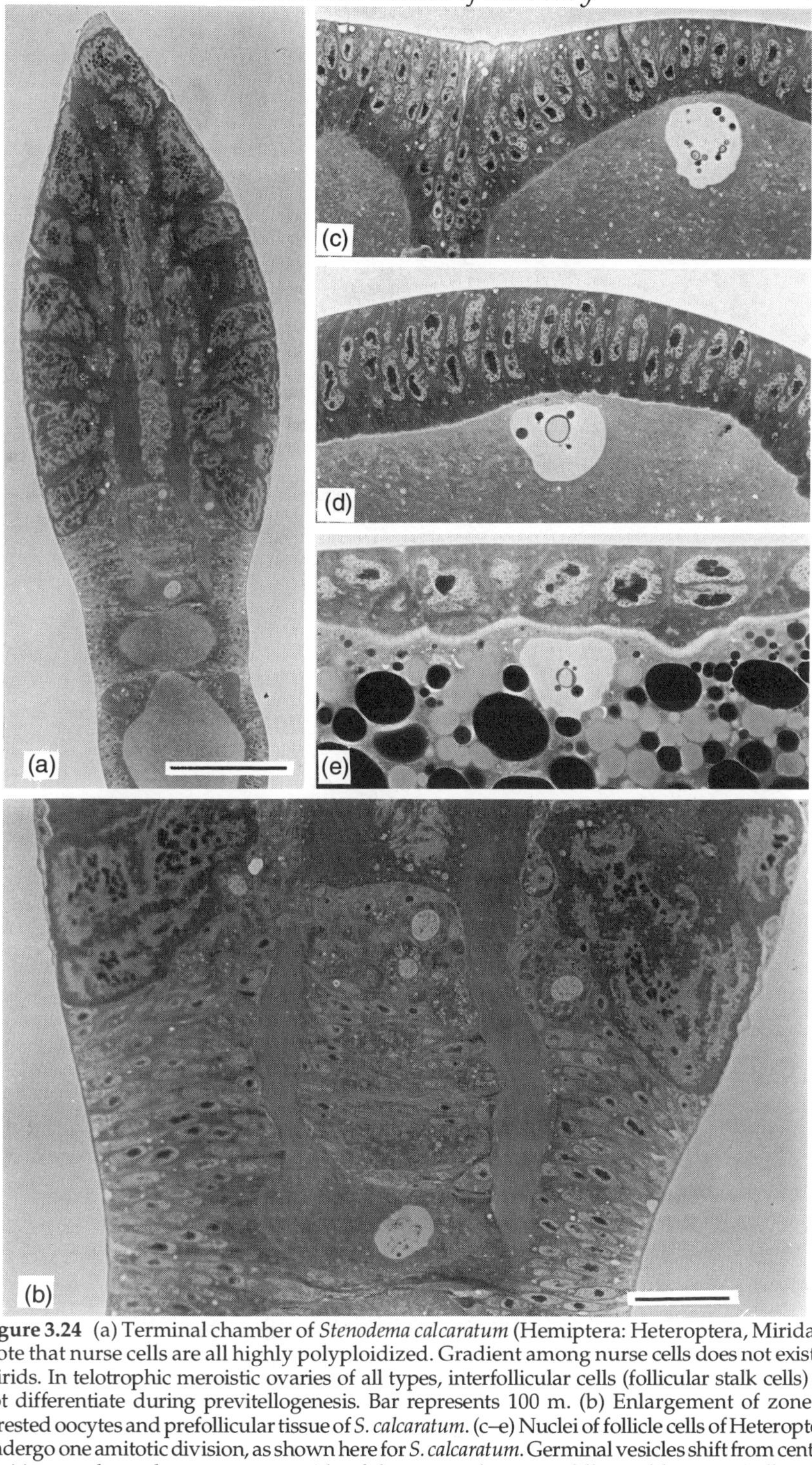

Figure 3.24 (a) Terminal chamber of *Stenodema calcaratum* (Hemiptera: Heteroptera, Miridae). Note that nurse cells are all highly polyploidized. Gradient among nurse cells does not exist in mirids. In telotrophic meroistic ovaries of all types, interfollicular cells (follicular stalk cells) do not differentiate during previtellogenesis. Bar represents 100 m. (b) Enlargement of zone of arrested oocytes and prefollicular tissue of *S. calcaratum*. (c–e) Nuclei of follicle cells of Heteroptera undergo one amitotic division, as shown here for *S. calcaratum*. Germinal vesicles shift from central positions at the early stages to one side of the oocyte during middle and late previtellogenic growth stages. They contain some vacuolated endobodies beside some other inclusions, which stain heavily with Azur B, indicating a high content of RNA. Bars represent 25 μm.

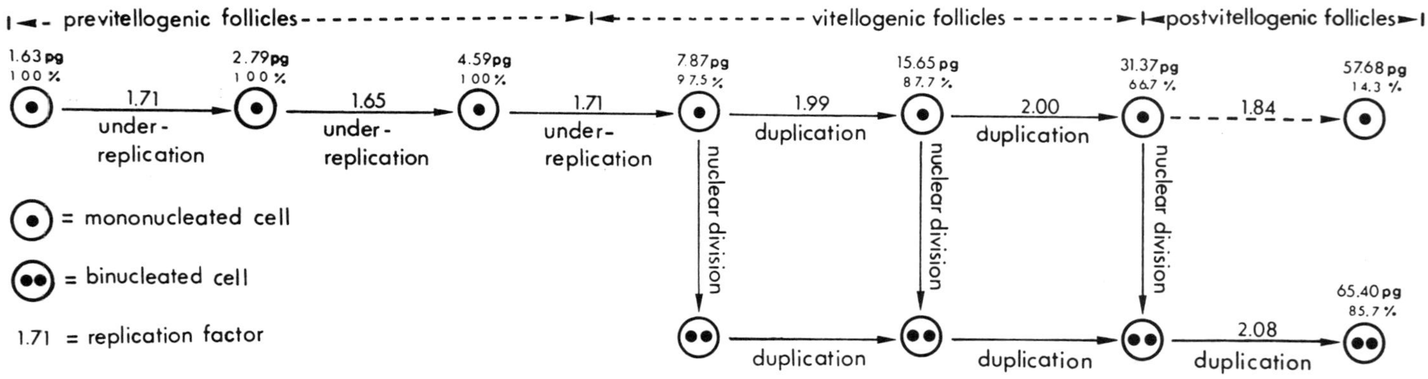

Figure 3.25 DNA content of follicle cells from *Dysdercus intermedius* ovarioles during growth periods. DNA measurements were made by Feulgen cytophotometry. (From Dittmann and Maier, 1987; with permission of Pergamon Journals Ltd.)

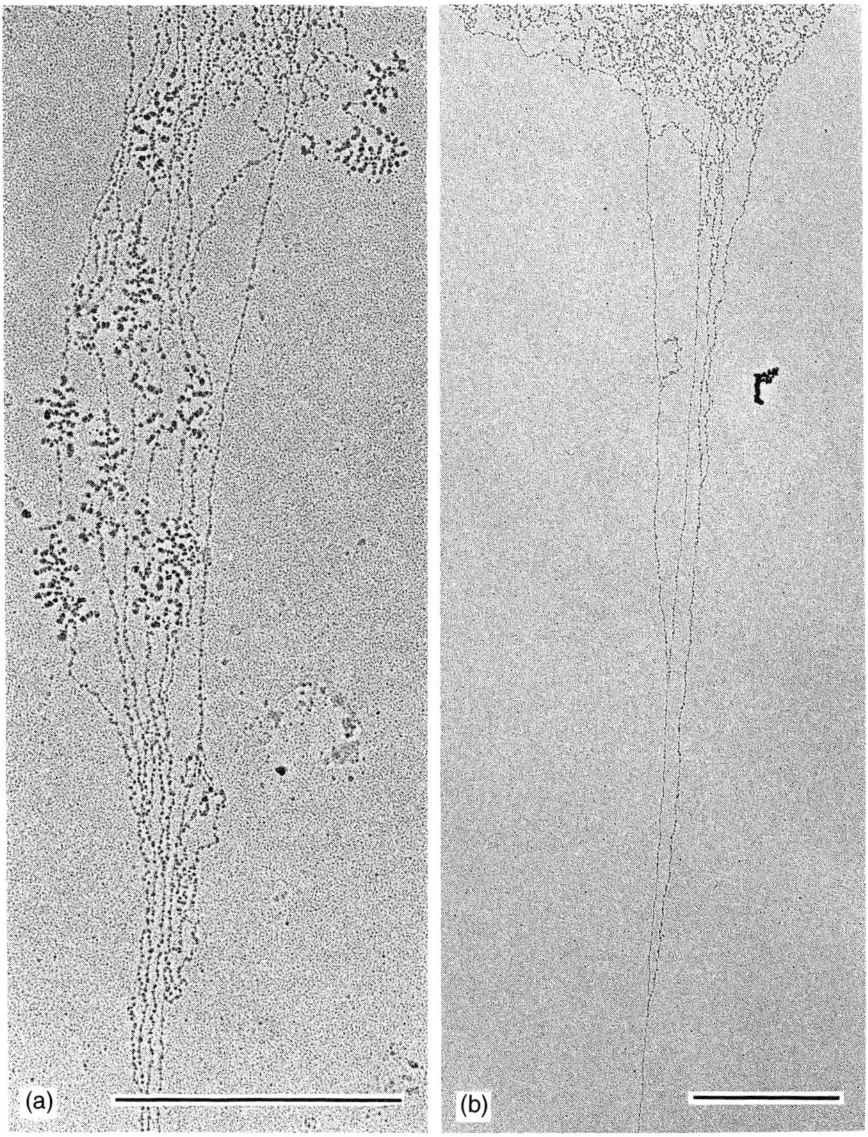

Figure 3.26 (a) DNA of follicular cells can polyploidize discontinuously, as shown here in a spread preparation of chromatin of *Drosophila melanogaster*. Note tandemly arranged active chorion genes. (b) Inactive DNA of follicular cells showing bifurcations, the measured distances of which are shown in Figure 3.27a. Bars represent 1 μm. (From Osheim and Miller, 1983; with permission of O.L. Miller, Jr. and Cell Press.)

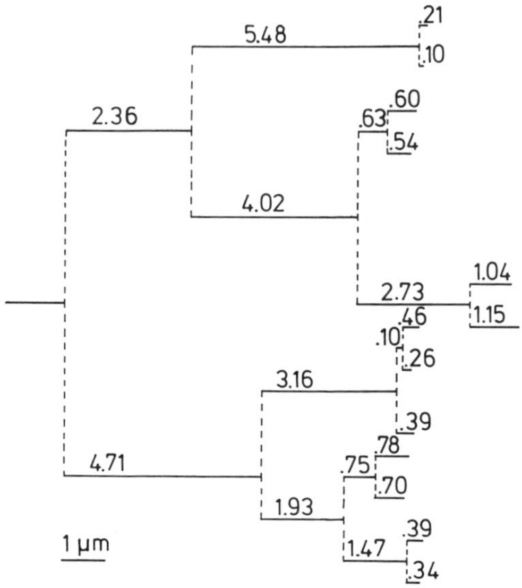

(a)

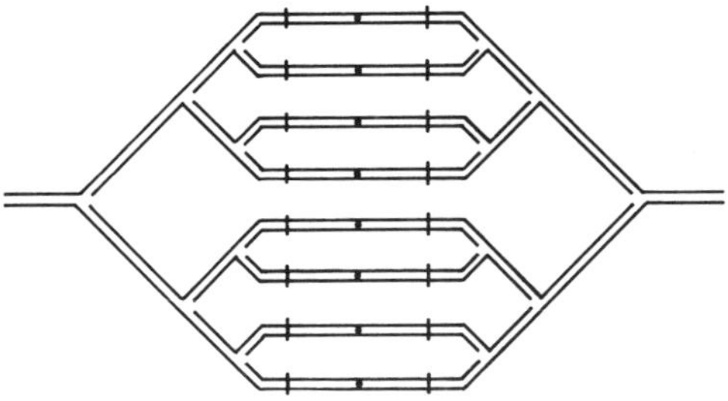

(b)

Figure 3.27 (a) Measurement of replication forks (from Osheim and Miller, 1983; with permission of O.L. Miller, Jr. and Cell Press). (b) Model describes general features of localized intrachromosomal amplification. Dots indicate origin of bifurcated replication. Small bars include regions of amplified genes (here 8-fold).

contains two major chorion proteins and four minor chorion proteins. The third chromosome cluster contains four major chorion proteins. Both clusters are amplified during stages 9–13 but to different degrees: the X chromosome cluster about 16–20-fold, the third chromosome cluster to 60–80-fold (for staging of *Drosophila* see section 3.5.2). Amplification of both is controlled by special regions nearby and inside the clustered genes. Although the clustered genes are regulated together in the amplification process, the temporal and tissue-specific transcription of genes is regulated singularly (Kalfayan *et al.*, 1985; Wakimoto *et al.*, 1986). The X chromosome clustered genes start early during stage 11 and reach their highest activity at stage 12, whereas the third chromosome gene transcripts accumulate during stages 13 and 14. Furthermore, the minor chorion genes of the X chromosome cluster are active only in special sets of follicle cells, corresponding to special structures (see below). An essential promoter element seems to be a highly conserved hexamer TCACGT, located approximately –60 bp from the transcription start site of all major chorion genes.

The architecture and ultrastructure of follicle cells are polarized in two ways: at first perpendicular to the long axis of the ovariole in each cell, as in prefollicular tissues; second along the ovarial tube (see below). Within each cell, vacuoles and mitochondria accumulate more in the apical region, whereas rough ER and Golgi complexes are found more often at basal and lateral sides (see for instance Huebner and Anderson, 1972c; or Dennis and Bradley, 1989). The lateral membranes of follicle cells communicate with each other by gap junctions, the number and distribution of which are stage specific. Tight junctions, septate junctions and desmosomes connect follicle cells tightly to each other on baso-lateral membranes which often interdigitate. Gap junctions and desmosomes remain contact regions even during patency, i.e. the rise of the follicular cleft, which separates the follicular cells and gives space for the haemolymph during pinocytotic uptake of vitellogenins when vitellogenesis is in process (Telfer *et al.*, 1982).

During previtellogenesis, follicle cell membranes facing the oocyte remain smooth, as shown for beetles (Bilinski and Petryszak, 1978; Bilinski *et al.*, 1985; Bilinski and Sklarzewicz, 1987), or become more and more folded, interdigitating with the oocyte membrane, as shown for bugs (Huebner and Injeyan, 1981). Thereby, gap junctions between oocytes and follicle cells are built (Figure 3.28) (Woodruff, 1979; Huebner and Injeyan, 1981; Telfer *et al.*, 1982; Bilinski *et al.*, 1985; Münz, 1988).

During patency, the oocyte develops numerous microvilli, while the follicle cell membrane facing the oocyte remains smooth, as found in most insect species and shown here for *Hystrichopsylla* follicles (Figure 3.16b). In bugs and in some other species, like *Hyalophora* (Lepidoptera), microvilli of follicle cells interdigitate with microvilli of the oocyte membrane. During secretion of the vitelline layer and during choriogenesis microvilli are abundant at the oocyte facing side of follicle cells (Margaritis, 1980, 1985).

Together with ultrastructural changes during previtellogenic, vitellogenic and choriogenic growth phases, the gross architecture of follicle cells changes. This depends on two antagonistic processes: the mitotic activity and growth of follicle cells on one hand, and the growth of oocytes which expands the follicular tissue on the other hand. Thus, in panoistic ovaries in which prefollicular cells are restricted to small nests, the first previtellogenic growing oocytes are enclosed by only some, highly flattened cells; whereas in telotrophic meroistic ovarioles of bugs (Heteroptera), cicadas (Cicadiformes), and polyphage beetles (Coleoptera, Polyphaga) the

Figure 3.28 Heterologous gap junctions between follicle cells and oocyte of *Tribolium destructor* (Coleoptera). Note that intermembrane particles are grouped to round or oval areas, according to attachment of oocyte microvilli to more or less flat membrane of follicle cells. Freeze-fracture preparation; bar represents 1 μm. (From Bilinski *et al.*, 1985; with permission of S. Bilinski and Springer Verlag.)

first previtellogenic growing oocytes are accompanied by many and therefore high columnar follicle cells. In middle previtellogenic stages, the follicle cells become more cube-shaped in all ovariole types. When follicular cells grow more intensely during late previtellogenic and early vitellogenic phases in which concomitantly the follicular cleft opens, the cells again become highly prismatic before they finally flatten during later stages.

Shape formation of follicle cells is accompanied by changes in the cytoskeleton. Gutzeit (1990a) showed that the pattern of F-actins (microfilaments) is polarized in follicle cells of *Drosophila* and undergoes dramatic changes during early and mid-vitellogenic stages (stages 8-11, King, 1970; for *Drosophila* stages of oogenesis see section 3.4). Sheath cells, bordering the nurse cells (see below), possess microfilament bundles not orientated to any axis of the cell itself or to the axis of the follicle but build an evenly distributed three-dimensional network. In follicle cells, encompassing the oocyte, microfilament bundles are orientated parallel to each other and to the basal lamina to which they tightly adhere. The apical part of follicle cells, facing the oocyte, as well as the lateral sides and the inner cytoplasm show only weak fluorescence with rhodamine-conjugated phalloidin, a substance which stains exclusively F-actin. At stage 10, when nurse cells have reached their maximum volume, follicle cells at the border between oocyte and nurse cells grow centripetally. These cells have orientated their basal microfilaments parallel to the latitude of the follicle, forming a circumferential belt. The number of basal microfilaments rises at stage 10, compared with stage 9. The pattern becomes slightly changed with respect to microfilament orientation, which keeps its array parallel to the basement membrane within a cell, but cell to cell orientation of microfilaments can differ. These basal microfilaments seem to be engaged in tightening the basal lamina to the outer follicular cell membrane (Gutzeit, 1990a). The same holds true for other elements of the cytoskeleton. Their three-dimensional pattern in follicle cells was investigated by several authors, such as Bilinski and Petryszak (1978) for *Fourcartia squamulata* (Coleoptera, Polyphaga), Bilinski and Szklarzewicz (1987) for *Gyrinus natator* (Coleoptera: Adephaga), Huebner (1981a) for *Rhodnius prolixus* (Hemiptera), Junquera (1983a,b) for *Heteropeza pygmaea* (Diptera, Cecidomyiidae), Gutzeit (1986a) for *Drosophila melanogaster* (Diptera, Drosophilidae), and Fleig *et al.* (1991) for *Apis mellifera* (Hymenoptera) and *Dysdercus intermedius* (Hemiptera). In *Gyrinus* bundles of microtubules are found, most of them parallel to the lateral sides of follicular cells, anchored at the oocyte facing side and opposite, on the basal lamina facing side.

Follicle cells are not unique but differentiate along oocyte growth into several subsets. The diverse differentiation processes are preceded and/or accompanied by mitotic programmes. In Diptera (*Drosophila, Musca, Calliphora, Aedes, Stomoxys*), Hymenoptera (*Apis*) and Mecoptera (*Boreus*) these mitoses are followed in part by incomplete cytokinesis giving birth to interconnected clusters, combined by small intercellular bridges (Figure 3.20) (Meola *et al.*, 1977; Ramamurty and Engels, 1977; Giorgi, 1978; King *et al.*, 1978, Mahowald *et al.*, 1979; Büning, unpublished). Geometry, constancy, exact timing, and therefore, functioning of these somatic clusters remain unknown, as well as their further distribution among insect orders. In *Drosophila*, Margaritis (1980, 1985) could detect 11 subpopulations of follicle cells, each of which has its own function and its own heritage (Figure 3.29). Possibly, the somatic clusters are involved in these differentiation processes.

A new access to study follicular cell gene expression patterns is given by P-element-mediated enhancer detection procedures as has been done by Grossniklaus *et al.* (1989). They found a variety of gene expression patterns, which have not been recorded morphologically before. Besides special expression of genes in border cells, polar cells and follicle cells at the oocyte nurse cell boundary, they also recorded so-called 'double gradients' among follicle cells, which show the highest expression of reporter genes at the dorso-anterior side of columnar follicular cells, whereas follicle cells in

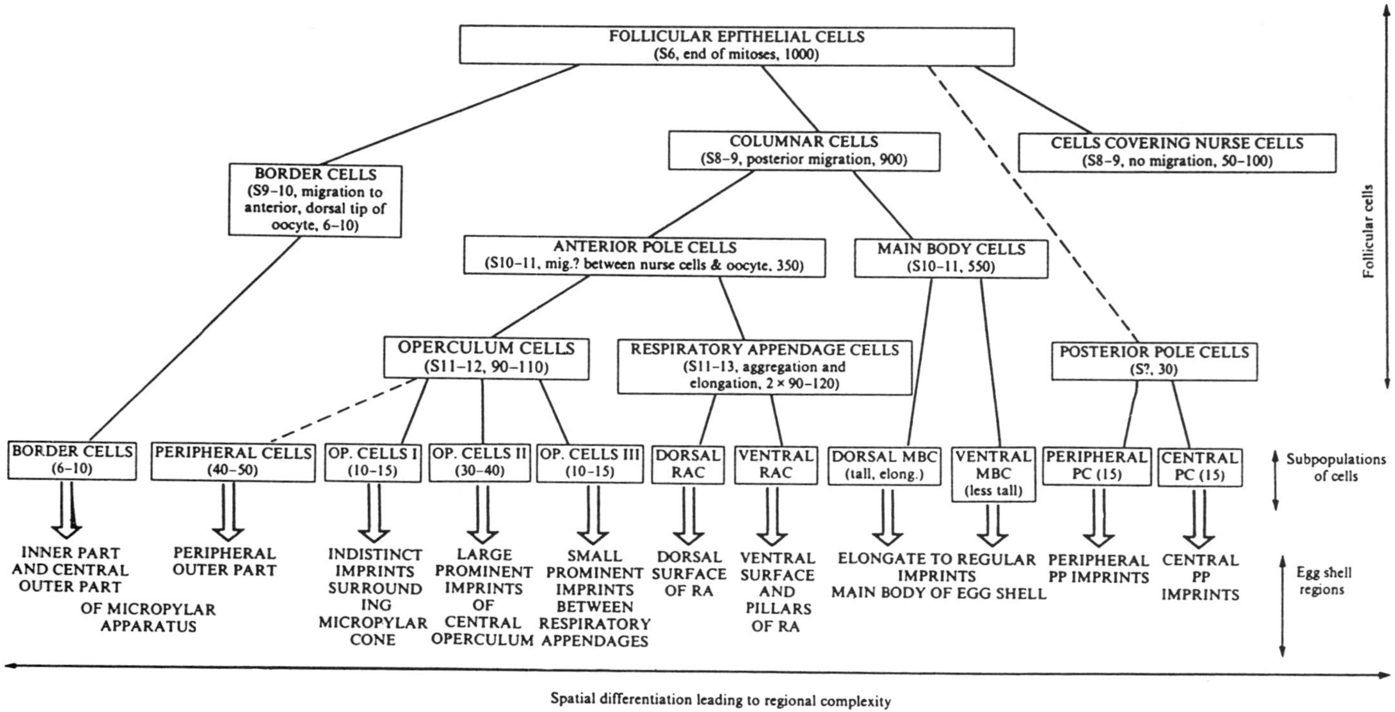

Figure 3.29 Morphological differentiation of follicular cells of *Drosophila melanogaster* during follicle formation, growth periods and egg shell production. Categories of cells are shown by boxes, their origin by lines (broken lines = still speculative). Thick arrows point to type of egg shell produced. (From Margaritis, 1985; with permission of L. Margaritis and Pergamon Press.)

ventro-posterior positions show decreasing activity. Such genes might play fundamental roles in establishing embryonal axes.

Among morphological subpopulations one is found exclusively in polytrophic meroistic ovaries, namely those follicle cells embracing the nurse cells of each follicle. These cells stop their mitotic programme early, become flattened during nurse cell growth and have reduced their RNA and protein syntheses as shown by autoradiography (Bier, 1963a,b; 1965a; Bier *et al.*, 1967). In this respect they are directly comparable to inner sheath cells of germaria and to terminal chambers of telotrophic meroistic ovaries (Büning, 1972; Mays, 1972; Ullman, 1973). This subpopulation is not developed in the aberrant polytrophic meroistic ovary of Hystrichopsyllinae. This subfamily is considered as advanced and includes the largest flea, the mole flea (Büning and Sohst, 1988, 1989). Here, follicle cells develop from small nests of quickly dividing mesodermal cells scattered among germ cell descendants as known from panoistic ovaries (Figures 3.21 and 3.30). The nurse cells remain small and become integrated into the upper part of the developing follicular epithelium of the oocyte during middle and late stages of previtellogenesis.

What are 'normal' follicle cells, surrounding the oocyte, actively doing during oocyte growth phases? The answer is complex and not as clear and simple as the question. One answer embraces all insect orders: follicular cells synthesize the shells of the insect egg (i.e. the vitelline envelope and the chorion). Choriogenesis has been studied intensively by many groups using all kinds of ultrastructural, biochemical and genetic methods available. Although the chorion of many insect species has been looked at with SEM studies, at least those which are kept in permanent cultures, most work is done in two orders, the Diptera (*Drosophila* species and few others) and the Lepidoptera (*Bombyx mori, Hyalophora cecropia, Antheraea polyphemus* and others). Thus, we will restrict ourselves mainly to these orders in a short review of choriogenesis.

Follicle cells, i.e. main body cells (Figures 3.29, 3.31 and 3.32) (Margaritis, 1985), synthesize subsequent layers around the egg at the end of vitellogenesis. The structure and thickness of the layers varies from species to species. However, some corresponding characters are shared by most insect chorions. Next to the vitelline layer (see below) a small (1–3 μm) trabecular layer (endochorion) is formed first, consisting of flocculent material embracing large cavities (Figure 3.31b). This layer is followed by the exochorion containing large amounts of proteinaceous fibres, mostly ordered in parallel to the surface of the egg and to each other. Only in Lepidopterans are the lamellae of fibres of the exochorion rotated in a helicoidal manner within each layer (Figure 3.32g) (Mazur *et al.*, 1982; Margaritis, 1985). This mode of construction is known from other exoskeletal elements, such as the cuticles of insects and crabs (Bouligand, 1972). The exochorion is disrupted by holes, serving different purposes. One is

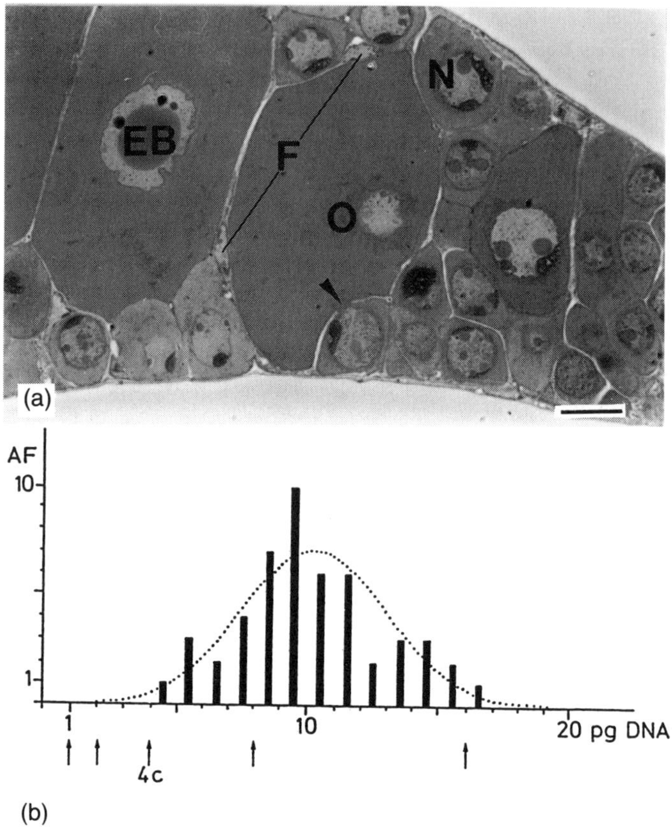

Figure 3.30 (a) Polytrophic meroistic ovary of *Hystrichopsylla talpae* (Siphonaptera). Nurse cells (N), follicular cells (F) and oocytes (O), in a region of early previtellogenic growth. Intercellular bridge between nurse cell and oocyte (arrowhead). Note big endobody (EB) in germinal vesicle. A nurse cell chamber is not formed, as known from all other polytrophic meroistic ovaries of ectognath insects. (b) Ovary of *H. talpae* (Siphonaptera); cytophotometric measurement of DNA from germ cell clusters during early previtellogenesis. Note that lowest amount is clearly above the 4C value and all values follow a Gaussian distribution, indicating extrachromosomal DNA replication in *all* cells. No accumulation around 8C or 16C is found. Therefore, endomitosis of nurse cells is restricted to only some cells and to only one round of replication, if at all.

the micropyle, used as an entrance for sperm, when eggs are inseminated just before egg deposition (Figure 3.32h). Others serve as aeropyles (Figure 3.32a–f), which communicate with the holes of the endochorion. Thus, the whole egg is able to respirate, but this architecture of the chorion provides desiccation of the egg. *Drosophila* eggs need specialized respiratory appendages (Figure 3.31a), since the main body of layed eggs sinks into the food, except these highly vacuolized appendages. The distribution of aeropyles and outer imprints of the chorion, follows the pattern of follicle cell architecture. Late in choriogenesis, the fibres of the exochorion become impregnated with a non-fibrous matrix, secreted by the follicle cells. The

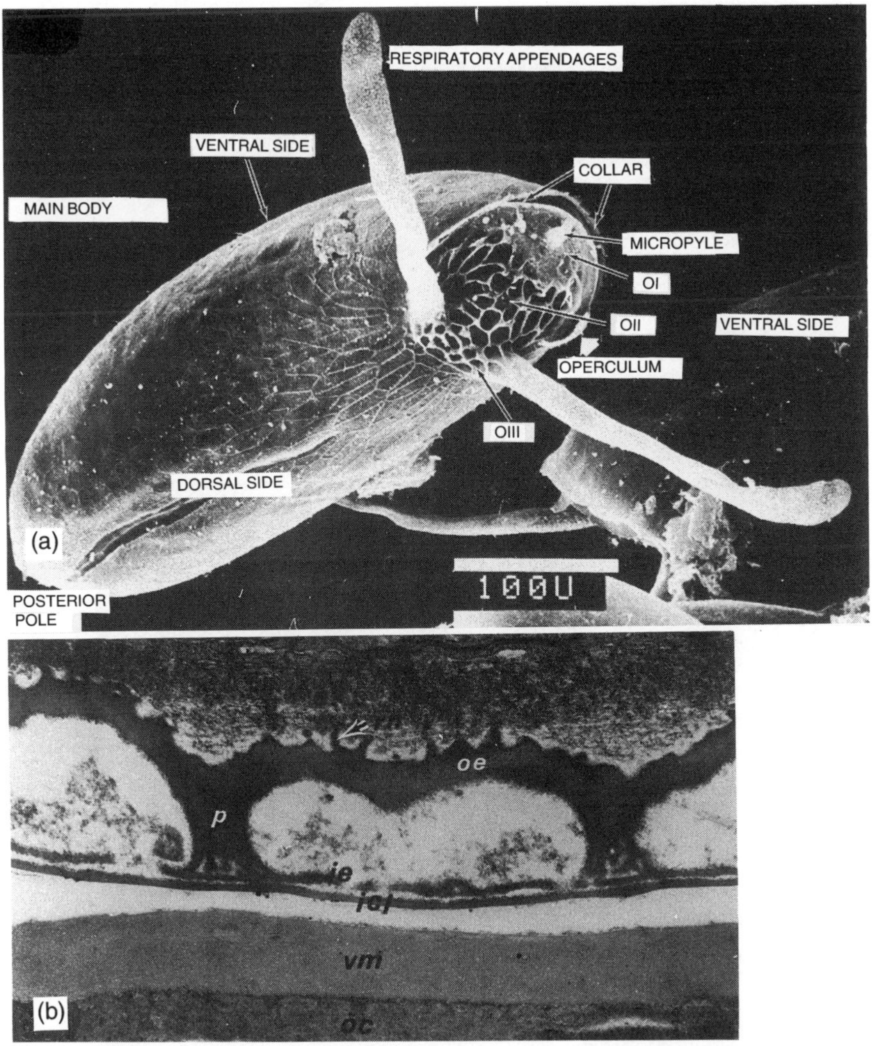

Figure 3.31 (a) Chorion of *Drosophila melanogaster* viewed by SEM. Note specialized regions like operculum, respiratory appendages, micropyle, collar, dorsal and ventral side. (b) Thin section through chorion of *D. melanogaster*. Note the different layers of egg shell material. Oocyte (oc); vitelline membrane (vm); inner endochorionic layer (icl); inner endochorion (ie); pillars (p) and roof network (rn) as parts of the outer endochorion (oe); parallel fibres are seen in the exochorion (ex). (From Margaritis, 1985; with permission of L. Margaritis and Pergamon Press.)

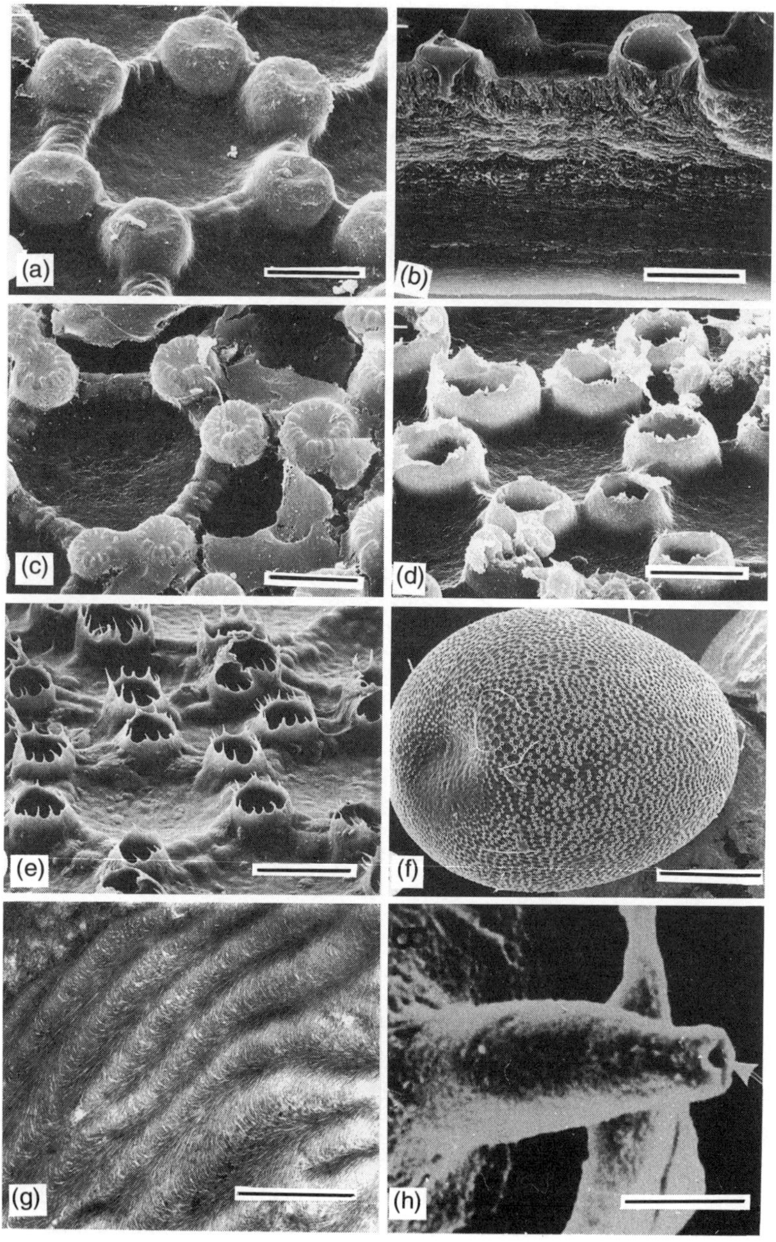
(a)
(b)
(c)
(d)
(e)
(f)
(g)
(h)

molecules of this matrix penetrate the endochorion and the exochorion before interacting with the fibre proteins. Ultrastructural, biochemical and molecular analysis of chorion genes clearly demonstrates that the chorion of different species are not, or at least not closely, related to each other and we must expect to find a variety of ways of constructing the chorion. Today, we are still far from answering questions on phylogenetic relationships between insect orders concerning the choriogenesis of the main groups of insects; however, chorion ultrastructure has been used for phylogenetic reasons within many orders (Kafatos *et al.*, 1977; Regier and Kafatos, 1985; Margaritis, 1985; Kambysellis, 1993; Mazzini *et al.*, 1993).

During the 1980s it became evident through genetic methods and molecular biology that the innermost layer of egg shells, the vitelline envelope, is also a product of the follicular cells, at least in *Drosophila* (Fargnoli and Waring, 1982, 1984; Higgins *et al.*, 1984; Mindrinos *et al.*, 1986; Burke *et al.*, 1987). These findings confirm older ultrastructural evidence gained from a variety of insect species of many insect orders (e.g. *Drosophila* (Diptera): King and Koch, 1963; Quattropani and Anderson, 1969; *Anopheles* (Diptera): King, 1964; *Hystrichopsylla* (Siphonaptera), *Hyalophora* (Lepidoptera): King and Aggarwal, 1965; *Ephestia* (Lepidoptera): Cruickshank, 1971, Cummings, 1972; *Nasonia* (Hymenoptera): Richards, 1968; *Chrysopa* (Neuroptera): Matsuzaki, 1978; *Bothrogonia* (Hemiptera): Matsuzaki, 1975; *Parthenothrips* (Thysanoptera): Büning, unpublished; *Sympetrum* (Odonata): Matsuzaki, 1971; *Aeschna* (Odonata): Beams and Kessel, 1969; *Gryllus* (Orthoptera): Matsuzaki, 1971; Favard-Sereno, 1971; *Bacillus* (Phasmatodea): Mazzini and Giorgi, 1985; *Habrophlebia* (Ephemeroptera): Mazzini and Gaino, 1988, Gaino and Mazzini, 1989, 1990). In all species, in which vitellogenesis and choriogenesis are not repressed, follicle cells increase their content of rough ER during the late stages of previtellogenesis and, concomitant with the onset of vitellogenesis, some vacuoles accumulate in follicle cell cytoplasm facing the oocyte as shown for follicle cells of *Hystrichopsylla talpae* (Siphonaptera) (Figure 3.16b). These vacuoles are of the same electron density as the droplets seen in the intercellular cleft shortly thereafter. Sometime later, the droplets begin to fuse into the

---

Figure 3.32 (a–f) Morphogenesis of aeropyles of *Antheraea pernyi*. SEMs of middle and late stages of development. Aeropyles develop in junctions of three neighbouring follicle cells. Filler material is shown in side view in (b) (arrow), which is used as a scaffold for crown formation (c–d). (e) Mature aeropyle crowns. (f) Whole surface of mature egg is covered with aeropyle crowns. (g) Contrasting, the exochorion of moths like *Bombyx mori* shows several layers of paralleled fibres with helicoidal orientation. (h) Micropylar apparatus of *Drosophila melanogaster* with opening (arrow), as shown by SEM (a, b, e, f and h from Margaritis, 1985; c, d and g originals courtesy of L. Margaritis; with permission of L. Margaritis and Pergamon Press.)

homogenous vitelline layer filling the whole cleft. The vitelline layer is still permeable to yolk precursors, since vitellogenesis via pinocytotic uptake by the oocyte membrane is undisturbed. Vitellogenesis is stopped by the onset of choriogenesis as shown for *Rhodnius* (Heteroptera) (Huebner and Anderson, 1972a; Telfer *et al.*, 1982), or by rising of occlusion zones between follicle cells near the apical (oocyte facing) side (Rubenstein, 1979; Telfer *et al.*, 1982). The role of both processes in the termination of vitellogenesis in other species has not been investigated so far.

Some other protein syntheses of follicle cells, not involved in the synthesis of egg shells, must be mentioned, which contribute to oocyte growth. The first is the synthesis of a vitellogenin accompanying protein (named paravitellogenin), which may be engaged in yolk uptake and is stored later on together with vitellin in the yolk droplets of the Cecropia moth (Anderson and Telfer, 1969, 1970; Bast and Telfer, 1976). The same protein is synthesized by the follicle cells of *Bombyx* follicles (Zhu *et al.*, 1986). Follicle cells of *Drosophila* are able to synthesize vitellogenin, which is exported from follicle cells into the oocyte (Bownes, 1979, 1980, 1982; Bownes *et al.*, 1983, 1993; Brennan *et al.*, 1982). Until now, *Drosophila* species have been the only insects in which yolk is synthesized by follicle cells as well as by fat body cells, which are the classic sources of vitellogenin synthesis (for vitellogenesis see for instance: Kunkel and Nordin, 1985; Postlethwait and Giorgi, 1985; Telfer and Pan, 1988; Giorgi *et al.*, 1989, 1993; Kulakosky and Telfer, 1990; Raikhel and Dhadialla, 1992; Ferenz, 1993). The third contribution to vitellogenesis was detected in *Hyalophora cecropia*, in which a sulphated proteoglycan is synthesized and/or modified by follicle cells and deposited in the intracellular cleft. This extracellular matrix is supposed to enrich vitellogenin molecules in the intracellular cleft, supporting their pinocytotic uptake by the oocyte (Telfer, 1979; Telfer *et al.*, 1982).

The former three syntheses occur during the vitellogenic stages. We will now focus on the previtellogenic stages, in which two genes have been identified, which are transcribed and translated by follicle cells of *Drosophila*. These are the genes for laminin and for collagen type IV (Garzino, 1989; Knibiehler *et al.*, 1990). Follicle cells first accumulate mRNA, which becomes destroyed after translation, at the beginning of vitellogenesis. The proteins accumulate in the oocytes on vitellogenic stages. Transport mechanisms have not yet been investigated. Both proteins are used during embryogenesis, when the collagen type IV may play a role in tissue recognition and differentiation as a source of the extracellular matrix. The last two examples corroborate the earlier findings of high RNA synthesis in follicle cells during previtellogenesis. Other genes will be found in the future, which will contribute their products to the growing oocyte.

Of even more interest are the activities of those somatic genes, the products of which may play a fundamental role in the construction of egg follicles. One of these is the gene *Notch*, also engaged in correct spermatogenesis

(Xu *et al.*, 1992). *Notch* encodes a transmembrane protein that has homologies with the epidermal growth factor of mammalian cells. It is expressed by prefollicular cells, border cells and follicular cells and its product accumulates at the apical side of follicular cells, facing the oocyte and nurse cell membranes. Highest accumulation occurs in those somatic cells which are just moving, such as border cells or the follicular cells at the anterior-dorsal and ventral end of the oocyte. At late stages of vitellogenesis, high activity is also found in those cells forming the dorsal appendages and the micropyle. In nurse cells, *Notch* is also active and the protein accumulates in the intercellular bridges forming the rings, which are next to the rings of F-actin (see section 3.4.1). The *Notch* gene product is thought to have a permissive role (i.e. it functions to make the cell receptive to specific signals, whether inhibitory or inductive) (Cagan and Ready, 1989; Artavanais-Tsakoras and Simpson, 1991). Other genes that encode to the same group of permissive molecules are *strawberry notch* (Coyle-Thompson and Banerjee, 1993) and *amardillo* (Peifer *et al.*, 1993). Other genes like *Hairless, Delta, groucho, serrate* and *deltex* may also be engaged in these cell–cell interactions and it may be a general method of developing differentiation among tissues. However, the cascade of interactions of these genes is not yet known. Other examples of follicular gene activities which are important for imprinting embryonal axes will be discussed in section 3.3.2.

3.2.6 Interfollicle cells, pedicel and lateral oviduct

Interfollicle tissues differentiate early during previtellogenic growth phases and isolate subsequent follicles from each other. The gross morphology of interfollicle cells varies between ovariole types and from species to species. In the polytrophic meroistic ovaries of many species, at least in Diptera and Lepidoptera, these cells produce a stalk, clearly visible between follicles in previtellogenic and vitellogenic stages as seen in *Drosophila* (Figures 3.33 and 3.34) (King, 1970) or *Hyalophora* (King and Aggarwal, 1965). Stalk cells (border cells) are involved in the formation of the anterior–posterior axis of *Drosophila* oocytes (see sections 3.3.2 and 3.3.4).

In the ovarioles of other species, this tissue is not separated as early thus the subsequent follicles have broad contacts to each other in previtellogenic stages (Figures 3.18, 3.24, 3.35–3.37).

Such a delayed separation of follicles is found throughout telotrophic meroistic ovarioles, irrespective of their different phylogenetic origins (see section 3.7) (Büning, 1972, 1979a,b,c; Huebner and Anderson, 1972c; Matsuzaki and Ando, 1977). Thus, broad contacting regions may be necessary at least during previtellogenic growth phases when nutritive cords are functioning (see section 3.5.3). However, in a variety of species of other ovariole types we notice broad contact zones, for example the polytrophic meroistic ovaries of adephage Coleoptera (Bier, 1965b), of Trichoptera

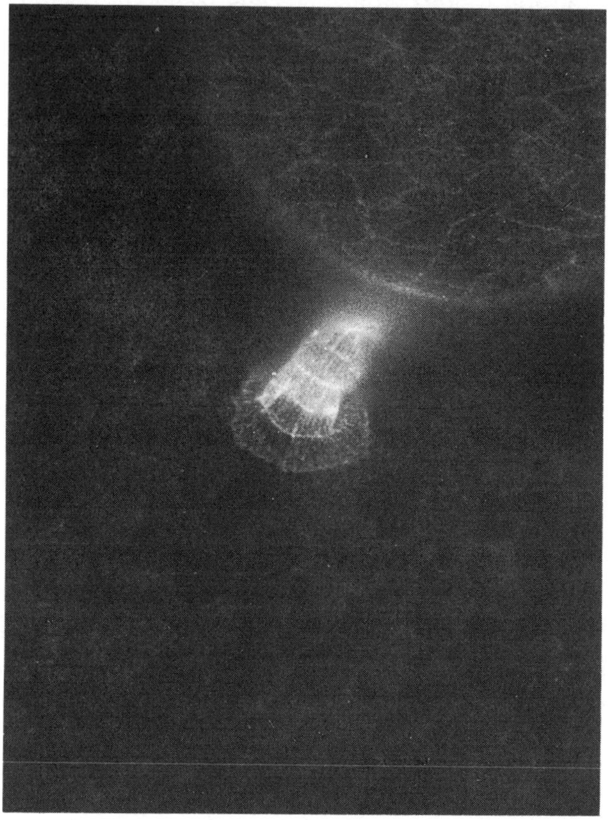

Figure 3.33 Stalk cells connecting two follicles of an ovariole of *Drosophila melanogaster.* The bright fluorescence comes from impregnation of whole mounts with anti PSb (integrin) antibodies. (Courtesy of H.O. Gutzeit, University of Freiburg, Germany.)

Figure 3.34 A wild type ovary (1) and ovariole (2) of *Drosophila melanogaster* (Diptera), representing a typical polytrophic meroistic ovariole. Three zones of the germarium are designated. Zone 1 contains stem cells and their direct siblings, the newly built cystoblasts; zone 2 contains clusters during cystocyte divisions; zone 3 houses the most basal cluster, which has already started previtellogenic growth and which is ensheathed fully with a monolayered follicular epithelium. Five follicles follow (S2–S6) according to stage definitions from Koch *et al.* (1967). (From Koch *et al.*, 1967; with permission of Alan R. Liss, Inc.)

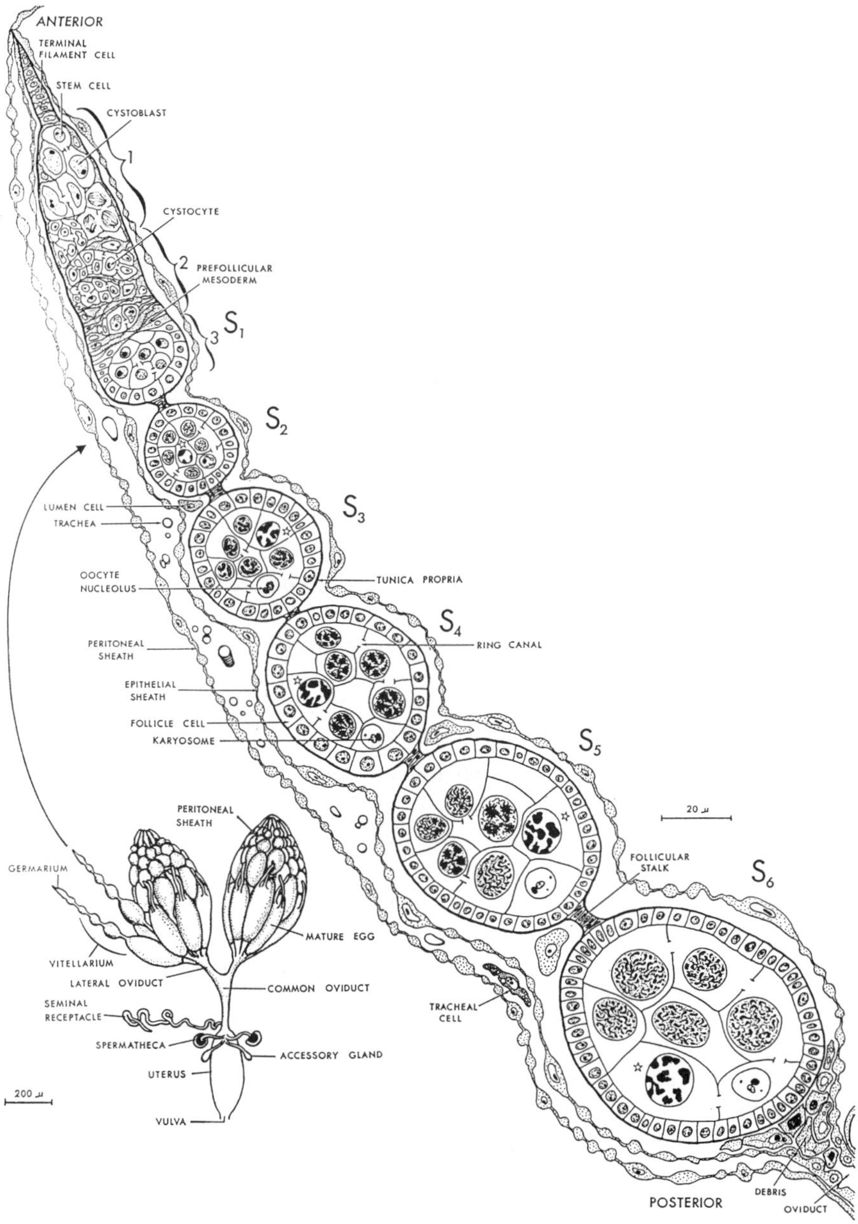
ANTERIOR
TERMINAL FILAMENT CELL
STEM CELL
CYSTOBLAST
1
CYSTOCYTE
2
PREFOLLICULAR MESODERM
3
S1
S2
S3
LUMEN CELL
TRACHEA
OOCYTE NUCLEOLUS
TUNICA PROPRIA
S4
RING CANAL
PERITONEAL SHEATH
EPITHELIAL SHEATH
FOLLICLE CELL
KARYOSOME
S5
20 μ
PERITONEAL SHEATH
GERMARIUM
VITELLARIUM
MATURE EGG
LATERAL OVIDUCT
COMMON OVIDUCT
SEMINAL RECEPTACLE
SPERMATHECA
ACCESSORY GLAND
UTERUS
200 μ
VULVA
FOLLICULAR STALK
S6
TRACHEAL CELL
POSTERIOR
DEBRIS
OVIDUCT

(Matsuzaki, 1972), of Mecoptera and Neuroptera (Matsuzaki, 1978), or of Hymenoptera (Paulcke, 1900; Hegner, 1915; Meng, 1968; Engels, 1973), as well as in panoistic ovarioles (Matsuzaki *et al.*, 1979; Gottanka and Büning, 1990). Possibly, these different aspects are caused by different timing strategies which, finally, lead to the same structure: a separating tissue during vitellogenesis. Figure 3.22b shows this tissue between two follicles of *Hystrichopsylla* at the beginning of vitellogenesis. Cells are isolated from each other and from the adjacent follicles by thick basal laminae. This construction may be helpful for the morphogenetic and/or physiological isolation of subsequent follicles. Electrical currents exist between each vitellogenic egg and the environment and these currents may play a prominent role during vitellogenesis (Woodruff and Telfer, 1980; Dittmann *et al.*, 1981; Kunkel, 1986). As shown above, follicle cells are electrically coupled to each other and to the oocyte by gap junctions and cannot serve as isolators, as is assumed for interfollicular tissues during vitellogenesis.

The basal part of each ovariole is called the pedicel. This region is clearly separated from normal follicle cells and rises from baso-lateral mesodermal cells after invasion of germ cells into the ovariole anlagen very early during ontogenetic development, throughout all insect orders investigated so far. Thereby, prefollicular tissue and pedicel tissue are not separated in the very beginning, demonstrating their common origin (Figures 3.7, 3.8 and 3.15) (see for instance King, 1970 for *Drosophila* (Diptera); Büning, 1979c for *Sialis* (Megaloptera); Huebner and Anderson, 1972 and Huebner, 1984a,b for *Rhodnius*, and King and Büning, 1985 for *Pyrrhocoris* (Hemiptera)). After invasion of germ cells into the ovariole anlagen, the tube will be closed by outgrowing cells of the pedicel. The pedicel plug will open again by autolysis of its innermost cells, which happens at the latest during

Figure 3.35 Development of ovarioles in Hemiptera, Heteroptera. (a) Ovariole anlage of early fourth larval instar of *Pyrrhocoris apterus* (Hemiptera, Heteroptera: Pyrrhocoridae). Note that all germ cells are ordered around a centre (arrowhead) and that they are embraced by a monolayer of inner sheath cells (outer margin is stippled). (b) In ovarioles of late fourth larval instar, four zones can be found: in the apical region (I), germ cells can divide mitotically (arrowheads). After a transitory zone (II) nurse cells become active in synthesis in zone III, as shown by intense staining with Azur B. In the centre of zone III a huge mass of fine threads and vesicles appear, the future trophic core, which are nothing else but the processes of peripheral nurse cells and of nutritive cords of resting oocytes (O). The latter are found in region IV at the bottom of the terminal chamber. (c) Apical region of an ovariole of adult *Dysdercus intermedius* (Pyrrhocoridae), showing the region of mitotic activity (arrowheads) in the anterior tip. (d) Overview of the terminal chamber and previtellogenic growing oocytes (O) of adult *D. intermedius*. Note apical–basal gradient with degenerating nurse cells in the basal region of the terminal chamber. Trophic core (TC); nutritive cords (NC); bars represent 50 μm.

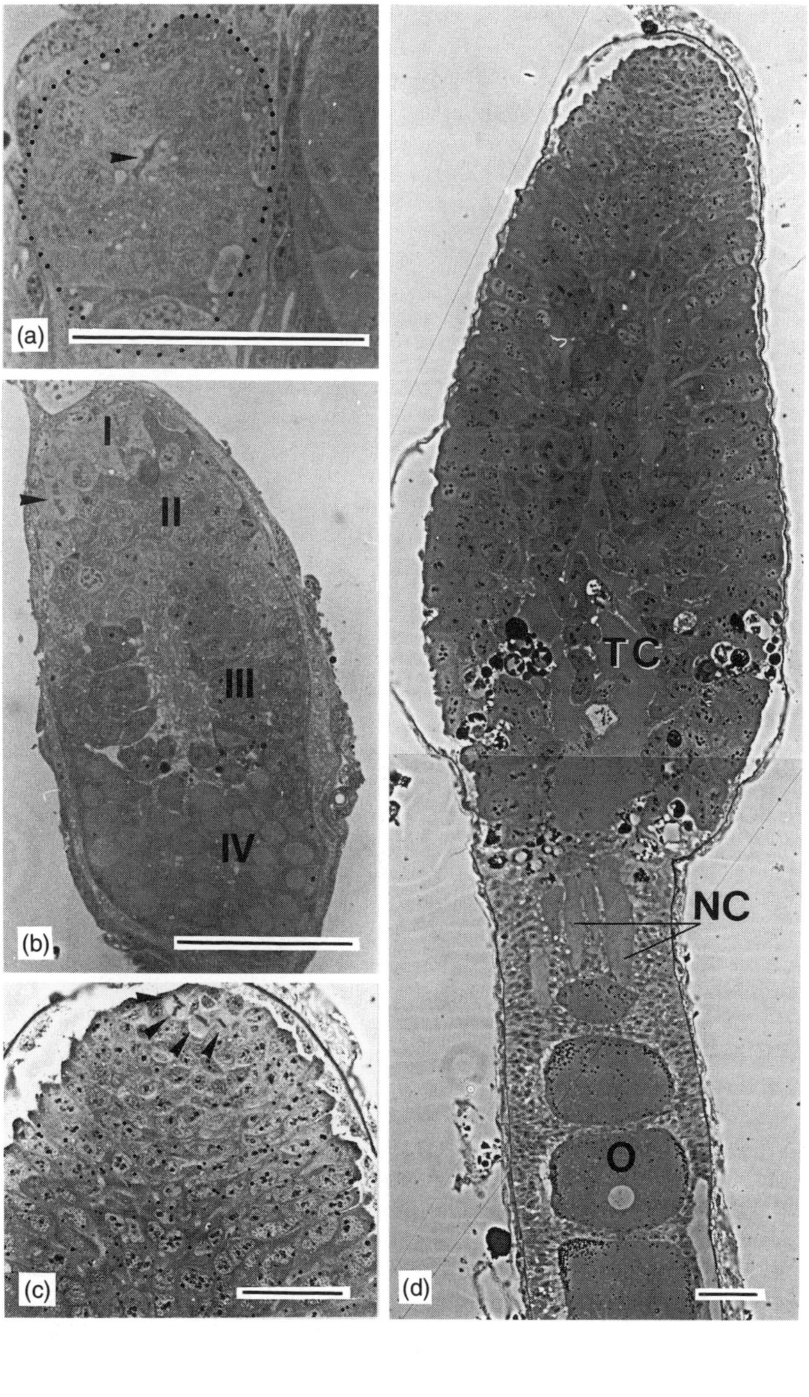
(a)
(b)
I
II
III
IV
(c)
(d)
TC
NC
O

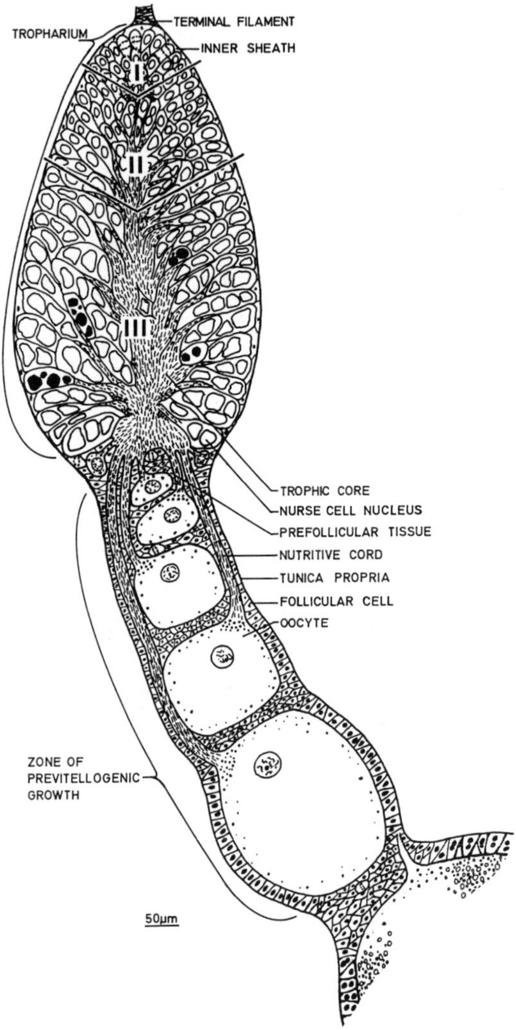

Figure 3.36 (a) Scheme of the telotrophic meroistic ovariole of advanced Heteroptera. Zone I is composed of few mitotic active germ cell descendants, producing nurse cells, which keep contact with the central core via intercellular bridges. Zone II houses nurse cell nuclei, the genomes of which multiply by endomitotic polyploidization. All cells open to the central core. Zone III contains nurse cells, which have divided amitotically, producing large areas containing the amitotic nuclei. Degeneration of nurse cell nuclei is shown by black dots. The vitellarium starts with the prefollicular tissue, which embraces all resting oocytes. Resting oocytes are situated as a collar around the previtellogenic growing oocytes. All oocytes are connected from the very beginning to the trophic core by nutritive cords. (Redrawn after King and Büning, 1985.)

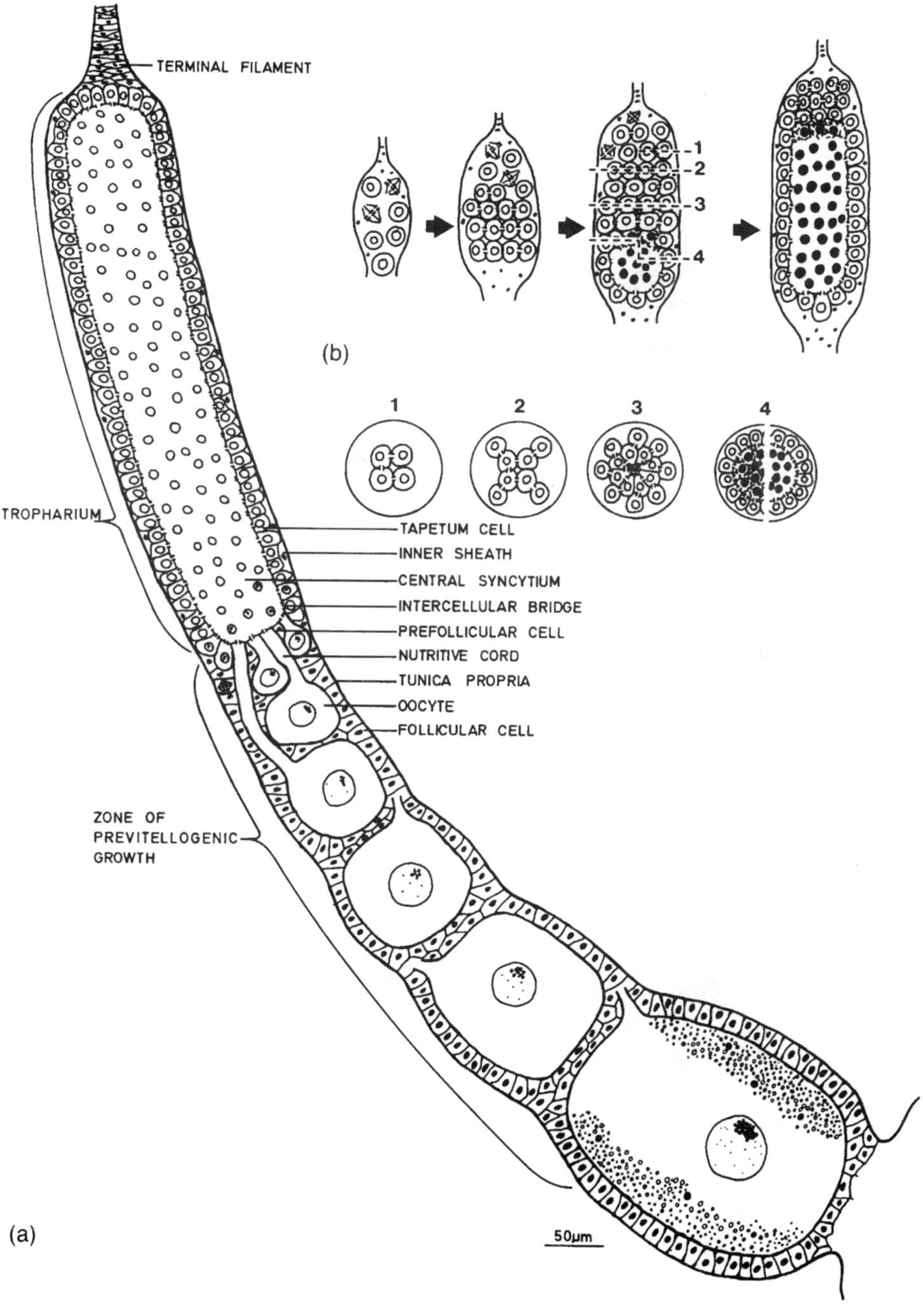

Figure 3.37 (a) Telotrophic ovariole of *Sialis*, just before final moult. (b) Hypothetical model of cluster formation in *Sialis*. Relatively small clusters are supposed, the cellular character of which is maintained only in tapetum cells, which are directly contacting the somatic inner sheath cells. Cell membranes of all other siblings degenerate totally, forming the central syncytium. Oocytes derive from basally located tapetum cells.

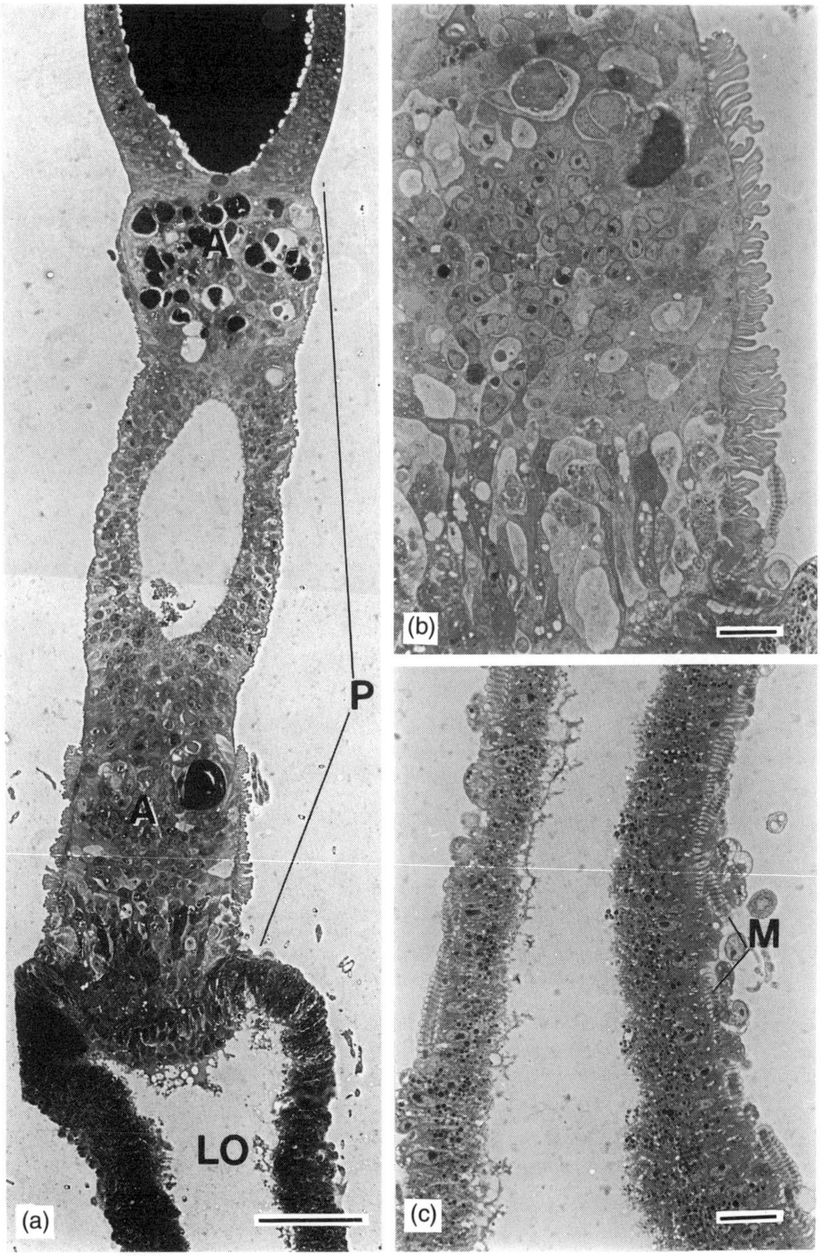

Figure 3.38 (a) Semi-thin section through pedicel (P) and beginning of lateral oviduct (LO) of an ovariole of *Dytiscus marginalis* (Coleoptera, Adephaga: Dytiscidae). Note autolysis of cells (A) within two yellow bodies. (b) Enlargement of basal zone of pedicel shows highly folded tunica propria. At the base, elongated cells connect pedicel region with lateral oviduct. (c) Lateral oviduct of the same species. Note striated muscles (M) and secretory vesicles in lateral oviduct cells.

choriogenesis of the first egg (Figure 3.38a). Besides autolysis of cells, indicated by large lysosomal vesicles, cells produce vacuoles of low electron density; these unknown products may help the ripe egg glide into the oviduct (Figure 3.38b,c).

When the first egg has passed, the remaining follicle cells, already in degradation, shrink to a globular body together with the encompassing tunica propria. This body is species-specific pigmented, mostly yellowish brown; however, greenish or red tones are also found. In analogy to mammals, this body is sometimes called the corpus luteum. However, hormone production is not proved and the ultrastructure gives no hints. In *Schistocerca* (Dictyoptera) the yellow colour comes from β-carotene (Goodwin, 1952; Lusis, 1963). In *Kalotermes* (Isoptera) post-ovulated follicle cells increase their number of lysosomal vesicles but keep their form for a while. Thus, one can distinguish two subsequent stages: an initial stage, in which the follicular tissue is still present, and an advanced stage, in which cells degenerate by autolysis. In this later stage, cells have lost their epithelial connections and are full of multilamellar bodies and of giant autophagic vacuoles (Griffith and Lai-Fook, 1986; Grandi, 1990).

In most species, the pedicel region is short and opens into the lateral oviduct. In lepidopterans, as in some other species of other orders, the pedicel is elongated and serves as a store for ripe, virgin eggs (Telfer, 1975). This function is carried out in most insects by the lateral oviduct or by the common oviduct. The lateral oviduct develops directly from the gonadal tube, which houses the germ cells, intermingled with some somatic cells during early developmental stages (Figures 3.7 and 3.8) (Heymons, 1895; King, 1970; Büning, 1979c). Thus, lateral oviducts are of mesodermal origin. However, in many species, distal parts of the lateral oviduct can also develop from the common oviduct, which is always of ectodermal origin (Weber and Weidner, 1974; Davey, 1985).

Differences in the gross anatomy of lateral oviducts have been used to establish diverse anatomical types of insect ovaries (Stein, 1847; Deegener, 1928; Weber, 1933). The types seem to be ordo-, sometimes family-specific. However, detailed investigations are lacking in most orders. Recently, Welch (1993) gave a review on ovary types in Staphylinidae (Coleoptera, Polyphaga). In this group, nearly all types are found, from elongated comb-like arrangements of ovarioles to lateral oviducts (ovaria pectinata), from those forms which have ramified lateral oviducts (ovaria racemosa; ovaria disticha) to those in which all ovarioles assemble (ovaria fasciculata). This example shows that these characters can be used within families, but its value for higher taxa is questionable. More investigations are needed, but the large amount of parallel development between different lines in insect phylogeny (see section 3.6) and the apparently low contribution of lateral oviduct anatomy and ultrastructure for ovary functions has made further investigations in the past two decades unattractive.

Lateral oviducts are always ensheathed with a net of striated muscles (Figure 3.38b,c). The epithelial cells are highly prismatic and are filled with secretion granules, which are thought to help ripe eggs glide down the tube. Lateral oviducts and the pedicel region are the only mesodermal parts of the insect ovary which are under direct nerve control, as shown by back-fills of abdominal nerves (Cook and Peterson, 1989).

3.3 GERM CELLS

As outlined in section 1.2 the idea of a germ line and its potential immortality goes back to the last century to the highly stimulating work of August Weissmann (1885, 1892), whose theories are still influencing the work of todays developmental biologists (Lawrence, 1988). In insects, the germ line is well documented in many species from different orders: in Diptera (*Miastor*: Kahle, 1908; *Chironomus*: Hasper, 1911; Sachtleben, 1918; *Drosophila*: Counce, 1961; Jura, 1964a,b), in Coleoptera (Chrysomelids: Hegner, 1908, 1911; Hirschler, 1909; Schnetter, 1965; Richard-Mercier, 1977), in Hymenoptera (*Pimpla*: Fleischmann, 1975) and others. Modern methods of molecular biology allow a precise documentation of the fate of pole cells as shown in Figure 3.5 for *Drosophila* embryos (Lasko and Ashburner, 1990).

However, not all insects possess clearly visible pole cells and there are some reports, which suggest late separation of germ cells, for instance in honey bees (Fleig and Sander, 1988). Günther (1971) could raise fertile individuals of *Pimpla* (Hymenoptera), after ablation of pole cells by X-raying posterior cells or by ligation experiments. To make sure that no pole cells have been overlooked, similar experiments were done by Fleischmann (1975), who gained the same results. Both results were interpreted as dual ways to get germ cells: normally, pole cells will be the source of germ cells, however, mesodermal cells of the inner cell mass of gonads can 'redifferentiate' to germ cells, presumably when the amount of former pole cells reaching the gonad anlagen is too low, or pole cells are absent (which may be the case in *Apis* (Hymenoptera)). Using modern molecular developmental biology it would be interesting to re-examine these data.

3.3.1 Differentiation into sperm and oocytes

(a) Metazoans and oogamy

When higher developed multicellular organisms occurred on earth about 0.7 billion years ago, sexual reproduction had been established long before, about 1.4 billion years ago (Stearns, 1987). However, the zygote , formerly representing the new generation directly in unicellular diploid species and, after reduction, in haploid species, will now divide, differentiate, and grow into the adult metazoan. In other words, embryonic development became an undeliverable prerequisite of life. During embryogenesis, the new organism cannot feed by itself but depends fully on the resources, which are invested by the gametes. The more resources the embryo can use, the more developed the young animal will be and the better start in life it will have. On the other hand, embryogenesis should be kept short, since embryos cannot defend themselves from predators for instance. Thus, one can assume an evolutionary tendency to raise a mass of gametes. This contradicts the need of karyogamy, to keep gametes mobile.

The conflict is solved by a division of labour between male germ cells, sperm, which are extremely small but highly mobile, and female germ cells, oocytes, which become immobile but can increase their mass up to 10^{12} times their original volume. This enables the later zygote to overcome the dangerous period of embryogenesis very quickly. This concept is based on evolutionary pathways, which lead from isogamy, via anisogamy to oogamy in lower, water-living eukaryotes, recently reviewed by Stearns (1987). Thus, we can say that oogamy is a prerequisite for the development of higher organisms, and all metazoans living today followed this type of evolution.

(b) Morphology of sperm development

The earliest stages of sperm development in insects are identical to the earliest stages in oogenesis from the morphological point of view (see also sections 3.4 and 3.6). First differences contribute more to the number of germ cells housed in early gonad anlagen than to their ultrastructure. However, our knowledge is based on a few investigations on *Drosophila melanogaster* (Mahowald, 1971a; Tokuyasu, 1974a,b,c, 1975a,b; Hardy *et al.*, 1979; Underwood *et al.*, 1980; Lifschytz, 1987; Gönczy *et al.*, 1992), *Drosophila hydei* (Hackstein, 1987; Hackstein *et al.*, 1987, 1990), locusts (Szöllösi, 1982) and moths (King and Akai, 1971a,b; Mandelbaum, 1980; Szöllösi, 1982), whereas there is much literature describing the histogenesis and ultrastructure of fully developed sperm, summarized by Jamieson (1987).

Insect testes have a similar anatomy to insect ovaries. Both organs are subdivided into homologous functional units, in ovarioles (ovaries) and follicles (testes). The male germ cells in a follicle are ensheathed by somatic mesodermal cells, homologous to the inner sheath cells of a female germarium. The apical end of a testes follicle bears no terminal filament as in ovarioles; however, one or some giant cells next to the apex of the follicle are found, first described by Verson (1889, 1894). He interpreted the singular apical cell found in testes of Lepidoptera as a primordial germ cell, which would produce primary spermatogonia. However, the somatic character of the apical cell was interpreted correctly a few years later by Toyama (1894), Tichomirow (1898) and others. The typical apical cells, as found in many insect orders, have a star-like shape with a centrally located spherical nucleus each (Figure 3.39).

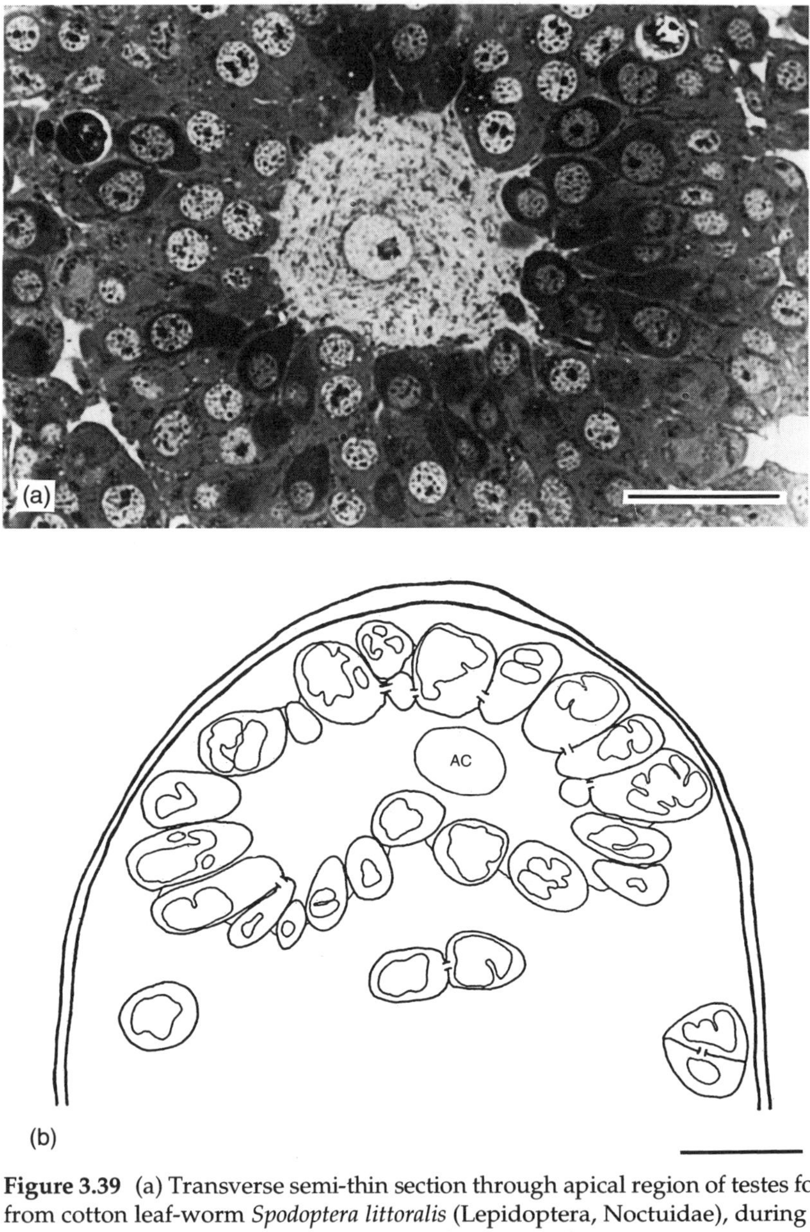

Figure 3.39 (a) Transverse semi-thin section through apical region of testes follicle from cotton leaf-worm *Spodoptera littoralis* (Lepidoptera, Noctuidae), during sixth larval instar. Radial protrusions interdigitate with primary spermatogonia. (Courtesy of Jacek Godula, Jagiellonian University of Krakow, Poland.) (b) Longitudinal section through apical region of testes follicle of *Locusta migratoria*. By ultrathin serial sectioning it can be shown that germ cells build clusters around the somatic apical cell (AC). The whole follicle is surrounded by a strong basal lamina and a tunica externa. Bars represent 20 μm.

Primary spermatogonia assemble around and cytoplasmic protrusions of the apical cell keep contact with the flat monolayer of somatic cells, which is equivalent to the inner sheath of ovarioles. The apical cells persist far into the adult stage in some species as in grasshoppers and in beetles, or are found only during late larval and pupal stages as in many butterflies (Knaben, 1931; Menon, 1969; Szöllösi, 1982). In beetles and flies, there are some apical cells in each follicle, forming a conus-like intrusion, which is also called the hub (Menon, 1969; Hardy *et al.*, 1979; Lindsley and Tokuyasu, 1980; Lifshytz, 1987; Gönczy *et al.*, 1992). In *Drosophila* the hub consists of 8–16 somatic cells to which about 18 stem cells of the germ line adhere. Mitotic activity of apical cells has never been reported. Thus, apical cells are functioning cells. However, the functioning of these somatic cells remains unclear. In older literature (Verson, 1889; Grünberg, 1903; Zick, 1911) the apical cell is considered as a nurse cell, which nourishes the surrounding spermatogonia. Muckenthaler (1964) found high RNA synthesis in the apical cell of *Melanoplus* (Orthoptera) by ^{3}H-uridine incorporation, and Szöllösi and Marcaillou (1980) found huge numbers of mitochondria, smooth ER, lysosomes and multilamellar (residual) bodies in an ultrastructural investigation of moth testes. These findings are consistent with other investigations on moths and locusts (Leclerck-Smekens, 1978; Szöllösi, 1982). Thus, endocytotic activity and mechanical support of germ cells and cysts is the favourite interpretation today; however, further experimental data are needed (Szöllösi, 1982). Recently, Gönczy *et al.* (1992) have tested over 700 viable and fertile autosomal P-element enhancer detector lines for expression in *Drosophila* testes. Among some specific lines, they found one which was specifically expressed in apical cells. Using this line in agametic testes, they could show that apical cells are enlarged and their position is altered, indicating that normal architecture in testes is controlled by interaction of somatic, here apical cells, and germ cell genes. For further examples of germ cell–somatic cell interactions see sections 3.3.2 and 3.3.4.

Male germ cells develop in an apical basal gradient in each follicle, whereby one or some stem cells divide by differential mitosis to cystoblast cells. By ultrathin serial sectioning of several apical regions of *Drosophila* testes, Hardy *et al.* (1979) could elucidate ontogenetic events during spermiogenesis. The number of stem cells decreases, but some of them remain functional throughout life. Each stem cell remains connected with its daughter cell, the cystoblast, via an intercellular bridge, the morphology of which is similar to that of intercellular bridges occurring during cystocyte divisions. A fusome stretches through the bridge. Later on, the daughter cell detaches and becomes settled in the inner sheath, where the four synchronous mitotic divisions occur, by which clusters of 16 interconnected cells arise. All cells of a cluster go through meiosis, forming 64 interconnected cells, enclosed by the two somatic cells from the final cyst. During these events, each cystocyte shows a 25-fold increase in volume, before morphogenesis transforms each spermatid into a ripe sperm. These events, shown here briefly for *Drosophila,* are similar in each insect testicular follicle; however, some differences may occur among insect groups. Preliminary results of an analysis of stem cell behaviour of *Locusta* male follicles clearly show that germ cells, which are still anchored to the apical cells, are connected to each other by intercellular bridges in the last larval stage, forming aggregates of more than two cells, as found in *Drosophila* (Figure 3.39b) (Büning, unpublished). Two interpretations are possible: (1) stem cells are combined to stem cell clusters; (2) the cystoblast cell will start its mitotic programme before the intercellular bridge, connecting the stem cell and the cystoblast, is split. Later on, these cells split and clones of interconnected cystocytes are created by synchronized mitotic cycles, the number of which is species specific and ranges among insects from four to seven (King and Akai, 1971a; King *et al.*, 1982). During cluster formation, intercellular bridges become stabilized during arrested cytokinesis; the cluster is regularly branched and a polyfusome stretches through all bridges. The bridges move to a centre, creating a rosette configuration of siblings (Carlson and Handel, 1988). These clones go through two meiotic divisions, thus, clusters of 64, 128, 256 or 512 interconnected spermatides occur, which can all be transformed into fertile sperms synchronously. Each cluster of germ cells becomes ensheathed by some somatic cells, forming a cyst. In *Drosophila,* each cyst is formed by only two somatic cells (Hardy *et al.*, 1979; Gönczy *et al.*, 1992). Since the morphological characters of germ cell divisions are identical in both sexes, they are outlined in more detail in section 3.6. Differential mitoses of stem cells, arrested cytokinesis, and the 2^n-rule are three, highly effective methods of producing masses of synchronized sperms within a short time. This concept is phylogenetically very old, since it is not restricted to insects but occurs in nearly all species of higher ranked monophyletic groups, except the nematodes (Wirth, 1984; King and Büning, 1985). What are the reasons for such evolutionary success? Some plausible answers may be as follows: (1) germ cells are needed in very high numbers at a very special point in time, the time of mating; (2) nearly all of these germ cells should be in the right physiological state to maximize sexual success. Both proposals are facilitated by interconnected clones of germ cells which divide synchronously and keep contact with each other as long as possible by cytoplasmic bridges. Looking at sperm, a third reason to conserve this mode to multiply can be added: in many species the sperm decides the sex of the offspring by two alternative subsets of genomes. Correct spermiogenesis in both types of sperm is guaranteed by the open cytoplasmic bridges, which allow a free flow of gene products (ribonucleoproteins (RNPs) and proteins) throughout the whole period of sperm development, which includes post-meiotic gene expression, as found in mice (Figure 3.40) (Erickson, 1990), and includes the period of DNA condensation when the sperm head is formed (e.g. Willison and Ashworth, 1987).

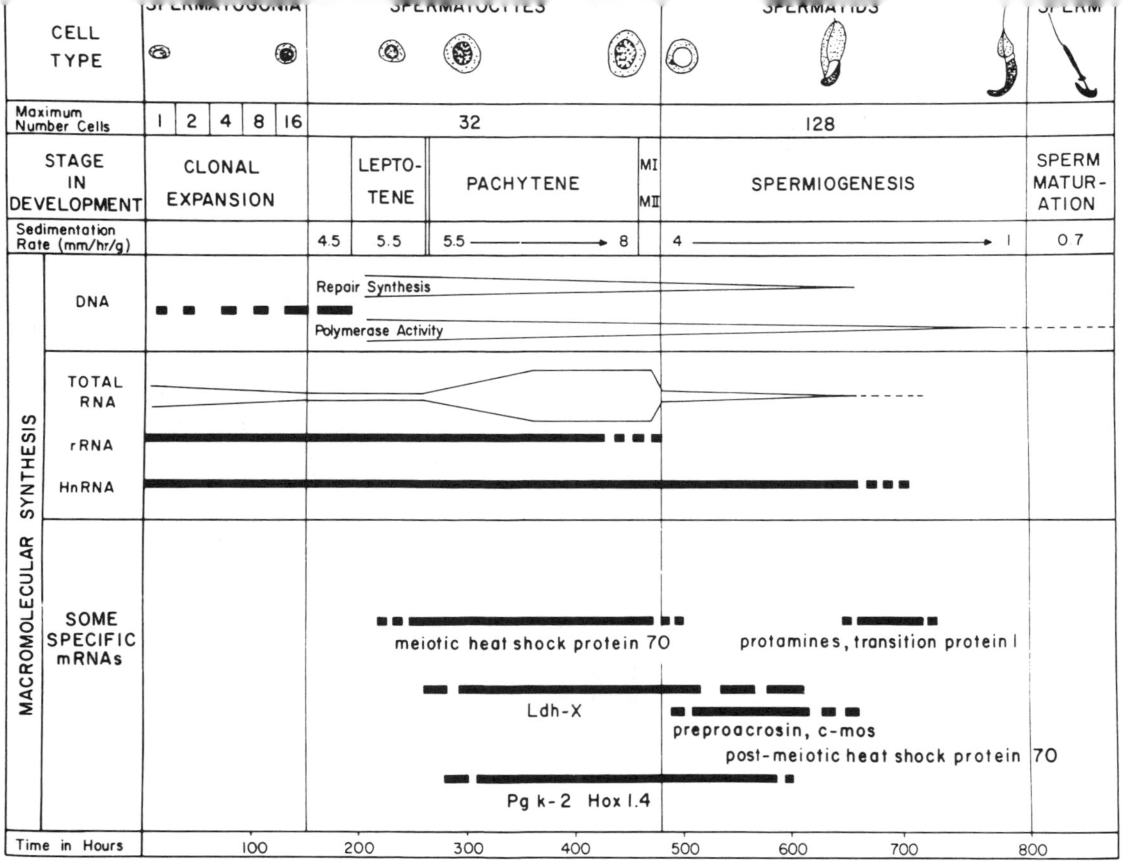

Figure 3.40 Synopsis of macromolecular syntheses during sperm development. Note that post-meiotic transcription characteristics are due to hnRNA and not to rRNA. (From Erickson, 1990; with permission of R.P. Erickson and Elsevier Science Publishers.)

(c) Sex differentiation in Drosophila

Oocytes are giant cells, which contribute to the embryo not only the genome, but also nearly all the cytoplasmic substances needed for successful embryogenesis (see section 3.3.2). This contradicts a germ cell dividing programme which leads to a very large number of small cells. Thus, the simplest way to overcome this conflict seems to be to stop this programme in females (i.e. the regulation of this gene set, 'on' in males and 'off' in females, has to be sex related).

Sex differentiation in insects depends on genetic factors, which can be influenced by environmental conditions (for review see Lauge, 1985). We will not go into the problem but will concentrate on *Drosophila*, the genetics of which have been investigated most thoroughly in insects.

Sex differentiation in *Drosophila* is a developmental process in which maternal and zygotic gene activities are lined in a cascade. This cascade differs in somatic and germ line cells. We will focus first on the cascade of events in somatic tissues and will later turn to germ line cell differentiation processes. Most events are shown in Figure 3.41 where those genes which are involved in the differentiation processes of somatic cells are in blank boxes and those genes for germ line cell determination are stippled.

Sex differentiation in *Drosophila* depends on the ratio of the large sex chromosome (X) to the total amount of non-sex chromosomes, the autosomes (A). As all metazoans, *Drosophila* is diploid, thus, a ratio of 1 (XX:AA) switches the female pathway, a ratio of 0.5 (X:AA) switches the male pathway (Figure 3.41; Bridges, 1925). The sex chromosome Y is not engaged in this basic decision but contributes some fertility genes necessary for correct spermiogenesis. Since meiosis will divide the sex chromosomes by chance to the haploid sperms, a 1:1 sex ratio will result normally. Such homogametic differentiation of females is found in many insect orders as in many other higher ranked taxa and in mammals. However, this is not always the case as a short look into insect and other taxa will show; heterogametic differentiation of females occurs in Lepidoptera, Trichoptera, in some dipteran families, and outside the insects in some families of frogs (Anura) and in birds (Aves) generally.

Another basic mechanism of genetic sex differentiation occurs among other phyla in insects: the haplo-diploidy, in which the female pathway is determined by diploidy, whereas the male pathway is followed by haploidy. Such a mechanism makes the sex ratio of the offspring variable and dependent on the physiology and behaviour of the mother, and/or on environmental conditions, as shown by Hymenoptera, white flies (Aleyrodidae), thrips (Thysanoptera), scale insects (Coccoidea) and others (Stearns, 1987).

Back to *Drosophila*: during the 1980s, we gained some insight into the cascade of sex differentiation (Figure 3.41), which, of course, has important

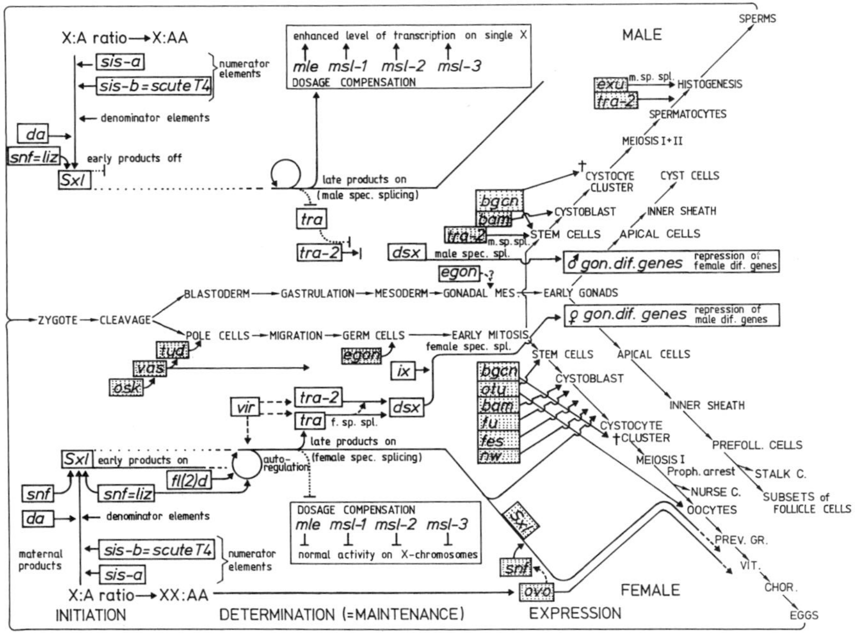

Figure 3.41 Diagram showing cascade of sex differentiation in germ line and somatic gonadal cells during life cycle of *Drosophila melanogaster*. Genes are shown in italics and are boxed. Blank boxes represent genes involved in the differentiation processes of somatic tissues, stippled boxes show genes engaged in the differentiation of germ line. Thick arrows indicate influence on following targets, which can be other genes or cells of special developmental stage. Broken arrows indicate influence in question. Thin arrows connect subsequent stages of development from zygote to eggs and sperms. Function of *egon* is not known but this gene is expressed during a short period of early gonad formation most probably in germ cells. *Vasa* is one of the genes necessary for maintenance of germ line character. For function of central gene *Sex-lethal* (*Sxl*) and other genes see text.

consequences for the morphology and function of the primary sex organ tissues, testes and ovaries (Belote *et al.*, 1985; Baker *et al.*, 1987; Boggs *et al.*, 1987; Lucchesi and Manning, 1987; Cline, 1988, 1989; Oliver *et al.*, 1988; Steinmann-Zwicky, 1988; Wolfner, 1988; Baker, 1989; Nöthiger *et al.*, 1989; Steinmann-Zwicky *et al.*, 1989, 1990; Amrein *et al.*, 1990; Bownes *et al.*, 1990; Pauli and Mahowald, 1990). A key role is played by a gene called *Sex-lethal (Sxl)* positioned on the X-chromosome, the mutant alleles of which cause lethality of males (Sxl^{M1}) or females (Sxl^{F1}) (Cline, 1978). The wild type gene (Sxl^{+}) shows high activity in females and low activity in males. Furthermore, different products occur by different splicing mechanisms. In males, ineffective products are built, which cannot interact with the *transformer (tra)* gene. In females, other RNAs are spliced out of the primary RNA, and are translated into proteins which, finally, can interact with *tra*. The *tra* gene product is spliced female-specifically under the influence of *Sxl* protein and will be translated into a functional protein which interacts with the gene *doublesex (dsx)*, the RNA of which will be spliced also in a female-specific mode. Finally, the female *dsx* proteins will repress the male-specific target genes, while the female-specific target genes are expressed. Female-specific target genes are found for instance in the three vitellogenin genes which act in fat body cells as well as in follicle cells in *Drosophila*. Another set of about 29 female-specific target genes are the genes for chorion proteins which act in follicle cells at a very special time (see also section 3.2.5).

In males, *Sxl* transcripts are spliced in a male-specific mode, most of which are not translated into protein. If they are translated, the final proteins cannot interact with *tra*, which keeps silent; thus the female pathway is blocked. However, the gene *dsx* is transcribed but its RNA is spliced in a default, male-specific mode which represses female-specific target genes. Female-specific regulation of *dsx* RNA processing occurs by promoting the usage of the female-specific splice acceptor site, rather than by repressing the usage of the alternative male-specific splice acceptor (Nagoshi and Baker, 1990).

Like *Sxl* and *dsx*, primary gene products of *tra-2* are also spliced differently (Mattox *et al.*, 1990; Mattox and Baker, 1991). However, differently spliced RNA of *tra-2* in the soma is not sex-specific and *tra-2* activity in somatic tissues must be regulated through a post-translational sex-specific interaction with the product of *tra* gene rather than through the expression of a female-specific *tra-2* gene polypeptide (Mattox *et al.*, 1990). As we have seen, differential splicing of RNAs is one key to sex differentiation. Recently, the differential splice products of Sxl^{+} in females and males have been characterized via hybridization of small, defined DNA probes of the *Sxl* gene to poly(A)$^{+}$ RNA probes derived from different developmental stages of male and female samples (Salz *et al.*, 1989). Four different RNAs were found in adult females, two of which occurred only when a functional germ line was present, indicating that these products are needed to maintain female determination of germ line cells. Three RNAs are found only during very early stages of embryogenesis, indicating initial steps in sex differentiation. In addition to these findings, Bopp *et al.* (1991) analysed the resulting proteins and showed that several minor *Sxl* proteins appear to be present in both sexes in specific tissues in head and thorax, indicating unknown functions. Early RNA spliced products of *Sxl* are translated into proteins only in female embryos, not in males. Using monoclonal antibodies against these early products the topology of proteins could be elucidated. Early proteins were found only in somatic nuclei whereas pole cells gave no anti-*Sxl* staining reaction.

As just indicated, differentiation is a complex process, which can be split into at least three steps (Gehring, 1976): the first step is the initiation event, in our case of sex determination it is the correct reading of the X:A ratio by the embryo. The second step is the commitment of a cell to a specific developmental pathway, the real determination; and the third step is the final expression, which enables us to correlate morphological findings with determinative processes. In initiation events (Figure 3.41), at least two other genes are involved, *sisterless-b (sis-b)* and *daughterless (da)*. *Sis-b* is localized on the X-chromosome and its gene product is synthesized very early, in preblastoderm or early blastoderm stages. At its time of synthesis, the X-chromosomes of male embryos have not yet hyperactivated their gene activity to compensate for the XX status found in females. Thus, only half of the product is synthesized in male embryos compared with female embryos. This 'counting' ability designates these genes as numerator elements, contrasting with yet unknown denominator elements the loci of which are on autosomes (Torres and Sanchez, 1989). The *da* gene is found on the second chromosome and is active in adult females. Thus, each egg contains the same amount of this gene product, irrespective of the genetic constitution of the later embryo. Meanwhile, the above genes have been cloned and putative proteins show domains, shared with other proteins, which are known to interact with promoter regions (Cronmiller *et al.*, 1988; Cline, 1989). Thus, products of both genes seem to be part of the promoter and/or enhancer region pattern of the *Sxl* gene, opening sex-specific splicing. The cascade of gene products via *Sxl*, *tra*, *tra-2* and *dsx* is influenced by other yet unknown genes. Recently, a new gene was found, *vir* (*virilizer*), the gene products of which act above and via the *tra* and *tra-2* genes (Hilfiker and Nöthiger, 1991). Another gene, called *female-lethal-2-d (fl(2)d)*, is part of the autoregulation of late female-specific splicing of *Sxl* RNA, similar to *snf*. Furthermore, *fl(2)d* has a vital function independent of *Sxl*. This additional function is needed during larval development in both sexes (Granadino *et al.*, 1990, 1991).

All that we have stated above, is due to sex differentiation of somatic cells. In *Drosophila*, the cascade of sex differentiation differs in the germ line (Figure 3.41) (Nöthiger *et al.*, 1989; Pauli and Mahowald, 1990). First evidence came from transplantation experiments in which pole cells (germ line cells) of special genetic constitutions were transplanted into embryos of other genetic background (Marsh and Wieschaus, 1978; Schüpbach, 1982, 1985; Steinmann-Zwicky *et al.*, 1989). Four genes are known, which are involved in sex differentiation of germ line cells. The first gene is again *Sxl*; however, regulation of its expression is not yet understood. It remains unclear by which mode the X:A ratio is measured by germ line cells, since *sis-b* seems not to be involved. Furthermore, it is not known if dosage compensation has any influence on germ cell sex differentiation. However, signals of somatic tissue for reception of germ cells seem to work via the gene *ovo* (Pauli and Mahowald, 1990). *Ovo*$^{+}$ is expressed during oogenesis in nurse cells, beginning at stage 8. Highest expression is at stage 10 and transcripts are transported to the oocyte and accumulate until stage 12. Transcripts are uniformly distributed in ooplasm and seem to be used during blastoderm formation. Transcripts seem to accumulate for a short period in the cytoplasm of early pole cells. At gastrulation *ovo* RNA is no longer detectable. The gene has been cloned meanwhile and it encodes a zinc finger protein (Mével-Ninio *et al.*, 1991). In sexual differentiation of germ cells, two other genes are involved which belong to so-called tumour mutants (see section 3.3.4): the gene *sans fille (snf)*, also known as *liz* (Steinmann-Zwicky, 1988) and identical to *fs(1)1621* (Gans *et al.*, 1975; Gollin and King, 1981), acts directly on *Sxl*, whereas the gene *oocyte tumour (otu)* acts upstream of *Sxl* (Pauli *et al.*, 1993). *Otu*-transformed germ cells splice *Sxl* pre-mRNAs in a male-specific mode. When coupled with *Sxl* alleles which express the feminizing activity *Sxl* constitutively, *otu* mutants can be rescued. These *otu* alleles work synergistically with those of *ovo*^{D} and with one allele of the *snf* locus. Those *otu* alleles that work on morphogenesis of ovarioles will be further described in section 3.3.4 when we will discuss oocyte–nurse cell differentiation.

In the male germ line *dsx* products are not involved in differentiation processes. However, *tra-2* RNA is spliced male specifically. Thereby, *tra-2* gene products regulate, besides their function in controlling *doublesex (dsx)* in the female soma, the splicing of their own product in the male germ line and this autoregulatory function may serve in negative feedback control of *tra-2* activity during spermatogenesis. In addition to the role of *tra-2* products in sex-differentiation of the male germ line, this gene also plays a role later on during histogenesis of sperms in a way not yet understood. The finding that *tra-2* controls multiple spicing decisions (its own splicing in the male germ line and the splicing of *dsx* in female soma) suggests that a variety of different alternative splicing choices could be regulated by a relatively small number of gene products (Mattox and Baker, 1991).

Recently, Hazelrigg *et al.* (1990) have shown that the *exuperantia (exu)* gene is also involved in spermiohistogenesis. The *exu* gene is known as one of those maternal genes, which are engaged in anteroposterior axis formation (see section 3.3.2). However, under germ line sex control, male-specific RNA is spliced from the *exu* gene, which is necessary for correct transformation of mitochondria into the nebenkern. This example shows clearly that specificity of a regulatory gene is always the product of environmental conditions.

In addition to this discussion on sex differentiation in *Drosophila*, other differentiation processes should be added which are close to this process. When looking for *knirps* homologous gap class segmentation genes via reduced stringency hybridization conditions, Rothe *et al.* (1989) isolated a gene called *embryonic gonad (egon)*. This gene is active during a short period of very early gonad formation. The involvement of this gene in regulatory functions is based on conserved homologous DNA-binding domains, which belong to zinc finger proteins.

Another determination process, related to sex differentiation, is the determination of pole cells. This process is tightly coupled with egg architecture and will be discussed in section 3.3.2.

At the end of our discussion of early sperm development, we stated that the programme of cystocyte divisions should be cut off in females; however, nature is more complex and in many insects we find synchronized syncytial germ cell cluster in females (Giardina, 1901; Hirschler, 1945; King, 1970; Telfer, 1975; King and Büning, 1985). In which way can females make use of this programme? The following sections will try to give some answers.

3.3.2 Growth of oocytes to giant cells

Insects are small animals with short generation times. Eggs are built in the reproductive phase, which occurs at the climax of imaginal life in most species. Even in those examples in which egg growth is shifted into preimaginal stages (see section 3.2.1 and Figure 3.9) a high evolutionary pressure must be assumed to shorten oocyte growth, which is time and energy consuming. Thus, optimized rates of RNA and protein synthesis must have evolved, together with effective mechanisms to protect the products against early degradation and to keep them in a dormant stage until they are activated during embryogenesis. Somatic cells, like gland cells, can increase their genetic activity by polyploidization of genomes. This way is not open to germ cells since they have to reduce their genome size to the haploid status during meiosis (section 3.3.3). This dilemma is solved by combinations of different strategies:

(1) The prophase of meiosis becomes extremely prolonged in females: thus, RNA can be synthesized from the four-fold genome size since the DNA has doubled during the preceding S-phase. This strategy is followed by all animals which have developed eggs.
(2) The rate of mRNA synthesis of a special gene, the products of which

will be needed in high amounts during early embryogenesis, may become enhanced by accumulation of active RNA polymerases as seen in lampbrush chromosomes.

(3) Extrachromosomal gene amplification of rDNA can be introduced, becoming visible by extra nucleoli. These genes, producing structural RNA, cannot amplify their information in the cytoplasm as happens with products of other genes during translation.

(4) The use of germ cell clusters in females. However, this needs a clear determination of the final oocyte(s) against those germ cells which become functionally 'somatic', the nurse cells.

(5) In some insects, the germ line information becomes separated from the somatic information by a special set of chromosomes, the E-chromosomes, which become eliminated from somatic cells during the first cleavages of the embryo. Somatic cells contain only S-chromosomes, whereas germ cells contain the whole set of chromosomes, E- and S-chromosomes. All these ways will be discussed in detail in section 3.3.4.

(6) Another way to circumvent the veto of polyploidization of oocyte genomes is the use of true somatic cells which fuse with germ cells, the genomes of which come under the register of the genomes of the oocyte and/or nurse cells to support their syntheses, as is found in some gall midges (Cecidomycidae) (sections 3.5 and 3.6).

(7) Insects as well as species of nearly all other phyla have split the syntheses needed for egg growth into those which are tightly coupled to group and species-specific information as is shown above (eucytoplasmic growth = euplasm), and those syntheses producing macromolecules, which serve as pools for anabolism and catabolism during embryogenesis, comprising vitellogenic growth. The macromolecules are stored in vacuoles (lipid and proteinaceous yolk spheres) or they assemble into characteristic aggregates (glycogen). Lipids are synthesized in oocytes and nurse cells throughout oogenesis in small amounts; however, most lipid yolk spheres occur at the onset of vitellogenesis (Bier, 1954; Bonhag, 1958; Telfer, 1965; Schreiner, 1977a,b) and its basic molecules are synthesized in other organs, transported at least in part by the haemolymph as shown for lipophorins (Gondim *et al.*, 1989). Glycogen is the last component synthesized and stored in the egg during vitellogenesis (Bier, 1954, 1967b; Bonhag, 1958; Engels, 1966, 1968, 1971, 1973; Schreiner, 1977a,b; Huebner and Anderson, 1972b). The main store products are globular proteins without enzymatic character, all named vitellin, which is the cumulative name for these store proteins, irrespective of their phylogenetic origins. The synthesis is not controlled by the germ cell genome directly. The corresponding genes become active in somatic cells as in fat body cells generally and the follicle cells of at least *Drosophila* species

(Bownes, 1982; Brennan *et al.*, 1982). The synthesized vitellogenins are exocytosed into the haemolymph, transported to the intercellular cleft, which opens under hormonal control between oocytes and the surrounding follicular cells, and, finally, these macromolecules become incorporated into the oocyte by adsorptive endocytosis (Roth and Porter, 1964; Kunkel and Pan, 1976; Ferenz, 1978; Röhrkasten and Ferenz, 1986a,b; Röhrkasten *et al.*, 1989). This last mode of oocyte growth cannot be discussed in detail here and we must refer to other publications (Telfer, 1965; Engelmann, 1979; Telfer *et al.*, 1982; Kunkel and Nordin, 1985; Postlethwait and Giorgi, 1985; Wyatt, 1991; Bownes *et al.*, 1993; Ferenz, 1993; Giorgi *et al.*, 1993).

(a) Timing and quantitative aspects

Euplasmatic growth precedes vitellogenesis in insects and goes on to some extent during endocytotic yolk formation. The preceding period is called the previtellogenic growth phase. During this phase the oocyte expands roughly 60-fold in volume, even in the smallest oocytes, as found in virginopare aphids. In normal insects, however, oocytes grow during previtellogenesis between 1000 or even 10 000 times the volume of the original germ cell.

Vitellogenesis can increase the original volume to about 10^6 or even 10^7. Thus, the euplasm amounts to about 1% (0.1–10%) of the final egg volume. When vitellogenesis begins, eucytoplasmatic growth continues for a while. This can be deduced qualitatively by autoradiography using labelled precursors of RNA and protein synthesis (Bier, 1963a,b; Ramamurty 1963; Engels, 1966, 1968; Bier *et al.*, 1967; Büning, 1972; Mays, 1972; Ullmann, 1973; Ray and Ramamurty, 1979). A quantitative comparison between euplasm synthesized during previtellogenic growth and that synthesized during vitellogenesis has not be done and seems of minor interest. Direct volumetric measurements are difficult to interpret, since the amount of euplasm between yolk spheres can vary in time and space, for instance by changes in the ionic milieu, by aggregation of proteins to paracrystalline associations and so on. As shown, euplasmatic growth contributes only to about 1% of the total volume of the final egg. However, beside its fundamental qualitative consequences it has some quantitative aspects, too. The amount of euplasm decides if the embryo will be constructed over the whole length of the egg which leads to a long germ band (langkeim) as found in *Drosophila*, an insect with polytrophic meroistic ovarioles, or if the embryo has to be constructed from low amounts of euplasm which results in a short germ band (kurzkeim) as found in *Gryllus*, which has panoistic ovarioles (Krause, 1939; Bier, 1968, 1970; Sander, 1976). These two types of embryogenesis correlate to some extent with the construction of insect ovaries. Generally, panoistic ovaries produce a relatively small amount of

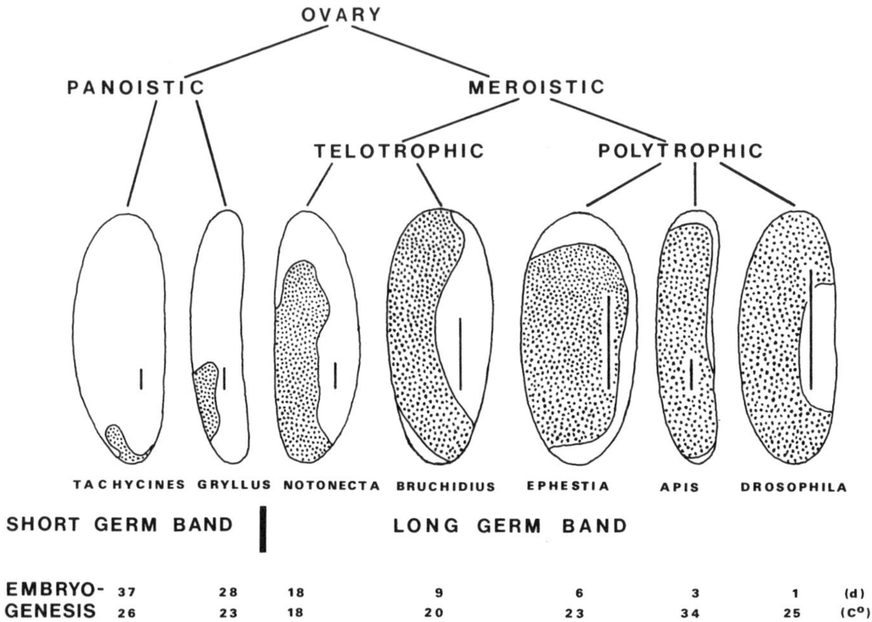

Figure 3.42 Synopsis of ovary types, germ band types, amount of euplasm and time of embryogenesis in insects. Each bar represents 0.2 mm. Embryonic areas are stippled, showing the embryo during early germ band formation. For further information see text. (After Krause, 1939 and Bier, 1969.)

euplasm and have short germ bands and embryogenesis takes a long time (several weeks), whereas meroistic ovaries produce more euplasm which is used to build long germ bands and embryogenesis is short (a few days) (Figure 3.42).

However, between these extremes, transitions exist, as is known from telotrophic meroistic ovarioles. These eggs contain a medium amount of euplasm, have large embryos and embryogenesis takes a week as in the coleopteran *Bruchidius* (Jung, 1966; Büning, 1972, 1983). Another example is the bug *Oncopeltus* (Hemiptera) in which embryogenesis takes 2 weeks; it also has telotrophic ovaries (Harris and Forrest, 1967). The inverse proportion between the amount of euplasm and the time of embryogenesis becomes more clear if we compare the pools of small metabolic molecules like ATP or UTP. The titres are high from the very beginning of embryogenesis in *Drosophila*, indicating satisfactory maternal equipment, whereas these titres have to first increase in *Bruchidius* embryos (Büning, 1983). Generally, the more the embryo has to synthesize out of its own resources via metabolic pathways, the longer embryogenesis will last.

(b) Quality of euplasm

As stated before, the cytoplasm of oocytes and eggs can be divided into two parts concerning their origin: the euplasm and the heteroplasm. Euplasm of oocytes includes all the substances synthesized under the regime of germ line DNA. This includes macromolecules bearing information for the embryo such as masked mRNA, RNPs and pure proteins of enzymatic character, as well as structural proteins like elements of the cytoskeleton or elements for protein synthesis like ribosomes. Furthermore, it includes self-replicating organelles, the mitochondria, as well as a basic stock of cellular membranes, vesicles and vacuoles of each kind, including rough and smooth ER, dictyosomes and annulate lamellae. Macromolecules contributing to each kind of euplasmatic structural component can be stored as ultrastructurally visible aggregates of similar dimensions as ribosomes as is assumed for cytoskeleton elements, such as tubulins or G-actins which polymerize later on to F-actins at the time they are used by the embryo (McKnight and Miller, 1976; Wolf, 1978, 1985; Davidson, 1986). Such aggregates can be structurally identified in the ooplasm when they assemble in greater compartments which withstand at least in part, those forces which mix the cytoplasm (Figures 3.21, 3.43b and 3.44b). Many of these compartments stain basophilic, are Feulgen-negative and are often described in older literature as emission bodies, coming from the germinal vesicle. Aggregates of this kind have sometimes been named as 'yolk nuclei' or 'Balbiani bodies' and have been found in a variety of organisms as in the ovaries of dragonflies (Kessel and Beams, 1969; Halkka, 1980), beetles (Büning, 1979b), fleas (Büning and Sohst, 1989), bark lice (Büning and Sohst, 1990) and snowfleas (Büning, unpublished). However, proof which identifies these aggregates biochemically, for instance by decoration with immunogold particles coupled with monoclonal antibodies against special proteins, is today known only for polar granules (Figure 3.45).

(c) Architecture of euplasm

As stated above, in special cases we can gain some insights into the timing and amount of cytoplasmic qualities by ultrastructural analysis. When we analysed the flea ovary (Büning and Sohst, 1988, 1989), we found at prepachytene stages a nearly homogenous cytoplasm which was very poor in free ribosomes (Figure 3.30). In corresponding nuclei, some small, proteinaceous endobodies (= Binnenkörper; Bier *et al.*, 1967; see also section 3.3.4) are found in the nucleoplasm (Figures 3.21, 3.30, 3.43 and 3.44). These endobodies rise to huge aggregates which can fuse to one big spherical body during the early and middle phases of previtellogenic growth. During this time, the cytoplasm shows only a small belt of ribosome-rich cytoplasm around the oocyte nucleus, which indicates a low rate of synthesis of rRNA

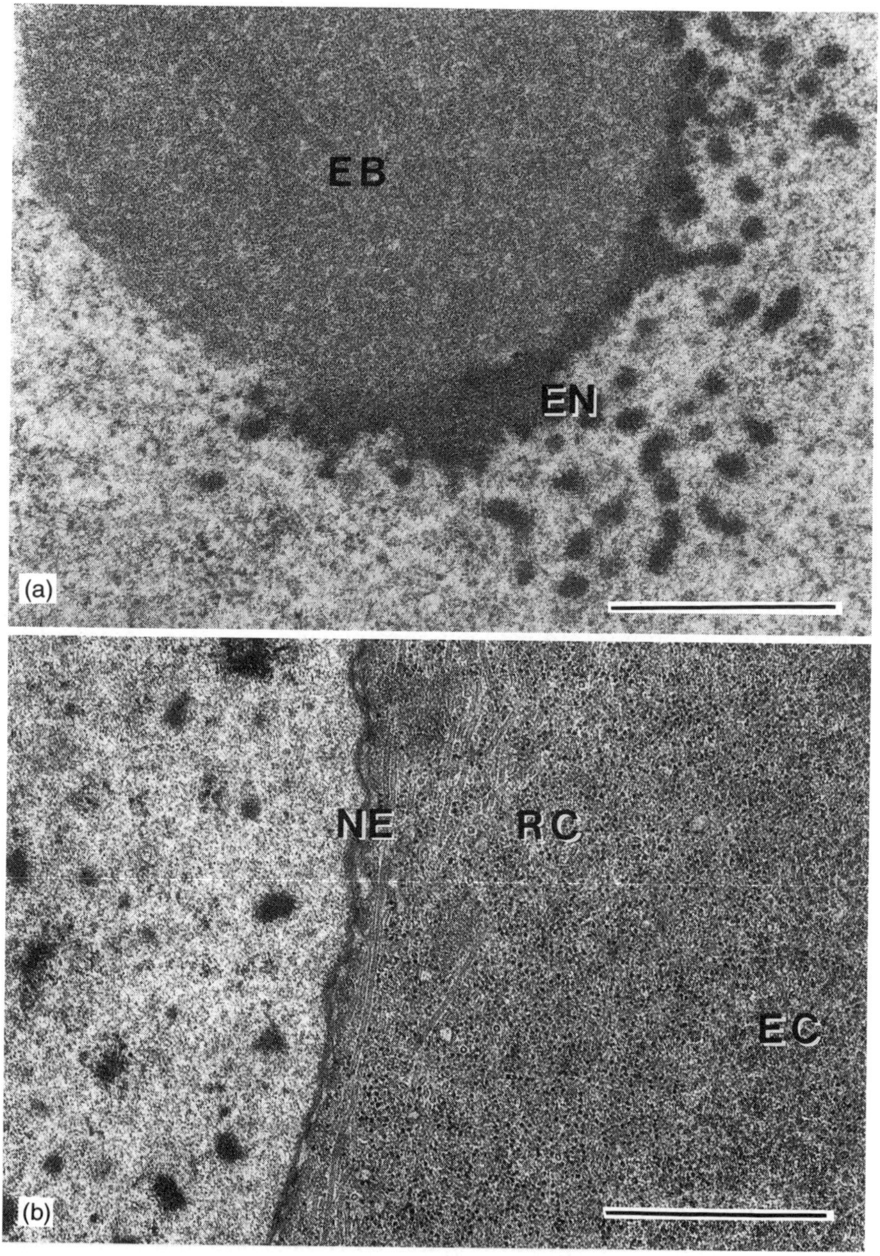

Figure 3.43 Polytrophic meroistic ovary of *Hystrichopsylla talpae* (Siphonaptera). (a) Oocyte nucleus at the end of previtellogenic growth. Extrachromosomal nucleolus material (EN) is associated in part to an endobody (EB). (b) Same stage as before. Newly built ribosomal-rich cytoplasm (RC) does not mix with early cytoplasm (EC). Note the belt of microtubules next to nuclear envelope (NE). Bar represents 1 μm.

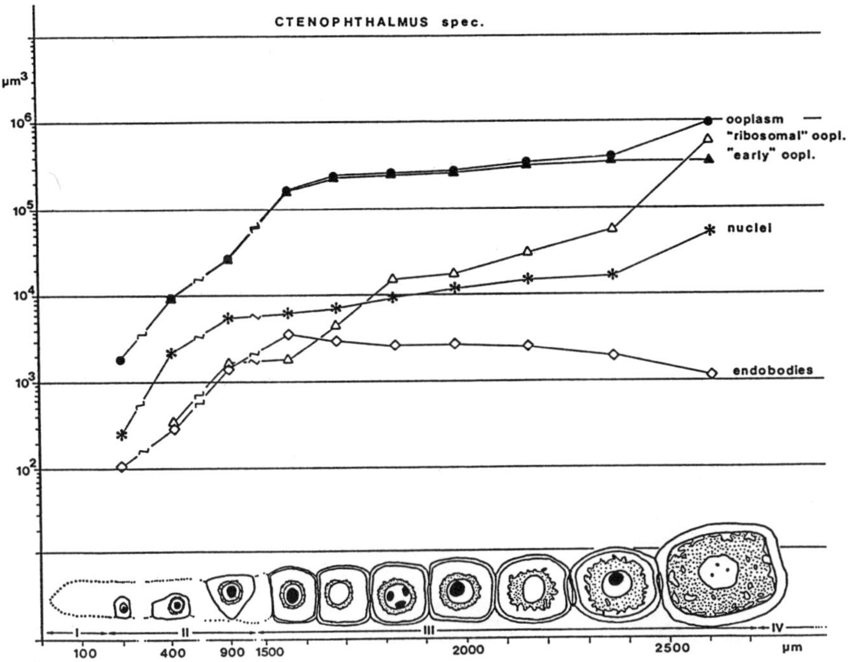

(a)

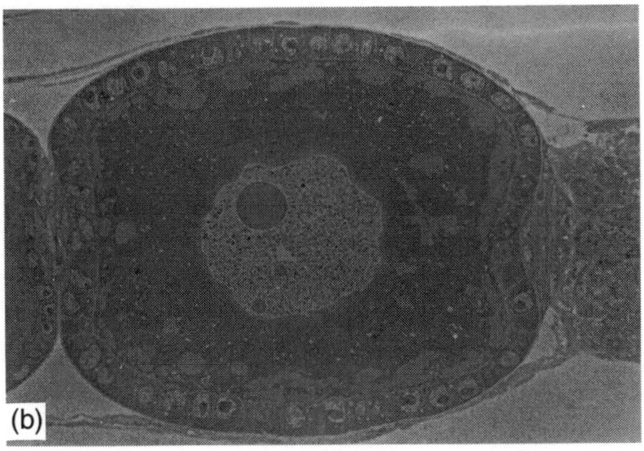

(b)

Figure 3.44 (a) Volume measurements during previtellogenic growth phases of the panoistic flea ovary (*Ctenophthalmus* sp.). Concomitantly with the rise of 'ribosomal' cytoplasm, the volumes of endobodies decrease. (b) Oocyte of *Ctenophthalmus* sp. (Siphonaptera) at the end of previtellogenesis. Note the numerous extrachromosomal rDNA beside decreasing endobody material. Ribosome-rich cytoplasm is now abundant, whereas early cytoplasm has been disrupted to small islands.

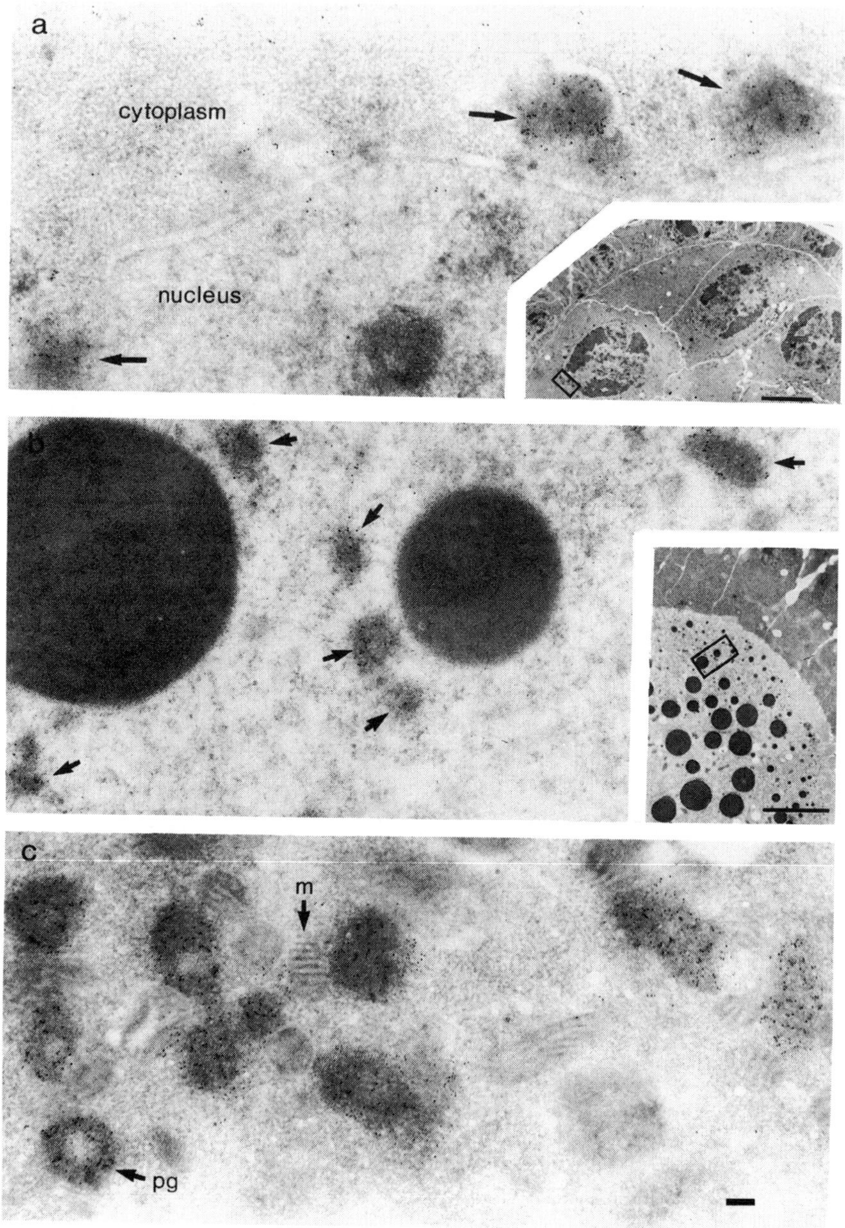
a
cytoplasm
nucleus
b
c
m
pg

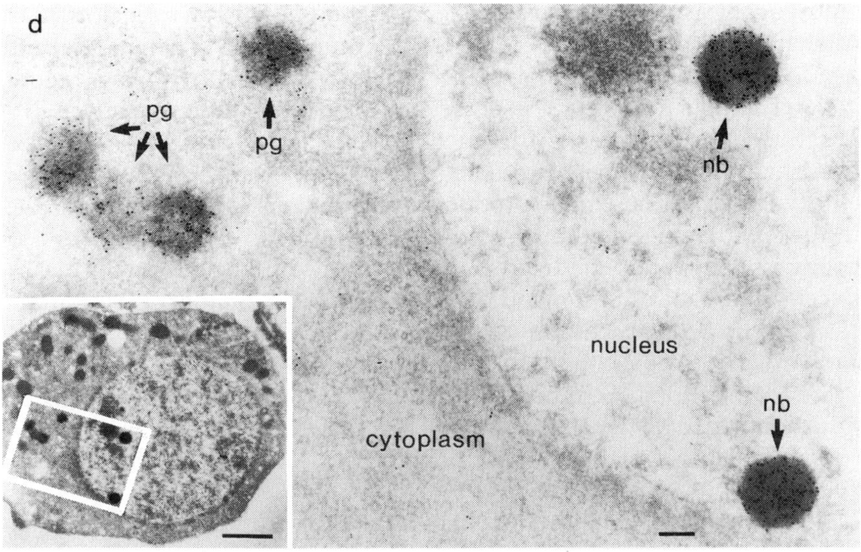

Figure 3.45 Fate of polar granules as visualized by gold labelling of polyclonal antibody (Mab46F11) during oogenesis and early embryogenesis. (a) Nuage material next to nuclear envelope of nurse cells from stage -3 or -4 shows labelling (arrows). Inset gives overview, boxed area is enlarged. (b) In oocytes of stage 10, polar granules have accumulated near the posterior pole. Inset shows posterior pole, boxed area is enlarged. (c) Posterior pole at stage 14. Note tight association of polar granules (pg) with mitochondria (m). Bars represent 100 nm. Inset bars represent 200 μm. (d) Polar granules (pg) and nuclear bodies (nb) of pole cells of *Drosophila* are labelled specifically. Inset shows overview of polar cell. Bar represents 100 nm; inset bar 1 μm. (From Hay *et al.*, 1988a; with permission of authors and The Company of Biologists.)

and ribosomes. This aspect changes dramatically during later stages, when extrachromosomal rDNA genes start their synthesis (Figure 3.44). The endobodies begin to shrink and thousands of extrachromosomal nucleoli spread over the whole nucleus and start their synthesis (see also section 3.3). Ribosome-rich cytoplasm contributes to more than 90% of the final euplasm at the end of previtellogenic growth. The rise of the endobody during the early phase of euplasmatic growth suggests that part of the early cytoplasm consists of nuclear proteins, which are imported into the nucleus and aggregate into endobodies. The endobodies are found closely associated to the granular component of nucleoli (Figure 3.43a). Thus, the proteins of endobodies may accumulate around rRNA precursors as structural and/or enzymatic components during the massive ribosome production at the end of euplasmatic growth.

The endobodies of flea oocytes described here may be functionally similar to the endobody described for cricket oocytes (Jaworska and Lima-De-Faria, 1973; Tröster *et al.*, 1985). However, there are some structural differences, which will be discussed in section 3.3.4.

During the 1980s we had our first insight into the regulation of insect embryogenesis; this work was done mostly on *Drosophila melanogaster*. In brief, insect embryogenesis is directed by communication of maternal and zygotic gene products, for instance nuclear proteins, with the zygotic genome and its followers (e.g. embryonic DNA) over time and space. The onset of embryogenesis is triggered by fertilization processes. Thereby, maternally derived gene products play an important role during early embryogenesis. These gene products imprint the axes of the embryo and influence the segmentation of the larval and adult body. Later on, products of zygotic gene activity are added step by step during the blastoderm stages, gastrulation and organogenesis (e.g. Sander, 1976; Sander *et al.*, 1985; Schüpbach and Wieschaus, 1986; Gehring, 1987; Nüsslein-Volhard *et al.*, 1987; Schüpbach, 1987; Driever and Nüsslein-Volhard, 1988, 1989; Manseau and Schüpbach, 1989; Jäckle *et al.*, 1989; Struhl, 1989; St Johnston *et al.*, 1989, 1991; Davidson, 1990; Driever *et al.*, 1990; Ingham and Nakano, 1990; Nüsslein-Volhard, 1991; St Johnston and Nüsslein-Volhard, 1992).

These maternal gene products are part of the euplasm. Some of them occur in the egg in the cortical cytoplasm, distributed homogenously or as gradients along the main axis of the body. Maternal gene products, visible under the electron microscope as dense granular aggregates which are distributed as gradients, are, for instance, the polar granules which are involved in determination of primary germ cells and the pole cells which are clustered around the vegetative pole (Figure 3.45) (Mahowald, 1962, 1971a,b; Illmensee *et al.*, 1976; Niki, 1986; Hay *et al.*, 1988a,b, 1990). Other maternal gene products, the functions of which are not known, occur as aggregates of different morphology in the ooplasm, as shown in the panoistic oocytes of a dragonfly, *Cordulia aenea* (Halkka and Halkka, 1968, 1975; Halkka, 1980). Similar structures are known to occur in the oocytes of other insects, as shown here for fleas (Figure 3.22a), psocopterans (Figure 3.23), aphids (Figure 3.46b) or mecopterans (Figs 3.47a,b). Other maternal information, not yet detectable in discrete morphological structures, is distributed anteriorly and is responsible for the development of head and thoracic structures of the embryo (Kalthoff, 1979; Lohs-Shardin, 1982; Sander *et al.*, 1985; Nüsslein-Volhard *et al.*, 1987).

We will discuss now some single maternal genes, which are engaged in pattern formation during early embryogenesis, the activation pattern of which during oogenesis might be important in respect of their deposition and integration into the egg cortex. This includes the time of transcription, the export and storage of such masked mRNAs in the cytoplasm, their translation into the final protein and storage of primary (RNPs) and/or secondary (proteins) products until use during early embryogenesis. Such analysis will give insights into the regulation of the cytoarchitecture of oocytes and it will answer questions on qualitative differences among the diverse groups of meroistic ovaries (see sections 3.5 and 3.6).

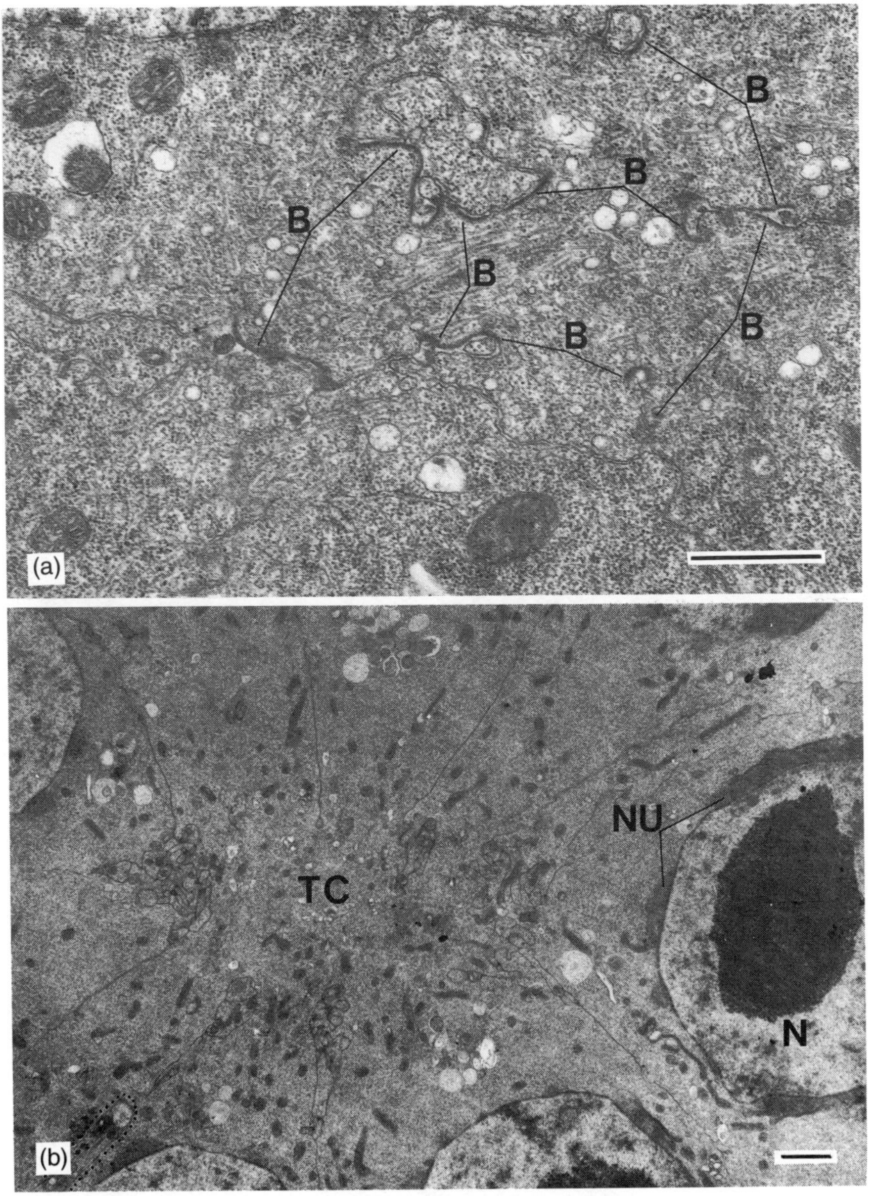

Figure 3.46 (a) Central part of an ovariole of *Drepanosiphum platanoides* during embryogenesis. Intercellular bridges (B) closely assemble in a centre. Bridges are filled with parallel microtubules. (b) Later in development, this area forms the trophic core region (TC) to which all siblings, nurse cells and oocytes are open via broad zones (nurse cells) or small nutritive cords (oocytes). Margin of nutritive cord is stippled. Note nuage material (NU) on nurse cell nucleus (N) envelope. Bars indicate 1 μm. (From Büning, 1985; with permission of Wiley and Sons.)

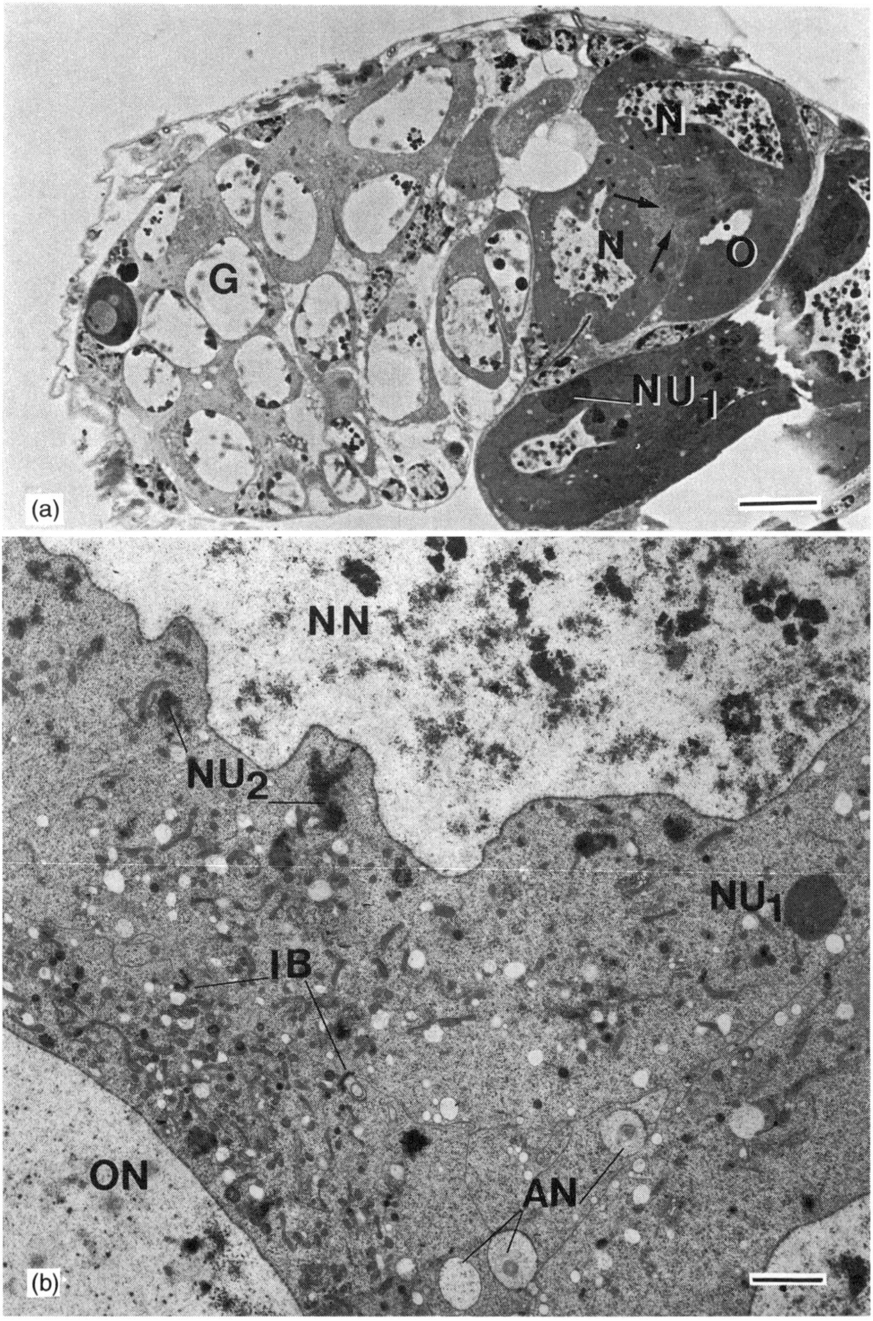
N
N
O
G
NU1
(a)
NN
NU2
NU1
IB
ON
AN
(b)

In panoistic ovaries, all maternal information comes from only two sources, the oocyte genome itself and the somatic follicle cell genome. Follicle cell information, imprinting the oocyte, is known only for ecdysone production in *Locusta*, as will be discussed in section 3.3.3. However, this lack of information among panoistic ovarioles seems to be more a matter of work not yet done, since follicle cell information is well documented in the polytrophic meroistic ovary of *Drosophila* (see below), and there is no reason to assume that regulation via follicle cells does not occur in panoistic (or telotrophic meroistic) ovaries.

However, the bulk of maternal information needed during early embryogenesis will come from the oocyte genome itself. Again, nothing is known about the temporal pattern of single gene expression during oogenesis in panoistic ovaries and we have to wait for interspecific experiments, using the analysed genes of *Drosophila*, which will give new insights into the general strategies of oogenesis.

Information on panoistic ovaries can be adopted for telotrophic meroistic ovaries with a few exceptions. Telotrophic ovaries differ from polytrophic ovaries by retention of all nurse cells in a terminal chamber, the trophic chamber. In advanced telotrophic ovaries, all developing oocytes are nourished by all nurse cells at the same time. This might have some impact on regulation during oogenesis, since nurse cells cannot supply the oocyte with time-dependent products, which may be stored one after another in the oocyte. If there is any need for such a mechanism, the time-dependent activity of the genes must be regulated in the oocyte genome itself, or in the encompassing follicular cells. Indeed, telotrophic oocytes show a remarkable amount of RNA synthesis in their germinal vesicles, as shown by autoradiography (Büning, 1972; Ullmann, 1973; Matuszewski *et al.*, 1977; Ray and Ramamurty, 1979 for Coleoptera, Polyphaga; Mays, 1972 for Hemiptera; for Megaloptera (Sialidae) see Figure 3.48).

The mayflies (Ephemeroptera) are the fourth group in which telotrophic meroistic ovaries are found (see section 3.6). Here, membranes of nurse cells

Figure 3.47 (a) Anterior region of an ovariole of *Panorpa communis* (Mecoptera: Panorpidae). Semi-thin section shows germ cells (G), all arrested in prophase of meiosis. Cluster formation into four cells (three nurse cells and one oocyte) has already happened. At the beginning of previtellogenic growth, nurse cell nuclei (N) polyploidize their genomes and nurse cells grow tremendously, whereas the oocyte (O) remains small. Intercellular bridge rims (arrows) are embedded in polyfusomal material, showing lesser contrast than surrounding cytoplasm. Note giant spherical nuage material (NU_1); bar represents 10 μm. (b) Electron micrograph shows intercellular bridge (IB) between nurse cell and oocyte of previtellogenic growing follicle. Note two different kinds of nuage material, one more spherical (NU_1) and one more diffuse (NU_2). Near to oocyte membranes, some accessory nuclei are found (AN). Nurse cell nucleus (NN); oocyte nucleus (ON). Bar represents 2 μm.

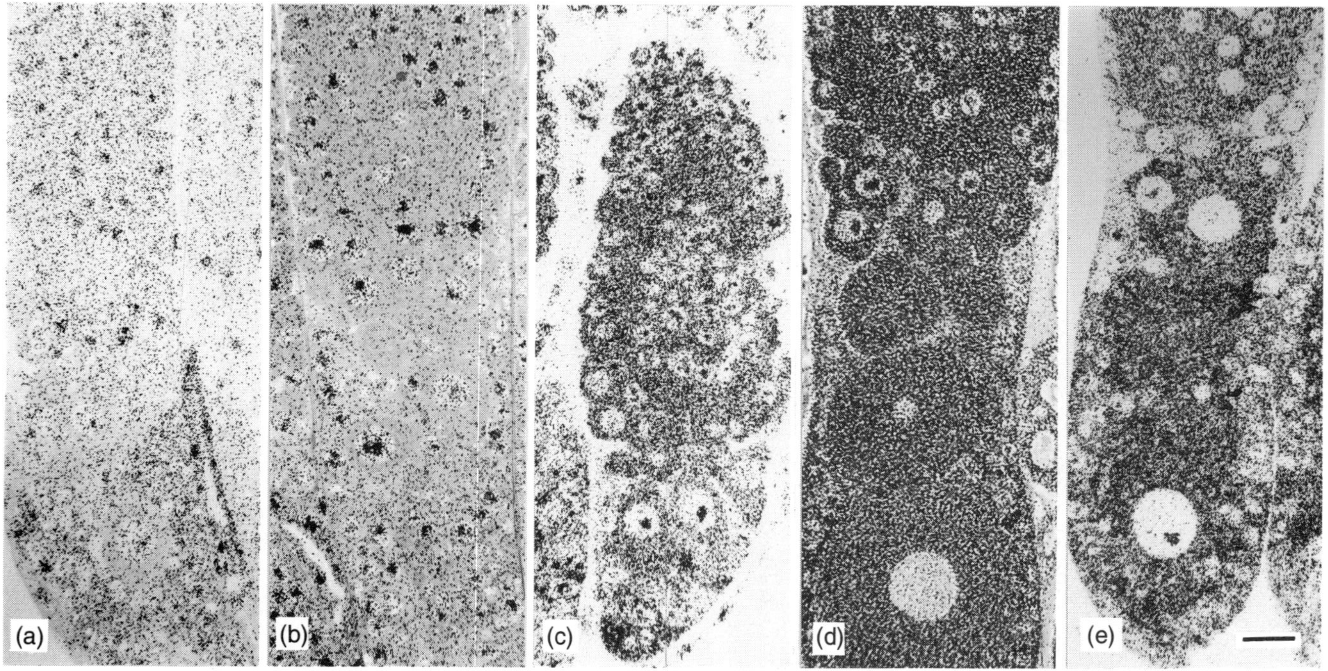

Figure 3.48 ^{3}H-uridine autoradiography in ovarioles of *Sialis flavilatera* (Megaloptera: Sialidae). Incorporation into the nuclei and export to the cytoplasm have the same pattern in nurse nuclei of central syncytium, in tapetum cells as well as in definitive oocytes. Incubation times: (a) 24 h; (b) 42 h; (c) 94 h; (d) 190 h; (e) 390 h. The pattern of ^{3}H-uridine autoradiography in *Sialis* ovarioles is similar to that observed in coleopteran and hemipteran ovarioles despite the fact that time is prolonged by a factor of 10 and apical–basal gradients in oocytes are not to be seen. Bar represents 20 μm.

remain stable as in primitive polyphagous coleopterans. RNA synthesis has not been investigated; however, the ultrastructure of oocytes and connected nurse cells is identical and we assume there is a measurable amount of RNA synthesis in the oocytes in this new type of telotrophic meroistic ovary.

Time-dependent oocyte genome activity has been reported for *Dysdercus intermedius* (Hemiptera). During previtellogenesis and early vitellogenesis, oocytes show a slightly reduced, but clearly above background level of RNA synthesis (Mays, 1972). This transcription activity cannot be distinguished from nurse cell activity, which produces the bulk of stable RNA (Winter, 1975). This RNA is transported as RNPs into the oocyte together with other cytoplasmic compounds via nutritive cords, which degenerate during vitellogenesis (see section 3.4). After choriogenesis, follicle cells degenerate and all syntheses must come from the oocyte itself. In the following hours, oocytes synthesize a class of short-living RNA, which becomes polyadenylated and translated in the ooplasm. The molecular weights of four main proteins (65, 48, 44, 40 kDa) have been determined; they are not in the range of *Dysdercus* chorion proteins. Winter *et al.* (1977) speculate on the effects of these proteins in early events of pattern formation in *Dysdercus*.

Today, we can report on the activity of some maternal genes during oogenetic growth known from *Drosophila melanogaster*. In future, however, the modern methods in genetics and molecular biology will allow us to test the pattern of transcription and translation of sympathetic maternal genes in other insect systems.

A special set of experiments, well probed for zygotic genes (e.g. Knipple *et al.*, 1985; Baldarelli *et al.*, 1988) to elucidate gene activity during early embryogenesis, has been adapted to maternal genes to demonstrate their activities during oogenesis. One, well-established method is the use of ^{35}S-labelled anti-sense RNA probes of defined genes, which can hybridize to insect ovarian tissues *in situ*, to demonstrate ovarian gene activities and localize their primary products. The gene *bicoid* is such a maternal gene, expressed only in germ line derived cells during oocyte growth (Figure 3.49) (Fronhöfer and Nüsslein-Volhard, 1986; Driever and Nüsslein-Volhard, 1988; Berleth *et al.*, 1988; Stephenson *et al.*, 1988; St Johnston *et al.*, 1989; Nüsslein-Volhard, 1991). *Bicoid* is engaged in the determination of anterior pattern formation during early embryogenesis of *Drosophila*. The stored mRNA is concentrated in the anterior region of the ripe egg, and is synthesized by nurse cells throughout previtellogenic and vitellogenic growth, with a maximum of synthetic activity at mid-vitellogenesis (stage 10; King, 1970; for *Drosophila* staging of oogenesis see section 3.5.2). The *bicoid* gene does not seem to be expressed by follicle cells, since anti-sense RNA does not hybridize specifically over these cells. Hybridization occurs in ooplasm, but is concentrated in anterior regions where intercellular bridges connect the oocyte with the nurse cells. Thus, most if not all *bicoid*

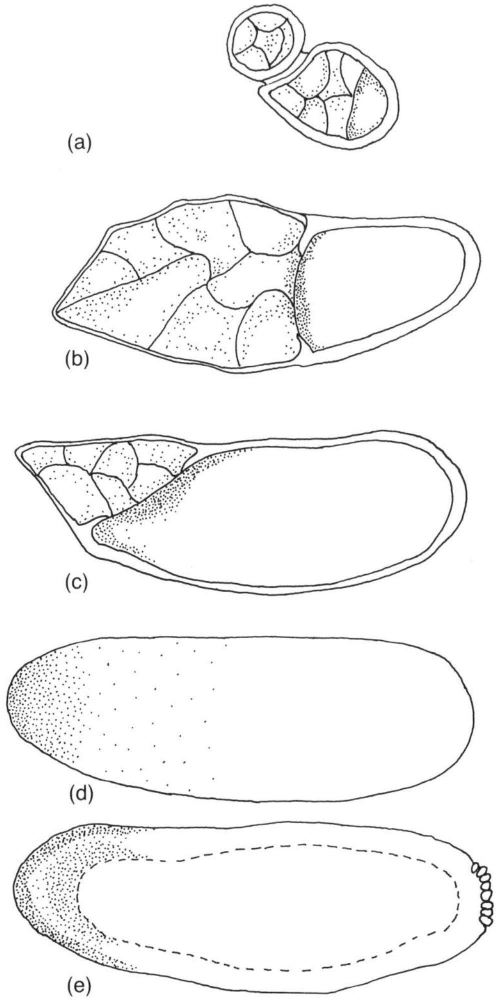

Figure 3.49 Distribution of *bicoid* transcripts in ovaries and embryos of *Drosophila melanogaster*. (a–c) The gene is transcribed in nurse cells and its products are stored in an apical–basal gradient in oocytes. (d,e) The gradient is maintained in egg cytoplasm and during early steps of embryogenesis. *Bicoid* RNA is shown by radiolabelled anti-sense RNA hybridization. (Drawings after colour pictures from Berleth *et al.*, 1988.)

RNA is built by nurse cells, whereas the oocyte genome seems to be silent. These experiments giving insights into the pattern of single gene activities, are in good agreement with the classical experiments of Bier (1963a,b), who found no specific incorporation of ^{3}H-uridine into RNA in the oocyte nucleus of the house-fly *Musca domestica*. Thus, most genes of the oocyte are silent, while nurse cell genes are transcribed with high efficiency. This classical conclusion is correct for most genes but not for all, as will be shown later, for instance for the *bicaudal-D* gene (section 3.3.4). Furthermore, highly repressed RNA synthesis of oocyte nuclei is found only in polytrophic meroistic ovaries, whereas telotrophic meroistic ovaries show somewhat reduced, but significant RNA syntheses in their oocyte nuclei (see above).

The deposition of *bicoid* transcript is influenced by other maternal gene products like *exuperantia*$^+$ or *swallow*$^+$ that are necessary for correct storage of *bicoid* transcripts. *Exuperantia*$^+$ is necessary for the patched storage in nurse cell cytoplasm, whereas *swallow*$^+$ is needed later for anterior localizaion in the oocyte (Frohnhöfer and Nüsslein-Volhard, 1986, 1987; Berleth *et al.*, 1988; St Johnston *et al.*, 1989; Stephenson and Pokrywka, 1993; Hedge and Stephenson, 1993). However, the functioning of both genes for the correct storage of *bicoid* transcripts is not well understood nor are the transport mechanisms which carry *bicoid* transcripts from nurse cells into the oocyte. However, the 3′-untranslated region of the *bicoid* transcript is essential for its localization and may consist of different elements, each responsible for different steps (Macdonald *et al.*, 1993). Transport may be achieved by cytoskeleton molecules (for transport mechanisms see section 3.4).

The transcript of *bicoid* becomes translated only in the embryo and not in the oocyte. The *bicoid* protein is localized in the embryo in an anterior–posterior gradient, which differs from the gradient of *bicoid* transcripts. The geometry of the protein gradient and its absolute amount decides the amount of anterior and posterior structures in cooperation with other maternal and zygotic genes.

As shown for the *bicoid* gene, information for the embryo can be stored asymmetrically in the egg cytoplasm. Other examples are maternal genes like *vasa*, the loss of which disturbs the organization of the abdominal segments. The posterior pole of insect eggs contains a cytoplasmatic quality which determines invading nuclei as germ cells during blastoderm formation. This germ cell cytoplasm is rich in ribosomes and polar granules, which are highly contrasting aggregates of smaller particles (Figure 3.45). These aggregates seem to be identical to highly contrasting aggregates found around nurse cell nuclei or oocyte nuclei during earlier stages of oogenesis (Mahowald, 1962, 1971a,b). Recently, one gene (*vasa*) involved in the differentiation of polar granules has been identified, the protein of which shows the same distribution as polar granules in embryos, oocytes and nurse cells (Figure 3.45; Hay *et al.*, 1988a,b, 1990). According to this gene it is its protein which is distributed asymmetrically in the oocyte. However, polar granules consist not only of proteins but also, at least in the beginning, of RNPs.

Such a transcript comes from the gene *nanos*. *Nanos* is transcribed in nurse cells, transported through the whole egg and deposited finally in the posterior pole plasm. *Nanos* belongs to the so-called posterior group genes, required for pole cell formation and for construction of terminal structures of the embryo. *Nanos* itself is not required for pole cell formation. However, its localization in posterior pole plasm is connected with the activity of other genes, such as *oscar, vasa, valois* and *tudor*. Other localization genes are *cappucino, spire, staufen, mago nashi* and *bicaudal-D. Oscar* mRNA and the proteins of *oscar, vasa* and *tudor* accumulate in the pole plasm and persist until pole cell formation or even longer. *Vasa* and *tudor* proteins are components of polar granules. Vasa is necessary for maintenance of polar granules and tudor is required after fertilization for correct functioning of polar granules. *Staufen* has been cloned and encodes a protein containing multiple copies of a domain which binds double-stranded RNA. *Bicaudal-D* also appears to act in localization and encodes a protein predicted to adopt a coiled structure, suggesting a structural role (Boswell and Mahowald, 1985; Hay *et al.*, 1988a,b, 1990; Wharton and Struhl, 1989, 1991; Lasko and Ashburner, 1990; Nüsslein-Volhard, 1991; Lehman and Nüsslein-Volhard, 1991; Lehman, 1992; Smith *et al.*, 1992; St Jonston *et al.*, 1992; Wilson and Macdonald, 1993).

Polar granules can be induced at the anterior pole by injection of posterior pole plasm (Illmensee and Mahowald, 1974). When UV irradiated posterior pole plasm is injected, pole cells do not form ectopically. However, this ability can be restored by coinjection of 16S rRNA from mitochondria. This mitochondrial rRNA is not sufficient to restore pole cell characters by itself but needs additional components of the UV irradiated pole plasm (Kobayashi and Okada, 1989). One of the posterior maternal gene products which may be important in the correct functioning of mitochondrial 16S rRNA may be the *tudor* protein, which has been found in mitochondria in all parts of the embryo as well as in the posteriorly localized polar granules (Bardsley *et al.*, 1993).

The gene *germ cell-less (gcl)* seems to be directly involved in germ line formation (Jongens *et al.*, 1992). Its mRNA is localized posteriorly and needs the function of the same localization and provision genes as known from *nanos*. Reduced *gcl* function results in sterile adult progeny lacking germ cells. *Gcl* protein specifically associates with nuclei of germ cell precursors, the pole cell nuclei. Thus only *germ cell-less, tudor* and *vasa* seem to have the potential to encode an independent germ cell determinant.

The population of pole cells is not uniform as could be demonstrated by differences in their splicing activity of the third intron of P-elements, which is known to occur only in germ line cells. Almost all pole cells having this splicing activity are able to penetrate the embryonic gonad and differentiate to primordial germ cells; the others fail and degenerate (Kobayashi *et al.*, 1993).

Pole cells differ from other blastoderm cells not only by polar granules but also because they have different timing for cell divisions. Mitosis is part of the cell cycle and the last growth phase of cells (G2) and its transition to mitosis is tightly coupled with slow accumulation and abrupt degradation of cyclin proteins. Cyclin proteins emanate from different splicing products. One of them, the long cyclin B transcript, is stored specifically in the posterior pole plasm and accumulates later in the perinuclear region of the pole cells (Dalby and Glover, 1992). Raff *et al.* (1990) assume that the specific localization of this splice product requires a polar granule component. However, the specific function of the long cyclin B transcript is not yet known. The specific localization needs the 3′-untranslated region of the transcript. This finding is consistent with findings on other genes, the transcripts of which are also specifically localized in the oocyte, like *bicoid* (Macdonald and Struhl, 1988; Macdonald *et al.*, 1993) or *K 10* transcripts, which are needed as dorso-ventral determinants (Cheung *et al.*, 1992). In both transcripts it is also the untranslated 3′ sequence which is necessary for correct localization.

These maternal gene products are synthesized by the oocyte itself or by nurse cells, and we should try to find the underlying mechanisms responsible for the cytoarchitecture of the final egg cytoplasm. There are some possibilities which should be mentioned.

Within each follicle an electrical gradient can exist, as shown in follicles of diverse species, including all types of insect ovarioles (Woodruff and Telfer, 1973, 1974, 1980, 1990; Telfer *et al.*, 1982; Huebner, 1984b; Woodruff and Anderson, 1984; Kunkel, 1986; Kunkel *et al.*, 1986; Münz and Dittmann, 1987; Woodruff *et al.*, 1988; Woodruff, 1989), by which cytoplasmic components may be ordered asymmetrically in the egg, according to their own charge (see also section 3.4).

The molecular architecture of the cortical cytoplasm differs in special regions (e.g. the anterior and the posterior part, or the dorsal and ventral side of the growing egg). This cortical cytoplasm is devoid of yolk spheres and houses cytoskeletal macromolecules and ribosome-like elements as well as large amounts of ribosomes (Figures 3.19, 3.22, 3.50 and 3.51) (Haget, 1970, 1972; King, 1970; Zissler and Sander, 1973, 1977, 1982; Miya, 1978).

In general, the cortical cytoplasm will be constructed similarly to the cortical cytoplasm of vegetative cells (Bray *et al.*, 1986; Alberts *et al.*, 1988), but it may be more complex. This includes special proteins on the outer face of the oocyte membrane, which can communicate with 'outdoor' molecules, which themselves can induce local specialization of the cortical cytoplasm. These signals may come directly from somatic follicular cells, or such molecules may be stored in the vitelline membrane, or they will come from other tissues via the haemolymph. As shown before (section 3.2.5), follicular cells differentiate into special groups and it seems possible that local signals of these cells can be used to differentiate the cortical cytoplasm of the egg.

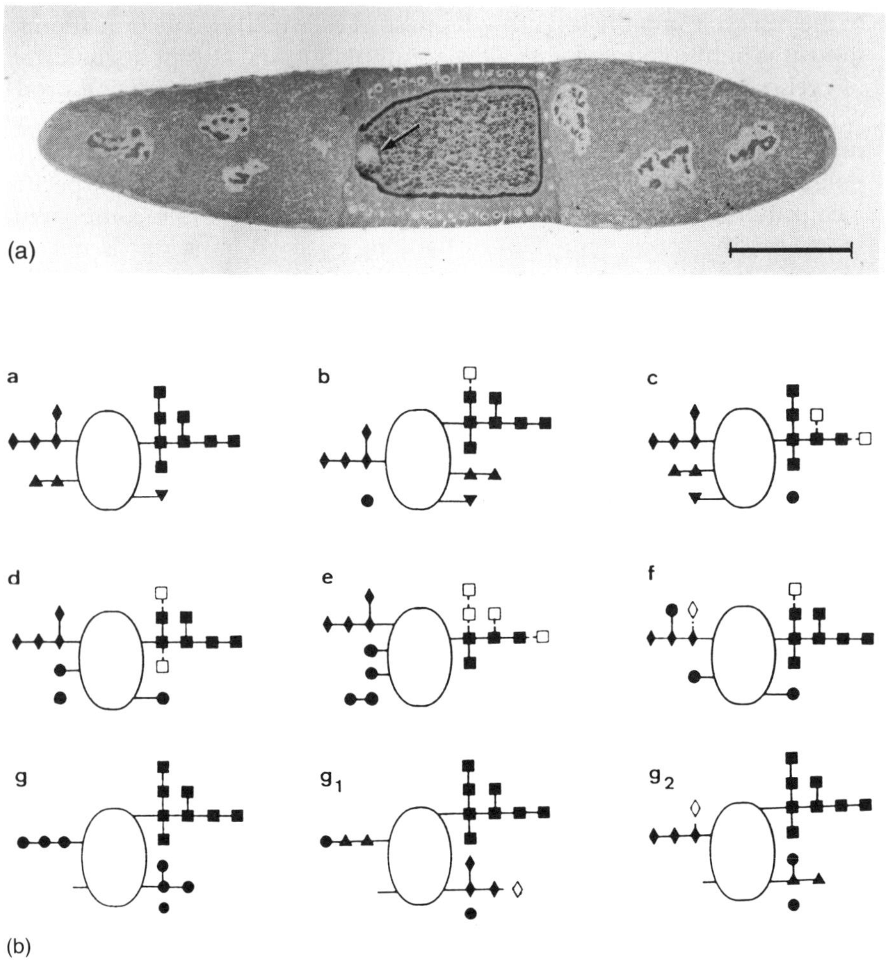

Figure 3.50 (a) Semi-thin section of *dicephalic* mutant follicle of *Drosophila melanogaster*. The trophic chamber is split and nurse cell cytoplasm is injected via intercellular bridges at two sides (anterior and posterior) into the oocyte, inducing anterior structures at both poles during embryogenesis. The arrow points to the germinal vesicle, which is located asymmetrically as in wild type follicles. (From Lohs-Shardin, 1982; with permission of Springer Verlag.) Bar represents 100 μm. (b) Synopsis of cluster formation with bipolar distribution of cell families as found in *dicephalic* follicles. Affiliation to cell family is indicated by shape of symbol. Open symbol shows place of missing cell; filled circle marks cell of uncertain identity. Cluster formation can be normal as shown in part a. However, many follicles show abnormalities in cluster formation (parts b–g). Two additional interpretations of cluster configurations are given in parts g_1 and g_2. (From Frey *et al.*, 1984; with permission of K. Sander and Springer Verlag.)

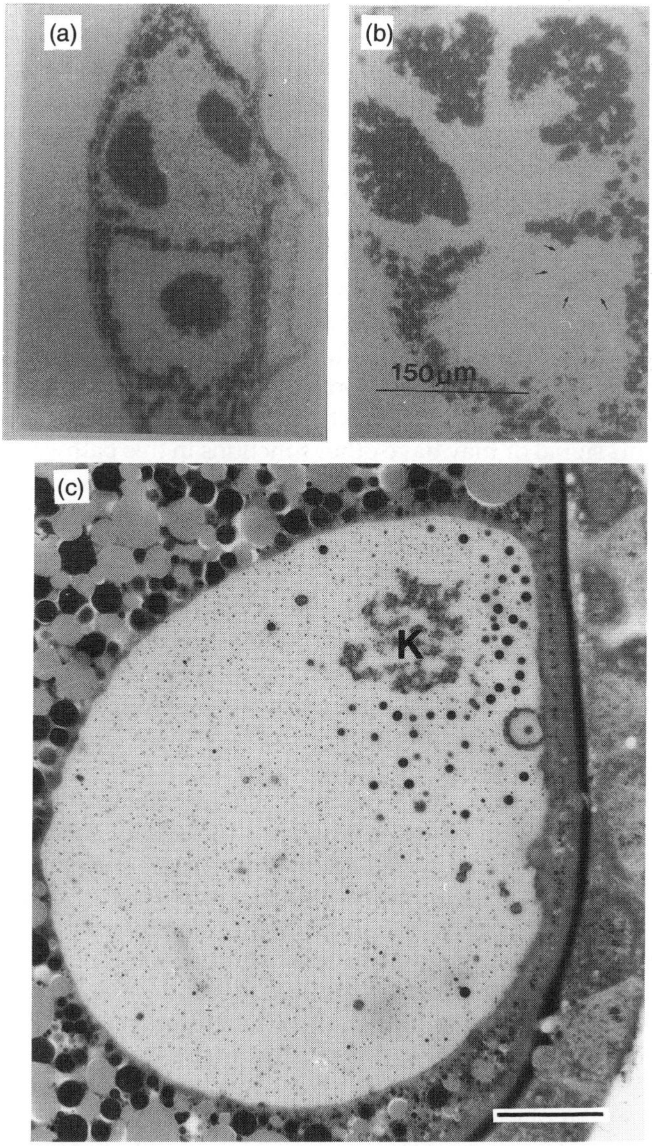

Figure 3.51 (a,b) Previtellogenic growing follicles of *Hyalophora cecropia* (Lepidoptera). Autoradiograms of follicles incubated *in vitro* with ^{3}H-uridine and overexposed to dramatize label of germinal vesicle. (a) Shows the situation before electrical coupling of oocytes with follicle cells. (b) Shows a follicle after electrical coupling. The germinal vesicle is not labelled; its position is marked by arrows. (From Woodruff and Telfer, 1990; with permission of R. I. Woodruff and Academic Press.) (c) Germinal vesicle during mid-vitellogenesis of *Anthaxia nitidula* (Coleoptera, Polyphaga: Buprestidae). The germinal vesicle is attached to lateral periplasm. Note that karyosome (K) and large extrachromosomal aggregates are found asymmetrically distributed next to lateral periplasm, whereas very small granules fill the whole germinal vesicle evenly. Bar represents 20 µm.

Insect eggs are not only polarized in the anterior and posterior regions but also at the dorsal and ventral sides. This dorsal–ventral polarity has been analysed best in *Drosophila* and we will discuss it here. The first visual sign of dorsal–ventral polarity begins in the oocyte by movement of the germinal vesicle from a central position to an antero-lateral position which is the future dorsal side (see also section 3.3.2). What follows is a cascade of gene activities (*fs(1)K10, squid, cappucino* and *spire*) which localizes the oocyte nucleus signals to the dorsal side of the oocyte. These dorsal signals are encoded by the genes *gurken* and *cornichon* and will be transmitted to follicle cells via the *Drosophila* epidermal growth factor receptor, encoded by *torpedo. Torpedo* will be activated only in dorsal follicle cells, as the ligand is predicted to be localized. *Gurken* is a dorsalizing signal coming from the oocyte nucleus, the level of expression of which is regulated negatively by the *fs(1)K10* product. Other genes such as *rhomboid* or *brainiac* may help to transduce this signal or may have other functions in this pathway (Prost *et al.*, 1988; Govind and Stewart, 1991; Schüpbach *et al.*, 1991; St Johnston and Nüsslein-Volhard, 1992; Goode *et al.*, 1992; Stewart and Govind, 1993; Forlani *et al.*, 1993; Ruohola-Baker *et al.*, 1993).

This dorsal–ventral polarity of the oocyte must be maintained in the ripe egg and will finally be imprinted on the embryo to establish its dorsal–ventral axis. Therefore, several components of the egg and egg coverings are used. Firstly the oocyte cortical cytoplasm with its cytoskeleton molecules, secondly the oocyte membrane with its transmembrane signalling pathways, thirdly the perivitelline fluid in which special proteins can be released or activated, and finally the vitelline membrane itself as an asymmetric deposit of special proteins synthesized by special sets of follicle cells during oogenesis. Such proteins are the products of *nudel, pipe* and *windbeutel*, which are deposited in the vitelline membrane on the ventral side, which ultimately results in the asymmetric activation of *torpedo*, as has been proposed by Schüpbach *et al.* (1991), Stein *et al.* (1991) and Stewart and Govind (1993). Proteins engaged in dorsal–ventral polarity of embryos are products of *spätzle, snake, gastrulation defective* and *easter*, which are all components required for the formation and maturation of the Toll ligand at the ventral side of the embryo. These proteins are therefore components of the perivitelline fluid. Toll itself is a transmembrane receptor, uniformly distributed in the plasma membrane of the oocyte. Its ventral activation results finally in the uptake of *dorsal* protein into the blastoderm nuclei on the ventral side. The *dorsal* protein has been stored uniformly in the egg cytoplasm previously (Hashimoto *et al.*, 1988, 1991; Steward, 1989; Roth *et al.*, 1989; Rushlow, 1989; Schneider *et al.*, 1991; Stein and Nüsslein-Volhard, 1992; Chasan *et al.*, 1992; Steward and Govind, 1993).

Localized gene activity of follicle cells is known from the gene *torsolike*, which is engaged in the terminal group of maternal genes. *Torsolike* is expressed only in a clone of terminal follicular cells at the posterior pole and in the small set of follicle cells situated anteriorly between nurse cells and oocyte. The *torsolike* protein is probably incorporated into the vitelline layer at these special regions and will interact with its receptor, the evenly distributed cortical *torso* protein (Sprenger *et al.*, 1989; Stevens *et al.*, 1990; Nüsslein-Volhard, 1991).

The mutant *dicephalic* (Lohs-Shardin, 1982) disturbs the normal distribution of germ cell cluster cells within a follicle (Figure 3.50). Normal follicles contain a 16 cell cluster in which a central oocyte is connected to its 15 nurse cells via intercellular bridges (see section 3.6). At the base of the germarium, the oocyte becomes orientated in a final position at the posterior pole of the recently formed follicle in which the whole cluster is ensheathed by follicle cells (King, 1970). In a series of transplantation experiments of pole cells, Frey and Gutzeit (1986) demonstrated that *dicephalic* mutation products can produce the *dicephalic* phenotype irrespective of their origin: soma versus germ cell, which clearly demonstrates cell to cell communication. The same mutant gives an example for nurse cell–follicle cell cooperation: a special sub-set of follicular cells, the border cells, construct the micropyle during choriogenesis (see section 3.2.5; Figures 3.29, 3.31 and 3.32). A second micropyle is established in the mutant *dicephalic* only when the nurse cell cluster splits into two parts of which one assembles at the correct anterior position and the other becomes positioned at the posterior pole of the oocyte. Thus, the latter nurse cells must induce associated follicular cells to construct the second micropyle, irrespective of their normal fate (Bohrmann and Sander, 1987).

During intercelluar communication, external signals cross the plasma membrane, either directly or indirectly, and provoke a celluar response. The response may be universal, i.e. seen throughout the oocyte cytoplasm, or it may be localized to particular areas of the cell, e.g. the cortical cytoplasm (cytoplasm which immediately underlies the oolemma), as shown above.

The inner cytoplasm can move and mix during oogenesis and early embryogenesis, as found in eggs rich in euplasm, which all derive from meroistic ovaries, whereas the cytoplasm of most panoistic oocytes seems to remain more static. Such movements have been visualized in *Heteropeza* (Diptera, Cecidomycidae) and in *Drosophila* (Gutzeit and Koppa, 1982). Whereas in young previtellogenic stages a plasmatic streaming was not observed, the breakdown of nurse cells during stage 10 (King, 1970) initiates a heavy streaming of cytoplasmic components from nurse cells via intercellular bridges into the oocyte. Up to the end of nurse cell breakdown the ooplasm shows cytoplasmic streaming, the direction, speed and intensity of which varies among the follicles. Cortical regions do not mix. After stage 12, the cytoplasmic streaming ceases (transport processes will be discussed in more detail in section 3.4).

However, we must consider that the oocyte membrane and the adjacent cortical cytoplasm will not be static but may have a highly dynamic fate during early embryogenesis as well as during oogenesis. Oogenetic changes of the cortical cytoplasm must be assumed when the vitelline layer will be deposited (Figures 3.16b and 3.51c) during the late phases of vitellogenesis. Now, the oocyte membrane becomes sealed with the vitelline layer and intermembrane communication will be stopped until external signals such as sperm entry activate the cascade of early embryonic events, which will induce new changes in the cortical cytoplasm.

(d) The germinal vesicle (remarks on accessory nuclei written by S. Bilinski)

Besides cell to cell communication, another source for local differentiation of cortical cytoplasm comes from the oocyte itself. As in somatic cells, the oocyte nucleus and its cytoplasm communicate during the cell cycle with each other via detailed information, such as mRNA, and via regulative molecules, such as nuclear proteins. This also happens in oocytes during the prolonged meiotic prophase. In general, the oocyte nucleus grows concomitantly with the ooplasm to a giant germinal vesicle (Figures 3.24, 3.30, 3.44, 3.48, 3.51–3.59) (Bier, 1967b; Matuszewski, 1968), which breaks down during late vitellogenic stages. Thereby, all components which have been stored in the germinal vesicle can mix and interact with the surrounding cytoplasm. The breakdown occurs when the germinal vesicle has reached a species-specific final, and in most cases, asymmetric position. In locusts the germinal vesicle is situated near the posterior pole and the breakdown is stimulated by a local increase of ecdysone in the egg, which is produced by surrounding follicle cells (Kappler *et al.*, 1986). In coleopterans the vesicle is mostly situated at 70% of the egg length, with the posterior pole defined as 0% egg length (Jung, 1966; Van der Meer *et al.*, 1982; Büning, unpublished). In *Gryllus (Acheta) domesticus*, the final position is at 40% egg length (Netzel, 1965), whereas in dipterans the germinal vesicle is found near the anterior pole (e.g. Bier *et al.*, 1967; King, 1970). Furthermore, the germinal vesicle has shifted from a medium position to an exocentric position contacting a broad area of the cortical cytoplasm. At this position reorganization of the nuclear envelope will occur and the meiotic divisions will be finished during fertilization. It is also this area which determines the future dorsal side of the embryo (Sonnenblick, 1950; Jung, 1966; Van der Meer, 1978; Sander *et al.*, 1985). Looking at the germinal vesicle when reaching this final exocentric position, we notice in many species an asymmetrical position of enclosed granules as shown here for *Cosmoconus meridionator* (Hymenoptera) (Figure 3.52) and for *Anthaxia nititula* (Coleoptera, Buprestidae) (Figure 3.51c). In *Anthaxia* very small granules are spotted over the whole area of the germinal vesicle, whereas large and heterogeneous granules assemble only at the side where the

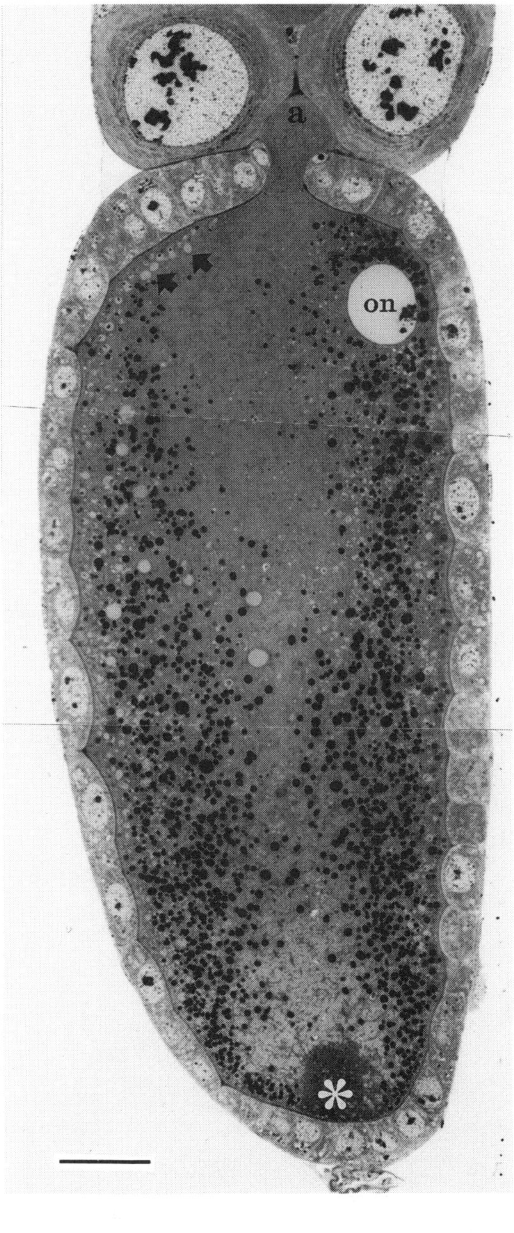

Figure 3.52 Light micrograph of vitellogenic growing oocyte of *Cosmoconus meridionator* (Hymenoptera) (longitudinal section). Oocyte nucleus (on); accumulation of accessory nuclei in the anteroventral periplasm (arrows). Oosome (*); nutritive appendix (a); × 600. (Courtesy of Sczcepan Bilinski, Jagiellonian University, Kraków, Poland.)

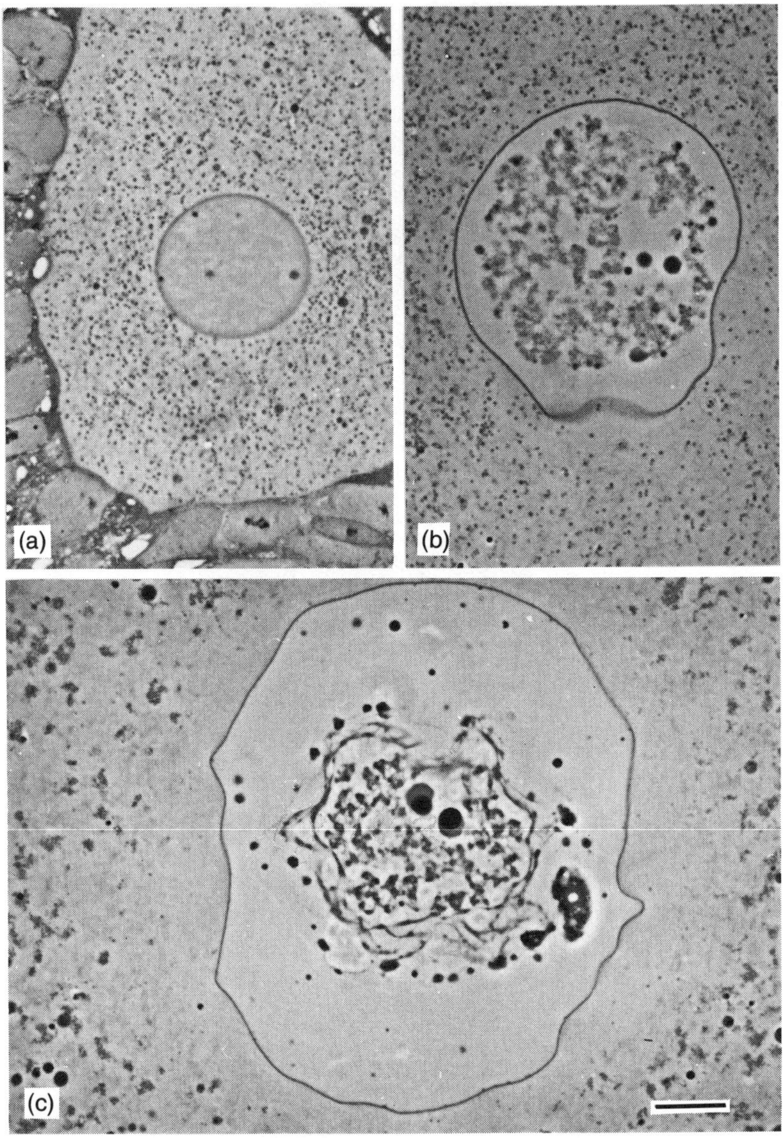

Figure 3.53 Germinal vesicle of *Psilothrix cyaneus* (Coleoptera, Polyphaga: Melyridae). (a,b) During previtellogenic growth, chromosomes of germinal vesicle stick together, forming a karyosome. (c) At beginning of vitellogenesis, karyosome is ensheathed by protein layers. Note numerous nuclear bodies of different size and different amounts of proteins and RNA.

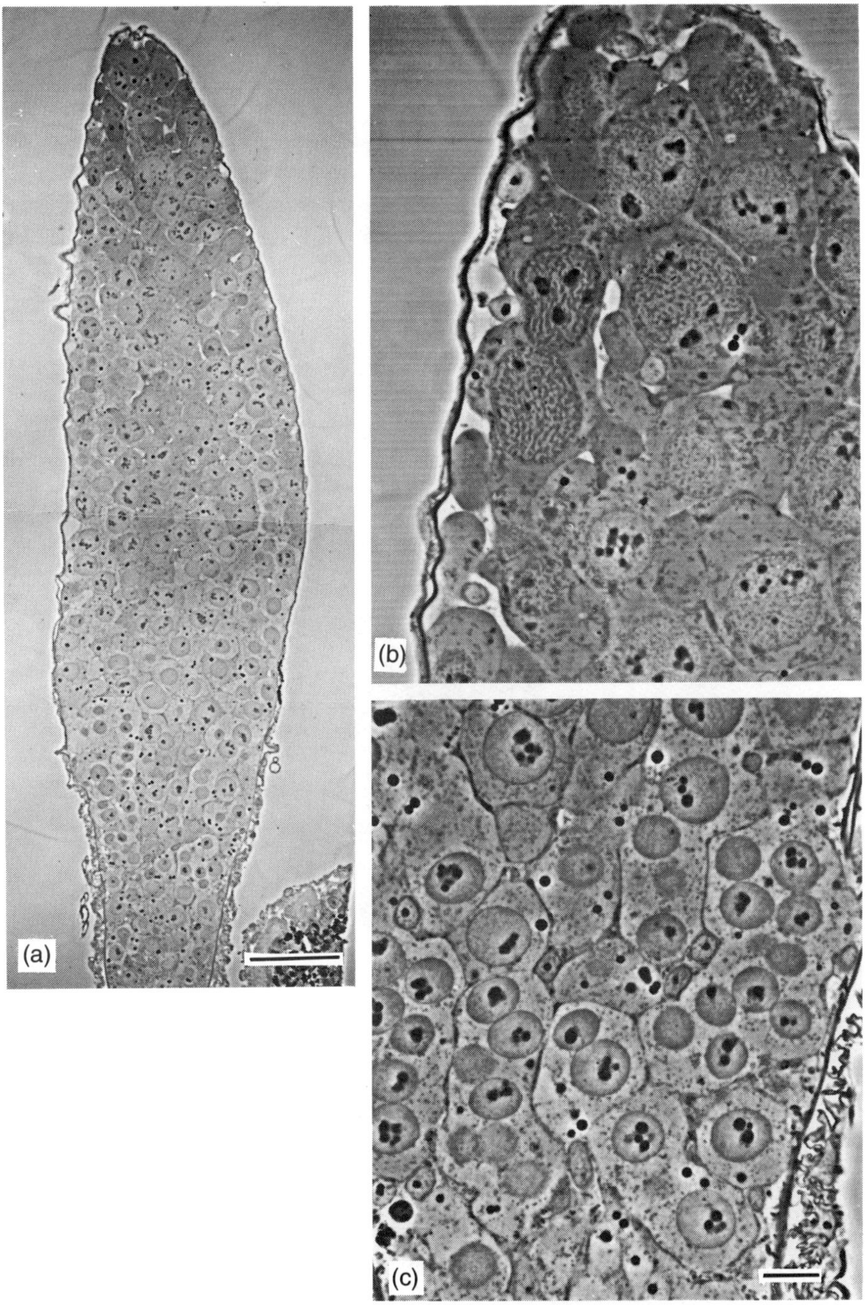

Figure 3.54 (a) Tropharium of ovariole of *Gaurotes virginea* (Coleoptera, Polyphaga: Cerambycidae). Bar represents 100 μm. (b) The anterior part of the tropharium contains large nurse cell nuclei, the structure of which clearly indicates endomitotic polyploidization. (c) In the posterior part nurse cell nuclei become smaller. On the other hand, syncytial areas of the same magnitude contain more nuclei in the posterior parts than in the anterior parts, indicating nurse cell nuclei divisions which must be amitotic, since mitoses have never been found in the tropharium of fully developed ovarioles. Bar represents 10 μm.

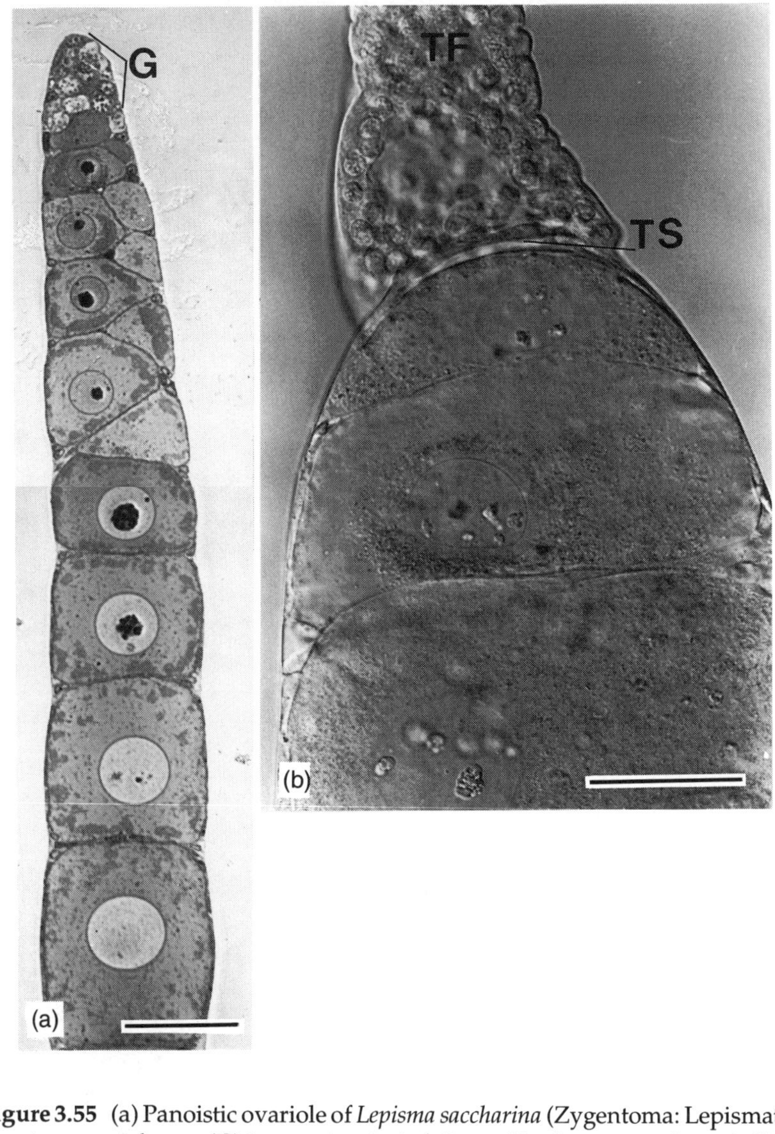

Figure 3.55 (a) Panoistic ovariole of *Lepisma saccharina* (Zygentoma: Lepismatidae). Short germarial zone (G) in anterior tip is followed by long area of previtellogenic growing oocytes (from King and Büning, 1985; with permission of Pergamon Press). (b) Anterior region of panoistic ovariole of larval ovary of *Isophya leonorae* (Orthoptera: Phaneropteridae). Terminal filament (TF) is separated by transverse septum (TS). A germarium does not exist. A maximum of about eight cells can develop as oocytes; however, some degenerate regularly and only three to five oocytes per ovariole undergo vitellogenesis. Bars represent 50 μm.

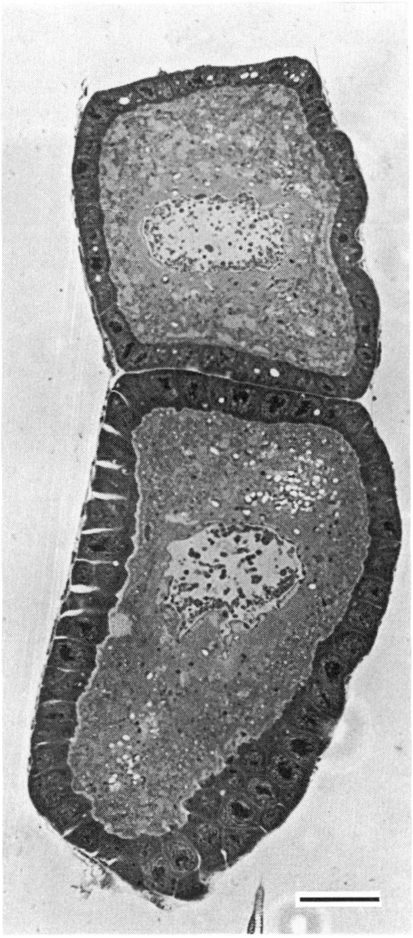

Figure 3.56 Two previtellogenic growing oocytes of *Boreus hiemalis* (Mecoptera: Boreidae). Germinal vesicles show numerous aggregates, most of them heavily stained with Azur B, indicating a high content of nucleic acids. Such a feature is known from germinal vesicles with extrachromosomal rDNA amplification. Note faintly staining cytoplasmic islands in oocytes. Bar represents 30 μm.

vesicle contacts the cortical cytoplasm tightly. Some of the large granules stain deep violet with Azur B, indicating a high content of RNA, whereas others stain faintly and are vacuolized in part, indicating a high degree of proteins. A similar disproportional distribution of granules in oocyte nuclei occurs in *Gryllus* (Netzel, 1965; Trendelenburg *et al.*, 1988). The asymmetric order of granules inside the germinal vesicle begins when the previtellogenic growing oocyte becomes ordered in a single line. Most of the granules are orientated towards the future dorsal side of the embryo, although this

Figure 3.57 (a) Ovariole of *Nebria brevicollis* (Coleoptera, Adephaga: Carabidae). The number of nurse cells is constant in this species and equals 15. Granules in the oocyte nucleus stain heavily with Methylene Blue, indicating their RNA nature and activity of the oocyte genome. (b) Ovariole of *Carabus obsoletus*. Note the different amount of nurse cells, serving the oocyte in each follicle. Bars represent 100 μm. (Courtesy of Mariuzs Jaglarz, Jagiellonian University, Kraków, Poland.)

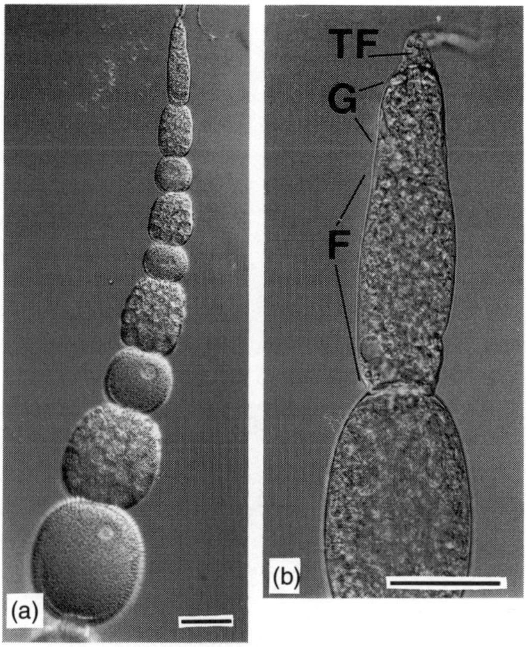

Figure 3.58 (a) Ovariole of *Athalia rosae* (Hymenoptera: Tenthredinidae). Follicles have different amounts of nurse cells. Bar represents 100 μm. (b) Enlargement of short germarial zone (G) and of following youngest follicle (F). Terminal filament (TF). Bar represents 50 μm.

asymmetry becomes visible only during late stages of vitellogenesis by differential growing of follicular cells on the dorsal and ventral side (Netzel, 1965). Thus, it seems possible that the germinal vesicle is engaged very early in dorsal–ventral axis determination. In *Gryllus*, the RNA-containing granules were identified as huge extra-chromosomal rDNA rings (see section 3.3.4). Additional support for special RNA synthesis, even in the germinal vesicles of polytrophic ovaries, comes from P-element-mediated reporter genes, which have been anchored in enhancer regions of *Drosophila* genes. Grossniklaus *et al*. (1989) found reporter genes active in germinal vesicles during previtellogenesis from stage 6 onwards. Thereby, the number of strains showing such activity increased until stage 10, which is a vitellogenic stage (King, 1970). This indicates selective activity of some genes during previtellogenesis. This selective activity may be connected with the 'double gradients' of reporter gene activities found by the same authors among follicle cells (see section 3.2.5).

The above findings correspond well with older findings from *Drosophila* and other fly ovaries, in which RNA synthesis was supposed to be repressed in oocyte nuclei at least during the vitellogenic stages (Figure 3.19) (Bier, 1963a,b; King, 1970; Mahowald and Tiefert, 1970; Davidson, 1986).

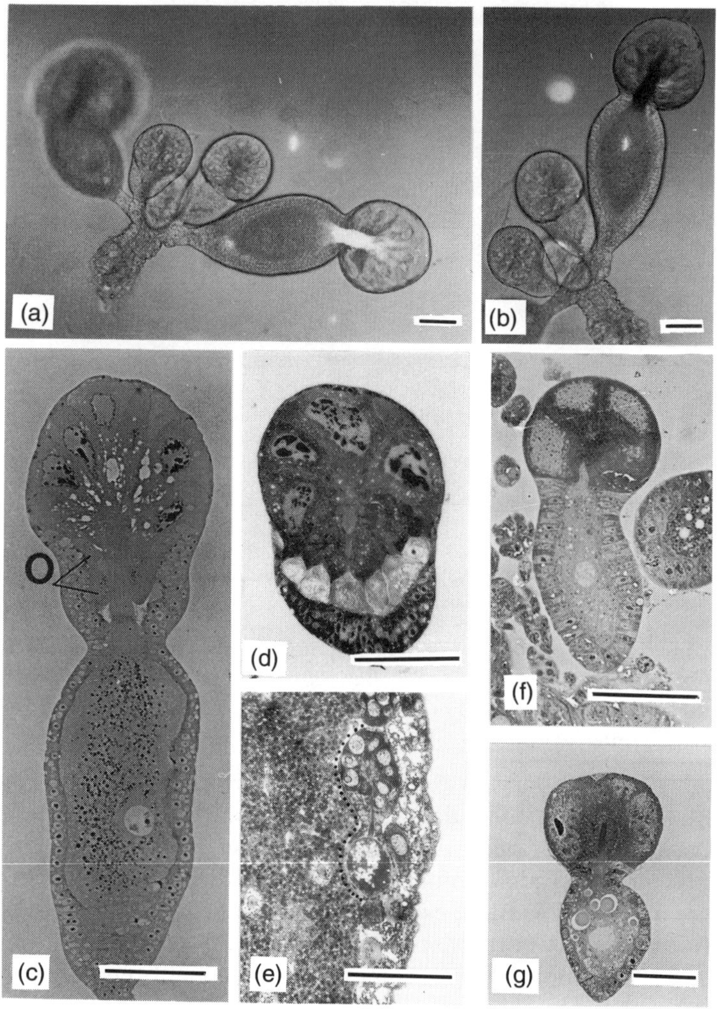

Figure 3.59 Ovarioles of Aphidimorpha. Note that all follicle cells are mononucleate. (a,b) Ovary of *Drepanosiphum platanoides* (Hemiptera, Aphidoidea: Callaphididae) viewed with polarized light. Birefringence occurs only in large nutritive cord. Note developmental difference between ovarioles. (c) Semi-thin section through oviparous ovariole of *D. platanoides*. Only one oocyte grows, all others are arrested (O). (d) Oviparous ovariole of same species before previtellogenic growth. Note basal layer of arrested, brightly stained oocytes. (e) Viviparous ovariole of *Macrosiphum rosae* (Aphidoidea: Aphididae). Margin from ovariole to mycetome is stippled. Same architecture occurs as in ovipare ovarioles, but nurse cells and growing oocyte remain small. (f) Ovariole of *Pseudococcus maritimus* (Hemiptera, Coccoidea: Pseudococcidae). Four nurse cells from a total of seven are shown in this section. (g) Ovariole of *Quadraspidiotus pyri* (Coccoidea: Diaspididae). This species has three nurse cells and one oocyte, as is found in most coccids. Bars indicate 50 μm. (Figures a–c from Büning, 1985; with permission of Springer Verlag. Figure g from King and Büning, 1985; with permission of Pergamon Press.)

However, a variety of small, uncharacterized granules occur in the germinal vesicle of *Drosophila*, concomitantly with a short pulse of RNA synthesis (Mahowald and Tiefert, 1970). A similar short phase of RNA activity in the germinal vesicle of *Musca* was found by Bier *et al.* (1969). Recently, Woodruff and Telfer (1990) reported measurable amounts of RNA synthesis in the germinal vesicles of previtellogenic growing oocytes of *Hyalophora* (Lepidoptera), which, as most orders of holometabolous insects, have polytrophic meroistic ovaries. The ^{3}H-uridine label was repressed, when electrical coupling of subsequent follicles became measurable, which occurs shortly prior to vitellogenic growth (Figure 3.51a,b). In *Drosophila* oocytes, the germinal vesicle will be shifted early in previtellogenesis to the dorsal side and keeps its dorsal-anterior position within the egg (Gutzeit, 1990b), similar to the dorsal positions of the germinal vesicles of house flies, crickets, grasshoppers or beetles (e.g. Netzel, 1965; Jung, 1966; Bier *et al.*, 1967; King, 1970; Büning, 1972).

Strong evidence for the engagement of the germinal vesicle in the process of axis determination of eggs and embryos comes also from laser-ablation studies in *Drosophila* (Montell *et al.*, 1991). The oocyte nucleus was eliminated during stages 6–9, i.e. during the last phases of previtellogenic and the early and middle vitellogenic growth phases. In all stages, the germinal vesicle has already shifted to the antero-dorsal side. The ablations result in a ventralization of the oocyte, similar to the effects known from some genes engaged in dorsal–ventral polarity, like *gurken* and *torpedo*. Accumulation of *vasa* protein at the posterior pole was also disrupted. This shows clearly that the germinal vesicle is needed for pattern formation in eggs.

Which forces can move the germinal vesicle? Ultrastructural analysis as well as immunotechniques clearly show that each nucleus is embedded in a tight net of cytoskeletal elements, comprising microtubules, actin filaments and intermediate filaments (Figure 3.43b) (Gutzeit, 1986a,b, 1990a; Büning and Sohst, 1988, 1989). In the small eggs of *Heteropeza pygmaea*, a parthenogenetic and paedogenetic gallmidge, moving of the oocyte nucleus could be demonstrated directly by time-lapse cinematography (Went *et al.*, 1978). The amoeboid pulsations and movement of the germinal vesicle through the ooplasm are forced by the cytoskeleton. No indications for autonomous movements by the germinal vesicle itself were found by ultrastructural analysis.

In some species, the oocyte nucleus becomes compartmentalized through the budding off of accessory nuclei during the previtellogenic and vitellogenic stages (Figures 3.52, 3.60 and 3.61).

These structures are surrounded by an envelope, identical to that of the oocyte nucleus, and often contain electron-dense, RNA-positive inclusions, referred to as 'pseudonucleoli' (Hopkins 1964; King and Fordy 1970; Cassidy and King, 1972; Meyer *et al.*, 1979; Bilinski, 1989, 1991a). Accessory nuclei have been described in developing oocytes of hymenopterans (Blochmann, 1886; Stuhlmann, 1886; Buchner, 1918; Hopkins, 1964; King and Richards, 1968; King and Fordy, 1970; Cassidy and King, 1972; Wolf and Sokoloff, 1984; Billen, 1985; Bilinski, 1991a,b), mallophagans (Ries, 1932; Ries and van Weel, 1933; Bilinski and Jankowska, 1987; Bilinski, 1989) and dipterans (*Chiromomus, Pryne*) (Sachtleben, 1918; Meyer *et al.*, 1979; Zissler and Sander, 1982). Similar structures occur also in the oocytes of *Panorpa* (Mecoptera) (Figure 3.47b; Büning, unpublished) and may be found also in other species. Besides oocytes, similar structures have also been reported in insect embryos (Zissler and Sander, 1982). Although the morphogenesis of accessory nuclei has been traced only in a limited number of species at the ultrastructural level, it has become evident that they arise by folding (budding) of the nuclear envelope (Meyer *et al.*, 1979; Bilinski, 1989, 1991a). Recent ultrastructural studies suggest that microtubules and nuclear lamina are involved in the formation and/or stabilization of accessory nuclei (Bilinski 1989, 1991a). Despite similar origins, accessory nuclei in mallophagans and hymenopterans differ in their structure and distribution in oocytes and may be involved in different cellular processes. This is in agreement with the opinion that accessory nuclei evolved independently in these not closely related orders (Meyer *et al.*, 1979).

In *Eomenacanthus stramineus* (Mallophaga), accessory nuclei arise during previtellogenic growth (Bilinski and Jankowska, 1987). Newly formed, small accessory nuclei contain accumulations of fibrillo-granular material (Figure 3.61) that are RNA-negative and stain positively with silver, when

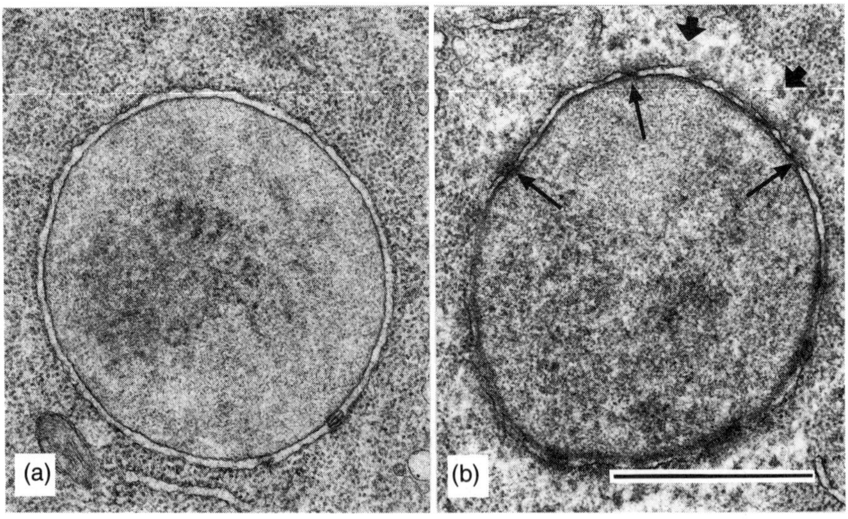

Figure 3.60 (a,b) Accessory nuclei in the ooplasm of the hymenopteran *Tenthredo olivacea* × 26 000. (b) Note close contact of nuclear pores (thin arrows) with accumulation of nuage material (thick arrows). (Coutesy of Sczcepan Bilinsky, Jagiellonian University, Kraków, Poland.)

the silver-nucleolus organizer region (Ag-NOR) method is used. In subsequent developmental stages, accessory nuclei grow markedly, often assuming a pear-like shape. They assemble, while growing, around the germinal vesicle and maintain contact with the nuclear envelope by tapered stems (Figure 3.61). Simultaneously, in each vesicle one dense, RNA-positive pseudonucleolus gradually develops. The above observations led to the suggestion that the accessory nuclei of this species may contain extrachromosomal rDNA (Bilinski, 1989). Consequently, pseudonucleoli can be interpreted as an equivalent of the multiple nucleoli described in germinal vesicles of other species (see section 3.3.4).

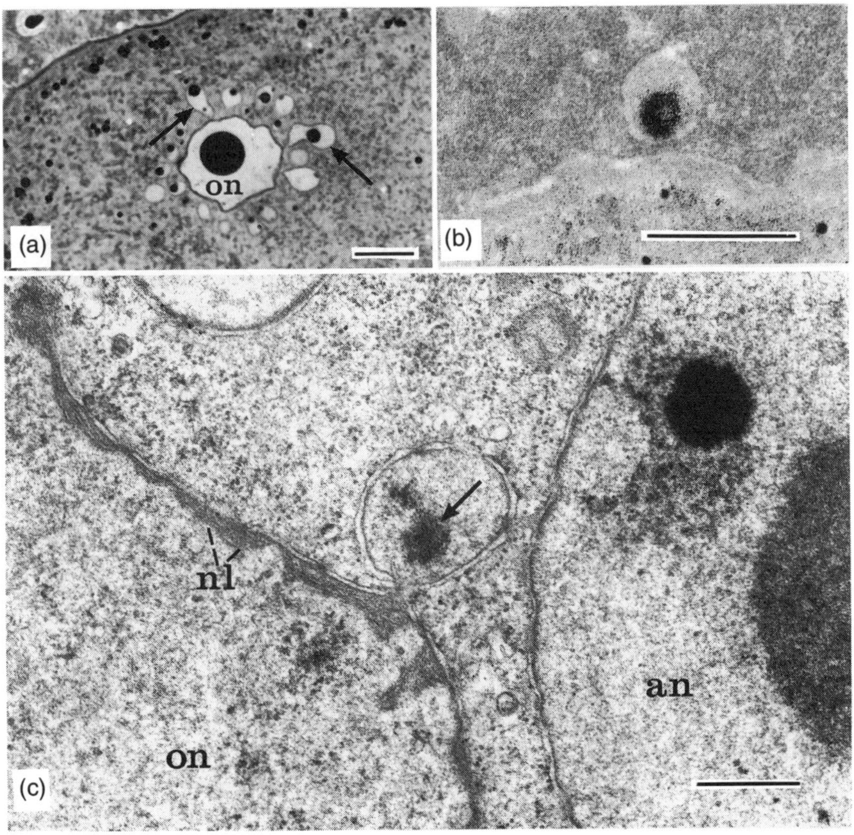

Figure 3.61 (a) Light micrograph of early vitellogenic oocyte of *Eomenacanthus stramineus* (Mallophaga). Oocyte nucleus (on); accessory nuclei (arrows) × 800. (b) Ag-NOR positive aggregations of fibrillo-granular material in newly formed accessory nuclei of *E. stramineus* × 20 000. (c) Electron micrograph showing newly formed accessory nuclei in oocytes of *E. stramineus*. Nuclear lamina (nl); oocyte nucleus (on); accessory nuclei (an); aggregation of fibrillo-granular material (arrow) × 14 000. (Courtesy of Sczcepan Bilinski, Jagiellonian University, Kraków, Poland.)

The morphogenesis of accessory nuclei in hymenopterans is apparently different. Here, newly formed accessory nuclei are transported rapidly from the place of origin, the perinuclear cytoplasm, to the peripheral ooplasm (periplasm), where they grow and multiply (Hopkins, 1964; Meng, 1968; King and Fordy, 1970; Bilinski, 1991a). It has been shown recently, that at least in some species (*Tentheredo olivacea, Cosmoconus meridionator, Lissonota* sp., *Coleocentrus* sp.; Bilinski, 1991a,b; unpublished) accessory nuclei are distributed unevenly and occur predominantly in the periplasm of the anterior pole. The extreme example of such spatial distribution has been described in the ichneumonid *Cosmoconus meridionator* (Bilinski, 1991b). In this species accessory nuclei are preferentially aggregated in the anteroventral periplasm, and along both (anteroposterior and dorsoventral) axes of the oocyte their number and size reduce in a gradient manner (Figure 3.51). Furthermore, ultrastructural and histochemical investigations support the idea that hymenopteran accessory nuclei contain RNA (Hopkins, 1964; King and Fordy, 1970; Cassidy and King, 1972; Demina, 1989). However, RNA-tracer experiments in *Pimpla* ovarioles failed, whereas proteins seem to accumulate in accessory nuclei (Meng, 1970). Accessory nuclei are also often connected to 'nuage material' of fine granular consistency, possibly extruded from them via nuclear pores (Figure 3.60) (Bilinski, 1991a).

These findings suggest that accessory nuclei of hymenopterans might transport RNPs and/or proteins to special regions of ooplasm and thus might be involved in the establishment of developmental gradients (morphogenic signals). This supposition is consistent with the discovery of accessory nuclei in deposited eggs of *Smittia* at the pole cell stage (Zissler and Sander, 1982).

The function and ultimate fate of accessory nuclei in both systems needs further investigation. Techniques such as immunohistochemistry and Miller spread preparations can provide new insights into the role of accessory nuclei in developing oocytes and/or embryos. From the last examples, the movement of the germinal vesicle to different defined positions in the oocyte during defined growth phases, squeezing, budding which forms accessory nuclei, growth and decay of so-called endobodies, and germinal vesicle breakdown, can all be looked at as a new function of the germinal vesicle, namely as an organelle responsible for the assembly, transport and specific deposition in time and space of nuclear RNPs or nuclear proteins which are needed during early embryogenesis.

Finally, we should mention a technique which will give some insights into the dynamic fate of nuclear proteins in the future. It has been probed in other organisms like *Xenopus* (Dreyer *et al.*, 1985; Wedlich *et al.*, 1985) and will be important in insects with their diverse structural and physiological differences in panoistic, polytrophic meroistic and telotrophic meroistic ovaries. Furthermore, it could be an ideal test for phylogenetic investigations: monoclonal antibodies should be raised against nuclear proteins, and labelled with fluorescent dyes or immunogold particles. Such prepared antibodies can couple with their targets in different tissues. Using different cells, for instance nurse cells, oocytes and follicle cells at different times during oogenesis, one would be able to gain information about the dynamics of these proteins, their (selective) importation into, or export from, nuclei, as well as their docking behaviour in cytoplasm and karyoplasm.

3.3.3 Meiosis in insects

No doubt, the two fundamental functions of germ cells are the reduction of the genome to the haploid status and the generation of new allelic combinations. Both functions are verified by meiosis in which a preceding S-phase is followed by two nuclear divisions, which finally gives rise to four haploid cells. New allelic combinations occur by two processes. The first is the exchange of DNA between homologous chromosomes, crossing-over, which happens during the prophase of the first meiotic division. The second process includes the mixing and separation by chance of the (remodelled) homologous chromosomes and occurs during metaphase and anaphase of the first meiotic division.

Both processes demand recognition by homologous chromosomes of one other, together with a tight physical pairing of homologues during genetic exchange (crossing-over). Such pairing takes time and its morphology makes meiotic prophase clearly discernible from mitotic prophase. We will discuss these morphological aspects of meiosis generally and with special reference to oogenesis. The ultrastructure has been elucidated during the past three decades. Numerous articles describing meiosis in both sexes have led to an unique model. The main subjects, however, have been *Drosophila melanogaster* (Diptera) and *Bombyx mori* (Lepidoptera), two species the genetics of which are well understood and in which crossing-over is restricted to the homogametic sex: to females in *Drosophila* and to males in *Bombyx*. This was very important for a correct interpretation of morphological and genetic facts (e.g. Moses, 1958, 1968; Roth, 1966; Koch *et al.*, 1967; Moens, 1968, 1978; Smith and King, 1968; King, 1970; King and Akai, 1971b; Fiil and Moens, 1972; Carpenter 1975, 1979a,b, 1981, 1982, 1989; Rasmussen, 1975, 1976, 1977a,b; Rasmussen and Holm, 1980, 1982; Holm and Rasmussen, 1981; for reviews see Roeder, 1990; Kleckner *et al.*, 1991; Giroux, 1992).

Meiotic prophase is turned on by as yet unknown processes. Meiotic prophase begins with the condensation of chromosomes which become visible as thin, long and often coiled strands; in this way sister-chromatids are kept tightly together on their whole length and thus behave as one unit. This first phase is called the leptonema or the leptotene stage. During this stage, nuclear proteins anneal asymmetrically at only one side along the coiled sister-chromatids giving a polar aspect to cross-sectioned chromosomes (Figure 3.62).

The recently formed protein axis is called the lateral element and has a diameter of 30–40 nm. The second morphological element which makes this stage clearly discernible from early mitotic prophases is the attachment of chromosomes, via their telomeres, to the nuclear envelope. These attachment plaques are specialized protein structures which seem to be necessary for the chromosomes to move and find homologous partners.

As soon as homologous chromosomes start intimate pairing, the zygotene stage (zygonema) is reached. In classical models of meiosis this intimate pairing, the synapsis, begins at the attachment plaques. At this time, the attachment plaques concentrate at one hemisphere or even at a smaller segment of the nuclear envelope. Now, a protein complex, the central element, anneals the lateral elements of homologous chromosomes

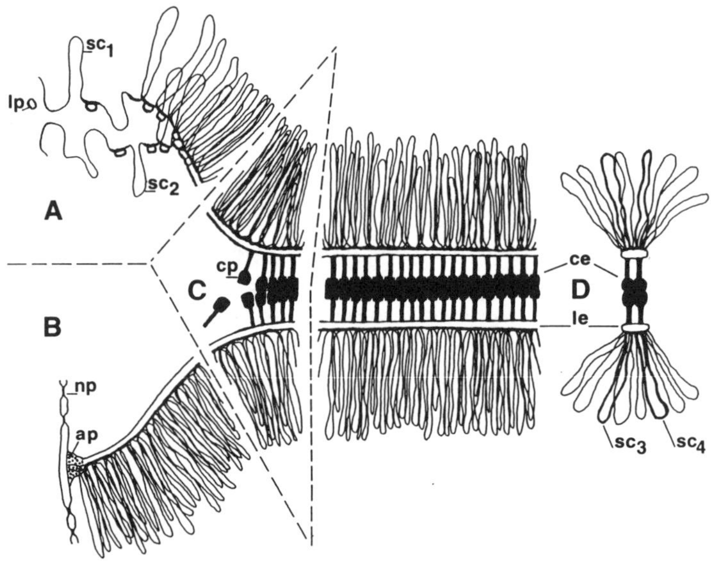

Figure 3.62 Early events during meiotic prophase. (A) Preleptotene stage: sister-chromatids (sc1, sc2) assemble with 'lateral' proteins (lp), which construct the lateral element (le). (B) Leptotene stage: the lateral element anneals the sister-chromatids by special proteins to the nuclear envelope, forming attachment plaques (ap). This region is free of nuclear pores (np). (C) Zygotene stage: 'central' proteins assemble at both lateral elements, forming the central element (ce). Especially in insects, the central proteins form ladder-like structures. The chemical nature and function of all these proteins are not known. (D) The pachytene stage is reached, when homologous chromosomes are fully synapsed. The cross-section through a tetrade shows the asymmetric binding of chromatin to the lateral elements. Sister-chromatids of the same plane are shown by thicker lines (sc3, sc4).

in a zippering manner, starting from the attachment plaques. In insects, this central element is ladder-like and has a diameter of about 100–120 nm. In detailed analyses of serial sectioned primary spermatocytes and primary oocytes, Rasmussen and coworkers (cited above) unravelled the dynamics of synaptonemal complexes (SC), giving some additional new aspects: at zygonema chromosomal interlocking was observed regularly, this is resolved during pachynema by yet unknown mechanisms. Both strands of a double helix must break simultaneously, since terminalization of interlocked regions was not observed but free ends were found regularly during zygonema (Figure 3.63).

When homologous chromosomes are fully synapsed, the pachytene stage (pachynema) is reached. The paired homologous chromosomes are also called dyads since we can see two elements combined into one. During pachynema, chromosomal exchange occurs. Some huge, proteinaceous

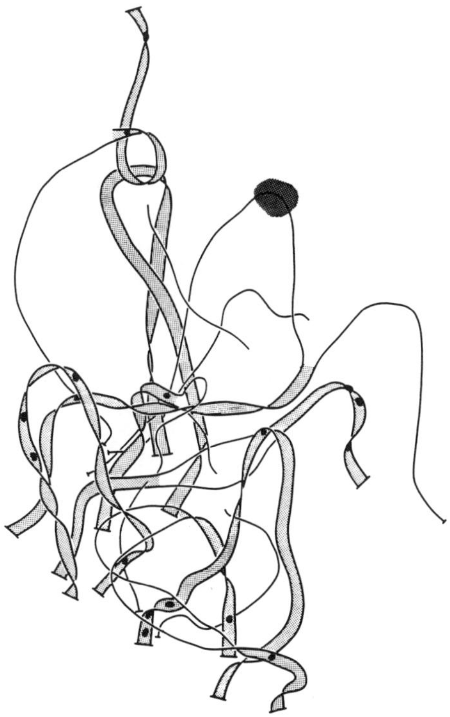

Figure 3.63 Partial reconstruction of *Bombyx mori* spermatocyte. Several interlockings of bivalents and univalents during late zygonema are shown. Attachment plaques are indicated by short bars; free ends of chromosomes indicate chromosomal breaks. Synaptonemal areas are shadowed. Recombination nodules are shown by black dots. (From Holm and Rasmussen, 1981; with permission of S.W. Rasmussen and Springer Verlag.)

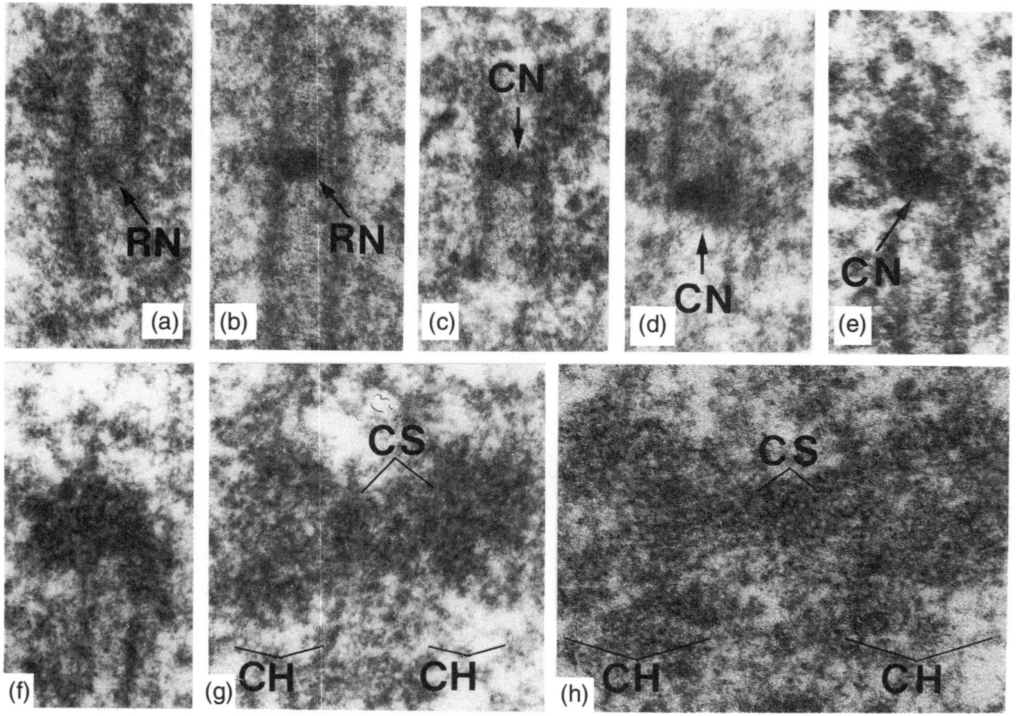

Figure 3.64 Recombination nodules (RN) during meiotic prophase stages in *Bombyx mori* spermatocytes. (a) Late zygonema; (b) mid-pachynema; (c) from mid-pachynema, more and more chromatin associates with the nodules, which are now called chromatin nodules (CN); (d) late pachynema; (e) increase of condensed chromatin around the nodule; (f) further chromatin association; (g) chiasma at late diplonema. A circular structure (CS) with a dense core is found between homologues (CH); (h) chiasma at mid-diakinesis. (From Holm and Rasmussen, 1981; with permission of S.W. Rassmussen and Springer Verlag.)

complexes arise along the synapsed chromosomes on the central element, called recombination nodules (Figure 3.64).

The association of these nodules with recombination processes is indirect: the number of recombination nodules is equivalent to the number of later chiasmata, which, themselves, are equivalent to the number of crossing-over events. Moreover, the distribution of nodules along the paired chromosomes is similar to the distribution of chromosomal exchanges. Both arguments are deduced from experiments with *Drosophila* mutants which have a reduced capacity for recombination. Recombination needs limited DNA synthesis to splice and anneal the homologous strands. Consequently, radioactive DNA precursors were incorporated near recombination nodules (Carpenter, 1975, 1979a,b, 1981, 1982, 1989). One, additional argument comes from the analysis of male and female germ cells of *Bombyx mori* (Rasmussen and Holm, 1980): crossing-over is found only in males and recombination nodules occur only in males, whereas synaptonemal complexes are found in both sexes. Thus, meiotic recombination needs synaptonemal complexes, but their existence is not sufficient for successful meiotic recombination.

During recombination of genetic material, the morphology and distribution of nodules changes (Carpenter, cited above). During early phases (zygonema) the number of recombination nodules seems to be greater than at late (pachynema) stages. The distribution of so-called early nodules is random, whereas the amount and distribution of late nodules is congruent to the amount and distribution of genetic exchange events (i.e. crossovers) measured along the chromosomes (Figure 3.65a).

There are more chromosomes with one and two exchanges (late nodules) than expected by a Poisson distribution. This effect is called positive interference (Muller, 1916; Laurie, 1980; King and Mortimer, 1990). The molecular mechanisms of crossover, chiasma and positive interference are still unknown. King and Mortimer (1990) proposed a speculative model based on computer simulation, which satisfies most of the available data for positive interference. A number of early nodules bind randomly at synaptonemal complexes, and some of these, starting with crossover reactions, initiate polymerization reactions (or vice versa) which spread over the synaptonemal complexes. The growing polymers eject early nodules, which have not initiated polymerization. Early free and ejected nodules can degrade, or reassociate only with those synaptonemal complexes which are not polymerized (Figure 3.65b).

At diplonema, desynapsis begins. The synaptonemal complexes become separated from sister-chromatids, and four elements can now be seen. Thus, the dyad of former stages transforms to a tetrad. The four chromatids of the tetrad still anneal in regions of crossover events. These points, the chiasmata, are the cytological expression of the genetic crossovers. In many species the proteins of the synaptonemal complexes dissolve; however,

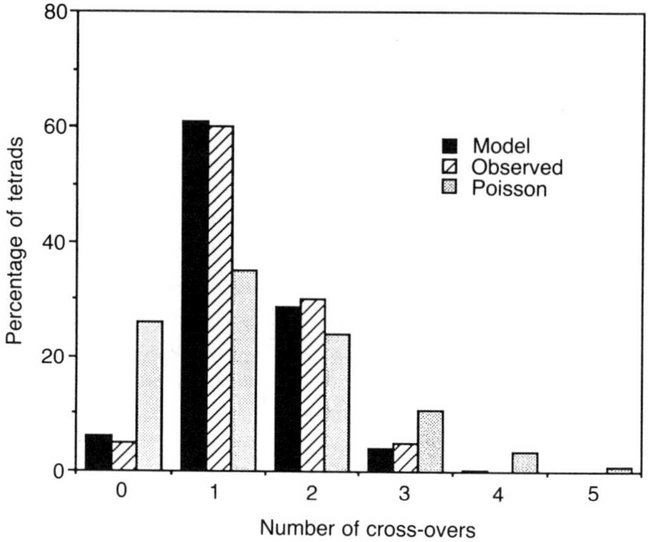
80
60
40
20
0
Percentage of tetrads
Model
Observed
Poisson
0
1
2
3
4
5
Number of cross-overs

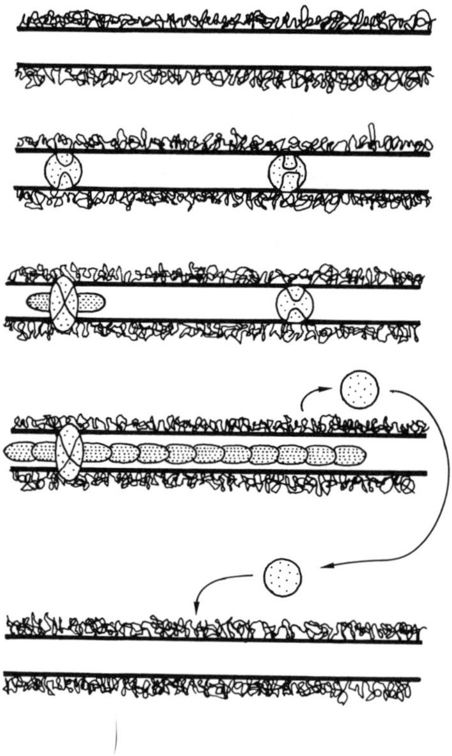

sometimes they form stable polycomplexes, not bound to chromosomal material (Roth, 1966; Moses, 1968; Fiil and Moens, 1972). The chiasmata keep their position during the final stages of meiotic prophase or they move to the ends of chromosomes as found first in *Primula sinensis* (Darlington, 1931) and in other plants and animals e.g. grasshoppers (summarized in Whitehouse, 1970). This process is called the 'terminalization of chiasmata'. Chiasmata keep homologous chromosomes together until metaphase I. They are thought to serve as additional elements ensuring correct segregation of homologues in anaphase I. This idea comes from experiments using mutations with decreased amounts of chiasmata (crossovers). Such mutations have increased non-disjunctions; homologous chromosomes are not separated equally to the daughter cells, but one cell becomes the whole tetrad, the other none.

The last stage of meiotic prophase I is called diakinesis. Here, chromatids condense more and more and the attachment plaques become dissolved. The tetrads move to the equatorial plate and metaphase I is reached. Sister-chromatids, although clearly discernible, behave as a unit and possess a single kinetochore region to which microtubules anchor in only one direction. The tetrads assemble by chance in the equatorial plate, thus segregation of homologous chromosomes occurs by chance during anaphase I. Telophase I ends the first meiotic division which results, in males, in two diploid cells which immediately undergo the second division which resembles a normal mitotic programme. Sister-chromatids have separated kinetochores to which kinetochore fibres point in opposite directions. Since an S-phase is omitted during the two meiotic divisions, the second division results in four haploid cells. In females, prophase I lasts until the entry of the chorionated egg into the lateral oviduct. The germinal vesicle becomes situated at the dorsal side, embedded in the cortical cytoplasm near to the oocyte membrane.

Figure 3.65 (a) Number of cross-over events per tetrad. Computer simulation is denoted as 'model'. *Drosophila* data from Weinstein (1936) are denoted as 'observed'. The corresponding Poisson distribution is denoted as 'Poisson'. The diagram demonstrates clearly that fractions of 0, 3, 4 and 5 events are under-represented, whereas fractions of 1 and 2 are over-represented. Mean number of cross-overs is 1.4 per tetrad. Computer simulation is based on 150 000 tetrads with an initial distribution of average 2,8 early nodules per tetrad and a polymerization start probability of 0.009. (From King and Mortimer, 1990; with permission of the Genetics Society of America.) (b) A speculative model of the regulation of positive interference during cross-over. Early nodules bind randomly to the synaptonemal complex (circular). Some initiate polymerization reactions thus becoming late nodules (oval). The growing polymers eject early nodules. Ejected early nodules are either degraded or bind to synaptonemal complex that is free of polymer. (From King and Mortimer, 1990; with permission of the Genetics Society of America.)

According to the physiological status of the oocyte nucleus, the morphology of the germinal vesicle changes. Whereas in early stages of previtellogenic growth oocyte chromosomes are diffuse and are considered to have the lampbrush chromosome aspect, during the later stages chromosomes behave differently. In panoistic ovaries, oocyte chromosomes keep their diffuse aspect during the entire previtellogenic growth. Condensation of chromosomes occurs just before germinal vesicle breakdown and condensed chromosomes become individually ensheathed by remnants of the nuclear envelope. In meroistic ovaries chromosomes condense earlier and in many cases all the chromosomes condense in one area of the germinal vesicle, into the karyosome. The karyosome has a strongly reduced capacity to synthesize RNA (Bier *et al.*, 1967). In some species like *Aedes aegypti* (Diptera) (Roth, 1966; Fiil and Moens, 1972), *Chrysopa perla* (Neuroptera) (Gruzova *et al.*, 1972), *Blabs mortisaga, Gaurotes virginea, Psilothrix cyanaeus* and *Phyllobius urticae* (Coleoptera, Polyphaga) (Büning, unpublished) the karyosome becomes ensheathed before germinal vesicle breakdown by proteinaceous sheaths (Figure 3.52). There is some morphological evidence that these sheaths may be transformed synaptonemal complexes (Fiil and Moens, 1972).

After germinal vesicle breakdown in late vitellogenesis, the karyosome, which includes all chromosomes in a tight association, becomes the karyosphere, enclosed by proteinaceous sheaths and remnants of the nuclear envelope. The oocyte nucleus is now ready for further meiotic events. The first meiotic nuclear division of the primary oocyte is triggered by as yet unknown processes. However, sperm entry in many species coincides with the final stages of the first meiotic division. On the other hand, unfertilized eggs of *Drosophila* complete meiotic divisions (King, 1970). The nuclear divisions occur next to the cortical cytoplasm and are not followed by cytokinesis in insects (Counce, 1972). Thereby, one nucleus remains in the cortical cytoplasm, while the other is shifted more centrally. In *Bruchidius,* as in other Coleoptera, the polar body nucleus is always found in the cortical cytoplasm of the future dorsal side of the embryo (Jung, 1966; Buning, unpublished), a position which is nearly identical to the former position of the karyosphere. This finding may be important for understanding dorso-ventral pattern formation (see section 3.3.4) and the role of the germinal vesicle for transport and deposition of maternal information (see section 3.3.2).

After the second meiotic nuclear division, the innermost nucleus becomes the female pronucleus while the polar body nuclei degenerate (King, 1970; Counce, 1972; Tremblay and Caltagirone, 1973). Thus, meiosis in females results in a single, giant, haploid cell.

Meiosis in males proceeds continuously during the early stages of spermiogenesis, i.e. immediately after the period of cystocyte divisions. In females, however, meiosis is arrested twice after the prolonged prophase of meiosis, the first time by the breakdown of the germinal vesicle, until mid-vitellogenesis, which resembles entry into the first meiotic metaphase. Reinitiation is under hormonal control in *Locusta migratoria* and depends on the ovarian titre of ecdysone, which works via the cAMP pathway (Kappler *et al.*, 1986; Lanot *et al.*, 1987, 1988, 1990). Some evidence for asymmetrically distributed ecdysone in *Drosophila* eggs has been found by Grau and Gutzeit (1990), who demonstrate that 2.5 h before completion of oogenesis, the uniform distribution of an antigen changes: it is concentrated in granules mostly found in cortical cytoplasm at the posterior pole. However, pole cells themselves are void of the antigen. The second meiotic arrest occurs in the ripe egg, mostly during interkinesis, and is broken by external modifications, such as sperm entry.

Recent molecular and genetic analysis of the *claret* (*ca*) region of *Drosophila melanogaster* has shown that the protein coded by the *non-claret disjunctional* (*ncd*) locus is strikingly similar to the microtubule motor protein kinesin (Komma *et al.*, 1991). However, it moves to the minus end of the microtubules. Some mutations of this protein have been analysed and one of them, ncd^D, has a change in one amino acid residue of a putative microtubule-binding region. This mutant only affects meiotic chromosome segregation (non-disjunction and loss of the fourth chromosome) and leaves mitotic segregation unaffected. This indicates different kinds of motors during mitotic and meiotic chromosome segregation. Furthermore, another protein coming from the gene *nod* has significant homology to the ATP-binding domain of kinesin. This protein may play a role as the microtubule motor that translocates chromosomes not engaged in crossing-over processes in *Drosophila* females. Such non-exchange chromosomes are the fourth chromosomes which never exchange, balancer chromosomes, in which multiple inversions hinder exchange processes, and finally the X chromosomes which fail to exchange in 5% of meiosis. All of these show a separate moving behaviour which is strongly dependent on the function of *nod* (Theurkauf and Hawley, 1992).

Entry into mitosis and meiosis are initiated by similar regulatory events occurring late in the G2-phase of the cell cycle. One of the key events is the abrupt degradation of cyclin B. Cyclin B builds a complex with $p34^{cdc2}$-kinase. The activation of this kinase requires the removal of inhibitory phosphate modifications in the ATP-binding site of the $p34^{cdc2}$-kinase. Therefore, a special phosphatase is needed. Two phosphate loci have been identified in *Drosophila*, one, *string*, is necessary for mitotic initiation, whereas the other, *twine*, initiates meiosis in both sexes (Courtot *et al.*, 1992).

3.3.4 Strategies to enhance oocyte gene activities

Growth of oocytes to giant cells within a short time requires adaptations to

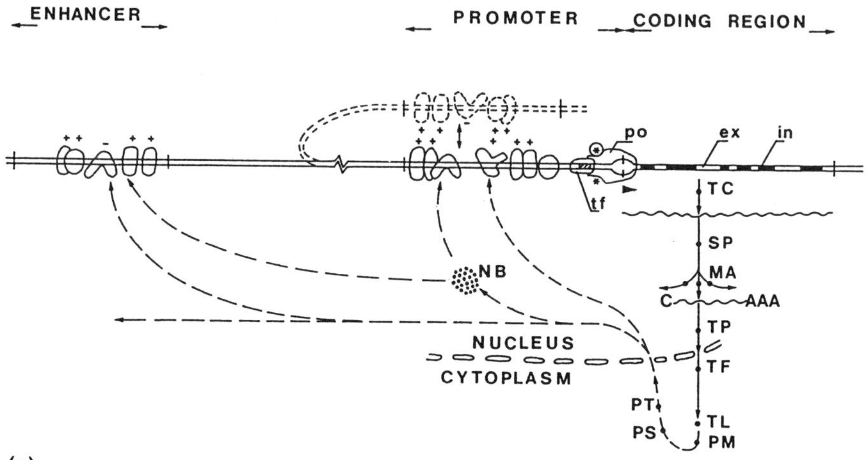

(a)

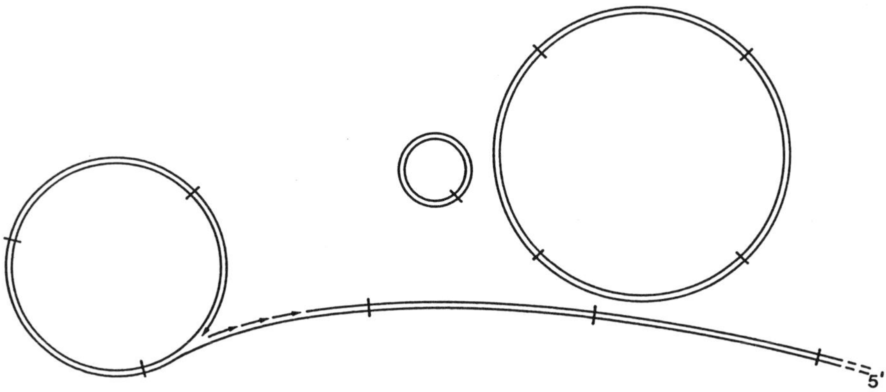

(b)

accelerate protein biosynthesis by a factor of about 1:1000 or more. This becomes apparent when comparing the growth of somatic cells with the previtellogenic growth of oocytes. As explained earlier, growth of oocytes will start during meiotic prophase, which gives a two-fold increase of all genes from 2n to 4n. However, other mechanisms must be added to establish giant growth within a sufficient time (i.e. within some days or weeks) (Bier, 1967b, 1970; Davidson, 1986).

If we look at the molecular level of transcription in eucaryotes, we learn that rates of transcription are regulated by the amount of RNA polymerase molecules transcribing a gene at a given time. The elongation rate of transcripts remains constant and RNA polymerases work at high and constant speed (i.e. about 30 nucleotides per second for all three RNA polymerase classes) if substrates (XTPs, transcription factors) are not rate-limiting (Davidson, 1986; Alberts *et al.*, 1988). Thus, RNA polymerase binding and the start of transcription are the main components regulating the transcription rate of a given gene (Figure 3.66). Consequently, the amount of RNA polymerases transcribing a gene would be directly proportional to the transcription rate of the gene.

However, it is not only the amount of actively transcribing RNA polymerases which makes the normal growth of somatic cells differ from the previtellogenic growth of oocytes. Other, stabilizing mechanisms are needed, which regulate and mostly reduce the decay of recently synthesized products. These mechanisms are involved in RNA processing to produce stable RNP complexes, and/or they are involved in protein processing to produce stable, inactive and storable forms. Some of these products assemble as detectable aggregates in the ooplasm (see section 3.3.2).

As we have learned from comparative embryology, the quality of the maternal message in the egg, stored in the euplasm, determines the pathway of early embryogenesis (i.e. in which way pattern formation is

---

Figure 3.66 (a) The hypothetical molecular architecture of a regulator gene. The coding region encloses exons (ex), bearing the information of the later regulator proteins, disrupted by introns (in). Regulator proteins can bind to promoter and enhancer regions of their own genes (self control), or they bind to such regions of other regulator and/or structure genes (stippled lines). At the start of transcription, DNA is opened (bulbous region) by the bound RNA polymerase II (po), which has some additional binding sites (*), one of which is just occupied by a binding factor (circled). The polymerase itself is closely bound to a transcription factor (tf), which binds to the TATA-box (striped box). Transcription of polymerase is under control of the three-dimensional structure of the promoter region, the regulator proteins of which interact with each other and with the DNA. These regulator proteins influence transcription positively or negatively (+,-) or are indifferent. The promoter region itself can interact by looping (stippled) with an enhancer region of similar structure, which can be situated some thousand base-pairs upstream (as shown) or downstream. All these structures influence the amount of transcription (TC). However, the quality of transcription (i.e. amount and structure of final mRNA) depends on splicing procedures (SP), not shown in detail, which can create different mRNAs in different amounts. During maturation (MA) transcripts are capped (C) at the 5′-end and poly A (AAA) sequences are added at the 3′-end. Transport (TP) through nuclear pores can be followed by cytoplasmic docking and/or cytoplasmic transformation (TF) or immediately by translation (TL). Protein maturation (PM) is followed by protein storage (PS) in cytoplasm or by protein transport (PT) into the nucleus. In the nucleus, regulator proteins reach their final destination immediately (enhancer or promoter regions) or are stored preliminarily in nuclear bodies (NB). (b) In this simplified rolling circle model direction of transcription is indicated by arrows. Two circularized tandem repeats are shown, bearing one and four repeat units, respectively.

influenced by the maternal genome). The amount and quality of maternal information decides the species-specific pathway between the extremes: the total 'regulative' or the total 'mosaic' development in the early embryo (Davidson, 1986). Consequently, we must assume that each species has its own, characteristic pattern of maternal gene activity during oogenesis.

(a) Lampbrush chromosomes

In oocytes, which synthesize euplasmatic components exclusively under register of their own genome, we can expect a maximized hnRNA synthesis. We will find many RNA polymerases, together with their transcripts, elongating along transcription units. This ultrastructural aspect is found in spread preparations of lampbrush chromosomes. Most details come from amphibian oocytes, whereas spread preparations of insect oocytes are rare (Figures 3.67–3.69).

The visualized polarity of transcription units fits well with the polarity of huge lateral loops observed with classical methods of light microscopy. The same can be said of the length of loops (Macgregor, 1980). For many cases we can assume that one transcription unit resembles one loop. However, in some spread preparations we notice relatively short spacer regions between transcription units which may indicate that more than one transcription unit can belong to one loop (Angelier and Lacroix, 1975; Scheer *et al.*, 1976b). In special cases, phase-contrast visible loops and corresponding spread preparations may represent sets of tandemly repeated gene families (Scheer *et al.*, 1976b; Macgregor, 1980; Scheer, 1981). Dot-like structures, known as chromomeres of lampbrush chromosomes as seen with the light microscope, have their counterparts in electron microscopic spread preparations as long spacer sequences, situated between transcription units. These spacers have lost their natural three-dimensional structure during the spreading procedure by dissociation of nuclear proteins, which are not further characterized (Angelier *et al.*, 1984).

Lampbrush chromosomes have also been identified in some panoistic insect ovaries using classical methods (Bier *et al.*, 1967; Kunz, 1967a,b, 1969; Bier, 1968). However, loops are very small and delicate and only in some subjects, like *Locusta migratoria* and *Decticus albifrons* (Kunz, 1967a) does the loop structure seem comparable to the situation found in amphibians. In other panoistic ovaries, as in the ovary of *Acheta (Gryllus) domesticus*, the loop structure is not as clear (Kunz, 1969). Trendelenburg and coworkers made spread preparations but failed to demonstrate transcription units, in which many nascent RNA fibrils are associated via transcribing RNA polymerases (Tröster *et al.*, 1985; Trendelenburg *et al.*, 1986, 1988). They found RNP transcript complexes by ultrathin serial sectioning, but could not find an association with the assumed central axis (DNA). Thus, lampbrush chromosomes may differ in respect to the association of transcripts to their matrix.

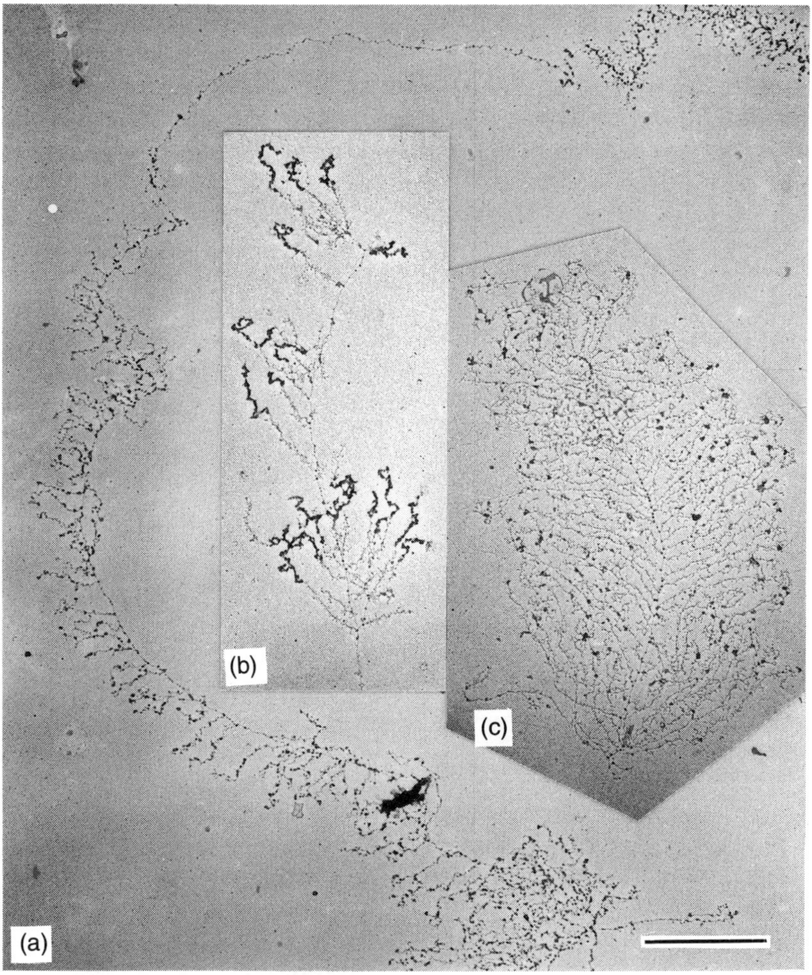

Figure 3.67 Lampbrush chromosomes of *Locusta migratoria*. Non-nucleolar transcription units, showing different architecture at different coding regions. Note that transcripts are generally not as densely packed as rDNA transcripts. (a) Polar organization of a transcript unit. Note that the neighbouring gene is transcribed in the opposite direction. (b) Part of a transcription unit, the transcripts of which are densely packed only in distal regions. (c) Transcription unit, the transcripts of which show many small hairpin structures and terminal globular aggregations. Bar represents 1 μm. (Courtesy of J. Gottanka, University of Erlangen-Nürnberg, Germany.)

However, during spermiogenesis of *Drosophila hydei* and some related species, several giant lampbrush loops have been identified on the Y-chromosome (Meyer *et al.*, 1961; Meyer, 1963; Hess and Meyer, 1963; Hennig, 1967). Spread preparations for electron microscopy have been done by

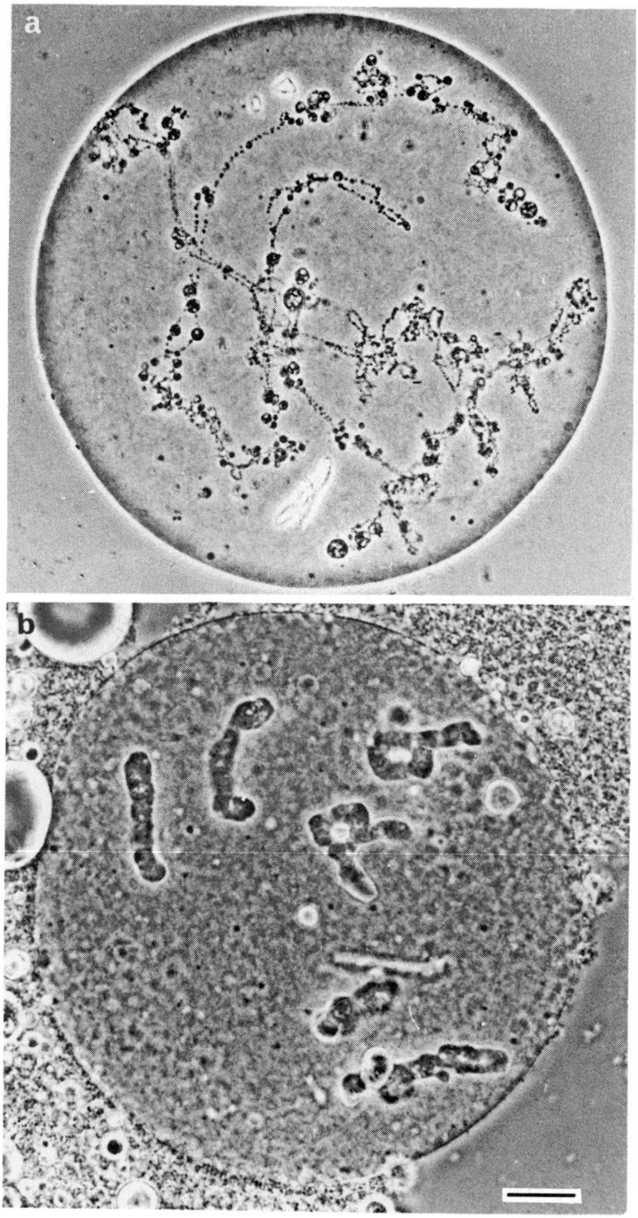

Figure 3.68 Isolated germinal vesicles from freshly prepared follicles of *Locusta migratoria* (Orthoptera). (a) Isolation at room temperature. Repetitive rDNA is shown as long threads loaded with small spheres (Perlschnurstränge). (b) After prolonged incubation of animals to low temperatures, threads shorten and three pairs of threads can be distinguished, indicating three NORs in the haploid genome. Bar represents 20 μm. (From Schäfer and Kunz, 1987; with permission of W. Kunz and Springer Verlag.)

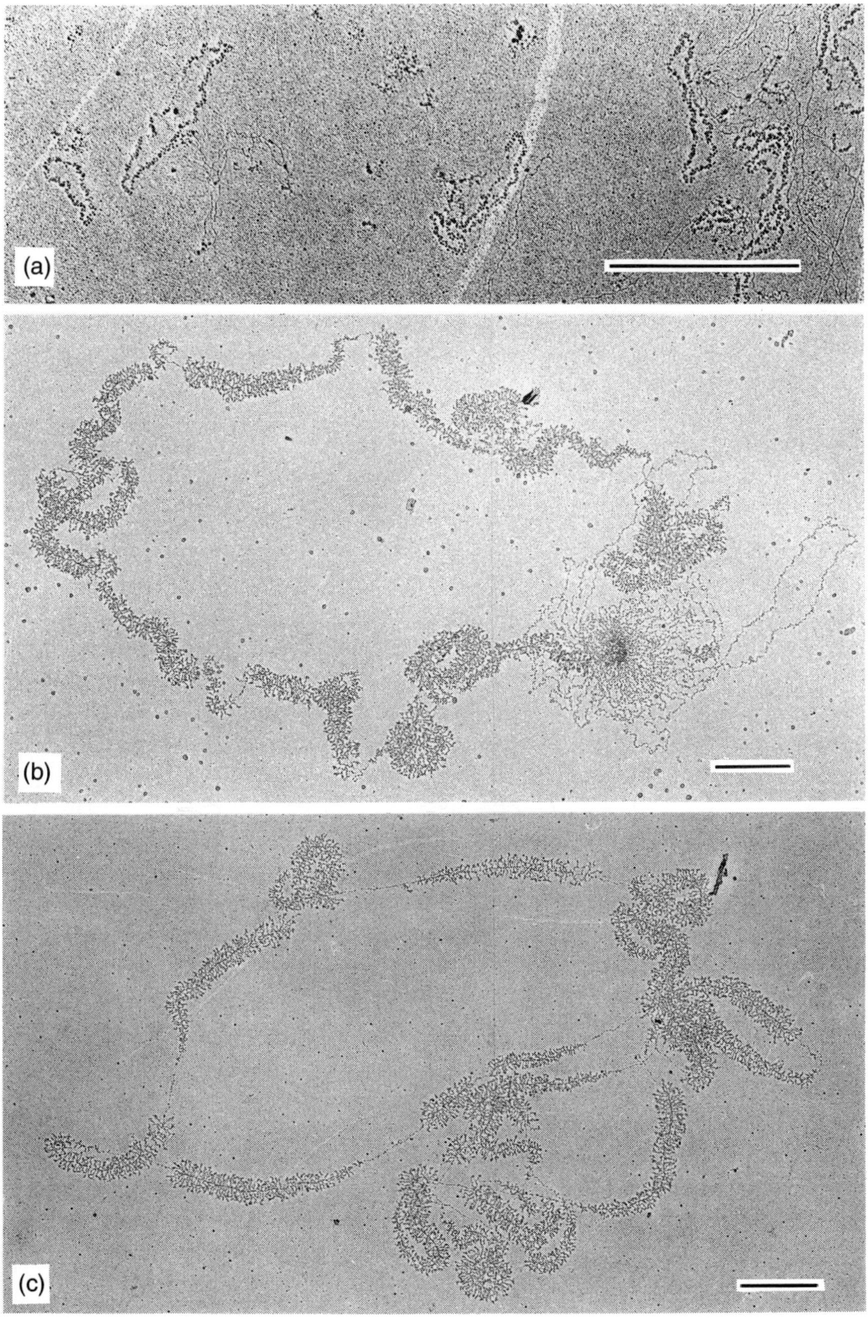

Figure 3.69 Spread preparation of extra-nucleoli of *Nosopsyllus fasciatus* oocytes (a–c). (a) In early and mid-previtellogenesis, extra-nucleoli are silent and show higher ordered structure. (b) During the last stage of previtellogenesis, extra-nucleoli occur, which are activated only in part; however, active tandem repeats are mostly neighbouring. Note that silent DNA loops form higher ordered structures and are beaded with nucleosomes. (c) At the beginning of vitellogenesis nearly all extrachromosomal genes are actively transcribing.

Glätzer (1979, 1980), Glätzer and Meyer (1981), Grond *et al.* (1983, 1984) and de Loos *et al.* (1984), summarized by Hennig (1987). In all investigations huge transcription units were found, which can be identified by their length and by characteristic folding of transcripts. In living cells however, this backbone of DNA and transcripts accumulates nuclear proteins, which interact specifically with the three-dimensional architecture of transcripts (Figure 3.70).

Our unpublished results on spread preparations of *Locusta migratoria* oocyte nuclei, taken from the last stages of previtellogenic growth, show typical transcript assemblies on transcription units as known from amphibian oocytes (Figure 3.67) (Gottanka, unpublished). Thus, the general concept of lampbrush chromosomes as elucidated with amphibian oocytes (Callan, 1986; Davidson, 1986) can be adopted to insect oocytes of panoistic ovaries; although we hardly need more investigations on other species to integrate the somewhat differing results gained on *Acheta* oocytes.

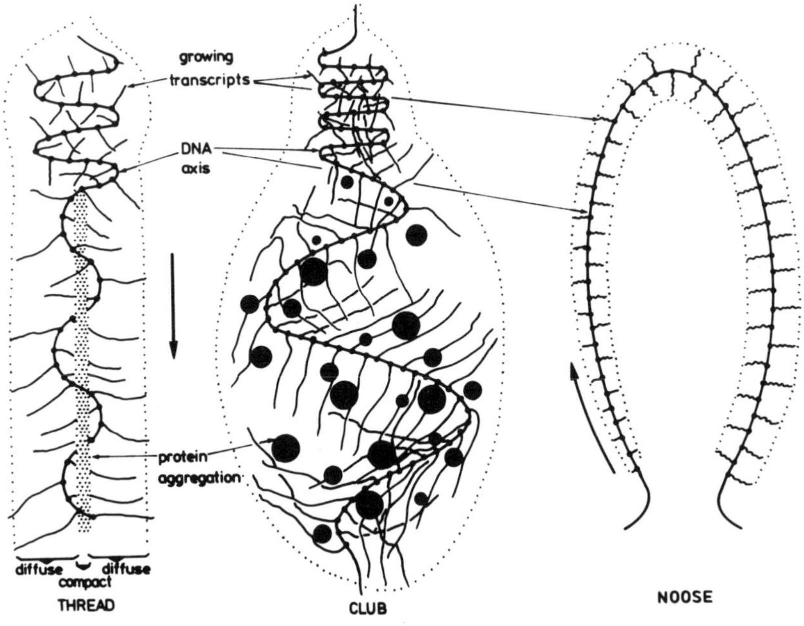

(a)

Figure 3.70 (a) Scheme of three lampbrush loop types of the Y-chromosome of *Drosophila hydei*. Different secondary structures of transcripts are omitted. Large arrows indicate direction of transcription. (From Hennig, 1987; with permission of W. Hennig and Springer Verlag.) (b) and (c) Lampbrush chromosome spread preparation of the Y-chromosome of *D. hydei*. Heavily branched RNP structures are found attached to a DNA axis, which is assumed to belong to 'thread' loops. Scale bar in (b) represents 1 μm and 5 μm in (c). (From Glätzer and Meyer, 1981; with permission of K.H. Glätzer and Biol Cell.)

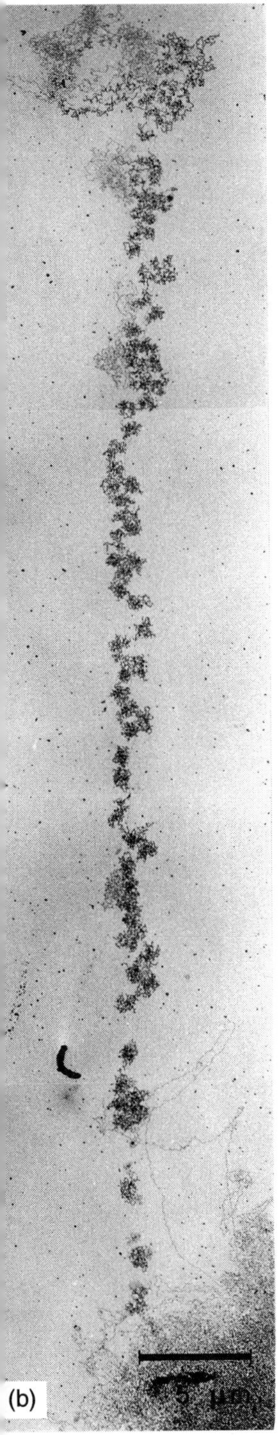
(b)
5 µm

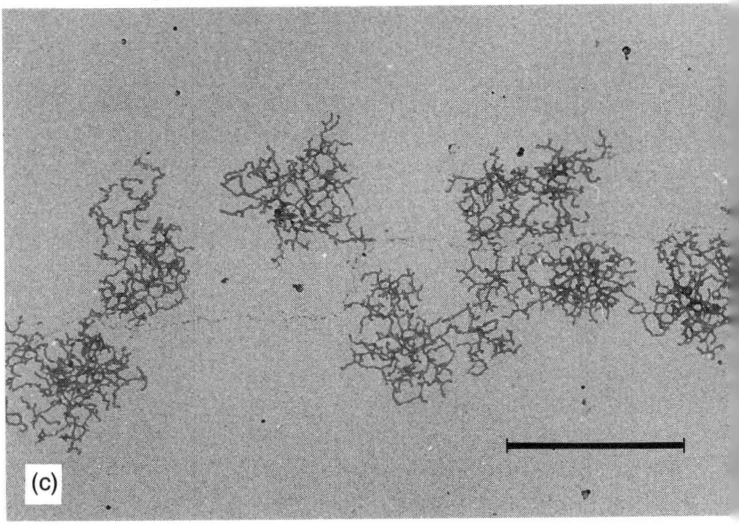
(c)

As shown for the giant loops of the Y-chromosome of *Drosophila hydei*, as well as from the analysis of lampbrush chromosomes of amphibia and *Locusta*, we must assume that lampbrush chromosomes are not only the morphological equivalent for enhanced gene activity, but may also serve for individual assembly, storage and release of nuclear proteins. These proteins will become very important during early embryogenesis since the onset of zygotic gene activity will be governed by nuclear proteins (Figure 3.66) (Davidson, 1986; Gilbert, 1988). In amphibia as well as in insects, proteinaceous bodies occur in oocyte nuclei during previtellogenesis. They are called 'Binnenkörper' or endobodies and their function remains obscure (Bier *et al.*, 1967; Bier, 1968; Troester *et al.*, 1985). One possible role of endobodies as storage for ribosomal proteins has already been discussed (section 3.3.2). However, there are good reasons to assume other functions for these heterogeneous endobodies. In crickets there is only one, relatively small endobody and its structure seems to be constant during previtellogenesis. Furthermore, this endobody is associated with DNA as was shown by Feulgen staining and DNAse controls (Jaworska and Lima-de-Faria, 1973). Recently, Gall and Callan (1989), Callan and Gall (1991) and Callan *et al.* (1991) have analysed similar proteinaceous bodies in *Xenopus*. Some of these 'sphere organelles' or 'snurposomes' are associated with special loops of lampbrush chromosomes, while others are not tightly associated and occur freely in the nuclear sap. Many spheres gave cross-reactions with autoimmune antibodies against snRNPs. Surprisingly, some of these spheres are located at or near the clustered histone genes and Callan *et al.* (1991) assume a functional connection although histone genes have no introns. The above antibodies give similar bright fluorescence with the endobody of *Gryllus*, indicating that this specific endobody may have the same function as the 'sphere organelles' in *Xenopus*: they may serve as intranuclear stores for spliceosomes, needed for intensive syntheses of RNA during the prophase of meiosis in oocytes. Thus, loops of lampbrush chromosomes may be used as individual landmarks for special nuclear proteins inside the oocyte nucleus (see below and section 3.3.2).

(b) rDNA gene amplification

Within the set of highly active genes during oogenesis, rRNA, the central structural element of ribosomes, is the most important since its primary gene product is needed in large amounts. In *Drosophila* each oocyte contains about 4.5×10^{10} ribosomes (Anderson and Lengyel, 1979; Davidson, 1986), which are synthesized during previtellogenic and early vitellogenic stages not by the oocyte itself, but by nurse cells, which are sister cells of the oocyte.

In those insects in which euplasmatic growth depends only on the activity of the oocyte itself, as found in panoistic ovaries, there are three general methods: (1) the amount of euplasm can be reduced; (2) the time of euplasmatic growth can be prolonged; (3) the amount of rDNA can be enhanced: (i) in all tissues; (ii) only in the germ line; (iii) only in oocytes during euplasmatic growth, which happens during the prophase of meiosis.

The first two strategies are seen in most insects which house panoistic ovarioles (Figure 3.42) (Bier, 1970). These insects develop short germ bands, the volumes of which are directly proportional to the small amounts of euplasm found in their eggs, when compared with equivalent large eggs of meroists, such as the eggs of *Notonecta* and *Apis*, which can develop long germ bands according to their higher amount of euplasm, synthesized by nurse cells. Furthermore, panoistic embryos need extended time for embryogenesis, since they have to synthesize much more on their zygotic genes, as do embryos which are derived from meroistic ovaries and which can make use of large amounts of maternal resources, derived from nurse cells.

The third method seems to be the most powerful, since tandemly repeated rDNA genes are found in all eukaryotic cells studied so far and additional multiplication of these genes in one of the three subtypes described above should be easy.

The oocytes of *Locusta* seem to resemble those insects in which the amount of rDNA has been raised in all tissues, including germ line cells (Kunz, 1983; Schäfer and Kunz, 1985, 1987). The total number of rRNA genes is about 3300 per haploid genome, approximately 14-fold the amount found in *Drosophila* haploid genomes of somatic cells (i.e. 235 genes, Tartof, 1973). Thus, during previtellogenic growth of oocytes, *Locusta* can activate about 13 200 rRNA genes. This activation of repeated rRNA genes becomes visible as pearl threads in living oocytes (Perlschnurstränge) (Kunz, 1967a,b) (Figure 3.68). A spread preparation of such rDNA genes is shown in Figure 3.71a.

About 120 polymerase molecules, together with their transcripts are coupled simultaneously to each transcription unit in a polarized manner. Furthermore, in chromatin spread preparations all nascent rRNA transcription units show terminal 'balls' at their ends, which contrasts to nascent mRNA transcripts which have no balls (compare Figures 3.71 and 3.67). Mougey *et al.* (1993) have analysed these terminal balls as a 5′ external transcribed spacer processing complex by comparing spread preparations of cloned genes of *Xenopus laevis*, each lacking different parts of the 5′ region. In *Locusta*, spacer sequences are slightly longer than transcription units and measure 11 kb, whereas the whole repeat measures 18 kb (Figure 3.71b) (Schäfer and Kunz, 1985, 1987). The same authors verified that oocyte tandem repeats do not differ in amount and composition from NORs of somatic cells. Thus, extrachromosomal rDNA amplification does not exist in *Locusta*. Furthermore, the non-transcribed spacer (NTS) does not contain a series of repeats as found in NTS of amphibia (see below). However, there

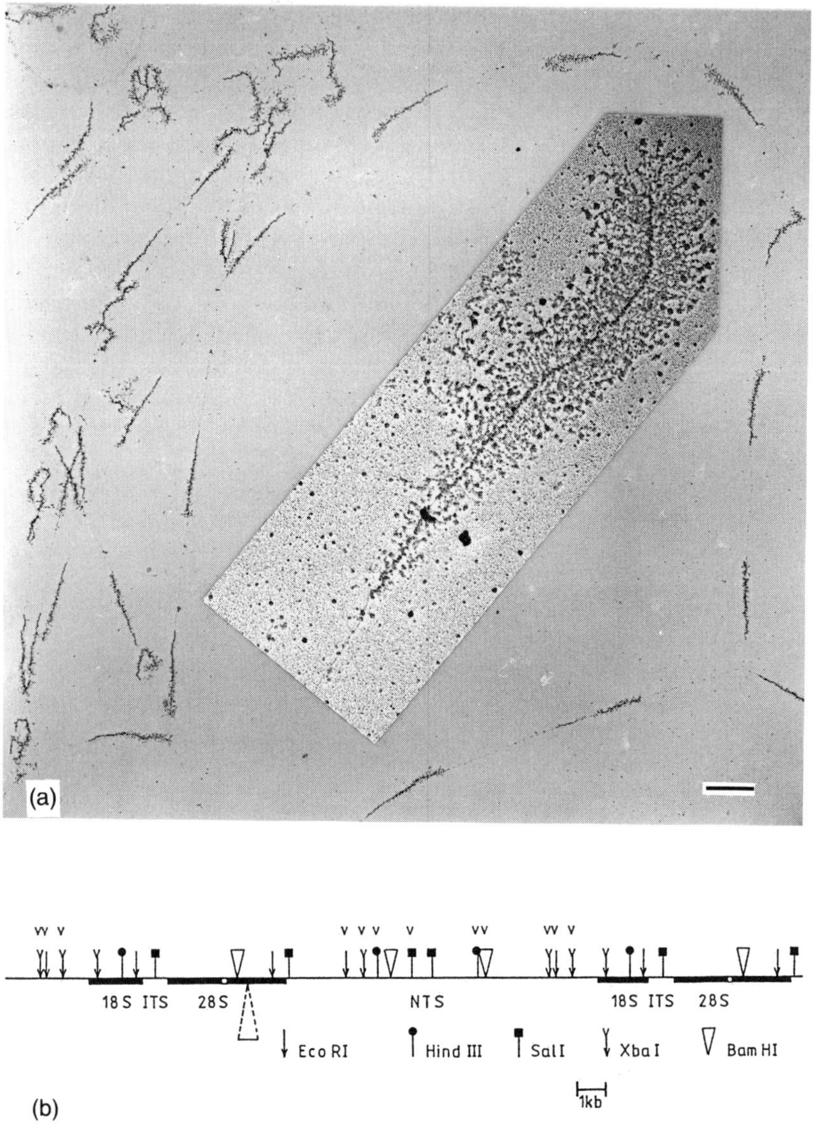

Figure 3.71 (a) Spread preparation of actively transcribing tandem repeats of NOR from germinal vesicles of *Locusta migratoria*. Geometry of tandem repeats is consistent with biochemical data of (b). Bar represents 2 μm. The inset shows a fully activated transcription unit. Bar represents 1 μm. (Courtesy of J. Gottanka, University of Erlangen-Nürnberg, Germany.) (b) Map of rDNA repeat (18 kb) unit from *L. migratoria*. NTS region is constant and measures about 11 kb. However, NTS region is variable at special sites, marked by arrowheads above symbols of restriction enzymes. Repeats within NTS are not found. (From Schäfer and Kunz, 1987; with permission of M. Schäfer and Academic Press.)

are some heterogeneous regions within the NTS, the function of which remains unknown (Schäfer and Kunz, 1985, 1987).

The high amount of repeated genes found in *Locusta* makes us ask how they are regulated in somatic tissues, in which much lower amounts of ribosomes are needed. The question is still unanswered and it is open to speculation whether all rRNA genes are activated with reduced rates of synthesis of each gene (i.e. reduced amount of polymerases per gene) or if subsets of rDNA genes are shut off, while others transcribe with full activity. One speculative method is discussed below. Another mode occurs in the regulation of 5S rRNA; however, regulation of gene activity may differ from gene to gene and/or from organism to organism (Davidson, 1986).

The second, possible way to enhance rDNA gene activity during oogenesis may be verified in those species that have developed germ line specific chromatin. This chromatin is organized in special chromosomes, often called E-chromosomes, which are active during oogenesis but become eliminated in somatic tissues during the early steps in embryogenesis, while somatic (S-) chromosomes are active. However, panoistic ovaries, in which E-chromosomes occur, are not known. Furthermore, in *Wachtliella* (Diptera), a species with polytrophic meroistic ovaries and possessing E-chromosomes, the nucleolus organisor is associated with an S-chromosome, the nucleolus of which becomes active only in nurse cells and not in oocytes where E-chromosomes are active and S-chromosomes are inactive (Kunz *et al.*, 1970). In a recent study on the oogenesis of another gall midge, *Aphidoletes aphidimyza* (Cecidomyiidae), Gruzova and coworkers showed, by ^{3}H-uridine incorporation studies, that S-chromosomes of oocytes are actively transcribing during the late stages of vitellogenesis, while E-chromosomes transcribe in all stages to low amounts. However, about 99% of transcription occurs in nurse cells. Silver impregnation studies give a strong indication for active NOR in E-chromosomes, showing similar intensity to the NOR regions of nurse cells, while S-chromosomes are poorly impregnated and their NORs seem to be silent. This shows that the germinal vesicle assembles NOR proteins to a certain extent but the quality of RNA synthesis must be shown for instance by hybridization experiments before a clear answer can be given (Batalova and Gruzova, 1987; Gruzova *et al.*, 1987, 1989; Gruzova, 1989; Batalova *et al.*, 1990; Gruzova and Batalova, 1990).

The third mode differs from the earlier ones, since rDNA is tandemly repeated at the NOR regions as in normal somatic cells, and is amplified extrachromosomally. Extrachromosomal gene amplification means that this special gene is no longer aligned to a centromere region of a chromosome and these genes are lost during the next division. This 'trick' allows the organism to circumvent the restriction to the 4C status during meiotic prophase.

Although extrachromosomal nucleoli are found in species of several phylogenetic groups, such as teleosts (Baumeister, 1973), echiurids (Dawid and Brown, 1970; Das, 1976; Kidder, 1976), and molluscs (Brown and Dawid, 1968), most information comes from two groups, the Amphibia and the Insecta. We will focus first on the experiments done with Amphibia, showing the general pattern of rDNA amplification combined with some special aspects typical for this group. Thereafter, we will focus on insect rDNA amplification and special adaptations.

How does extrachromosomal DNA come into being? The favoured model today is that of the rolling circle, first developed by Gilbert and Dressler for virus replication in 1968 and proposed for extrachromosomal rDNA replication of frog oocytes. *Xenopus*, as are other amphibia, is known to have amplified, extrachromosomal rDNA (Brown and David, 1968; Gall, 1968, 1969), besides tandemly repeated genes at the nucleolus organizer region. The rolling circle hypothesis was developed by Hourcade *et al.* (1973), who demonstrated lariat structures in spread preparations. Further confirmation was given by Rochaix *et al.* (1974) by a combination of pulse-label experiments with spread preparations. As shown (Figure 3.66b), an excised ring molecule bearing at least one repeat unit (i.e. a non-transcribed spacer and a transcription unit) will be opened at one strand and DNA polymerase will synthesize unidirectionally along the circle without termination. This mode results in an 'endless' growing linear double helix and a double helical ring, both annealed by the replication fork together with the synthesizing DNA polymerase. This extrachromosomal replication occurs during pachytene stages in amphibia. Non-transcribed spacer sequences are often of identical length in extrachromosomal nucleoli, whereas transcribed units of intrachromosomal nucleoli are separated by nontranscribed spacers of heterogeneous length. This finding supports the idea that extrachromosomal nucleoli are derived from only one tandem repeat out of the NOR by a rolling circle mechanism and not from large parts or even the whole NOR (Wellauer *et al.*, 1976a,b; Scheer *et al.*, 1977).

In Amphibia, large, linear tandem repeats are not found as intermediate molecules, as proposed by the above model. They seem to circularize spontaneously by an as yet unknown mechanism forming large ring molecules, consisting of many identical tandem repeats. The number of extrachromosomal nucleoli varies between 500 and 2500 in oocytes of the same *Xenopus* female, and the number of rDNA genes varies between 500 and 11 000 per extra nucleolus; however, the total amount of amplified rDNA genes is relatively constant and is in the range of 2.5×10^6 genes, which is equivalent to 30–35 pg (Perkowska *et al.*, 1968; Thiébaud, 1979; Davidson, 1986).

Transcription of extrachromosomal rDNA starts early in meiotic prophase but remains low: 0.01–0.5% the rate of fully active extrachromosomal nucleoli during mid-vitellogenesis as measured in spread preparations of *Triturus alpestris* oocytes. In contrast, in vitellogenic stages about 95% of extrachromosomal rDNA is fully active (Scheer *et al.*, 1976a).

Extrachromosomal rDNA transcription seems to be regulated in several ways, in amphibia as well as in insects (Figure 3.69): totally silent extrachromosomal rings are beaded, the knobs of which seem to be nucleosomes and which shortens the rings to about 40% of its unbeaded and fully active status. Furthermore, part of the inactive rings shows higher ordered structures, which shortens the rings to about 8% of active rings (Scheer *et al.*, 1976a; Scheer and Zentgraf, 1978; Trendelenburg and McKinnell, 1979). Rings can be activated in part or fully. Furthermore, each gene seems to be regulated autonomously as is suggested by different amounts of transcribing rRNA polymerases on tandem repeats of the same ring (Scheer *et al.*, 1976a).

In insects, extrachromosomal multiple nucleoli are proposed for a variety of species, evidence for which came from morphological, histochemical and biochemical investigations (Bier, 1965b, 1968; Lima de Faria *et al.*, 1968; Allen and Cave, 1969, 1972; Gall *et al.*, 1969; Lima de Faria, 1974; Gall and Rochaix, 1974; Cave, 1982) (Tables 3.1, 3.3 and 3.4) (data from Cave, 1982; and Büning, unpublished). However, spread perparations, which were introduced by Miller and Beatty (1969), and which give a direct view on structure and synthetic capacity of extranucleoli are investigated in only four orders: Orthoptera: Gryllidae: *Acheta domesticus*; Coleoptera, Adephaga: Dytiscidae: *Dytiscus marginalis* (both analysed by Trendelenberg, Scheer and coworkers); Siphonaptera: Ceratophyllidae: *Ceratophyllus* sp., *Nosopsyllus fasciatus* (Figure 3.69) and *Hystrichopsylla talpae* (Büning, unpublished), and in different species of *Tipula* (Figure 3.68b,c) (Diptera, Tipulidae) (Büning and Domazer-Schmidl, unpublished).

The rise of extrachromosomal bodies in female germ cells was first studied in *Dytiscus marginalis* by Giardina (1901). Although he could not know the composition and function of that body at this time, he noted that the body arises very early, during germ cell divisions. The body rises in only one cell, the future oocyte, whereas the sister cells lack such bodies. However, in other species like *Tipula lateralis* (Diptera: Tipulidae), similar bodies rise in the two innermost cells after the second circle of synchronized sister cell mitoses and are transmitted to only these two cells, the later oocyte and its oldest sister cell, which transforms to a nurse cell, as do the rest of the 16 cell cluster. The DNA body is associated with the heterochromatic sex chromosome. During anaphase of each mitotic cycle, the body remains associated to only one chromatid and reaches only one sibling (Bayreuther, 1952). In fleas (Siphonaptera), extrachromosomal DNA is also found (Table 3.4) (Bayreuther, 1957; Büning and Sohst, 1988, 1989), irrespective of which type of ovaries, panoistic or meroistic, will be developed. In *Hystrichopsylla talpae*, a flea with clusters of 31 nurse cells and one oocyte (see section 3.6.2), all siblings will develop extrachromosomal DNA (Figure 3.60b). Thus, in insects, we can conclude that extrachromosomal rDNA replication occurs long before meiosis, during oogonial mitoses and, when

Tables 3.1 – 3.4 Synopsis of ovary characters of insects

Table 3.1

				Somatic cell characters														
				Gross morphology							Germarium				Follicle cells			
Genus	S	O	No 1	2	3	4	5	6	7	8	9	10	11	12	13	14	15	T
Machilis		A	1	7 c	≈5	1		+						–			–	P
Pretobius		A	2	8 c													–	P
Lepisma		Z	3	5 c	≈10	1			+		+			–	+		–	P
Thermobia +1		Z	4	3 c	few	1			+		+						–	P
Habrophlebia		Ep	5	>100 c	≈4	1–2	+		+		+			–			–	TM
Cloëon		Ep	6	c			+		+		+			–			–	TM
Ephemerella		Ep	7	c	1–2	≈2		+	+		+			–	–		–	TM
Siphlonurus		Ep	8	c	≈4	1–2			+	–	+			–			–	TM
Aeschna		Od	9	c	≈10			+	+		+			–	+		–	P
Anax		Od	10	c	≈20		+	+	+		+			–	+		–	P
Cordulia		Od	11				+	+	+		+			–			–	P
Libellula		Od	12				+	+	+	–	+			–			–	P
Sympetrum		Od	13	≈275 c	≈15	1–2		+	+	+	+			–	+		–	P
Nemoura		Pl	14	≈70 c			+	–	–		+			–			–	P
Perla		Pl	15	≈130 c	3–6	1–2	–	–	–		+			–			–	P
Haploembia		Em	16	5 c	≈10	1–2		+	+	–	+			–			–	P
Bacillus		Ph	17	≈36 c	≈16	1–2	–		+		+			–			–	P
Diapheromera		Ph	18	c	≈10	1–2			+		+			–			–	P
Sipyloidea		Ph	19	≈15 c	≈10	1			+		+			–			–	P
Carausius		Ph	20	c	≈6	1			+		+			–			–	P
Isophya		Or	21	≈23 b	5–6	1	+	+	+	+	+			–			–	P
Tachycines		Or	22	≈45 b	≈10	1–2	+	+	+					–	+		–	P
Locusta		Or	23		≈15	1–2		+									–	P
Gryllus		Or	24	≈150 r	≈15	1–3			+	–	+						–	P
Galloisiana		Gr	25	14 c	≈10	1–2			+					–			–	P
Kalotermes	Is	Dc	26	7 b	≈50	≈6		+	+	+	+			–				P
Odontotermes	Is	Dc	27	70–3200 be					+	–					+			P
Blatta + 2		Dc	28	8 b	≈20			+	+	+	+			–				P
Periplaneta		Dc	29	8 b	≈30	1–2		+	+	+	+			–	+			P
Anisolabis		De	30	5 c			+	+	+	–	+						+	PM
Forficula		De	31	≈30 c	≈4	1	+	+	+	–	+			–	–		+	PM

Table 3.1 continued

			Stem cells				Cystocyte cluster								Germ cell characters				
No	16	17	18	19	20	21	22	23	24	25	26	27	28	29	30	31	32	33	34
1																			
2																			
3																			
4																			
5	L		1																
6	L	0	1	–	56				+			+	–	+	+	–	–	+	–
7	L		1	–		2		≈10	+	–	–		–			–	–	+*	–
8	L	0	1	–	186	5	1	≈35	+	–	–		–			–	–	+*	–
9	L,I	f	1	–	–	–	–	–	–	–	–	–	–	–	–	–	–	–	–
10	L, I	f	1	–	–	–	–	–	–	–	–	–	–	–	–	–	–	–	–
11	L, I	f	1	–	–	–	–	–	–	–	–	–	–	–	–	–	–	–	–
12	L, I	f	1	–	–	–	–	–	–	–	–	–	–	–	–	–	–	–	–
13	L, I	f	1	–	–	–	–	–	–	–	–	–	–	–	–	–	–	–	–
14	L	f	1	–	≈16	16	16	–	–	+	+	s	+	+	+	–	–	+*	–
15	L	f	1													–	–	–	–
16	L, I	f	1	–	–	–	–	–	–	–	–	–	–	–	–	–	–	–	–
17	L, I	f*	1	–	–	–	–	–	–	–	–	–	–	–	–	–	–	–	–
18	L, I	f*	1	–	–	–	–	–	–	–	–	–	–	–	–	–	–	–	–
19	L, I	f*	1	–	–	–	–	–	–	–	–	–	–	–	–	–	–	–	–
20	L, I	f*	1	–	–	–	–	–	–	–	–	–	–	–	–	–	–	–	–
21	L, I	0	1	–	–	–	–	–	–	–	–	–	–	–	–	–	–	–	–
22	L, I	f	1							–	–	–	–	–	–	–	–	–	–
23	L, I	f	1							–	–	–	–	–	–	–	–	–	–
24	L, I	f	1		–*	–	–	–	–	–	–	–	–	–	–	–	–	–	–
25	I															–	–	–	–
26	I	m	u							–	–	–	–	–	–	–	–	–	–
27	I															–	–	–	–
28	E, L, I	f	1													–	–	–	–
29	L, I	f	1													–	–	–	–
30	L, I			+	8	4	1	1	–	+		s	+	+	+	–	–	+*	–
31	L, I	f	1	–	2	–	1	1						+		–	–	+	–

Table 3.1 continued

No.	Germ cell characters Nurse cells 35	36	37	38	39	40	41	42	43	44	45	46	Oocyte 47	48	49	50	51	52	T	References
1																			P	1
2																			P	2
3																			P	2–6
4																			P	7,8
5			–	–		–	–	–	–	+	–	–	–	+			+		TM	9
6	–	–	–	–		–	–	–	–	+							+		TM	9
7	–	–								+									TM	10–12
8	–	–	–	–		–	–	–	–	+									TM	9
9	–	–	–	–	–	–	–	–	–	–	–	–	–	+		+	+	–	P	6, 13
10	–	–	–	–	–	–	–	–	–	–	–	–	–	+		+	+	–	P	14
11	–	–	–	–	–	–	–	–	–	–	–	–	–	+	+	+	+	–	P	15
12	–	–	–	–	–	–	–	–	–	–	–	–	–	+	+				P	9, 16
13	–	–	–	–	–	–	–	–	–	–	–	–	–	+		+	+	–	P	17
14	–	–	–	–	–	–	–	–	–	–	–	–	–						P	6,18
15	–	–	–	–	–	–	–	–											P	19
16	–	–	–	–	–	–	–	–	–	–	–	–	–	+			+	–	P	9
17	–	–	–	–	–	–	–	–											P	20, 21
18	–	–	–	–	–	–	–	–											P	22
19	–	–	–	–	–	–	–	–				–	–	+					P	23
20	–	–	–	–	–	–	–	–						+					P	20, 24, 25
21	–	–	–	–	–	–	–	–				–	–	+		+	+	–	P	5
22	–	–	–	–	–	–	–	–				–	–	+		+	+	–	P	26
23	–	–	–	–	–	–	–	–				–	–	+	+		+	–	P	27–31
24	–	–	–	–	–	–	–	–				+	–	+	+	+	+	–	P	6, 32–39
25	–	–	–	–	–	–	–	–					–	+					P	40
26	–	–	–	–	–	–	–	–				–	–	+	+	+	+	–	P	41–45
27	–	–	–	–	–	–	–	–											P	45–47
28	–	–	–	–	–	–	–	–				–	–	+					P	5, 48, 49, 56
29	–	–	–	–	–	–	–	–				–	–	+		+	+		P	50–52, 57
30	–	–	–	–		+	–		–	–	–	–	+						PM	53
31	–	–	–	–		+	–		–	–	–	–	+	–	–	+		–	PM	5, 54, 55

Abbreviations of the column headings: S = suborders or families; O = order; No = consecutive numbers of species; 1–52 = number of characters; T = ovariole typus.

Abbreviations of families, suborders and orders: A = Archaeognatha; Ad = Adephaga; Al = Aleyrodidae; An = Anoplura; Ap = Aphididae; Bo = Boreidae; Cc = Coccidae; Ci = Cicadiformes; Co = Coleoptera; Cr = Corydalidae; Cy = Cyclorrhapha; Dc = Dictyoptera; De = Dermaptera; Di = Diptera; Em = Embioptera; Ep = Ephemeroptera; Gr = Grylloblattaria; He = Hemiptera; Ht = Heteroptera; Hy = Hymenoptera; Is = Isoptera; Le = Lepidoptera; Ma = Mallophaga; Mc = Mecoptera; Me = Megaloptera; Ne = Neuroptera; Nm = Nematocera; Od = Odonata; Or = Orthoptera; Pa = Panorpidae; Ph = Phasmida; Pl = Plecoptera; Po = Polyphaga; Ps = Psocoptera; Pt = Phthiraptera; Py = Psyllidae; Ra = Raphidioptera; Si = Siphonaptera; Sl = Sialidae; St = Strepsiptera; Tr = Trichoptera; Z = Zygentoma; Zo = Zoraptera.

Characters of somatic cells (1–15): gross morphology, 1–4; outer epithelia, 5–8; germarial region, 9–12; growth region, 13–15. 1 = number of ovarioles per ovary; n* = numerous and new, can be built during reproductive phase; 2 = architecture of ovary (= ovariole alignment): c = comb like (= pectinate), i.e. ovarioles insert serially to the lateral oviduct; b = bunchy, ovarioles insert to an end-standing calyx; be = bunchy, but calyx is elongated; r = ramified, lateral oviduct is ramified; 3 = amount of previtellogenic follicles/ovariole; –* = only one follicle exists/ovariole; 4 = amount of vitellogenic follicles/ovariole; * = cyclic growth; 5 = ovary envelope; 6 = ovariole envelope; +* = reduced; 7 = terminal filament; 8 = transverse septum; +* = in some species; 9 = inner sheath; +* = these somatic cells serve as follicle cells; 10 = apical cells; 11 = interstitial cells: + = found regularly in amounts equal to amounts of inner sheath cells; ± = only few cells; * = only the subset of border cells, passing through nurse cell chamber; 12 = prefollicular tissue (= continuous layer of

continued from previous page
somatic cells at the base of germarial region, showing high mitotic activity); ± = reduced prefollicular tissue; 13 = amitotic division of follicle cell nuclei; +* = deep division furrow; 14 = intercellular bridges of follicle cells; 15 = differentiated cells of nurse cell chambers; +* = these cells arise from the former inner sheath of the germarium which has now transformed into the tropharium.
Characters of germ cells and their progeny (16–52): 16 = main periods in which germarial characters (17–30) occur: G = characters occur in gonadal tubes, not in ovarioles, i.e. ovarioles are devoid of germaria from the very beginning; E = characters occur during embryonal phases; B = germ cell clusters occur freely in the body cavity; L = characters are found in larval ovarioles, not in ovarioles of pupae or imagos; P = characters are found in pupal, not in imaginal ovaries; I = characters occur in imaginal ovaries.
Stem cell characters: 17–19; cystocyte characters: 20–30; nurse cell characters: 31–42; oocyte characters: 43–52, germinal vesicle 46–52. 17 = amount of stem cells/ovariole; f = few; f* = few, but no sign of mitotic activity; m = many; 0 = no stem cells; 18 = stem cell mitoses: 1 = the amount of stem cell mitoses is limited; u = unlimited; 19 = stem cells and their progeny are temporarily combined to a stem cell cluster; 20 = Cystocyte cluster; numbers announce amount of combined cystocytes; c = constant; v = variable; –* = some 2-cell clusters in gonad anlagen which degenerate later on; 21 = secondary detachment to subclusters; numbers indicate amount of subclusters; 22 = amount of oocytes/cluster; c = constant; +s = somatic cells fuse to oocyte; 23 = amount of nurse cells/cluster; c = constant; v = variable; +s = somatic cells fuse to nurse cells; 24 = linear architecture of clusters; 25 = clusters ramified regularly after the 2^n-rule; 26 = clusters ramified irregularly; 27 = timing of cystocyte divisions: s = synchronous; s* = synchronous or metasynchronous in single clusters or subclusters; 28 = rosette formation; 29 = fusome; 30 = polyfusome; 31 = trophic core; 32 = masses of microtubules in parallel arrangement; ± = reduced; 33 = permanent intercellular bridges; +* = those will be closed, at which subclusters detach; 34 = cystocyte membrane reduction: p = partly reduced, only among cystocytes of same subcluster; t = total reduction; 35 = central syncytium; 36 = tapetum cells; 37 = gradient in nurse cell chambers which become apparent by decrease or increase of nuclei volumes and other structural modifications; 38 = apical mitoses; 39 = meiosis: + = in all presumptive nurse cells; +* = only in some; 40 = endomitoses; 41 = amitosis; ± some binucleate cells; 42 = rDNA amplification: u = under-replication, o = over-replication; 43 = nutritive appendix; 44 = nutritive cord; 45 = accessory nuclei; 46 = extrachromosomal rDNA amplication; 47 = karyosome; 48 = lampbrush chromosomes; +* = only in early phases of previtellogenic growth; 49 = RNA synthesis of germinal vesicle; +* = only in special phases; 50 = endobody (proteinaceous, globular aggregates in germinal vesicles); 51 = nuclear bodies (RNA-rich globular aggregates in germinal vesicles); 52 = nuclear lamina (proteinaceous sheaths, surrounding karyosome); + = character is found; – = character is not found; * = see footnote, according to character number; no sign = not known.

T = ovariole type: P = panoistic ovariole; PM = polytrophic meroistic ovariole; TM = telotrophic meroistic ovariole; M* = reduced polytrophic meroistic type.

1: Weyda, 1989; 2: Birket-Smith, 1974; 3: Heymons, 1897; 4: Jura, 1972; 5: Büning, unpublished; 6: Gross, 1903; 7: Klag, 1977, 1978, 1982; 8: Szlarzewicz, 1989; 9: Gottanka, 1992; 10: Mazzini and Gaino, 1988; 11: Soldan, 1979; 12: Soldan, 1981; 13: Beams and Kessel, 1969; 14: McGill, 1905; 15: Halkka, 1980; 16: Kessel and Beams, 1969; 17: Matsuzaki, 1971; 18: Gottanka and Büning, 1990; 19: Rosciszewska, 1976; 20: Pehani, 1925; 21: Taddei *et al.*, 1992; 22: Marshall, 1908; 23: Pijnacker, 1967; 24: Pijnaker, 1966; 25: Pijnacker and Godeke, 1984; 26 Von Kraft, 1960a,b; 27: Kunz, 1967a,b; 28: Kunz, 1969; 29: Bier *et al.*, 1967; 30: Bier *et al.*, 1969; 31: Schäfer and Kunz, 1987; 32: Buchner, 1909; 33: Netzel, 1965; 34 Favard-Sereno and Durand, 1963; 35: Lima de Faria *et al.*, 1968; 36: Allen and Cave, 1969; 37: Alan and Cave, 1972; 38: Cave, 1982; 39: Dennis and Bradley, 1989; 40: Matsuzaki *et al.*, 1979; 41: Truckenbrodt, 1966; 42: Truckenbrodt, 1970; 43: Grandi *et al.*, 1988; 44: Grandi, 1988; 45: Grandi, 1990; 46: Truckenbrodt and Amelung, 1986; 47: Banerjee, 1964; 48: Heymons, 1891; 49: Cave, 1976; 50: Bonhag and Arnold, 1961; 52: Anderson, 1964; 53: Yamauchi and Yoshitake, 1982; 54: Brauns, 1912; 55: Engels, 1969; 56: Matsuda, 1976; 57: Bonhag, 1956.

Table 3.2

				Somatic cell characters															
				Gross morphology								Germarium				Follicle cells			
Genus	S	O	No	1	2	3	4	5	6	7	8	9	10	11	12	13	14	15	T
Zorotypus		Zo	1	4–6	b	≈10	1–2												P
Peripsocus + 1		Ps	2	4–5	c	≈8	1–3		+	+	+	+	+	–	–	+*		+	PM
Eomenacanthus	Ma	Pt	3	5	c	2–3	1		+	+					–	+		+	PM
Pediculus	An	Pt	4	3–5	b	2–5	1–3	–		+	+*	+			–	+		+	PM
Psylla	Py	He	5	40–100	b	1	1	–	–	+		+	+	–	±	–		+*	TM
Aleurodes	Al	He	6	≈15	b	1	1	–	–	+	+	+	+	–	±	–		+*	TM
Porphyrophora	Cc	He	7	n*	be	–*	1			–									TM
Pseudococcus	Cc	He	8	n*	be	–*	1	–	–	–		+	+	–	±	–		+*	TM
Icerya	Cc	He	9	n*	be	–*	1			–									TM
Aspidiotus	Cc	He	10	n*	be	–*	1			–		+	+	–	±	–			TM
Lecanium	Cc	He	11	n*	be	–*	1			–									TM
Drepanosiphum	Ap	He	12	5, 6	b	1–2	1	–	–	+	+	+		–	±	–		+*	TM
Metopolophium	Ap	He	13	5	b	1–2	1	–	–	+	+	+		–	±	–			TM
Macrosiphum	Ap	He	14	6	b	1–2	1	–	–	+	+	+	+	–	±	–			TM
Aphis	Ap	He	15	5	b	1–2	1	–	–	+		+		–	±	–			TM
Bothrogonia	Ci	He	16	12	b	3–4	2			+	+	+			+			+*	TM
Cercopis	Ci	He	17	9	b					+	+	+	+						TM
Typhlocyba	Ci	He	18	6	b	2–4	1–2			+	+	+		–	+				TM
Gerris	Ht	He	19	4	b	≈7					+					+		+*	TM
Saldula	Ht	He	20															+*	TM
Notonecta	Ht	He	21			≈4	2			+	+							+*	TM
Sphaerodema	Ht	He	22	5	b	≈5	5*	+	+	+	+				+	+		+*	TM
Cimex	Ht	He	23												+	+		+*	TM
Rhodnius	Ht	He	24	7	b	2–3	1	+	+	+	+	+			+	+		+*	TM
Stenodema	Ht	He	25		b	4	1								+	+		+*	TM
Oncopeltus	Ht	He	26	7	b	≈6	≈3		+	+	+			–	+	+		+*	TM
Dysdercus	Ht	He	27	7	b	≈7	7*	–	+	+	+	+		–	+	+		+*	TM
Pyrrhocoris	Ht	He	28	7	b	4–6	4–6*		+	+	+	+		–	+	+		+*	TM
Eurygaster	Ht	He	29	7	b														TM

References: 1: Zaffagnini, 1969; 2: Büning and Sohst, 1990; 3: Bilinski and Jankowska, 1987; 4: Ries, 1932; 5: Ries and van Weel, 1934; 6: Gross, 1906; 7: Büning and Beyrich, unpublished; 8: Witlaczil, 1885; 9: Büning and Nowak, unpublished; 10: Magakyan *et al.*, 1976; 11: Emeis, 1915; 12: Büning, unpublished; 13: Hughes-Schrader, 1925; 14: Hughes-Schrader, 1927; 15: Ksiazkiewicz, 1980; 16: Büning, 1985; 17: Blackman, 1978; 18: Matsuzaki, 1975; 19: Ksiazkiewicz-Kapralska, 1985; 20 Ksiazkiewicz-Kapralska, 1989a,b; 21: Eschenberg and Dunlap, 1966; 22: Choi and Nagl, 1976; 23: Choi and Nagl, 1977a,b,c; 24: Hwang and Choi, 1989; 25: Bilinski *et al.*, 1990; 26: Stebbings, 1988; 27: Sahai, 1975; 28: Jawale and Ranade, 1990; 29: Davis, 1956; 30: Vanderberg, 1963; 31: Huebner and Anderson, 1972a,b,c; 32: Lutz and Huebner, 1980, 1981; 33: Huebner, 1984a,b; 34: Wightman, 1973; 35: Bonhag and Wick, 1953; 36: Wick and Bonhag, 1955; 37: Bonhag, 1955a,b; 38: Schreiner, 1977a,b; 39: Brunt, 1970; 40: Brunt, 1971; 41 Dittmann *et al.*, 1984; 42: Masner, 1968; 43: Mays, 1969; 44: Mays, 1972; 45: King and Büning, 1985; 46: Gross, 1901; 47: Kozhanova *et al.*, 1976; 48: Matsuda, 1976; 49: Winter *et al.*, 1977.

Table 3.2 continued

No	16	17	Stem cells 18	19	20	21	Cystocyte cluster 22	23	24	25	26	27	28	29	Germ cell characters 30	31	32	33	34
1																			
2	L, I	f	1	–	4	–	1	3		2	–	s	+	+	+	–	–	+	–
3		f			8	–	1	7		3						–	–	+	–
4		f			8	–	1	7				s				–	–		–
5	L, I	0	1		≈310		≈35	≈285	–	≈8	+	s	+	+	+	–	–	–	t
6	L, I	0	1		64	–	8	56	–	6	–	s	+	+	+	+	±	–	–
7	G, L	0	u		16	–	1	15	–	4			+			+	+		–
8	G, L	0	u		8	–	1	7	–	3			+			+	+		
9	G, L	0	u		8	–	1	7	–	3			+			+	+		
10	G, L	0	u		4		1	3		2			+			+	+	–	–
11	G, L	0	u		4		1	3		2			+			+	+		
12	G, L	0	1		32	–	16	16	–	5	–	s	+	–	–	+	+	–	–
13	G, L	0	1		32	–	8	24	–	5			+	–	–	+	+	–	–
14	G, L	0	1		32	–	11	21	–	5			+	–	–	+	+	–	–
15	G, L	0	1		32											+			
16	G, L	0	1				c		–							+			–
17	G, L	0	1		v		c	v	–		+			+		+	+	+	–
18	G, L	0	1				c		–							+	+		–
19	G, L	0	1		v		c	v			+					+	+		–
20	G, L	0	1		v		c	v			+					+	+	+	–
21	G, L	0	1		v		c	v			+					+	+		
22	G, L	0	1		v		c	v			+					+	+		
23	G, L	0	1		c		c	c								+	+		
24	G, L	0	1		v		c	v	–	+	+		+			+	+		
25	G, L	0	1		c		c	c								+	+		
26	G, L	0	1		v		c	v			+					+	+		
27	G, L	0	1		v		c	v			+					+	+		
28	G, L	0	1		v		c	v	–		+		+			+	+		
29	G, L	0	1		v		c	v	–		+		+			+	+		

No	Germ cell characters Nurse cells 35	36	37	38	39	40	41	42	43	44	45	46	Oocyte 47	48	49	50	51	52	T	References
1																			P	1, 48
2	–	–	–	–	+	+	–		–	–									PM	2
3	–	–	–	–		+			+	–	+								PM	3,4
4	–	–	–	–	+	+	–			–	+			+		+	+		PM	4–6
5	+	–	–	–	+	–	–												TM	7, 8
6	–	–	–	–	+	–	–		–	+			–						TM	9
7	–	–	–	–						+									TM	10
8	–	–	–	–						+			+						TM	11, 12
9	–	–	–	–						+									TM	13, 14
10	–	–	–	–	+	+	–		–	+									TM	15, 16
11	–	–	–	–	+					+									TM	11
12	–	–	–	–		+	–		–	+									TM	16
13	–	–	–	–		+	–		–	+									TM	16
14	–	–	–	–		+	–		–	+			+						TM	16
15	–	–	–	–		+	–		–	+			+						TM	17
16	–	–	+		+	+	+		–	+									TM	18
17	–	–	+	+	+					+									TM	19, 20
18	–	–								+									TM	12
19	–	–	+	+	+	+	+			+			+			+	+		TM	21–24
20	–	–	+	+			–		–	+									TM	25
21	–	–	+	+						+									TM	26
22	–	–	+	+						+									TM	27, 28
23	–	–	+	–						+									TM	29
24	–	–	+	+						+							+		TM	30–33
25	–	–	–	–						+									TM	12, 34
26	–	–	+	+						+							+		TM	35–38
27	–	–	+	+		+			u,o	+					+	+	+		TM	12, 39–41, 49
28	–	–	+	+		+				+					+				TM	42–46
29	–	–	+	+		+				+									TM	46, 47

Table 3.3

Genus	S	O	No	1	2	3	4	5	6	7	8	9	10	11	12	13	14	15	T
					Somatic cell characters Gross morphology							Germarium				Follicle cells			
Elenchus		St	1		–	–	–	–	–	–	–	+*	–	–	–	–	–	–	M*
Dytiscus +2	Ad	Co	2		b					+								+	PM
Hygrobia	Ad	Co	3		b					+								+	PM
Gyrinus	Ad	Co	4	≈20	b					+						–		+	PM
Cicindela	Ad	Co	5	13	b			+	+	+								+	PM
Agonum +6	Ad	Co	6		b					+								+	PM
Carabus +6	Ad	Co	7		b					+				+		–		+	PM
Scaphidium	Po	Co	8	5					+	+		+		+	+	–		+*	TM
Cercyon +3	Po	Co	9	11	c				+	+		+		+	+	–		+*	TM
Creophilus +1	Po	Co	10	4	c				+	+		+		+	+	–		+*	TM
Silpha +5	Po	Co	11	12	c				+	+		+		+	+	–		+*	TM
Elmis	Po	Co	12	11					+	+		+		+	+	–		+*	TM
Sericus	Po	Co	13	≈30	r				+	+	–	+		+	+	–		+*	TM
Psilotrix	Po	Co	14	16					+	+		+		+	+	–		+*	TM
Dermestes	Po	Co	15						+	+		+		+	+	–		+*	TM
Hedobia	Po	Co	16	23					+	+		+		+	+	–		+*	TM
Aphodius +5	Po	Co	17	≈6				+	+	+	–	+		+	+	–		+*	TM
Coccinella +3	Po	Co	18	≈60	b			+	+	+	+	+		+	+	–		+*	TM
Ischnomera	Po	Co	19	29	b				+	+		+		+	+	–		+*	TM
Meloe +1	Po	Co	20	≈1000	be	1–2	1		+	+		+		+	+	–		+*	TM
Anaspis	Po	Co	21	12	b				+	+		+		+	+	–		+*	TM
Isomina	Po	Co	22	7					+	+		+		+	+	–		+*	TM
Tenebrio +2	Po	Co	23	12	b			+	+	+	+	+		+	+	–		+*	TM
Byturus	Po	Co	24	7	b				+	+		+		+	+	–		+*	TM
Uleiota	Po	Co	25	6	b				+	+	+	+		+	+	–		+*	TM
Gaurotes	Po	Co	26	9	b				+	+		+		+	+	–		+*	TM
Leptinotarsa +5	Po	Co	27	≈34	b	≈3	1	+	+	+		+		+	+	–		+*	TM
Bruchidius	Po	Co	28	6	b	≈5	2	+	+	+	+	+		+	+	–		+*	TM
Ips	Po	Co	29	2	b			+	+	+	+	+		+	+	–		+*	TM
Phyllobius	Po	Co	30	2	b	≈30	≈15	+	+	+	+	+		+	+	–		+*	TM
Raphidia		Ra	31	≈50	b				+	+				–	±			+*	TM
Sialis	Sl	Me	32	≈150	c	2–4	1	+	+*	+	–	+		–	±			+*	TM
Protohermes	Cr	Me	33	≈150	c	≈14	≈10	+	+	+	–			–	–				P
Chrysopa		Ne	34		c	≈14		+	+	+	–	+		–				+	PM
Eumantispa		Ne	35	≈160															PM
Hagenomya		Ne	36	10															PM
Asculaphus		Ne	37	10															PM
Protidricerus		Ne	38																PM
Sisyla		Ne	39	≈17	c	≈10	≈6			+				+				+	PM

Table 3.3 continued

No	16	17	Stem cells 18	19	20	21	Cystocyte cluster 22	23	24	25	26	27	28	29	Germ cell characters 30	31	32	33	34
1	B, L, P				≈128	+	1	≈127	–	7		s	+	+	+	–	–	–	p
2	L, P				16	–	1	15	–	4	–	s	+	+	+	–	–	+	–
3					16	–	1	15	–	4	–								
4	P				8	–	1	7	–	3	–	s	+	+	+	–	–	+	–
5	L, P, I				32	–	1	31	–	5	–	s							
6	L, P, I				16	–	1	15	–	4	–	s							
7	L, P, I				18–182	–	1	17–181	–	7	+	s	+	+	+	–	–	+	–
8	L, P, I		1		+	–	c	c	+							–	–	+	–
9	L, P, I		1		+	–	c	c	+		+					–	–	+	–
10	L, P, I		1		+	–	c	c	+		+	s*		+		–	–	+	–
11	L, P, I		1		+	–	c	c	+		+					–	–	–	p
12	L, P, I		1		+	–	c	c	+							–	–	+	–
13	L, P, I		1		+	–	c	c	+							–	–	+	–
14	L, P, I		1		+	–	c	c	+		+					–	–	–	p
15	L, P, I		1		+	–	c	c	+		+	s*		+	+	–	–		
16	L, P, I		1		+	–	c	c	+							–	–	+	–
17	L, P, I		1		+	–	c	c								–	–	–	t
18	L, P, I		1		+	–	c	c	+		+	s*				–	–	+	–
19	L, P, I		1		+	–	c	c	+		+					–	–	+	–
20	L, P, I		1		+	–	c	c	+		+					–	–	+	p
21	L, P, I		1		+	–	c	c	+							–	–	+	–
22	L, P, I		1		+	–	c	c	+		+					–	–	+	–
23	L, P, I		1		+	–	c	c	+							–	–	+	–
24	L, P, I		1		+	–	c	c	+							–	–	–	p
25	L, P, I		1		+	–	c	c	+							–	–	–	p
26	L, P, I		1		+	–	c	c	+		+					–	–	–	p
27	L, P, I		1		+	–	c	c	+							–	–	–	p
28	L, P, I		1		+	–	c	c	+		+	s*		+	+	–	–	–	t
29	L, P, I		1		+	–	c	c	+		+					–	–	–	t
30	L, P, I		1		+	–	c	c	+		+					–	–	–	t
31	L, P, I	f	1				c	c								–	–	p	p
32	L, P	f	1				c	c								–	–	p	p
33	L, P	f	1																
34	L, P, I				12–14	–	1	11–13	–	3	+								
35	L, P, I				12–14	–	1	11–13	–										
36	L, P, I				12–14	–	1	11–13	–										
37	L, P, I				12–14	–	1	11–13	–										
38	L, P, I				13–15	–	1	12–14	–										
39	L, P, I				12–14	–	1	11–13	–										

Table 3.3 continued

	Germ cell characters																			
	Nurse cells												Oocyte							
No	35	36	37	38	39	40	41	42	43	44	45	46	47	48	49	50	51	52	T	References
1		–	–	–	–	–	–	–	–	–	–								M*	1–3, 46
2		–		–	+	+	–		+	–	–	+	+						PM	4–6
3																			PM	6
4		–				+				–	–	+							PM	7–10
5				–					+	–									PM	11, 12
6				–	+				+	–									PM	13, 15
7		–		–	+	+	+		+	–									PM	14
8		–	+	–		+	+		–	+			+			+	+		TM	16
9		–	–	–		+			–	+			+			+	+		TM	16
10		–	+	–		+	–		–	+		+	+		+	+	+	+	TM	17–21
11		–	–	–		+	+		–	+			+			+	+	+	TM	13, 16
12		–	–	–		–			–	+			+			+	+	–	TM	16
13		–	–	–		+			–	+			+			+	+	–	TM	16
14		–	+	–		+			–	+			+			+	+	+	TM	16
15				–		+			–	+			+						TM	20
16		–	–	–		+			–	+			+			+	+	–	TM	16
17		–		–		+			–	+			+			+	+	+	TM	13, 16, 22
18		–	–	–		+	±		–	+			+			+	+		TM	23
19		–	+	–		+	±		–	+			+			+	+		TM	16
20		–		–		+			–	+			+			+	+		TM	16
21		–	+	–		+			–	+			+			+	+		TM	16
22		–		–		+	+		–	+			+						TM	16
23		–	+	–		+	+		–	+			+	+*	+	+	+	+	TM	16, 24–27
24		–	–	–		+	+		–	+			+			+	+		TM	16
25		–	–	–		+	+		–	+			+			+	+		TM	16
26		–	+	–		+	+		–	+			+			+	+		TM	16
27		–	+	–		+			–	+			+	+*	+	+	+		TM	16
28		–	–	–	+	+			–	+			+	+*	+	+	+		TM	13, 16, 30, 31
29		–	+	–		+	+		–	+			+						TM	13, 16
30		–	+	–		+			–	+			+	+*	+	+	+	+	TM	13, 16, 32–34
31	+	+	–	–		–	–		–	+				+					TM	35
32	+	+	–	–	+	–	–		–	+			–	+	+	+	+		TM	36, 37
33														+			+		P	3, 38
34				–	+	+	–		–	–	–	+	+	–	+	+	+	+	PM	39–42
35																			PM	43
36																			PM	43
37																			PM	44
38																			PM	44
39									+									+	PM	45

References: 1: Brues, 1903; 2: Baumert, 1957; 3: Büning, unpublished; 4: Giardina, 1901; 5: Günthert, 1910; 6: Urbani, 1970; 7: Matuszewski and Hoser, 1975; 8: Bilinski and Sklarzewicz, 1987; 9: Bilinski, 1986; 10: Hegner and Russel, 1916; 11: Bilinski and Jaglarz, 1987; 12: Jaglarz, 1988, 1992; 13: Gross, 1903; 14: Kern, 1912; 15: Ribbert and Weber, 1970; 16: Büning, 1979a,b; 17: Kloc, 1976; 18: Matuszewski and Kloc, 1976; 19: Kloc and Matuszewski, 1977; 20: Matuszewski *et al.*, 1985; 21: Matuszewski *et al.*, 1977; 22: Willimzik, 1930; 23: Kozhanova and Pasichnik, 1979; 24: Saling, 1907; 25: Schlottman and Bonhag, 1956; 26: Ullmann *et al.*, 1973; 27: Gruzova, 1979a,b; 28: Gerber *et al.*, 1978; 29: Ray and Ramamurty, 1979; 30: Büning, 1972; 31: Büning, 1978; 32: Bielanska-Osachowska, 1957; 33: Bielanska-Osuchowska, 1960; 34: Kurihara and Matsuzaki, 1989; 35: Büning, 1980; 36: Matsuzaki and Ando, 1977; 37: Büning, 1979c; 38: Matsuzaki *et al.*, 1985; 39: Gruzova, 1966; 40: Gruzova *et al.*, 1972; 41: Matsuzaki, 1978; 42: Rousset, 1978a,b; 43; Matsuzaki and Satoh, 1986; 44: Matsuzaki, 1987; 45: Matsuzaki and Enomoto, 1990; 46: Gu *et al.*, 1992.

Table 3.4

Genus	S	O	No	1	Somatic cell characters Gross morphology 2	3	4	5	6	7	8	Germarium 9	10	11	12	Follicle cells 13	14	15	T
Tenthredo +1		Hy	1							+								+	
Atalia +1		Hy	2	≈14	be	≈9	1	+	+	+								+	PM
Pimpla +2		Hy	3	≈13	b	≈4	≈5	+	+	+	−	+			−	−		+	PM
Habrobracon		Hy	4	2	b	≈11	≈10	+	+	+	−	+			−	−		+	PM
Formica		Hy	5	≈30	b			+	+	+	−	+			−	−	+	+	PM
Eciton		Hy	6	≈1300	b	≈25	1*	+	+	+	−							+	PM
Polistes		Hy	7	3	b									±					PM
Solenius		Hy	8																PM
Apis		Hy	9	≈180	b	≈8	≈5	+	+	+	−	+		±	−	−	+	+	PM
Platyphylax		Tr	10							+								+	PM
Parastenopsyche		Tr	11	≈130		≈4	≈4		+	+								+	PM
Pieris +1		Le	12	4	b			+	+	−		+	+		−			+	PM
Macrothylacia		Le	13	4	b			+	+			+			−			+	PM
Galleria		Le	14	4	b	≈17	≈17	+	+	−		+	+	±	−			+	PM
Bombyx		Le	15	4	b			+	+	−		+	+		−			+	PM
Hyalophora		Le	16	4	b			+	+	−		+	+		−			+	PM
Panorpa	Pa	Mc	17	≈12	c	≈8	1	+	+	+	−	+		+	−			+	PM
Boreus	Bo	Mc	18		c										−	−	+	−	P
Nosopsylla		Si	19	3–5	c				+	+					−	−		−	P
Hystrichopsylla		Si	20	≈4	c	≈9	1–3		+	+	+	+			−	−		−	PM
Stenoponia		Si	21																PM
Tipula +10	Nm	Di	22		be	1–2	1	+	+									+	PM
Psychoda	Nm	Di	23	≈60	be	1–2	1	+	+									+	PM
Culex +1	Nm	Di	24			1	1	+	+	+							+	+	PM
Chironomus	Nm	Di	25	≈100	be	1–3	1	+	+									+	PM
Sciara	Nm	Di	26	≈150	be	0–2	1	+	+									+	PM
Heteropeza	Nm	Di	27				1											+	PM
Mycophila	Nm	Di	28				1											+	PM
Miastor	Nm	Di	29				1											+	PM
Mikiola	Nm	Di	30				1											+	PM
Drosophila	Cy	Di	31	10–30	b	≈6	≈4	+	+	+	−	+	+	*	−	−		+	PM
Chryromya	Cy	Di	32	≈80	b			+	+	+								+	PM
Musca	Cy	Di	33	≈70	b	≈3	1–2	+	+	+				*				+	PM
Calliphora	Cy	Di	34	≈80	b	≈3	1	+	+	+				*				+	PM
Melophagus	Cy	Di	35	2	b	1–2	1	+	+	+								+	PM
Glossina	Cy	Di	36	2	b		1	+	+	+								+	PM
Hippobosca	Cy	Di	37	1	b		1											+	PM

Table 3.4 continued

		Stem cells				Cystocyte cluster	Germ cell characters											
No 16	17	18	19	20	21	22	23	24	25	26	27	28	29	30	31	32	33	34
1 P, I				32	–	1	31	–	5	–					–	+		–
2 P, I				20–84	–	1	19–83	–	4–6	+					–			–
3 P, I				32	–	1	31	–	5		+				–	+		–
4 P, I				32	–	1	31	–	5	–	+	+	+	+	–	+	+	–
5 P, I					–	1		–	5		+	+	+	+	–		+	–
6 P, I					–	1	>20	–							–			–
7 P, I															–			–
8 P, I				16	–	1	15	–	4	–	+	+	+	+	–			–
9 P, I				40–60	–	1	39–59	–	5	+	+	+	+	+	–	+	+	–
10 P, I		1		8	–	1	7	–	3	–	+	+			–			–
11 P, I				3–2	–	1	1–2	–	1	+	+				–	–	+	–
12 L, P		1		8	–	1	7	–	3	–					–			–
13 L, P		1		8	–	1	7	–	3	–	+	+	+	+	–			–
14 L, P		1		8	–	1	7	–	3	–	+	+	+	+	–			–
15 L, P		1		8	–	1	7	–	3	–	+	+	+	+	–			–
16 L, P		1		8	–	1	7	–	3	–	+	+	+	+	–			–
17 L, P				4	–	1	3	–	2		+	+	+	+	–	–	+	–
18 L, P								–							–	–	–	–
19 L, P								–							–	–	–	–
20 L, P				32	–	1	31	–	5	–					–	–	+	–
21 L, P				10–13	–	1	9–12	–	3	+					–	–		
22 E, L, P	0	1		16	–	1	15	–	4	–					–	–	+	–
23 E, L, P	0	1		16	–	1	15	–	4	–					–	–		–
24 E, L, P	0	1		8	–	1	7	–	3	–					–	–		–
25 E, L, P	0	1	?	2	–	1	1	–	1	–					–	–		–
26 E, L, P	0	1	?	2	–	1	1	–	1	–					–	–	+	–
27 E, L	0	1		2	–	1 + s	1 + s	–	1	–					–	–	+	–
28 E, L	0	1		2	–	1 + s	1 + s	–	1	–					–	–		–
29 E, L	0	1		2	–	1 + s	1 + s	–	1	–					–	–		–
30 E, L	0	1		2	–	1 + s	1 + s	–	1	–					–	–		–
31 P, I	≈2	u		16	–	1	15	–	4	–	+	+	+	+	–	–	+	–
32 P, I				16	–	1	15	–	4	–					–	–		–
33 P, I				16	–	1	15	–	4	–					–	–	+	–
34 P, I				16	–	1	15	–	4	–					–	–	+	–
35 P, I				8	–	1	7	–	3	–					–	–		–
36 P, I				16	–	1	15	–	4	–					–	–		–
37 P, I				3	–	1	2	–	4	–					–	–		–

Table 3.4 continued

	Germ cell characters																			
	Nurse cells										Oocyte									
No	35	36	37	38	39	40	41	42	43	44	45	46	47	48	49	50	51	52	T	References
1	–	–		–		+	–		+	–	+		+			+			PM	1, 2
2	–	–		–		+	–			–	+		+			+			PM	3, 4
3	–	–	+	–		+	–		+	–	+		+			+			PM	5, 6
4	–	–		–	+*	+	–		+	–	+								PM	7, 8
5	–	–		–		+	–			–	+								PM	9, 10
6	–	–		–		+	–			–									PM	11
7	–	–		–	+*	+	–				+								PM	12
8	–	–		–		+	–			–	+								PM	13
9	–	–	+	–	+*	+	–		+	–	+		+						PM	14–19
10	–	–		–	+	+	–			–									PM	20
11	–	–		–		+	–			–									PM	21
12	–	–		–	+	+	–			–									PM	22, 23
13	–	–		–		+	–			–									PM	24, 25
14	–	–	+	–		+	–			–									PM	26, 27
15	–	–		–	+	+	–			–									PM	23, 28–30
16	–	–		–		+	–			–					+*				PM	31–34
17	–	–		–	+	+	–			–	+		+			+	+		PM	35–38
18	–	–	–	–	–	–	–			–		+				+	+		P	38, 39
19	–	–	–	–	–	–	–			–		+			+	+	+		P	38, 40
20	–	–	–	–	–	–	–	o	–	–		+			+	+	+		PM	41, 42
21	–	–	–	–						–									PM	40, 43
22	–	–	–	–	+*	–	–			–		+			+				PM	38, 44–46
23	–	–	–	–			–			–									PM	47
24	–	–	–	–		+	–			–			+						PM	48–51
25	–	–	–	–		+	–			–			+						PM	52–54
26	–	–	–	–		+	–			–									PM	38, 55, 56
27	–	–	–	–			–			–									PM	57–61
28	–	–	–	–			–			–									PM	62, 63
29	–	–	–	–			–			–									PM	64, 65
30	–	–	–	–			–			–									PM	66
31	–	–	+	–	+*	+	–	u		–		–	+				+		PM	67–70, 84
32	–	–		–		+	–			–									PM	71
33	–	–	+	–		+	–			–			+		+*				PM	72–74
34	–	–	+	–		+	–			–			+						PM	38, 75–78
35	–	–		–		+	–			–							+	+	PM	79
36	–	–		–		+	–			–									PM	80–82
37	–	–		–		+	–			–									PM	83

References: 1: Bilinski, 1991a; 2: Bilinski, 1991b; 3: King and Fordy, 1970; 4: Hatakeyama *et al.*, 1990; 5: Meng, 1968; 6: Meng, 1970; 7: Cassidy and King, 1972; 8: Henschen, 1928; 9: Billen, 1985; 10: Bier, 1952; 11: Hagan, 1954a,b,c; 12: Marshall, 1907a; 13: Buchner, 1918; 14: Paulcke, 1900; 15: Hegner, 1915; 16: Bier, 1954; 17: Engels, 1966; 18: Engels, 1968; 19: Ramamurty and Engels, 1977; 20: Marshall, 1907b; 21: Matsuzaki, 1972; 22: Zick, 1911; 23: Grünberg, 1903; 24: Knaben, 1934; 25: Hirschler, 1942; 26: Duveau-Hagege *et al.*, 1963; 27: Gruzova *et al.*, 1990; 28: Machida, 1926; 29: Miya, 1959; 30: Miya, 1975; 31: King and Aggarwal, 1965; 32: Mandelbaum, 1980; 33: Telfer, 1975; 34: Cardoen *et al.*, 1990; 35: Grell, 1942; 36: Ramamurty, 1964a,b; 1967; 37: Gruzova, 1967; 38: Büning, unpublished; 39: Steiner, 1937; 40: Kunitzkaya, 1960; 41: Büning and Sohst, 1988; 42: Büning and Sohst, 1989, 43: King and Teasley, 1980; 44: Bauer, 1933; 45:: Bayreuther, 1952, 1956; 46: Lima de Faria and Moses, 1966; 47: Koch, 1929; 48: Roth and Porter, 1964; 49: Meola *et al.*, 1977; 50: Fiil, 1976, 1978; 51: Ramadan *et al.*, 1986a,b; 52: Balbiani, 1885; 53: Hasper, 1911; 54: Sachtleben, 1918; 55: Berry, 1941; 56: Gutzeit, 1985; 57: Reitberger, 1940; 58: Hautscheck, 1962; 59: Camenzind, 1966; 60: Madvahan, 1973; 61: Went *et al.*, 1978; 62: Schüpbach and Went, 1983; 63: Schüpbach and Camenzind, 1983; 64: Kahle, 1908; 65: Mahowald and Stoiber, 1974; 66: Matuszewski, 1968; 67: Painter and Reindorp, 1939; 68: Meyer, 1961; 69: King, 1970; 70: Wieschaus and Szabad, 1979; 71: Avancini and Pires do Prado, 1986; 72: Verhein, 1921; 73: Bier, 1963a,b; 1970; 74: Bier *et al.*, 1969; 75: Bier, 1957; 76: Cardoen *et al.*, 1986; 77: de Loof *et al.*, 1990; 78: Sahai, 1989; 79: Ulrich, 1963; 80: Hardenberg, 1929; 81: Hagan, 1951; 82: Huebner *et al.*, 1975; 83: Ansari and Murad, 1981; 84: Renkawitz-Pohl and Kunz, 1975.

combined with meroistic ovaries, its propagation following cluster mitoses is species specific. Specific, but yet unknown, interactions of the extra DNA body to chromosomes seem to decide if extra DNA is carried in equiproportional or disproportional amounts to the next sibling generation.

The rDNA nature of this extra-DNA was determined for cricket oocytes by *in situ* hybridization experiments using ^{3}H-labelled rRNA, which was synthesized from isolated satellite rDNA *in vitro* (Ullman *et al.*, 1973). According to the unique and characteristic structure of active rDNA in spread preparations of other species, it is commonly accepted that all these genes resemble rDNA genes.

The length of spacers, which is more or less constant in amphibian extranucleoli, shows greater variations in the extranucleoli of *Dytiscus marginalis* and *Acheta domesticus* (Trendelenburg, 1974; Trendelenburg *et al.*, 1976). Such variation is interpreted as a reflection of intrachromosomal spacer length diversity as has been shown to exist in Amphibia (Franke *et al.*, 1978; Davidson, 1986). Thus, a model is preferred in which rolling circles, presumably derived from different regions of the NO, are used to establish extrachromosomal nucleoli (Trendelenburg *et al.*, 1986). Recently, Tröster *et al.* (1990) analysed tandem repeats of NORs and extrachromosomal nucleoli of *Acheta domesticus* by restriction analysis of 16 genomic clones of a conventional DNA library and one clone from micro-dissected NORs. They found tissue (germ line versus soma) and sex-specific modifications in NTS regions of *Acheta* rDNA, which is different from the findings of Schäfer and Kunz (1985, 1987) on *Locusta* NOR regions. In *Acheta*, NTS regions are variable in their length, whereas coding sequences are stable. The pattern of length diversity of NTS found in NORs of somatic tissues is similar to the pattern of extrachromosomal amplified NTS. However, some tissue and sex-specific variations occur, the function of which remains speculative.

In *Acheta* and *Dytiscus* most extranucleoli contain only a few, 1–5, rDNA genes, whereas hundreds of rDNA genes are found in each amphibian extranucleolus. In fleas, we found similar compositions as in the other two insect species; however, larger ring molecules occur in greater amounts, bearing up to 50 rDNA genes. They were found especially in oocyte nuclei derived from late previtellogenic stages. Corresponding to the pattern of rDNA activation described for *Triturus alpestris* and *Rana pipiens* (Amphibia) (Scheer *et al.*, 1976a; Trendelenburg and McKinnell, 1979) we note in the flea ovary an increasing amount of rDNA activation during late stages of previtellogenic growth (Figure 3.69a–c). In the early to middle stages there is a large population of silent DNA rings which are beaded to higher order structures (Figure 3.69a). Later on, these rings become activated only in part, especially when they are large (Figure 3.69b). During the early and middle stages of vitellogenesis, nearly all rings are activated (Figure 3.69c). At these stages the number of large rings decreases. This activation pattern corresponds to the increase of cytoplasmic free ribosomes during previtellogenic growth, which can be estimated by the increasing amounts of ribosomal cytoplasm, which do not mix with cytoplasm synthesized previously (Figures 3.21, 3.43b and 3.44) (Büning and Sohst, 1988, 1989; Büning, 1990, and unpublished).

Meanwhile, evidence has come from molecular and genetic work that the NTS region is engaged in regulating rDNA gene activity. In spread preparations, the spacers between transcription units appear smooth (i.e. no particles are associated with the DNA) especially when the spreading medium is enriched with detergents. In other preparations, in which no or low amounts of detergents are used, some aggregates are associated, from which short transcripts can sometimes emerge (Franke *et al.*, 1978). Thus, at least some of these aggregates are RNA polymerases, transcribing spacer sequences, the products of which may dissociate and/or become degraded at once. Molecular analysis studies of cloned rDNA gave additional support to this idea. Different constructs of transcription units and aligned spacer sequences derived from *Xenopus laevis*, are injected into living oocyte nuclei of *Xenopus borealis*. This bioassay allows transcription of injected genes since oocyte nuclei are provided with vast amounts of RNA polymerases, nucleotides and other substances needed for successful transcription. Using restriction enzymes and DNA sequencing methods, a vast pattern of deletions in spacer regions can be tested in order to elucidate transcription of rDNA genes by RNA polymerase I. The first spacer region analysed came from *X. laevis* (Figure 3.72) (Davidson, 1986 and references therein). The gene promoter region includes the nucleotides from –142 to +6, measured from the start of the transcript. Thereby, the initiation side includes 13 nucleotides (–7 to +6), needed for correct transcription of the rDNA gene. At position –27, which is occupied by the TATA-box in messenger genes, six T are found, possibly preventing a docking of polymerase II. The remaining region of the gene promoter houses a 42 bp sequence, which enhances transcription rates. This motive is found also in 22 repeats, which occur upstream of the gene promoter, including two spacer promoter sequences, which differ slightly from the gene promoter. Futher upstream, two sets of other repeated sequences are found, which differ strongly from the 42 nucleotide pair elements.

Spacer promoters as well as repeats function as binding sites and enhancers. This has been shown by analysis of spacer sequences of *X. laevis* and *X. borealis*, the spacers of which have only four repeats. In competition tests, *X. laevis* nucleoles are always superior in transcription to those of *X. borealis* (Labhart and Reeder, 1984; Reeder and Roan, 1985). To ensure correct transcription, a terminator site in front of the gene promoter cuts all spacer transcripts. The gene terminator at the 3′ end disrupts the gene product, the pre-rRNA. Nearby spacer promoters may catch the polymerases

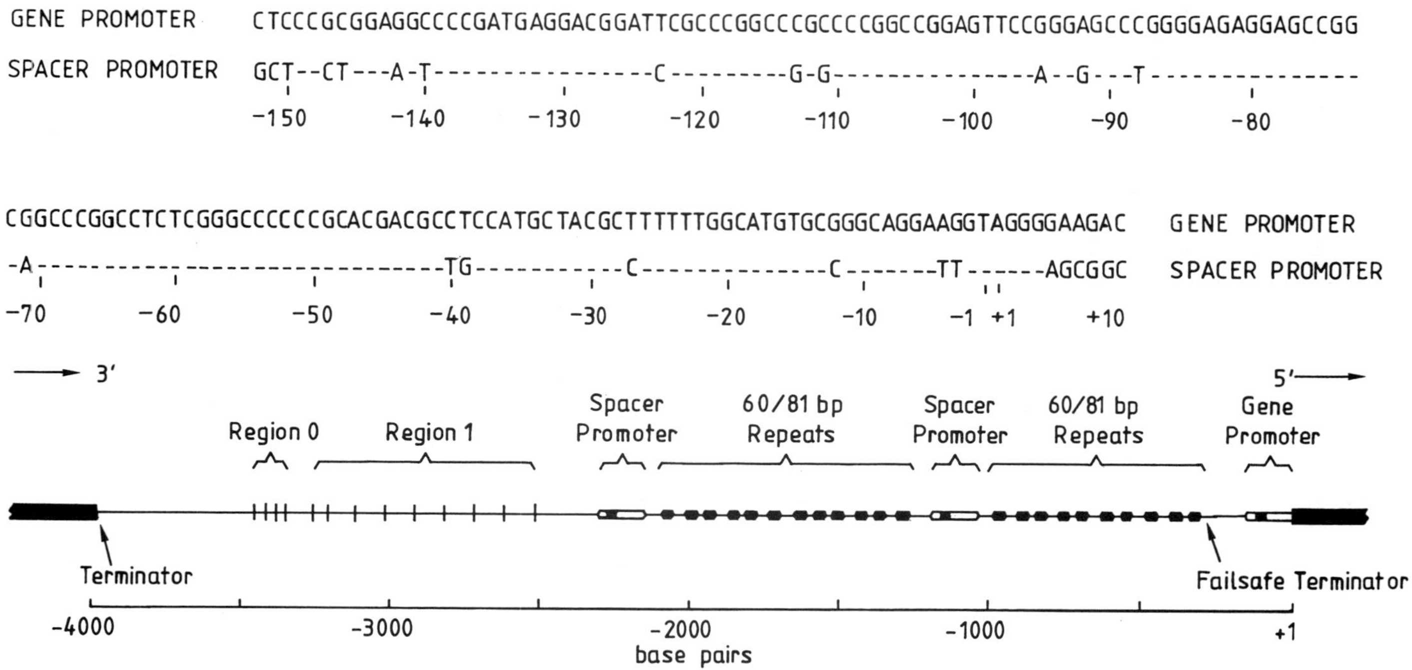

Figure 3.72 (a) Gene and spacer promoters of *Xenopus laevis* rDNA. Identical nucleotides are indicated by dashes. The 42 nucleotide pairs region reaches from -73 to -114. For further information see text. (b) Model of spacer region. Preceding and following transcription units are shown by heavy bars. The 42 nucleotide pairs repeat is shown by black squares, promoters are blank. Regions 0 and 1 contain other repeats. The fail safe terminator stops spacer transcription. For further information see text. (After Davidson, 1986; slightly modified.)

just terminated, and the whole spacer region seems to work as a trap for polymerases of class I.

Another mode of regulation in tandem repeats, which have already started rRNA synthesis, comes from careful spread preparations of *Acheta domesticus* NORs. Garlands occur, in which promoter regions of a given transcription unit are connected directly with terminator sites of the preceding repeat unit, whereas the included spacer region loops out (Trendelenburg, personal communication). Similar garlands are also known from spread preparations of *Dytiscus* extrachromosomal rDNA (Trendelenburg, 1974, his Figure 7c). Thus, one can speculate that RNA polymerases of type I do not enter the free pool of polymerases in the nuclear sap after transcription but remain connected to the DNA and can start again immediately by a 'short-circuit'.

Thus, one can postulate at least four, subsequent steps in rRNA synthesis:

(1) Nucleosomes have to be removed before transcription of rDNA is started.
(2) The transcription rate is first balanced by the pool of free RNA polymerases of type I and by the individual architecture of the spacer region.
(3) At termination of rDNA transcription, specifically bound RNA polymerases of type I will not leave the templates but remain bound to the next spacer region, which still works as a trap for free polymerases to maximize transcription rates.
(4) Short circuits would allow the transcription rates to be kept constant for each gene and at any level, as long as the short circuit exists.

(c) 5S rDNA genes

Ribosomes are built by equimolar amounts of 18S, 5.8S, 28S and 5S RNAs, together with a stock of some 70 ribosomal (r-)proteins. Whereas the former three rRNAs are synthesized in the NO region and derive by splicing from a long precursor molecule of about 40S, the pre-rRNA, 5S RNA is not integrated into the nucleolus. Furthermore, its genes are not transcribed by RNA polymerase I but by RNA polymerase III, which transcribes all short RNA classes, including tRNAs. Thus, synthesis of ribosomes needs a well balanced and finally equimolar amount of 5S RNA, which must be regulated during the process from synthesis to self-assembly. As the rRNA genes of the NO region, 5S RNA genes are also multiplied (about 160 genes per haploid genome) and their total number is correlated roughly 1:1 with the number of multiple rDNA genes of the nucleolus (about 200 genes per haploid genome) in *Drosophila melanogaster* (Ritossa and Spiegelman, 1965; Tartof and Perry, 1970; Procunier and Dunn, 1978; Jacobs-Lorena, 1980). In this species, all 5S RNA genes are clustered as tandem repeats on the second chromosome in the 56EF region, situated on the right arm (Ritossa and Spiegelman, 1965; Wimber and Steffensen, 1970). However, such a singular cluster is more the exception than the rule: in *Drosophila virilis* as well as in closely related species there are multiple clusters (Wimber and Wimber, 1977) and in the house cricket, *Acheta domesticus*, about 300 5S RNA genes are spread over all chromosomes at multiple sites (Benes and Cave, 1985). In insects, regulation of these genes follows at least in part the lines known from *Xenopus* 5S RNA genes, in which specific transcription factors (TFIII A, B, C) are needed to initiate a special cluster of oocyte-type 5S RNA genes, whereas somatic-type clusters of 5S RNA genes can be initiated alone by RNA polymerase III. Thus, in somatic cells, oocyte-type 5S RNA genes are silent and only the somatic-type cluster of about 400 genes transcribes short-living 5S RNAs, whereas in oocytes about 25 000 5S RNA genes, slightly differing in nucleotide sequence, will produce sufficient and stable copies for the 10^{12} ribosomes, stored in the amphibian egg (for reviews see Davidson, 1986; Alberts *et al.*, 1988; Amaldi *et al.*, 1989 and references therein). The typical oocyte 5S RNA is synthesized during previtellogenic stages before activation of extrachromosomal rDNA rings. The molecules are stored first in 7S ribonucleoprotein particles (Dixon and Ford, 1982a,b; Pieler and Erdmann, 1983). Each particle contains one molecule of 5S RNA and one molecule of TFIIIA, the 5S RNA typical transcription factor (Picard and Wegnez, 1979; Pelham and Brown, 1980; Honda and Roeder, 1980). In fully grown oocytes, the 7S particles are composed of two subsets, one composed as before, the other built by 5S RNA and the L5 r-protein. Both classes are transported via nuclear pores into the cytoplasm. A third class, a construct of 5S RNA and the La-protein, is found as a transient nuclear complex (Guddat *et al.*, 1990).

Such different gene sets of 5S RNA genes, oocyte specific and soma specific, are not found in *Drosophila* (Benes and Cave, 1985; Davidson, 1986). In *Acheta* two size classes of 5S DNA repeats exist, measuring 3.0 and 2.1 kb; the former includes about 95% of the total 5S DNA. However, their possible role in the regulation of transcription is unknown (Benes and Cave, 1985).

(d) Ribosomal proteins

Ribosome production of eukaryotic cells is regulated by coordinated production of ribosomal RNAs and some 70 r-proteins. When, during oocyte growth, the amount of rRNAs, including 5S RNAs, increases to more than 10 000-fold, ribosomal protein production also needs altered pathways to compensate. However, ribosomal protein synthesis is much more flexible than rRNA production, since it can make use of the regulation of its own mRNAs (rp-mRNA) and their subsequent translation. Thus, regulation of these, mostly single copy genes, must not differ generally from regulation of any other genes involved in the synthesis of other structural proteins, which are needed in large amounts during embryogenesis. In *Drosophila*, rp-mRNAs of two of the some 70 r-proteins were isolated and their synthesis and translation were recorded in oocytes and embryos (Al-Atia *et al.*, 1985; Kay and Jacobs-Lorena, 1985, 1987). Abundance of these mRNAs is high during vitellogenesis, in mature eggs and in the early and late stages of embryogenesis. However, their association with ribosomes and polysomes shows striking differences, indicating translational control during embryogenesis. Polysomes of this kind are found in high amounts only in oogenesis and late embryogenesis but not in early embryogenesis, where ribosomal production is absent. This parallelism of translation of rp-mRNA with rRNA production indicates connections in the control of both.

One of these r-protein genes (rp49) was analysed in transformed flies, bearing extra copies of this gene in order to test if each r-protein gene is controlled independently or not (Tamate *et al.*, 1990). In transgenic flies, the gene transcript was overrepresented while other r-protein genes were not affected. The rp-mRNA was associated with polysomes in the same pattern as noticed in untransformed flies, indicating an overproduction of the final gene product. This shows that the basis of coordinate r-protein gene expression lies in the intrinsic properties of r-protein genes, requiring a common regulatory factor rather than individual regulation of each r-protein gene.

Transposons are powerful probes to study the developmental processes; they become integrated into a genome, carrying a special gene, which is combined in an anti-sense direction with a heat-shock promoter. Such anti-sense inhibition experiments have been carried out with one of the rp-genes of *Drosophila* (Qian *et al.*, 1988, 1989). Expression of these anti-sense rp-genes during oogenesis results in the production of small, abnormal eggs; but its expression does not affect embryogenesis, larval development or the development of male adults. Temperature shift experiments clearly indicate that oogenesis is viable only at the beginning of vitellogenesis. The mechanisms of this inhibition are still unclear since measurable amounts of sense–anti-sense hybrids could not be detected. A pleiotropic effect is obvious. Total mRNA synthesis and/or storage are affected and transport from nurse cells into the oocyte seems blocked directly or indirectly by delay in nurse cells, which uncouples the programmes of RNA and/or protein synthesis between nurse cells and follicle cells, which may then synthesize egg shells too early, disrupting nurse cells from their oocyte.

(e) Differentiation of oocytes and nurse cells

The most effective way to circumvent the restriction to the 4C status of oocytes is the transformation of their sister cells to nurse cells, which can then multiply their genomes. This method is followed with great success among insects (see sections 3.6 and 3.7). Here, we focus on oocyte–nurse cell differentiation, a process which is apparently necessary, but is verified by unknown strategies. Common to all meroists is the starting point, the germ cell cluster. Another common feature is the orientation of nurse cells and oocytes: in all cases, the oocyte (or the oocytes) of one cluster is (are) orientated to the pedicel, whereas nurse cells of the same cluster are always orientated towards the terminal filament of the ovariole. In *Drosophila*, this polarization is caused among other genes by the gene *dicephalic*, the mutant allele of which disturbs the correct orientation of clusters in the ovariole in such a way that part of the nurse cells will support their oocyte from the distal end (Figure 3.49) (Lohs-Schardin, 1982; Sander *et al.*, 1985) (see section 3.3.2). The amount of euplasm introduced in the posterior part of the egg by these wrongly orientated nurse cells will decide the expression of the anterior structures on the posterior pole, including a second micropylar apparatus on the posterior pole of the egg, and a mirror image of anterior larval structures. We do not know the final gene product; however, we can speculate that this product may be associated with asymmetric intercellular communication between prospective oocytes and prefollicular cells, by which an intracellular polar orientation of oocytes along the axis of the ovariole may be established (Frey *et al.*,1984; Frey and Gutzeit, 1986; Bohrmann *et al.*, 1986c; Bohrmann and Sander, 1987). Such asymmetric cell to cell communication between prospective oocytes and prefollicular cells may be used for asymmetric growth of oocyte membranes, by which oocyte growth proceeds only in that part which is in contact with prefollicular tissues. Thus, all intercellular bridges will automatically be situated around the anterior pole, when oocyte growth proceeds. Such a theoretical model is in concordance with ultrastructural findings, which clearly show that branched clusters during cluster formation or immediately after cluster genesis are not yet orientated to the longitudinal axis of the ovariole from the beginning, but their nurse cells are orientated by chance, while oocytes keep a central position (Figure 3.72b) (King *et al.*, 1982; Büning and Sohst, 1988, 1990; Storto and King, 1989; Büning, unpublished) (for cluster genesis see section 3.6).

Another gene found in *Drosophila*, which is called *bicaudal-D*, produces mirror images of abdominal segments when dominant alleles are investigated. A recessive allele, *PA66*, produces 16 nurse cells in each follicle but no oocyte, when homozygous. This gene has been cloned and its wild type RNA accumulates only around oocyte nuclei from the beginning of previtellogenesis from stage 1 up to the end of pure previtellogenic growth at stage 7 egg chambers. About stage 8 chambers, from the beginning of vitellogenesis until its end in stage 11, this RNA is synthesized in increasing amounts in nurse cells, is transported to the oocyte and becomes concentrated at the anterior pole, to which nurse cells are connected via intercellular bridges. After vitellogenesis, the primary gene product is distributed uniformly in the egg (Figure 3.73) (Suter *et al.*, 1989). The

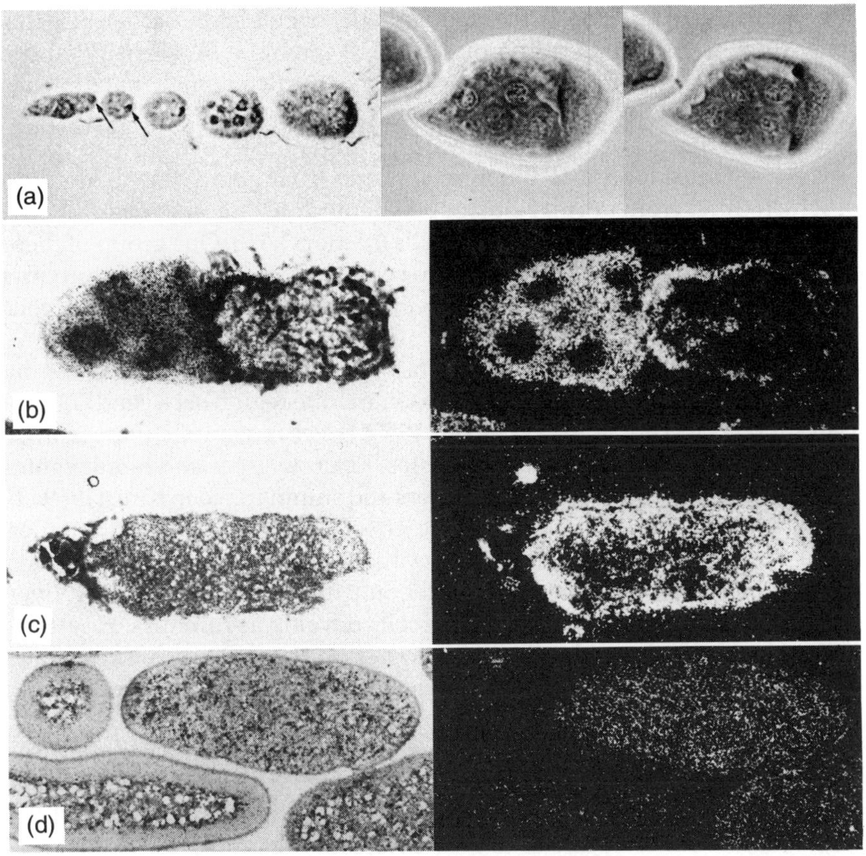

Figure 3.73 (a) The *bicaudal-D* gene product is distributed as a gradient in the *Drosophila* egg, when synthesized by the oocyte during early previtellogenesis (stage 1 and 2; arrows); whole mount preparation. (b) When nurse cell *bicaudal-D* genes are turned on during stage 10b–11, the morphogen is still distributed asymmetrically (hybridization with 35S-labelled RNA, derived from a *bic-D* specific DNA). (c) At stage 13, the *bic-D* RNA hybridizes uniformly in the entire oocyte. (d) In embryos, the *bic-D* RNA remains uniformly distributed and fades during cellular blastoderm formation. (From Suter *et al.*, 1989; with permissions of R. Steward and Cold Spring Harbour Press.)

predicted final gene product, a protein of 782 amino acid residues, shows similarities to the structural proteins of intermediate filaments, especially to the myosin heavy chain tail. Thus, the protein may play a role in the architecture of the oocyte cytoskeleton. The cytoskeleton itself can communicate with other cells via cell membrane proteins, which brings us back to the first hypothesis. However, this analysis clearly shows that in *Drosophila* germ cells, oocyte determination occurs very early and before previtellogenic growth is turned on. On the other hand, activation of *bicaudal-D* in oocytes seems to be necessary to maintain the oocyte status, otherwise this cell will be transformed into a nurse cell. It might be of interest that the onset of transcription of this gene occurs when fusomal material has disappeared from intercellular bridges.

When we discuss oocyte–nurse cell differentiation from the genetic point of view, we must look at all those *female sterile (fs)* mutants, the phenotypes of which differ from normal cluster behaviour in *Drosophila* regarding the determination of oocytes and nurse cells (section 3.6.2). One group of these mutants are tumour mutants, the germ cells of which can produce tumours in female gonads by aberrant division programmes of germarial stem cells and/or cystocytes (Figure 3.41).

The first tumour mutant, *fused (fu)*, was described and isolated by Bridges and mapped on the X-chromosome in the region between 17B6 and 17F1 (Bridges and Brehme, 1944; Green, 1953). Seven mutant alleles are known, the expression of which in homozygotes, heterozygotes and hemizygotes was investigated by King and coworkers and summarized in King (1969a,b; 1970). The mutant shows pleiotropic effects on wing venation, bristles, ocelli and maternal effects on embryogenesis during gastrulation and organogenesis, but mutant alleles differ only in tumour rates per day. Even in homozygous females, not all germ cells develop as tumours. However, if a germ cell decides to develop tumorously, mitotic activity of such clones does not stop. Each division is completed by cytokinesis; thus, the tumour contains single cells. The tumours do not spread into other tissues; hence follicle cells are not affected. Gene dosage effects and temperature sensitivity indicate that *fused* mutants produce an aberrant product. Meanwhile, the gene has been localized precisely on the X-chromosome (17C4-17D2 bands) and a set of 14 alleles is now available (Busson *et al.*, 1988). *Fused* belongs to the segment polarity class genes, responsible for the correct architecture of the posterior halves of thoracic and abdominal segments in embryos. Mutations show deletions of posterior halves and mirror images of anterior halves of segments (Nüsslein-Volhard and Wieschaus, 1980; Martinez-Arias, 1985; Perrimon and Mahowald, 1987). Molecular cloning has also been done (Mariol *et al.*, 1987) and the protein product is characterized as a putative serine/threonine kinase (Préat *et al.*, 1990).

The second *female sterile* mutation (*fes; or fs(2)B*) is located on the second chromosome and shows no effects on the morphology of males and females. Males are fertile, females sterile when homozygous. Follicle cells of tumorous follicles of these females show aberrant large carbohydrate deposits, not known from wild type follicular cells. Germ cells divide into interconnected cystocytes with mostly reduced cell numbers in each cluster. Since each ovariole contains several, subsequent chambers containing tumorous clusters in increasing amounts, King and coworkers came to the conclusion that cystocytes can sometimes complete cystocyte divisions. However, at least some of these subclusters can continue with mitotic activity, producing moderately growing tumours which do not spread into other organs (reviewed in King, 1970; Johnson and King, 1972). In old tumour chambers an apical basal gradient occurs, in which basal cells degenerate and adjacent cells develop into nurselike cells. Nurselike cells can develop polytene chromosomes. This development is temperature dependent: high temperatures (25°C) increase the amount of nurselike cells; chambers occur with synchrone developing nurselike cells, while others contain two or more differently developing nurselike cell populations, which may have developed from more than one stem cell or cystoblast (see section 3.5). Analysis of ovarioles of young females (2 days) compared with older females (9 days) clearly demonstrates that nurselike cells develop from formerly dividing cystocytes. Thus, the capacity of mutant cells to transform into nurse cells is not severely affected although oocyte determination is repressed to 2% in nurselike cell chambers. These few oocytes are capable of going through vitellogenic growth; however, only on two occasions were fertilized eggs laid, which died before hatching or died at the larval or pupal stages (Koch and King, 1964; King, 1970).

The *narrow* (*nw*) mutation, located at position 83 on the second chromosome (Bridges and Brehme, 1944), has pleiotropic effects in males and females but causes tumours only in the female germ line. Here, too, clusters of aberrant cystocytes are generated, which grow moderately. The tumour does not spread into other organs. Three alleles are known, one is dominant over the wild type (King, 1970).

In 1975, Gans *et al.* isolated two additional mutants, affecting germ cell divisions in females, producing tumours. These are named *female sterile (1) 1621* (equal to *sans fille (snf)* and to *liz* (see section 3.3.1) and *female sterile (1) 231*. The latter mutant, *fs231*, renamed *otu*, has been genetically and cytologically characterized by King and coworkers during the last 15 years (King *et al.*, 1978, 1986; King, 1979; Dabbs and King, 1980; King and Riley, 1982; Rasch *et al.*, 1984; Storto and King, 1987, 1988, 1989; reviewed by King and Storto, 1988). King's studies involved 17 mutant alleles of *otu*, all induced by ethyl methane sulphonate. These mutants were tested in their expression as homozygotes, hemizygotes and in all heteroallelic combinations. The resulting phenotypes could be grouped into three developmental classes: the first class (*quiescent: QUI*) is built by eight alleles, which are generally unable to generate mitotically active germ cells, resulting in ovarioles lacking chambers (Figure 3.74a).

The second class (*oncogenic: ONC*) consists of four alleles. Females homozygous for *ONC* alleles contain ovaries in which cystocytes continue to divide and rarely differentiate into either nurse cells or oocytes. Thus, hundreds or thousands of cells are generated, enclosed by a follicle epithelium (Figure 3.74a). The third class (*differentiated: DIF*) contains five alleles. Females carrying *DIF* alleles can form chambers containing nurse cell/oocyte syncytia or syncytia containing only nurse cells. In cases where nurse

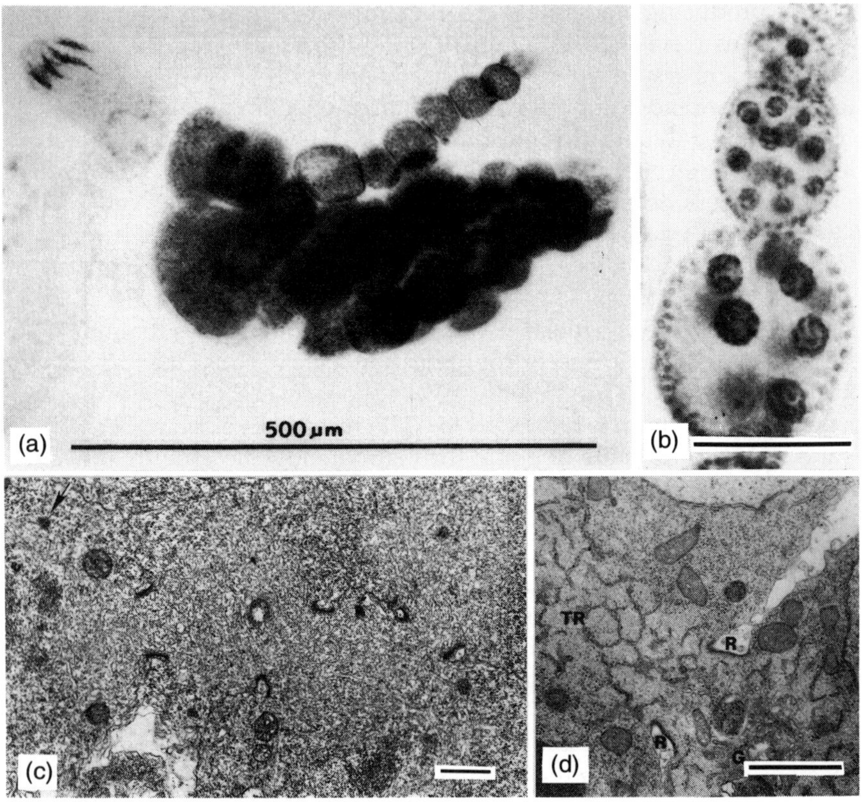

Figure 3.74 Feulgen-stained whole mount ovaries from *otu*^{16}/*otu*^{16} female. (a) Left ovary contains 11 chamberless ovarioles (*QUI* class) while the right contains six tumorous ovarioles (*ONC* class). (b) Part of ovariole from *otu*^{1}/*otu*^{11} ovary housing pseudonurse (PNC) cells (*DIF* class). Smallest follicle contains seven PNCs, medium follicle has 16 PNCs and large follicle contains 10 nurse cells with polytene chromosomes and one oocyte. Bar represents 100 µm. (From King *et al.*, 1986; with permission of R.C. King and Wiley and Sons.) (c) Wild type phenotype occurs in heterozygotes as shown here (*otu*^{7}/*otu*$^{+}$, *FM3*). Polyfusome stretches through three intercellular bridges. (d) Polyfusome and intercellular bridge from *otu*^{1}/*otu*^{1} female. Note network of ER in fusomal material. Bars represent 1 µm. (c and d from Storto and King, 1989; with permission of R.C. King and Wiley and Sons.)

cells/oocyte clusters form, the oocyte cannot complete development, but stops at a prior 12 stage. The number of nurse cells of each cluster is generally reduced with a mean of 10 instead of 15 as found in wild type clusters. Furthermore, in clusters developing into pure nurse cell clusters, the nurse cells show an increased and temperature-sensitive tendency to form polytene chromosomes. Thereby, these cells reach levels of about 8000 C, which is four times that found in the largest salivary gland chromosomes. As in polytene chromosomes of 'normal' somatic cells like salivary gland cells, polytene chromosomes of nurse cells show somatic pairing, which allows us to identify chromosome aberrations, such as inversions (King *et al.*, 1981; Rasch *et al.*, 1984). Since these cells do not deliver their synthesized products to an oocyte, they are named pseudonurse cells (PNCs) (Figure 3.74b).

The banding patterns of PNCs and polytene chromosomes derived from somatic tissues, like salivary glands, are strikingly similar (Sinha *et al.*, 1987; Heino, 1989). The same result was found earlier in *Anopheles stephensi* (Diptera, Culicidae), in which ovarian nurse cells contain polytene chromosomes (Redfern, 1981). However, constant banding patterns in the polytene chromosomes of different tissues of the same species are not always identical. Ribbert (1979) found a different banding pattern in temperature-induced polytene chromosomes of nurse cells compared with the polytene chromosomes of pupal trichogene cells in *Calliphora erythrocephala* (Diptera, Calliphoridae). Similarly, different banding patterns have been found in *Ceratitis capitata* (Diptera, Tephritidae) when comparing pupal trichogene and larval salivary gland polytene chromosomes (Bedo, 1987). The cause of these different patterns remains unknown.

Back to the role of the *otu* gene during oocyte nurse cell differentiation. King and coworkers have developed a model, which is in concordance with all the above genetic and morphological findings (King *et al.*, 1986; King and Storto, 1988; Storto and King, 1988, 1989): the *otu* gene will generate a product, which is used throughout oogenesis but in increasing amounts, from the germ cell stage, via the cystoblast, cystocytes, nurse cells and young oocyte, to the final oocyte (Figure 3.75).

The alleles of the *QUI* class fail to produce a functional product or the titre of functional products does not reach the level needed to transform into stem cells. Thus, those mutants are arrested and germ cells do not proliferate.

The alleles of the *ONC* class made sufficient amounts to enter the stem cell niveau but miss the niveau needed for oocyte–nurse cell determination. The pattern of cluster mitosis becomes altered: instead of 16 cells combined by incomplete cytokinesis, *ONC* class alleles show supernumery divisions, which, about half the time, complete cytokinesis. This results in single cells or small and irregular clusters of 2–4 cells, all distributed at random in the germarium. Generally, these alleles fail to form polyfusomes regularly and the morphology of fusomal material is altered (Storto and King, 1989). In wild type females, each germ cell cluster consists of 16 interconnected cells of which one develops as an oocyte and all the others as nurse cells. All cells are combined by a continuous polyfusome, the amount of which is highest in the two pro-oocytes, of which one develops finally into the oocyte. Polyfusomes of wild type females show a random array of small vesicles and fibrous material, whereas the rare polyfusomes and singular fusomes of *otu* mutants have a tubular network within the fibrous material, which may be derived from the remnants of the nuclear envelope of the preceding mitotic division (Figure 3.74c,d; Storto and King, 1989).

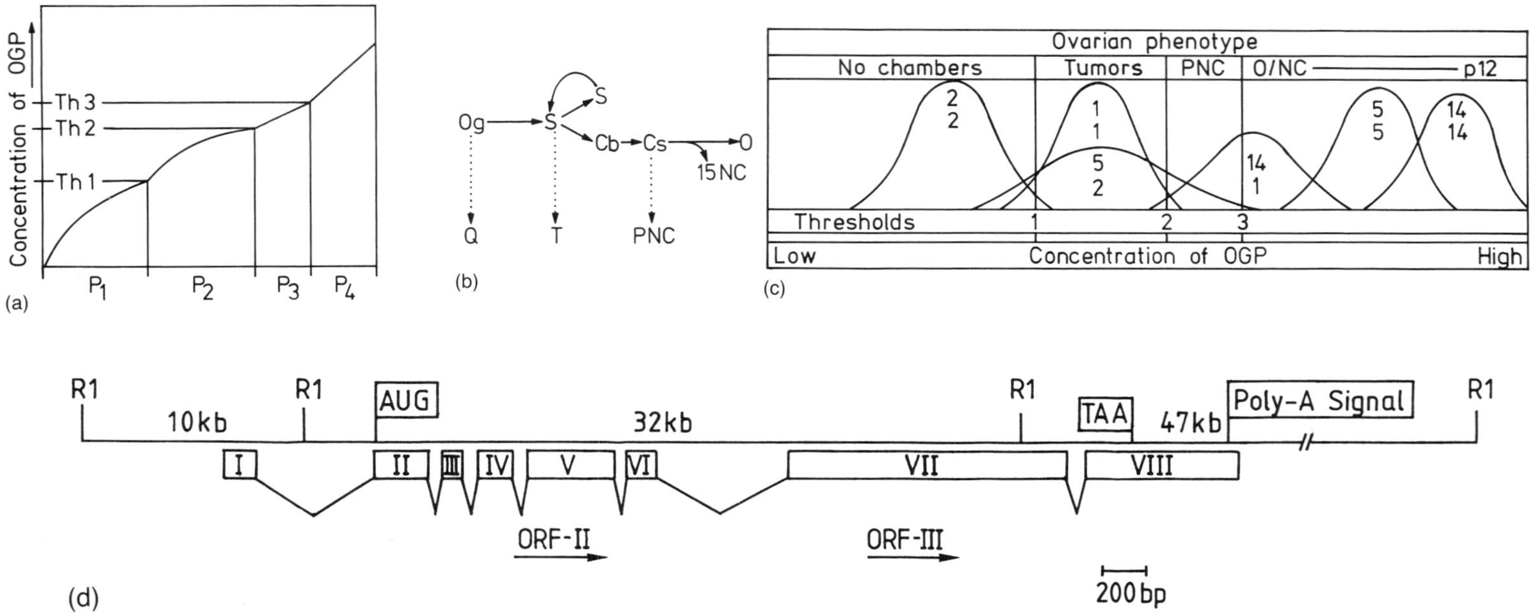

Figure 3.75 (a) The hypothesized phenocritical levels of an *otu* gene product (OPG) utilized during consecutive developmental periods (P1–P4) of oogenesis. Thresholds (Th) indicate concentration levels, which must be passed to reach the next developmental period. (b) The morphological events that happen during these stages: Q, quiescent stage, germ cells cannot divide; T, stem cells and/or cystoblasts cannot differentiate and undergo irregular mitoses, producing tumours; PCN, pseudonurse cells, cystoblasts cannot differentiate into oocytes; 15 NC, nurse cell–oocyte differentiation is regular, but development stops at stage 12. (c) Illustrates the hypothesized ranges of maximum levels of functionally active OPG produced by germ cell populations of some homozygotic and heteroallelic *otu* genotypes. (With permission of R.C. King and Company of Biologists.) (d) Structure of the *otu* gene. Genomic *Eco* RI restriction map, showing exons (open boxes, I–VIII) and introns (crooked lines). Translation initiates at AUG codon boxed above exon II and terminates at codon TAA boxed above exon VIII. Two additional reading frames (ORF-II; ORF-III) are indicated below. Direction of translation is the same as that of the main ORF. (After Steinauer *et al.*, 1989; with permission of American Society for Microbiology.)

The *DIF* class alleles produce higher amounts of the *otu* gene product by which the level for oocyte–nurse cell differentiation is reached. However, mutant products cannot communicate with other gene products sufficiently, thus, different phenotypes occur as PNC, and/or as reduced clusters, and/or as the failure to proceed in the final steps of oogenesis.

Complementation tests between heteroallelic combinations proceeded further in development than homozygous allelic combinations, when alleles were tested from the *QUI* class. In most other combinations, irrespective of their class, heteroallelic combinations are of medium expression in respect to their homozygous combinations. One allele of the *ONC* class, otu^{11}, differs in respect to heteroallelic combinations: when combined with one of eight other alleles (8, 1, 3, 9, 4, 7, 5, 14), oogenesis continues until stage 14, i.e. eggs are fully completed and are able to be fertilized. The expression is very high (97%) in the allelic combination otu^{11}/otu^{14}. This finding has suggested that *otu* may produce more than one gene product, which could be generated by differential splicing of a primary mRNA (King and Storto, 1988; Storto and King, 1988).

The *otu* gene has been cloned, sequenced and its structure has been published (Champe and Laird, 1989; Steinhauer *et al.*, 1989). The gene has seven introns spanning 4.5 kb (Figure 3.75d). With a long open reading frame in front, the putative translation start signal (TCGCC*AUG*) is found at position 1333 and the stop signal (*TAA*) occurs at position 4656. Two other, putative start signals are indicated, which would give rise to different mRNAs and different putative proteins. The main putative protein contains 811 amino acid residues, is highly hydrophilic and is enriched in proline (12%). Consensus sequences are not found with analysed sequences of, for instance ATP-binding sites, helices, RNA-binding sites, leucine zippers, or DNA binding sites like zinc-finger or GCN4 class motifs. Hydrophobic areas, as found in membrane spanning domains, are also missing. Two proteins from *Drosophila* ovary extracts were recognized by anti-*otu* antibodies with apparent molecular masses of 98 and 104 kDa (Steinhauer and Kalfayan, 1992). Sequence analysis of cDNA from *otu* suggests that these proteins are translated from two mRNAs generated by alternative splicing. The 104 kDa isoform appears to be required for efficient differentiation of germ cells. Immunostaining of *otu* proteins is restricted to the cytoplasm of germ cells. At stage 11, the oocytes will lose their staining while nurse cells keep staining up to the end of oogenesis. Thus, further analysis is needed to elucidate the role of *otu* products during oogenesis and other developmental processes, since the *otu* gene is not silent in males, or during embryogenesis (Mulligan *et al.*, 1988; Steinhauer *et al.*, 1989; see also section 3.3.1).

Whereas *otu* mutants show morphogenetic effects only in females, another tumour mutant, *benign (2) gonial cell neoplasm (bgcn)*, produces aberrant germ cell division programmes in both sexes (Figures 3.76 and 3.77) (Gateff,

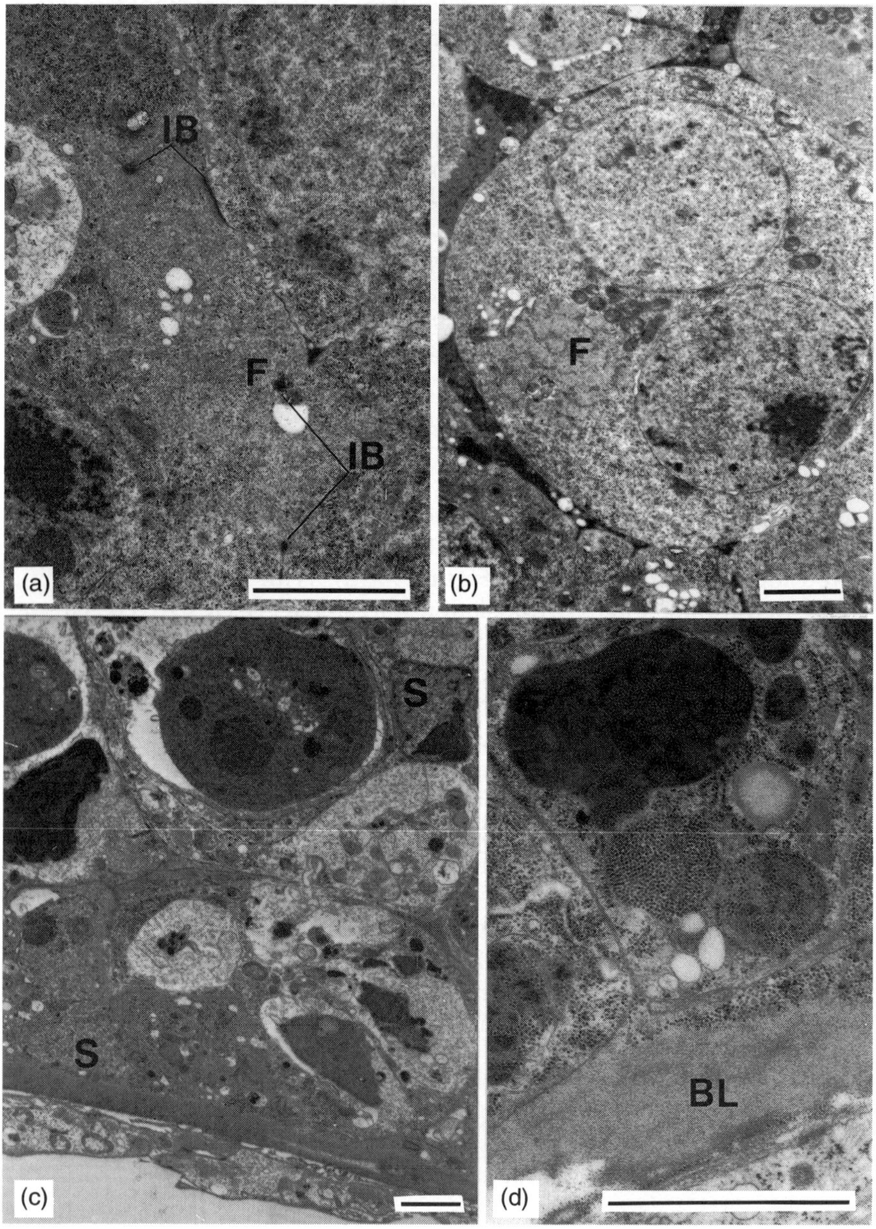
IB
F
IB
(a)
F
(b)
S
S
(c)
BL
(d)

1982; Gutzeit, 1990b). Similar to female-specific tumour mutants, the neoplastic growth of *bgcn* mutants is non-invasive and not lethal to adults. As the *otu* mutants, *bgcn* mutants show pleiotropic effects. In females, germ cells enter the phase of cystocyte division; however, clusters are aberrant in respect to the number of connected cystocytes, the number of which is reduced. Furthermore, intercellular bridge walls are reduced. Finally, cystocytes occur containing two nuclei. The ultrastructure of the fusomal material does not seem to be affected and polyfusomes are built (Figure 3.76a–d)(Büning, unpublished). Neither nurse cells nor oocytes will develop. The cells remain embryonic and do not enter previtellogenic or vitellogenic growth. A gradient is established from the apical to the basal parts of the germarium. Germ cell mitoses are found only in apical parts, while at the base the aberrant clusters degenerate. Thereby, the mitotic activity of apical clusters overrules the degeneration processes and death of posterior clusters and the aberrant female germaria grow throughout life (Figure 3.76a–d).

Similar effects are found in the male follicles of young adults (Figure 3.77a–f). However, at least mutant males stop the dividing programme of stem cells later on, whereas cystoblast mitoses seem not to be affected. Germ cell clusters will degenerate and cystocytes develop huge multilamellar bodies. Male germaria of older adults house huge masses of long, slender, curved rods which are clearly thicker and of higher contrast than microtubules. All germ cell membranes as well as all inner membranes, such as ER, Golgi vesicles and nuclear envelopes, become degraded, except the rods and some masses of chromatin (Figure 3.77f). Thus, spermiohistogenesis (Tokuyasu, 1974a–c; 1975a,b; Hackstein, 1987, Hackstein *et al.*, 1990) is highly aberrant. In the follicles of older testes no germ cells can be found, but the above mentioned rods and degrading chromatin are present. Thus, *bgcn* affects germ cells during cluster formation independently of sex differentiation, and results finally in germ cell death. The effect on cystocyte divisions mimics the effect of some *otu* mutant alleles, which produce similar

---

Figure 3.76 The tumour mutant *bgcn* of *Drosophila* influences life-time and morphological differentiation of germ cell clusters of both sexes in similar ways. (a–d) In females, oocyte–nurse cell differentiation does not happen; sibling cells remain undifferentiated and clusters undergo pycnosis at the bottom of each germarium. Germ cells cannot interact with somatic tissues and follicles are not formed. (a) Even in young siblings (apical part of germarium) intercellular bridge (IB) rims are shortened, polyfusome (F) looks normal. (b) In middle parts of germarium cell membrane reduction occurs and polyfusomes (F) can show network of interdigitating ER, as known also from *otu* mutants. (c) In basal parts, germ cell clusters become pycnotic and die, whereas somatic cells (S) of inner sheath survive. (d) However, somatic cells of basal area show huge masses of paracrystalline aggregations of globular particles, apparently bigger than ribosomes. Basal lamina (BL). Bars indicate 2 μm.

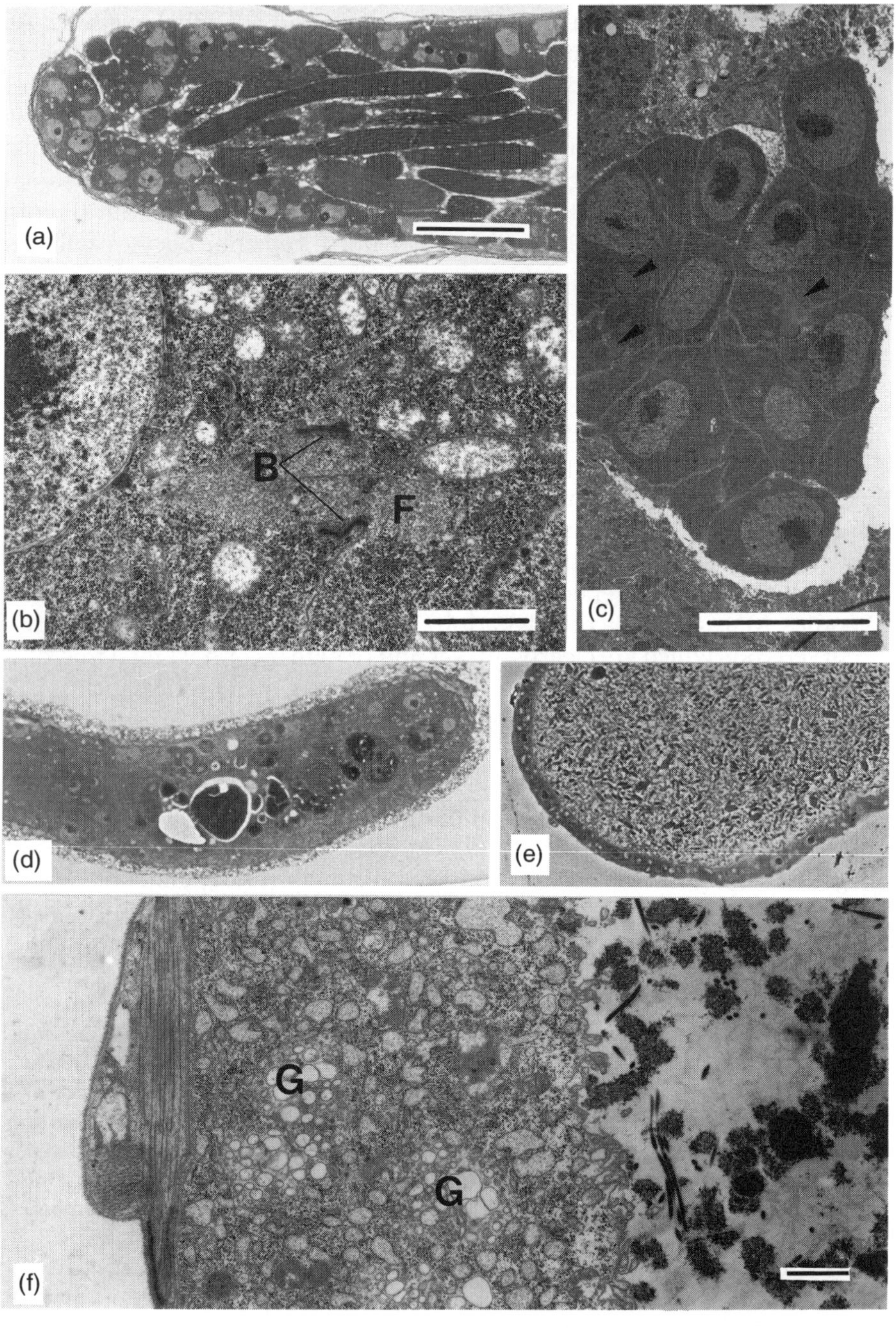
(a)
(b)
B
F
(c)
(d)
(e)
(f)
G
G

tumours as just described, but its product(s) are essential only in females, indicating that *otu* works only in females (i.e. after the primary events of sex decision). Even the ultrastructure of fusomal material seems similarly altered at least in female clusters. In both mutants the fusomal material is interrupted by some ER structures, producing a network (Figures 3.74d compared with 3.76b), which is not found in wild type female germ cell clusters (Storto and King, 1989; Büning, unpublished).

Recently, another gene was found, mutation of which affects both sexes in *Drosophila*, called *bag-of-marbles (bam)* (McKearin and Spradling, 1990). The normal programme of cystoblast divisions is disturbed in both sexes. The resulting cystocytes are unable to grow in either sex; thus spermiohistogenesis and oogenesis cannot proceed. Hybridization with anti-sense RNA in wild type ovaries shows specific activity of this gene during cluster mitosis and for a second time in nurse cells of stage 10B egg chambers. This RNA is shifted to oocytes and is found homogenously distributed in early embryos (about 1.5 h after fertilization). Shortly thereafter (about 3.5 h) the mRNA is degraded. This shows that the *bam* gene product is used at two points, in the cluster development of germ cells and in early cleavage. The *bam* gene protein shows similarities to the *otu* product, indicating that both genes belong to the same family.

All the above mutants show phenotypes which can produce pseudonurse cells, i.e. all cluster cells develop in a somatic pathway. No alleles are known by which cluster determination will follow the oocyte pathway; i.e. in *Drosophila* no mutants are known which change the $2^n - 1$ nurse cell : 1 oocyte pattern in favour of oocytes. However, we know some examples from nature in which the above pattern is changed in favour of oocytes.

In stone flies (Plecoptera) we found germ cell clusters from which all cells develop into oocytes (section 3.6.1) (Gottanka and Büning, 1990). In earwigs (*Anisolabis maritima*) clusters produce a 1:1 quotient of oocytes to nurse cells, which will split later on into subsets bearing polytrophic

---

Figure 3.77 In *bgcn* males, morphogenesis of clusters is also disturbed. (a) Semi-thin section through region of cluster growth in wild type testicular follicle. (b) Intercellular bridge (B) and fusome (F) during cystocyte divisions in testes of young *bgcn* mutants. Both structures look normal. (c) Low power TEM through germ cell cluster of young *bgcn* male. Cystocyte divisions seem to be normal, polyfusome is built (arrowheads) and rosette formation occurs. However, high contrast of germ cells indicates beginning of pycnosis. (d) Semi-thin section through *bgcn* testes follicle of similar age as in (b and c). Basal area is shown with highly pycnotic germ cell clusters. (e) Apical part of testes follicle from 1 week older male. Lumen is filled with numerous rods and chromatin aggregates. No living germ cells are to be seen. (f) Same stage as in (e). Outer envelope shows striated muscles; inner sheath cells show numerous Golgi vesicles (G). Lumen of testes is filled with rods and chromatin residues. Bars in a, d, e represent 50 μm; bars in b, f represent 1 μm; bar in c represents 10 μm.

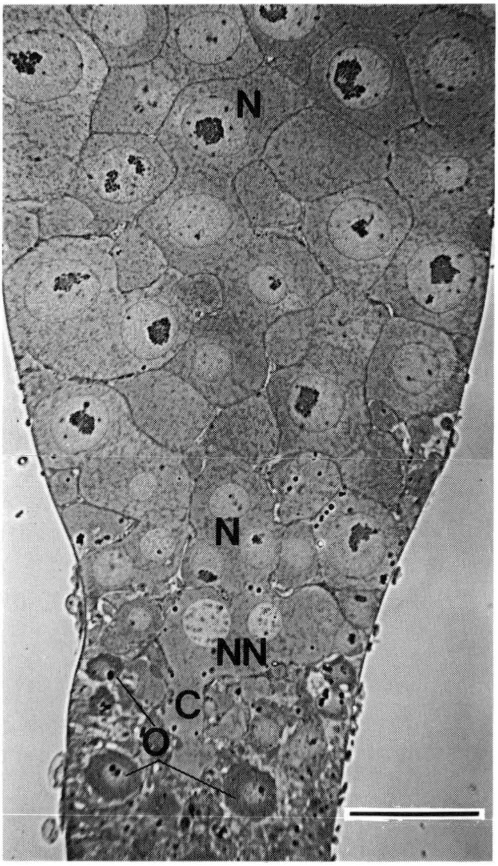

Figure 3.78 Ovary of polyphagous Coleoptera. Lower part of the tropharium of *Gastroidea viridula* (Chrysomelidae). Two nurse cell nuclei (NN) with distinct chromosomes are seen in common cytoplasm at bottom of tropharium. Arrested oocytes (O) are clearly distinguishable by their position and intense staining with Azur B/Toluidine Blue. Note also the decrease in volume of nurse cells (N) in the lower parts compared with the upper parts. Nutritive cord (C). Bar represents 50 µm.

meroistic follicles (section 3.6.2) (Yamauchi and Yoshitake, 1982). In aphids and presumably all other hemipteran groups, more than one oocyte develops from a single germ cell cluster, forming telotrophic meroistic ovaries (section 3.6.3) (Büning, 1985; King and Büning, 1985; Nowak and Büning, unpublished; Beyrich and Büning, unpublished). In most of these examples, presumed nurse cells begin to differentiate with steps out of the oocyte pathway: they enter meiotic prophase. A restriction point must exist during the pachytene stage. When cells can pass, they continue as oocytes. If not, they will be transdetermined to pseudo-somatic cells, the nurse cells. In polyphagous coleopterans the oocyte pathway and the nurse cell pathway seem to be dependent on the position of the cystocytes, by their interaction with basal interstitial cells or prefollicular cells, and/or on some internal signals which might be distributed as gradient(s) with their highest concentration in true oocytes. Figure 3.78 shows the transition zone between the tropharium and the previtellogenic zone of the vitellarium of *Gastroidea viridula* (Chrysomelidae). Apparently, some posterior nuclei of the nurse cell syncytium are designated as oocyte nuclei, as one can see by their individual chromosomal character, indicating the meiotic pathway of oocytes. Normal nurse cells became much bigger and the polyploidization of their genomes is assumed (see below).

(f) Nurse cell genomes

Nurse cells will be transdetermined from the oocyte status to soma-like cells. However, their synthesis remains oocyte-like. In 'primitive' nurse cells, we will find no morphological and presumably no biochemical differences between nurse cells and oocytes except their final destination, which may be caused by germ cell–somatic cell interactions. This implies that nurse cells and oocytes do not differ in their capacity to synthesize. In other words, the oocyte genome will be active during all stages of oogenesis, as known from panoistic ovaries, and nurse cells will not amplify their DNA either by endomitotic processes, or by selective under- or over-replication of specific genes. Candidates for such primitive organization are found in a variety of orders. Recent investigations in Ephemeroptera have shown that these insects develop a telotrophic meroistic ovary (section 3.5.3) (Gottanka and Büning, 1993). Nurse cells and oocytes cannot be distinguished by morphological criteria, except by position.

White flies (Hemiptera: Aleyrodidae) have telotrophic meroistic ovaries of the hemipteran type (section 3.5.3). However, in this family, oocytes and nurse cells are not distinguishable by morphological criteria except by their position and the rise of nutritive cords. This example is thought to reflect an ancestral status (Nowak and Büning, unpublished). The same can be said of the nurse cells of psyllids (Hemiptera: Psyllidae). In this ovary, nurse cell nuclei do not grow after they have stopped at the prophase of meiosis (see section 3.5.3; Beyrich and Büning, unpublished).

The endoparasitic insects, Strepsiptera, have reduced their somatic tissues, except a follicular layer (Figure 3.79; section 3.5.2). Germ cells form rosettes by incomplete cytokinesis. One of the morphologically equal siblings becomes the oocyte, all others will be incorporated by lysis of cell membranes and decay of nurse cell nuclei. This ovary resembles a highly reduced and deviated, but in its basal characters an ancestral polytrophic meroistic ovary (Kathirithamby *et al.*, 1990; Gu *et al.*, 1992; Büning, unpublished).

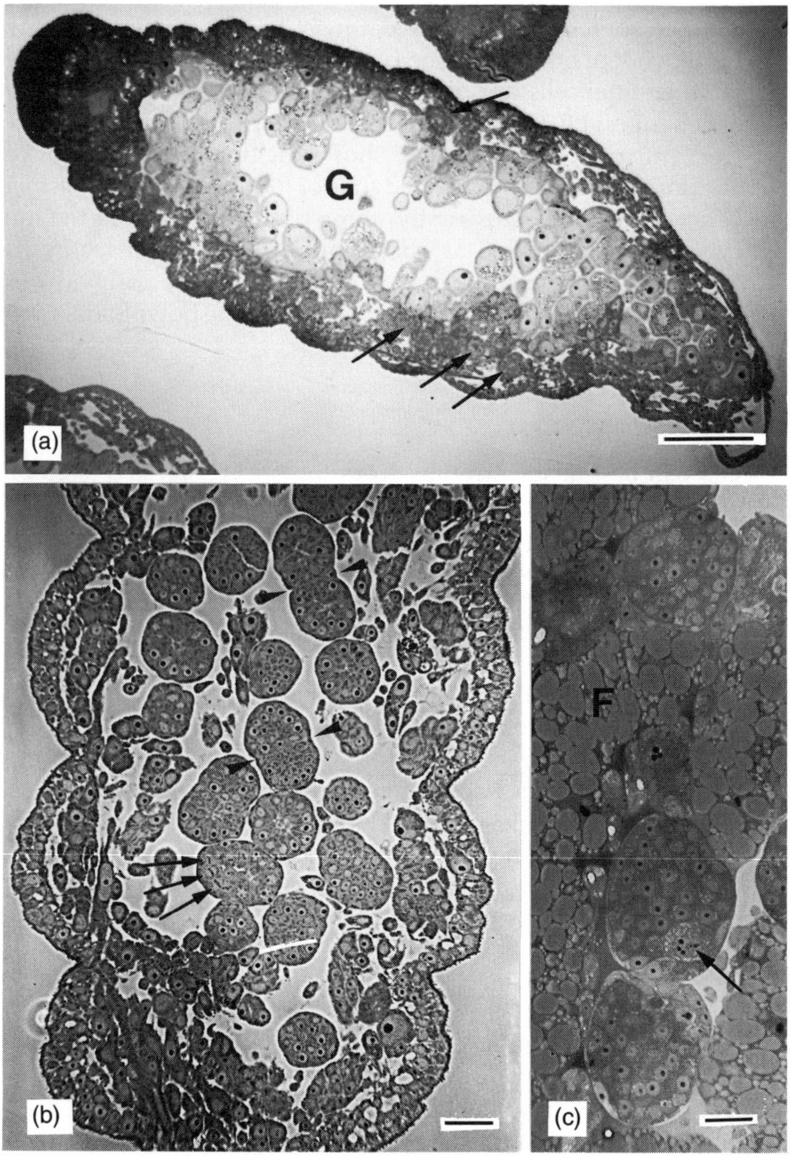

Figure 3.79 (a) Semi-thin section through larval stage 2 of *Elenchus tenuicornis*. Germ cells during cluster formation (arrows). Gut lumen (G). Bar represents 100 µm. (b) During larval stage 3 the number of clusters is raised by fission of clusters (arrowheads). One cluster shows half of siblings in mitotic division (arrows), the other half in interphase. Note that outer somatic epithelia of gonads are reduced totally, germ cells are ensheathed by some flat follicular cells. Germ cell clusters are floating freely in the body cavity. (c) In larval stage 4, final magnitude of clusters is reached. Rosette formation is lost and one of the peripheral germ cell siblings becomes the oocyte, indicated by incorporation of yolk granules and vacuoles (arrow). Fat body (F).

In Rhaphidioptera and Megaloptera (Sialidae) telotrophic meroistic ovaries, in which nurse cell nuclei of the central syncytium and nuclei of the tapetum cells from which oocytes come, are indistinguishable by morphological criteria. The same can be said of their RNA synthesis pattern (Figure 3.48; section 3.4) (Büning, 1979c, 1980; King and Büning, 1985).

The small group of advanced fleas, Siphonaptera: Hystrichopsyllinae, have developed a polytrophic meroistic ovary, the cluster formation and previtellogenic growth of which have been studied by Büning and Sohst (1988). Nurse cells and oocytes behave equally; however, oocytes always develop from the innermost and therefore oldest siblings, as known from most other polytrophic meroistic ovaries (section 3.5.2). Nurse cells remain small; however they contain extrachromosomal rDNA as do the oocytes (Büning and Sohst, 1989).

In all other meroistic ovaries, nurse cells undergo additional replications, which probably double whole genomes during each replication cycle, resulting in endomitotic polyploidization. The number of replication rounds seems to be measured by yet unknown mechanisms, which 'count' the whole content of nurse cell DNA in respect to the number of oocytes which will be served in one cluster. Thus, large clusters of germ cell siblings contain fewer rounds of endoreplications than small clusters. The cluster of *Drosophila melanogaster* (Diptera) contains 15 nurse cells and its highest amount of endoreplication rounds is 12. In silk moths, as in most lepidopterans, seven nurse cells contribute to an oocyte with highest amounts of endoreplication rounds of 16 (Berry, 1982; Rasch *et al.*, 1984). Recent investigations in *Hyalophora cecropia* demonstrate that endoreplication continues until vitellogenesis (Cardoen *et al.*, 1990). This showed that endoreplication rounds are not synchronized in a given follicle, resulting in different levels of polyploidy among nurse cells of the same cluster. Similar differences are known from the nurse cells of *Drosophila* follicles (Bohrmann *et al.*, 1986c). In the wax moth, *Galleria melonella*, these differences are extreme; in those nurse cells which are in direct contact with the oocyte, endoreplication rounds are normal and lead to high levels of polyploidy; however, some apical nurse cells do not share this process and remain small (Gruzova *et al.*, 1990). A heterogeneous replication pattern also occurs in the nurse cells of fleshflies (Cardoen *et al.*, 1986; De Loof *et al.*, 1990) and of blowflies (Avancini and Pires do Prado, 1986) and was also found in the nurse cells of aphids, which have telotrophic ovaries of the hemipteran type (section 3.5.3) (Büning, unpublished). These measurements are consistent with older findings derived from different species, including polytrophic and telotrophic meroistic ovaries, in which the DNA content of nurse cell nuclei was estimated by volume measurements, assuming that the volume of nuclei and DNA content are directly proportional (King *et al.*, 1956; Bier, 1957, 1958, 1959; Matuszewski, 1968; Büning, 1979a; King and Büning, 1985; Cardoen *et al.*, 1986).

As shown, the amount of polyploidization of nurse cells is dependent on their position inside their follicles in polytrophic ovaries or on their position inside the trophic chamber in telotrophic ovaries. Thereby, nurse cells, which are situated next to the oocyte show higher amounts of DNA than those far away (i.e. an anterior–posterior gradient is established). This has been shown in flies and moths by cytophotometry as well as by volume measurements of nuclei (Bier, 1958; Jacob and Sirlin, 1959; Brown and King, 1964; Dapples and King, 1970; Cardoen *et al.*, 1986, 1990). In the telotrophic ovaries of most bugs, an anterior–posterior gradient is established by subsequent unequal mitotic activity of some anterior nurse cells. Functioning nurse cells polyploidize their genomes in a more posterior region of the tropharium and finally they degenerate in the most posterior part (Figures 3.24 and 3.35; see also section 3.5.3) (King and Büning, 1985; Dittmann and Maier, 1987; Dittmann *et al.*, 1990).

In telotrophic ovaries of polyphage coleopterans such a gradient in nurse cell polyploidization is not as obvious. In most species the volume of nurse cells is more or less constant. However, in some species the trophic chambers have slightly larger nurse cell nuclei in posterior parts than in anterior regions. In other species like *Psilothrix cyaneus* (Cantharidae), *Gaurotes virginea* (Cerambycidae) or *Scaphidium quadrimaculatum* (Scaphidiidae) the nurse cell nuclei volume decreases next to arrested oocytes (Figures 3.53 and 3.78). Thereby, amitotic divisions of nurse cell nuclei in the posterior part of the tropharium are suggested. However, cytophotometric measurements by which the topological position of nurse cells remain identifiable are still missing (see also section 3.5.3) (Büning, 1979a; King and Büning, 1985).

In addition, special genes may be over-replicated as is proposed for rDNA in nurse cells of the bug, *Dysdercus intermedius* (Dittmann *et al.*, 1990), whereas in other bugs like *Oncopeltus fasciatus* no over-replication of rDNA could be detected (Cave, 1975). These apparent differences may influence the time needed for euplasmatic growth: after ecdysis, *Dysdercus* females have finished their first egg batch after some days, whereas *Oncopeltus* females need weeks (Rankin and Jäckle, 1980; Dittmann *et al.*, 1985). It is not known in which way over-replication occurs in *Dysdercus*: intra or extra-chromosomal. In other species, rDNA may be under-replicated as suggested for *Drosophila virilis* (Renkawitz-Pohl and Kunz, 1975). In *Drosophila melanogaster*, no under-replication was found by cytophotometric measurements during the 12 endocytotic cycles, which occur in pseudonurse cells of *otu* mutants (Rasch *et al.*, 1984).

In *Drosophila melanogaster*, Chooi (1976) investigated rDNA transcription units of nurse cells and embryos. Using different mutants of the rDNA complex, she could demonstrate that wild type nurse cells transcribe rDNA only from those NORs which contain long NTS, whereas transcribed embryonic NORs contain only short NTS. These data clearly show that regulation of NOR activity follows different pathways in embryos and nurse cells (see also section 3.3.4). Using ferritin-labelled antibodies against rat ribosomal proteins, Chooi also demonstrated directly on spread preparations for the first time that ribosomal proteins assemble in the endstanding knobs of growing rDNA transcripts (Chooi, 1976).

Under-replication of chromatin not involved in NORs has been reported to occur in *Anopheles stevensi* (Redfern, 1981). Giant chromosomes of salivary glands and of nurse cells miss heterochromatic areas of the short arm of the X-chromosome, which occur usually in somatic tissues and which correspond to a satellite DNA of lower density, not found in polytene DNA.

Recently, reporter genes have been introduced via P-elements into the genome of *Drosophila melanogaster*. This technique allows us to follow the expression of the reporter gene (e.g. β-galactosidase) when the P-element becomes stably integrated into or near the enhancer and/or promoter structures of *Drosophila* genes. Grossniklaus *et al*. (1989) have screened the ovaries of about 600 different strains carrying single copies of P-element enhancer detectors. Among many strains in which all nurse cells are marked, they obtained one strain, which showed a differing pattern: only anterior nurse cells show the staining of β-galactosidase, whereas in all posterior nurse cells the reporter gene remains silent. This differential pattern can be interpreted as different activation of some genes in different subsets of nurse cells. The use of reporter genes to unravel the complex pattern of gene activation will give new insight into the regulation processes during *Drosophila* development.

3.4 TRANSPORT FROM NURSE CELLS TO OOCYTES

As shown in the preceding sections, the growth of oocytes is supported to the greatest extent by nurse cells in meroistic ovarioles. Such support with euplasmatic substances raises questions of transport from the place of synthesis to the place of final deposition within the egg. When discussing these problems the architecture of interconnected siblings must be considered, which might have a great impact on transportation itself. Therefore, polytrophic meroistic ovaries will be discussed first, and the four different types (section 3.5.3) of telotrophic meroistic ovaries will be examined later on.

3.4.1 Transport in polytrophic meroistic ovaries

As will be shown, there are good arguments for a singular rise of the polytrophic meroistic ovary among the ancestors of today's living insects (section 3.5.2). Thus, one can assume for all polytrophic ovaries the same basic events, whatever they are, are responsible for the transport of macromolecules like RNPs, proteins and greater aggregates like ribosomes or organelles such as Golgi complexes, ER, vacuoles of each kind and mitochondria.

Most work on transport from nurse cells into oocytes has been done on fly and moth ovaries. First impressions about the dynamics of transport of macromolecules came from the pioneering work of Bier (1963a,b), who labelled newly synthesized RNA with tritiated precursor molecules and followed the path of marked macromolecules from the nurse cells into the oocyte by autoradiography of tissue sections. He noticed that these macromolecules assemble first in nurse cell nuclei, but after some minutes they are found in increasing amounts in the cytoplasm. Sometime later, at least in these follicles, the oocytes of which have started growth, label was detected in a gradient manner in the oocytes, showing the highest concentrations first in the connecting region of intercellular bridges (Figure 3.19). This transport could be prevented when flies were kept in anoxic atmospheres during incubation periods (Bier, 1965a). These experiments indicated that transport mechanisms depend on the metabolic activity of nurse cells. From today's point of view, Bier's experiments can be interpreted as an example of the dynamics of rRNA pathways, since precursors used in these labelling experiments had low specific activity and the resulting intensity of messenger RNA labelling must be assumed to be in the range of background activity and should be overruled by rRNA labelling.

Similar experiments were done in the following years by Bier's group and others, showing the pathway of RNA synthesis, mainly due to rRNA, in a variety of polytrophic meroistic insects of different orders, such as Hymenoptera (Engels, 1968; Meng, 1970), Dermaptera (Engels, 1969; Zinsmeister, 1973), other Diptera (Zalokar, 1965; Ribbert and Bier, 1969; Muhlach and Schwalm, 1977), Lepidoptera (Pollak and Telfer, 1969; Hughes and Berry, 1970; Paglia *et al.*, 1976), Mecoptera (Ramamurty, 1963), Neuroptera (Gruzova *et al.*, 1972), Coleoptera, Adephaga (Bier, 1965b; Matuszewski and Hoser, 1975). In all these examples, the transport of newly synthesized RNA from nurse cells to oocytes is quick and takes only a few hours (Table 3.5).

The transport of mRNA in presumed RNP complexes was demonstrated first in the telotrophic ovary of *Oncopeltus* via *in situ* hybridization of labelled poly U with the poly A-tails of mRNAs (Capco and Jeffrey, 1979). Similar dynamics were found as were known from unspecified RNA labelling experiments (see above). Today, one can follow the path of single mRNA products directly via anti-sense hybridization when the gene is cloned or at least indirectly, since reporter genes, which are mostly bacterial genes and normally not found and expressed in eukaryotes, can be established permanently in the germ line of *Drosophila* via P-element transfer. Thereby, reporter genes come under the control of promoter regions to which the P-elements have become integrated. Thus, reporter genes will be coexpressed with the normal gene or will substitute for the gene in question. The activity of reporter genes can be shown in whole mounts and/or tissue sections by staining procedures of reporter gene products. Using this technique, Cheung *et al.* (1992) have shown that the mRNA of the gene *fs(1) K10* is synthesized in nurse cell nuclei as well as in oocyte nuclei during the very early stages (stage 1 and 2 after King, 1970). Later on, label accumulates only in oocytes, not in nurse cells. Such accumulation in oocytes is interpreted as selective transport of *K10* mRNA from nurse cells into oocytes during previtellogenic stages, i.e. before the masses of ribosomes and other constituents of the euplasm are shifted to the oocyte at the beginning of vitellogenesis. Such selective transport of *K10* mRNA, perhaps together with other mRNAs, gained strong support from experiments using different *K10-lac Z* fusion constructs, which enabled Cheung *et al.* (1992) to analyse this selective transport in more detail. When both flanking regions of the *K10* gene were used, but the structure domain of *K10* was replaced by the *E. coli lac Z* gene, the construct behaved as the natural *K10* gene: i.e. label accumulated only in the oocyte, whereas nurse cells stained only faintly during early stages 1–7 (for staging see below). When, in addition to the structure domain, the 3′-end was also substituted, the synthesis in nurse cells continued, but no transport into the oocyte occurred. In constructs in which the 5′-end was additionally substituted by the *hsp26* nurse cell enhancer linked to the *Sgs3* (glue) promoter-translation start side, synthesis in nurse cells continued but no transport in oocytes occurred. When by substitution of the 5′-end the *hsp26* region was omitted, synthesis stopped in all germ cells. This analysis shows that the 5′-flanking region is needed for synthesis of *K10* products and that the 3′-region is needed for transport and accumulation in the oocyte. Although label is first distributed homogeneously in all parts of the ooplasm, at stage 8 (i.e. before tremendous oocyte growth begins) the label becomes concentrated in the anterior margin, just beneath the anterior border cells and nurse cells. This indicates a second transport process, which results in localized fixation of RNPs at special regions of cortical cytoplasm. During stages 9 and 10, nurse cells again transcribe the *K10* gene locus, the primary product of which will then be co-transported with the bulk of euplasm into the oocyte at stage 11. The conclusions of these experiments, which gave some insight into the dynamics of transport processes, are as follows:

(1) There is an active transport from nurse cells to oocytes in most systems, since simple diffusion cannot explain the speed by which marked molecules are shifted from nurse cells to oocytes.
(2) There is a strong indication of selective transport of special mRNAs from nurse cells to oocytes during the early phases of previtellogenic growth.
(3) RNPs can become sorted in the oocyte at different times to different places.

Before discussing the transport problem more deeply, it will be helpful to take a short look at the development of ovarian follicles of polytrophic meroistic ovaries during oocyte growth phases. This will be done using the

Table 3.5a RNA synthesis and transport in polytrophic meroistic ovarioles

	Strong silver-impregnation over							
	Nurse cells				Oocyte			
Genus	Nucleolus (h)	Cytoplasm (h)	Nucleus ≤ cytoplasm (h)	Grains passing IB into oocyte (h)	Nucleus	Temperature (°C)	Oogenesis (d)	References
Dermaptera								
Forficula	3–6	6–10	20–24	20	± at	21	45–60	Engels
					early			(1969)
Labidura	3	3		17	stages	21	30	
								Herter
								(1965)
Coleoptera								Bier
Adephaga								(1965b,
Carabus		0.5–1		1–4	+++ in			1967)
					extra			Bier *et*
Acilius				6–8	rDNA			*al.* (1967b)
Hymenoptera								
Formica	0.75			2.5	± in			
Vespa			(5–12)	(2)	early			Engels
Apis	1–3	1–3	24–48	3–5	stages	31	7–10	(1968)
Mecoptera								
Panorpa	2.5	0.5–1		2.5–3	+ in			Rama-
					early			murty
					stages			(1963)
Lepidoptera								
Malacosoma				2–5	+ in			Bier
					early			(1965b)
Diptera					stages			
Drosophila		0.25–1	4	4	–			Zalo-
								kar (1965)
Calliphora		0.25–0.5		0.75–1		21		Bier
								(1963a,b)
Musca	0.16–0.25	0.25–0.5	1.5–5	0.75–1	+ in	21	7	Bier
					early			(1963a,b)
					stages			(1965a)
		4	4	4	–			Trepte
								(1981)
								Zalo-
								kar (1965)

Autoradiography of sectional ovaries was done mostly after injection of ^{3}H-uridine into abdomen of females. Most data from Engels (1969).

Table 3.5b RNA synthesis and transport in telotrophic meroistic ovarioles

Genus	Strong silver-impregnation over: Nurse cells: Nucleolus (h)	Nurse cells: Cytoplasm (h)	Nurse cells: Grains in nutritive cord (h)	Nurse cells: Grains passing from nutritive cord into oocyte (h)	Oocyte: Nucleus	Temperature (°C)	Oogenesis (d)	References
Hemiptera								
Pyrrhocoris	0.25	1	4	5 early 30 late = main fraction	+	21		Mays, (1972)
Coleoptera, Polyphaga								
Tenebrio	1.5	≈ 5	48	48	++	21	6–7	Ullman (1973)
Bruchidius	0.1	4	8	8	++	21	6–7	Büning (1972)
Crynodes	0.25	6	6	6	++			Ramamurty (1979)
Megaloptera, Sialidae								
Sialis	24	24	48 no gradient	no gradient	++	21	≈ 60	Büning (this report)

Autoradiographic data after incubation in ^{3}H-uridine.

ovary of *Drosophila,* since most of our knowledge about transport comes from it. King and coworkers have staged these developmental processes very thoroughly and the staging they introduced has been accepted worldwide (Figure 3.34 and Table 3.6) (King *et al.*, 1956; Cummings and King, 1969; King, 1970; Mahowald and Kambysellis, 1980). As shown, previtellogenesis starts with equal growth of oocyte and nurse cells, i.e. the proportion of oocyte volume to the volume of an average nurse cell remains roughly 1:1 from stage 1 to stage 8. However, most oocyte growth during these early phases will also be supported by nurse cells, since the oocyte genome assembles during stage 3 into a karyosome, in which most genes are kept silent. Vitellogenesis begins during stage 8. This shows that the bulk of euplasm, synthesized by nurse cells, will be transported into the oocyte, while growth becomes enhanced via endocytosis of yolk proteins. Nurse cells grow until stage 10. Thereafter, the nurse cell cytoplasm becomes shifted into the oocyte very quickly, and the nurse cell chamber breaks down. Thus, three phases of euplasmatic transport can be assumed:

(1) An early transport phase, in which growth of oocytes is supported by nurse cells showing a constant slope of logarithmic growth and in which selective transport of RNPs may contribute to growth of oocytes

to a certain extent. This phase takes about 50 h (stages 2–7 inclusive), in which nurse cells will enhance their synthesizing capacity via polyploidization of genomes (section 3.3.4 and Table 3.6). The relative volume of the oocyte expands from 1 to about 200 by the beginning of vitellogenesis at stage 8. Although this volume contributes quantitatively to only 0.3% of the final volume, one can expect that about 2% of the final volume at stage 13 is euplasmatic. This assumption is made by extrapolation of the previtellogenic slope to stage 13, which can be determined between stage 2 and 8. Furthermore, it shows that in *Drosophila* the main transport from nurse cells to oocyte occurs during vitellogenesis.

(2) A middle transport phase starts with the onset of vitellogenesis, showing a higher slope (upward inclination) than before. Now, masses of ribosomes, mitochondria, vacuoles and other compartments of cytoplasm will be shifted to the oocyte. During this phase, nurse cells will reach their highest amount of DNA. The phase lasts about 19 h and most of the euplasm, about 87%, will be synthesized in this period (stages 8–10).

(3) A late transport phase, in which the already synthesized euplasm, stored in nurse cells, will be pushed into the oocyte. This phase shows the highest constant slope but it is the shortest phase and lasts about 2 h (stages 10–11).The amount of euplasm transported in this phase is about 95% of the final euplasmatic volume.

What forces can be used to establish these transport processes during the three phases and what main forces may be used for mass transport and for selective transport? Today, we are far from answering these questions correctly; however, there are some possibilities which might be used by nature:

(1) Transport via electrical currents between nurse cells and oocytes.
(2) Transport via osmotic gradients between the sites of synthesis (nurse cells) and the site of storage (oocyte).
(3) Transport via gross changes of the inner cytoskeleton of the syncytium, especially of the nurse chamber.
(4) Transport via special proteins, e.g. microtubule-associated proteins (MAPs), along cytoskeletal elements like microtubules.

The middle transport phase in which, among other constituents, masses of ribosomes will be passed from nurse cells to the oocyte was the first transport phase which was analysed in detail. In 1973, Woodruff and Telfer published their measurements on electrical currents within an insect follicle. They measured different electrical properties between nurse cells and oocytes in follicles of the giant silk moth, *Hyalophora cecropia*, during the early and middle stages of vitellogenesis. Whereas nurse cells had a resting potential of about 40 mV to haemolymph or physiological salt solution, when measured with implanted electrodes, the oocyte averaged only 30 mV. If fluorescein-labelled rabbit serum globulin was injected into nurse cells, the label was recorded sometime later in the oocyte. When injected into the oocyte, the label could not cross the barrier between nurse cell and the oocyte, although no morphological barrier exists in the wide intercellular bridges (diameter about 30 μm) between nurse cells and oocyte. However, when the potential difference of 10 mV was reversed by a second pair of stimulating electrodes, the fluorescence crossed the bridges and could be recorded in nurse cells after a while. These experiments indicated transport of negatively charged molecules from nurse cells into the oocyte using the 10 mV gradient, which was measured in these experiments between nurse cells and the oocyte at that developmental stage (mid-vitellogenesis). The results could be confirmed by non-invasive measurements of electrical currents, using a vibrating probe (Jaffe and Woodruff, 1979). Recently, these experiments were re-examined and extended (Woodruff and Telfer, 1990; Woodruff *et al.*, 1992). Electrical polarity of *Hyalophora* follicles rises with the onset of yolk deposition by approximately 20 mV hyperpolarization that affects nurse cells more strongly than oocytes. A steady-state ionic potential of about 5 mV, more negative in nurse cells than in attached oocytes, results. The hyperpolarization appears to be due to the superimposition of a vanadate-sensitive metabolic pump onto a –20 mV Donnan-type membrane potential. This steady-state potential depends mainly on the K^+ concentration of the surrounding medium. K^+ channels were maximally opened at 300 mOsm, which is the normal osmolarity of the haemolymph of *Hyalophora*, which has an unusually high K^+ and low Na^+ content (Woodruff *et al.*, 1992).

Similar experiments were done with *Drosophila* follicles by different groups. Whereas extrafollicular currents could be found in *Drosophila* follicles by others (Bohrmann *et al.*, 1984, 1986a; Jaffe and Overall, 1985), only Woodruff could measure intrafollicular potential differences between nurse cells and oocytes of about –2.5 mV significantly (Woodruff *et al.*, 1988; Woodruff, 1989). Other groups failed (Bohrmann *et al.*, 1986a,b; Sun and Wyman, 1987, 1989, 1993; Bohrmann and Gutzeit, 1987). Recent investigations by Bohrmann (1991a) have shown that the incubation medium has an enormous impact on the electrophysiology of isolated follicles as well as on protein synthesis and further development. However, even in better culture media like Robb's medium, all the tests are made with excised and isolated follicles *in vitro* and it remains highly questionable if these conditions can mimic sufficiently the *in vivo* situation. Further experiments are needed on current-mediated flow of cytoplasmatic components from nurse cells to oocytes, at least in *Drosophila*.

When looking at the final molecules that are synthesized by nurse cells, as well as the metabolites needed for this synthesis, it becomes evident that

Table 3.6 Principal characteristics of oogenetic stages of *Drosophila melanogaster* (adapted from King, 1970; table from Mahowald and Kabysellis, 1980)

Stage	Duration (h)	Length of follicle	Identifying features	Oocyte	Nurse cells	Follicle cells
1	9.56		16 cell cyst within region 3 of germarium	oocyte–nurse cell size ratio is 1:1 from stage 1 to 8		monolayer of ~80 cells
2	9.56		follicle enters vitellarium; spherical to oval chamber until stage 7		begin polyploidization 8C	follicle cells divide 4× between stages 2 and 5
3	9.56		beginning of karyosome in oocyte	nucleolus disappears; endobody and karyosome form	16C	
4	9.16		bulbous chromosomes in NC	karyosome (3μm) and endobody (1 μm)	32C	
5	2.61		bulbous chromosomes lost		some ploidy difference appears; posterior nuclei: 64C	
6	8.45		chamber still oval;		64C	cessation of cell division
7	8.69	110–185 μm (145 μm)[a]	first elongate chamber; last previtellogenic stage		anterior nuclei: 128C; posterior nuclei: 256C	cells enlarge; begin polyploidization; cells over oocyte begin to be columnar
8	5.21	140–270 μm (190 μm)	yolk formation begins	15–65 μm (30 μm)[a]	256C 512C	same as stage 7
9	5.61	215–320 μm (275 μm)	oocyte is ~1/3 of the size of the chamber	45–130 μm (75 μm) karyosome enlarges to 5 μm, some RNA synthesis occurs	256C	most follicle cells move over oocyte. Begin secretion of vitelline membrane
9				oocyte nucleus becomes located adjacent to the dorsal follicle cell-NC surface	512C 1024C	cluster of border cells migrate between nurse cells to anterior surface of oocyte
10	5.13	300–490 μm (430 μm)	oocyte is 1/3 to 1/2 of chamber; follicle cell layer is columnar over oocyte, and squamous over nurse cells	110–280 μm (150 μm)	512C 1024C cytoplasm becomes opaque due to lipid droplets	begin centripetal migration at anterior edge of oocyte
11	0.40	450–530 μm (490 μm)	oocyte is 1/2 to 3/4 of chamber	250–400 μm (320 μm)	nurse chamber rapidly pours into oocyte	follicle cell layer thins over oocyte and begins secretion of endochorion
12	1.90	500–690 μm (540 μm)	shrunken nurse cells form anterior cap	360–520 μm (420 μm) vesicle; oocyte is fully grown	nuclei have condensed to spherical shape	formation of chorionic appendages begins; reach ploidy of 45C
13	0179	550–730 μm (650 μm)	chorionic filaments visible; oocyte elongate	430–540 μm (490 μm) first appearance of glycogen	gradual disappearance of nurse cell nuclei	secretion of endochorion and exochorion
14	1.98	600–750 μm (700 μm)	mature oocyte; slightly shrunken	420–550 μm (510 μm) meiotic division arrests at metaphase I		completion of chorion; degenerate

[a]These values represent the range of lengths for living follicles or oocytes for the P2 strain of Oregon-R. The medium length is given in parentheses. Between 25 and 50 chambers were measured at each stage. No attempt was made to distinguish stages 1–6. (With the permission of Academic Press, New York.)

negatively charged molecules are more abundant than positively charged molecules. This implies that the number of cations compensating for these negative charges must increase in nurse cells to a higher extent than in oocytes. This can be used to establish negative potential differences, as are found in silk moth follicles and in dipteran follicles to a lesser extent. Thus, both osmotic (see below) and electrical forces can be used for active transport during the middle stages of synthesis, i.e. stage 8 to stage 9. Cations have been measured in *Sarcophaga* follicles by Verachtert *et al.* (1989) as well as by Bohrmann (1991b) in *Drosophila* follicles. In *Sarcophaga*, a potential difference between oocyte and nurse cells was measured at the beginning of vitellogenesis, showing nurse cells more negatively charged. This current depends on the ionic composition of the Ringer solution. Changing the amount and composition of Ringer solutions, Verachtert *et al.* (1989) could show that K^+ and Cl^- concentrations influence the potential difference between nurse cells and oocyte. The authors discuss a hypothetical model of asymmetry of ionic channels and ionic pumps, which will stabilize a negative potential between the nurse cell chamber and the oocyte at the beginning of vitellogenesis. Bohrmann (1991b) did more extended experiments to elucidate cation exchange (presumably that of K^+) in *Drosophila* follicles *in vitro* during vitellogenesis. He used rubidium-86 as a probe and tested several ion-channel inhibitors to obtain information about active and passive transport of ions. He showed that total K^+ uptake is stage specific and increases up to mid-vitellogenesis. However, dissected halves of follicles clearly show that the total uptake of the oocyte half exceeds that of the nurse cell half. This difference is due to ouabain-insensitive K^+-pumps, whereas (Na^+, K^+)-pumps are equally active in anterior and posterior halves and show no stage-specific variations. The active K^+ uptake increased during nurse cell regression (stages 11 and 12). Active and passive K^+ uptake increase with increasing pH. In the presence of ouabain or at alkaline pH the follicles develop normally up to stage 14 in the *in vitro* system. These results do not support the idea of electrical current as a main force for active transport from nurse cells to oocytes in *Drosophila.*

Gutzeit (1986b) wrote a critical review on the transport of molecules and organelles in the meroistic ovaries of insects. He highlighted another force by which transport of molecules can be explained and which had not been focused on in preceding studies. During the early growth phases of polytrophic meroistic follicles, each follicle becomes more and more polarized by synthesis of nurse cell genomes, which polyploidize and, when open, synthesize large amounts of their products. The oocyte nucleus, which has to remain in the 4C state, synthesizes only at low levels. Primary gene products (RNA) and secondary gene products (proteins and RNPs) are all molecules which have positive and negative charged ligands that attract ions (Donnan's law). In addition, high synthetic activity demands greater pools of metabolites like NTPs, NDPs, NMPs and amino acids, which are all charged molecules. This fraction, too, attracts ions (Donnan's law). Thus, the osmotic forces of nurse cells must be higher than those of the oocyte. If there is no force against, nurse cells must swell or active transmembrane transport must be established, or cell walls must be able to withstand the osmotic pressure as in plant cells. Insect follicles are enclosed by the tunica propria, also known as the basal membrane or basal lamina (section 3.2.3). Additionally, germ cells, oocytes as well as nurse cells, are enclosed by somatic cells, the follicular cells. The cytoskeleton of these somatic cells, together with the basal lamina, could serve as a counterpart for osmotic pressure in the same way as cell walls act in plant cells. During the early stages of follicle growth up to stage 7, the whole follicle of *Drosophila* is ensheathed by squamous follicle cells, the architecture of which shows no difference in the anterior (i.e. nurse cell facing) or posterior parts (i.e. oocyte facing). During this time, the oocyte volume grows concomitantly with the nurse cell volume. During stages 7 to 9 follicle cells encompassing the nurse cell compartment of the germ cell syncytium become flattened and the nurse cell compartment grows tremendously, while the oocyte compartment, which includes the follicular cells, now becomes highly prismatic and grows slowly. This changes during stage 9, when the maximum volume of the nurse cell compartment is reached. Now, concomitantly with the onset of vitellogenesis, follicle cells of the oocyte compartment change their structure again. The cells become cuboidal and later, with continuing vitellogenesis, become squamous, the intercellular clefts rise and the formerly closed follicular epithelium loses its strength. Thus, the osmotic pressure of the whole system can expand into this compartment. The pressure of the flattened follicular cells of the nurse cell compartment together with its basal lamina is now greater than that of the oocyte compartment.

Arguments supporting this idea come from morphological investigations of the cytoskeleton of follicular cells and basal lamina of *Drosophila* (Gutzeit, 1986a, 1990a,b; Gutzeit *et al.*, 1991; see also sections 2.3.3–5) and this analysis has been recently extended to two other polytrophic meroistic insects, *Bradysia* (Diptera, Nematocera) and *Apis* (Hymenoptera)(Gutzeit and Haas-Assenbaum, 1991). The F-actin pattern of submembrane cytoplasm of follicular cells contacting the basal lamina is orientated circumferential to the long axis of a follicle. The same orientation is found among laminin molecules of the basal lamina itself. The number of F-actin filaments increases during vitellogenesis, which is interpreted as a consequence of very tight connections between the intracellular cytoskeleton structures and the basal lamina to form one physiological unit. Comparative investigation of the mutant *kugel* shows that this circumferential orientation of F-actin and laminin is lost and eggs round up. Furthermore, when the basal lamina of wild type follicles become destroyed by collagenase or pronase E, eggs round up as do nurse cells. Similar experiments have been done with the neotenic fungus gnat, *Hetereopeza pygmaea* (Diptera:

Cecidomyidae). Here, too, denuded follicles round up, demonstrating that the shape of follicles and eggs depends on the development of follicle cells (Went and Junquera, 1981; Junquera 1983a,b). All this shows clearly that basal lamina and follicular cells contribute some pressure on the whole follicle and the final shape of the eggs is affected by the synergistic forces of the follicular sheath and basal lamina.

During mid and especially late vitellogenic stages (stages 10a, 10b and 11), other forces may contribute to the very fast and final transport of all nurse cell cytoplasm into the oocyte. During this late transport phase, also known as the nurse cell breakdown, the streaming of nurse cell cytoplasm into the oocyte can be recorded with time-lapse photography (Gutzeit and Koppa, 1982; Gutzeit, 1986a). During this mass flow of cytoplasm from nurse cells into the oocyte, a strong network of F-actin filaments occurs in the nurse cell compartment, orientated more or less parallel to the long axis of the follicle. Cytoplasmic streaming is first recorded in the oocytes of stage 10a, before cytoplasmic streaming of nurse cell cytoplasm into the oocyte is seen. This ooplasmatic streaming can be stopped by colchicine, whereas the mass streaming from nurse cells into the oocyte cannot be prevented by the same drug (Gutzeit, 1986a). When inhibitors (cytochalasins) of F-actin were tested, the opposite effect occurred: the mass streaming of nurse cell cytoplasm into the oocyte was prevented, but the oocyte streaming continued (Gutzeit and Huebner, 1986). Thus, mass streaming during nurse cell breakdown occurs independently of microtubule-mediated cytoplasmic streaming in oocytes (Gutzeit, 1986a). In addition to osmotic and/or electro-osmotic forces, which are the main forces during the middle phase of transport, mass streaming of nurse cell cytoplasm depends to a large degree on the architecture of the cytoskeletal elements of the nurse cell chamber, especially F-actin filaments. Gutzeit and coworkers have shown that during stage 10, F-actin molecules accumulate in the nurse cell chamber and are orientated mainly in the anteroposterior direction. These molecules might contract during the nurse cell breakdown phase (Gutzeit, 1986a,b, 1990a,b; Gutzeit and Huebner, 1986; Gutzeit and Haas-Assenbaum, 1991; Gutzeit *et al.*, 1991).

Transport from nurse cell to nurse cell and from nurse cell to the oocyte is possible via intercellular bridges. These structures are remnants of preceding incomplete cytokinesis which undergo ultrastructural changes during the growth of the follicles. The rims of intercellular bridges become impregnated with unknown electron-dense substances. The cytoplasm in the bridge is filled at first with the midbody, which is the remnant of the preceding mitotic spindle. During cystocyte divisions this midbody will be replaced by the fusome, which appears to be a homogeneous mass of middle electron density, which encloses some vesicles and sometimes ER-like structures. Fusomes and polyfusomes contain F-actin, at least in some species, as was shown for queen bees, (*Apis mellifera*) germarial region

(Gutzeit *et al.*, 1993) as well as for bumble bees and for the tiger beetle (*Cicindela hybrida*) (see Figure 3.80c; Rübsam and Büning, unpublished).

Transport of larger aggregates like mitochondria through intercellular bridges will be prevented by the fusome, and even ribosomes are extremely rare in the fusomal area. One can conclude that the flow of ribosomes through bridges is reduced to almost zero at these stages. The fusome exists until the end of cystocyte divisions. A circumferential microfilament ring is attached to the bridge rims, the main substance of which seems to be F-actin, as was shown by the bright fluorescence of inner ring structures with rodamine-phalloidin, shown in whole mounts of *Drosophila* follicles (Warn *et al.*, 1985; Storto and King, 1988) and in *Apis mellifera* follicles (Gutzeit *et al.*, 1993). The same structural component is also found in other flies (Figure 3.81c; Rübsam and Büning, unpublished; Carpucino *et al.*, 1992) as well as in social hymenopterans where they seem to be strongest (Cassidy and King, 1969; King *et al.*, 1982).

However, in other insects like *Chrysopa, Cicindela* and small solitary wasps, F-actin seems to be absent or below recognition by rhodamine-conjugated phalloidin in whole mounts (Rübsam and Büning, unpublished; Bilinski, unpublished). This inner ring persists during all stages of follicle growth until breakdown of nurse cells (Theurkauf *et al.*, 1992). This F-actin ring needs another protein encoded by the gene *hu-li tai shao (hts)* (i.e. too little nursing) which is similar to adducin, a membrane skeletal protein of mammals (Yue and Spradling, 1992) for correct assembly in intercellular bridges. The mutant alleles lack these F-actin rings and cytoplasmic transport from nurse cells into the oocyte is affected. Furthermore, this gene is involved in correct oocyte determination processes, as is assumed by strong alleles which reduce the number of oocytes in favour of pure nurse cell follicles. Such phenotypes can be explained also by disturbances of transport processes especially for special mRNAs or RNPs into the oocyte, during and after cluster formation. Sometimes, vesicular structures are attached to the inner ring of microfilaments, as was shown in the intercellular bridges of whirligig beetles (Bilinski, 1986). Apart from mechanical support, which keeps the bridge open, the role of intercellular bridge coating remains unknown.

The role of microtubules in oocyte differentiation and transport processes within the *Drosophila* follicle has been analysed, mainly using whole mounts stained with antibodies against α-tubulin (Theurkauf *et al.*, 1992, 1993). In the germarial region, microtubules are organized in the central region of developing clusters where intercellular bridges and polyfusome are assumed. During the early stages of follicle growth (stages 1–7) a polarized structure of microtubules exists, running from the oocyte through intercellular bridges into the nurse cells. Thereby, the centrosomes of pro-nurse cells become inactivated and their centrioles will be carried on into the oocyte later (Mahowald and Strassheim, 1970; Theurkauf *et al.*, 1993). During stages 7 and 8, microtubules in the oocyte cortex reorganize

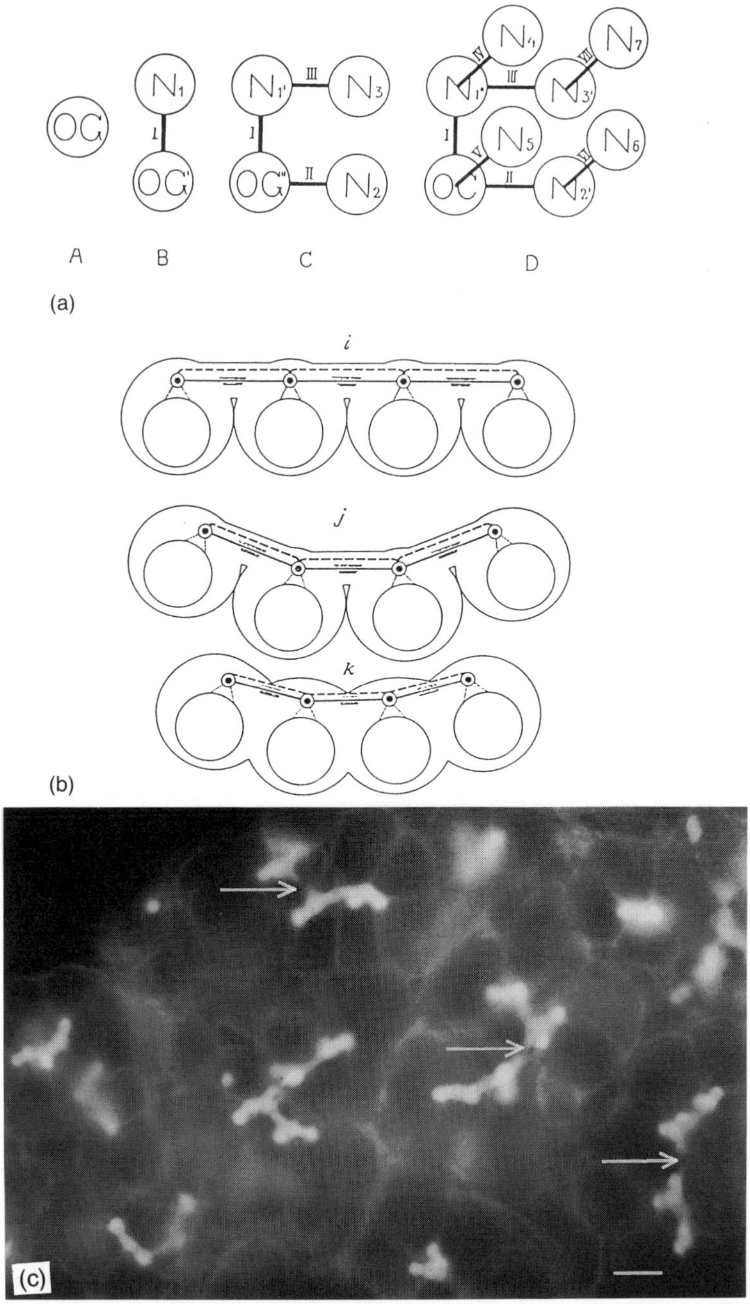

Figure 3.80 Cluster formation and fusomes in insect ovarioles. (a) This figure from Knaben (1934) shows the sister cell connections of *Tischeria angusticolella* (Lepidoptera: Tischeriidae). (b) This figure from Hirschler (1945) demonstrates the beginning of rosette formation by shortening of fusomes and by movements of intercellular bridges in *Macrothylacia rubi* (Lepidoptera). (c) The whole mount demonstrates bright F-actin fluorescence by specific coupling of rhodamine-conjugated phalloidin in polyfusomes of *Cicindela hybrida* (Coleoptera, Adephaga) germarial region. Note the symmetric arrangement within each cluster. Arrow points to the centre in which the content of F-actin decreases first. Upper part of germarium to the left. Bar represents 10 μm. (Courtesy of R. Rübsam, University of Erlangen, Germany.)

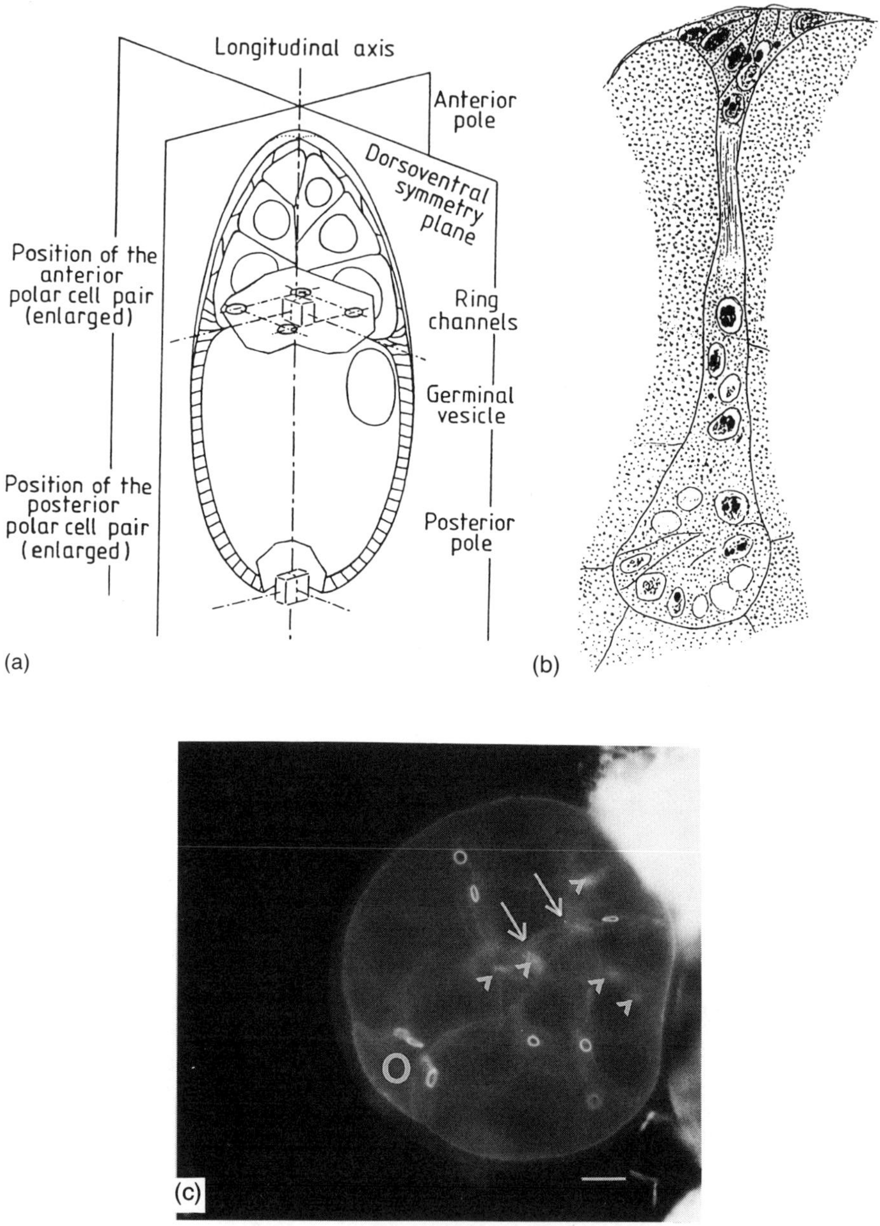
Longitudinal axis
Anterior pole
Dorsoventral symmetry plane
Position of the anterior polar cell pair (enlarged)
Ring channels
Germinal vesicle
Position of the posterior polar cell pair (enlarged)
Posterior pole
(a)
(b)
(c)

into an anterior-posterior gradient which is supposed to play an important role in the localization of morphogens like *bicoid*. During nurse cell breakdown, microtubules assemble all over the cortical region of the oocyte and the authors suggest an important role for cytoplasmic streaming in the oocyte, mediated by vesicle transport along microtubules. During this last phase actin may have a minor role (Theurkauf *et al.*, 1992).

However, Cooley *et al.* (1992) have analysed the *Drosophila* mutant *chickadee*, expression of which blocks the final transport of nurse cell cytoplasm into the oocyte during stages 10b and 11. *Chickadee* encodes a protein of the profilin family which is thought to be engaged in the architecture of the strong actin network of nurse cells. This network stretches from the cortical cytoplasm around the intercellular bridges up to a strong belt of actin, surrounding the nurse cell nuclei. This strong actin network arises during stage 10. In mutants, this network is absent and the nurse cell nuclei become displaced from their normal place in the middle of the nurse cells and are shifted to the rims of intercellular bridges. Thus, the streaming of cytoplasm is blocked by these displaced nurse cell nuclei rather than by contractile forces using the strong actin fibres as was proposed by Gutzeit (1986b,c).

Summarizing, one can say that transport in polytrophic meroistic ovaries starts with more or less selective transport of mRNAs or RNPs during the early phases of nurse cell growth; later on, during phases of intensive nurse cell growth, which in *Drosophila* are stages 4–7, an osmotic gradient becomes established, by which ribosomes and other euplasmatic constituents begin to flow into the oocyte. Finally, during nurse cell breakdown (stages 11–12), all nurse cell cytoplasm is shifted into the oocyte, possibly mediated by the contraction of nurse cells, which in part occurs through contraction of cytoskeletal elements. These three phases of transport must be looked at not as separate elements following one after another, but as phases and forces which will be established in addition to the preceding stages and forces. Selective transport will go on during the rise of osmotic/electric

---

Figure 3.81 Architecture of dipteran follicles showing the *Drosophila* type (*Calliphora erythrocephala*). (a) Border cells move down a median line through the nurse cells from the apical region of each follicle and settle between the nurse cells and oocyte (from Verhein, 1921). (b) Out of these border cells, two anterior polar cells derive, which might be involved in anterior–posterior and dorso-ventral axis formation in Diptera (*Sarcophaga bullata*). (With permission of Geysen *et al.*, 1988.) (c) The whole mount shows the distribution of intercellular bridges in a previtellogenic follicle by fluorescence of inner F-actin rings which couple specifically with rhodamine-conjugated phalloidin. Note the four intercellular bridges of the oocyte (O). Arrowheads show positions of intercellular bridges not in focus. Arrows indicate areas of somatic border cells on their way to the oocyte–nurse boundary. Bar indicates 20 μm. (Courtesy of R. Rübsam, University of Erlangen, Germany.)

gradients and this gradient will not cease at once, when the final breakdown of nurse cells occurs.

3.4.2 Transport of euplasm in telotrophic ovarioles

In all telotrophic ovarioles, the nurse cells remain in an anterior and endstanding position, whereas oocytes will be finally ordered in one row in the vitellarium. Nurse cells are connected via intercellular bridges to each other, at least at the beginning of ontogenesis; later on nurse cell membranes can degrade between sister cells, or even among nurse cells of other clusters (see section 3.5.3). Finally, one common syncytium is built, with the remarkable exception of Ephemeroptera (see below). In all other orders, the endstanding syncytial trophic chamber nourishes all growing oocytes at the same time. This may be important for the endogenous synthesis of RNPs of the oocyte itself (see sections 3.3.2 and 3.5.3).

Oocytes are connected to the endstanding trophic chamber by nutritive cords. These structures mediate transport to the oocyte of RNAs, RNPs and proteins synthesized by nurse cell genomes. Nutritive cords are elongated parts of the oocyte and not part of the nurse cell system. This can be deduced directly from the position of the permanent intercellular bridges in the orders of mayflies (Ephemeroptera), alder flies (Megaloptera: Sialidae), snake flies (Raphidioptera) and some beetles (Coleoptera: Polyphaga). In these species, intercellular bridges are always found at the anterior end of a nutritive cord, connected to the bottom of nurse cell syncytia. Intercellular bridges are never found at the opening of nutritive cords into the oocyte. In other species this can be deduced indirectly by the position of intercellular bridges during ontogenetic development (other polyphage Coleoptera; Hemiptera) (see section 3.5.3). The diameter of the nutritive cords can be less than 1 µm, when arrested oocytes are connected to the nurse cell syncytium, as found in many Hemiptera and some polyphage Coleoptera. The diameter can be more than 20 µm, when connecting growing oocytes with the trophic chamber. The length of nutritive cords varies with the amount and position of the growing oocytes in the trophic chamber. Thus, nutritive cords can be short, as found in aphids, scale insects or some chrysomelid beetles, in which only one oocyte will grow in a given ovariole, or nutritive cords can be very long (some 100 µm to several mm) as found in some Hemiptera or in some weevils (Coleoptera: Curculionidae).

As just shown, the four different types of telotrophic meroisitic ovarioles have some morphological findings in common, which may have some impact on the physiology of transport. In all systems a single and unbranched nutritive cord connects only one oocyte to the nurse cell syncytium. Constructs, in which two or more growing oocytes are supported via the same nutritive cord, are not reported. This shows that the oocyte needs its final and polarized position and this polarized position towards the nurse cell syncytium may be highly important for transport phenomena. In addition, one can assume that the oocyte itself is one of the main 'motors' of transport.

Principally, the same forces which work in polytrophic ovaries will work in telotrophic ones with one remarkable exception. Except in the ephemeropteran telotrophic ovary, there is no breakdown of nurse cells at the end of oocyte growth since nurse cells must give their resources to more than one oocyte. Thus, only two transport phases and/or forces can work in telotrophic ovaries. These are a possible selective transport of mRNAs or RNPs and a mass transport, preferentially of ribosomes, vacuoles and mitochondria. Mass transport has been shown first by ^{3}H-uridine autoradiography in *Oncopeltus* ovarioles (Zinsmeister and Davenport, 1971) and by Mays (1969, 1972) using the same methods in *Pyrrhocoris* ovarioles. Some hints for selective transport were given by the same author who was able to demonstrate a fast front of silver impregnation over nutritive cords, long before the main front of RNA labelling occurs (Table 3.5). *In situ* hybridization experiments with ^{3}H-marked poly-U, which hybridizes with poly-A tails of mRNA, indicated some sorting during late stages of oocyte growth in *Oncopeltus fasciatus* (Hemiptera). Silver grains accumulate during these stages preferentially at both oocyte poles (Capco and Jeffery, 1979). This indicates, as do older reports on oosome formation in polyphage Coleoptera, that differential sorting phenomena in oocytes take place in all oocytes, irrespective of their substitution by nurse cells and irrespective of ovariole types.

At this point, another phenomenon must be mentioned by which telotrophic ovarioles seem to differ from polytrophic ones. The germinal vesicle remains much more active in RNA synthesis than in most polytrophic ovarioles (see section 3.3.2 and Table 3.5).

Recent analysis of ovarioles of Ephemeroptera has shown that telotrophic ovarioles occur in this order (section 3.5.3; Gottanka, 1992; Beyrich and Gottanka, 1992; Gottanka and Büning, 1993). Some findings are unique among telotrophic ovarioles and might have some impact on transport phenomena. In contrast to the method in all other telotrophic ovaries, each oocyte is served by only one subcluster, which has lost its contact with the other subclusters. Clusters are linear and the topology of a given nurse cell inside a cluster is highly variable. Therefore, flow of material from the farthest nurse cell to the oocyte follows, at least in part, not the main polarity from anterior to posterior (oocyte) positions but must strictly follow the unbranched chain of nurse cells, which can be orientated against the main polarity (Figures 3.82–3.84).

Since there are no microtubules orientated in parallel to the main direction of the nurse cells, mass transport cannot be facilitated by these cytoskeletal structures. The only 'motor' of transport seems to be the oocyte itself and its storage behaviour of macromolecules, which does not differ from that

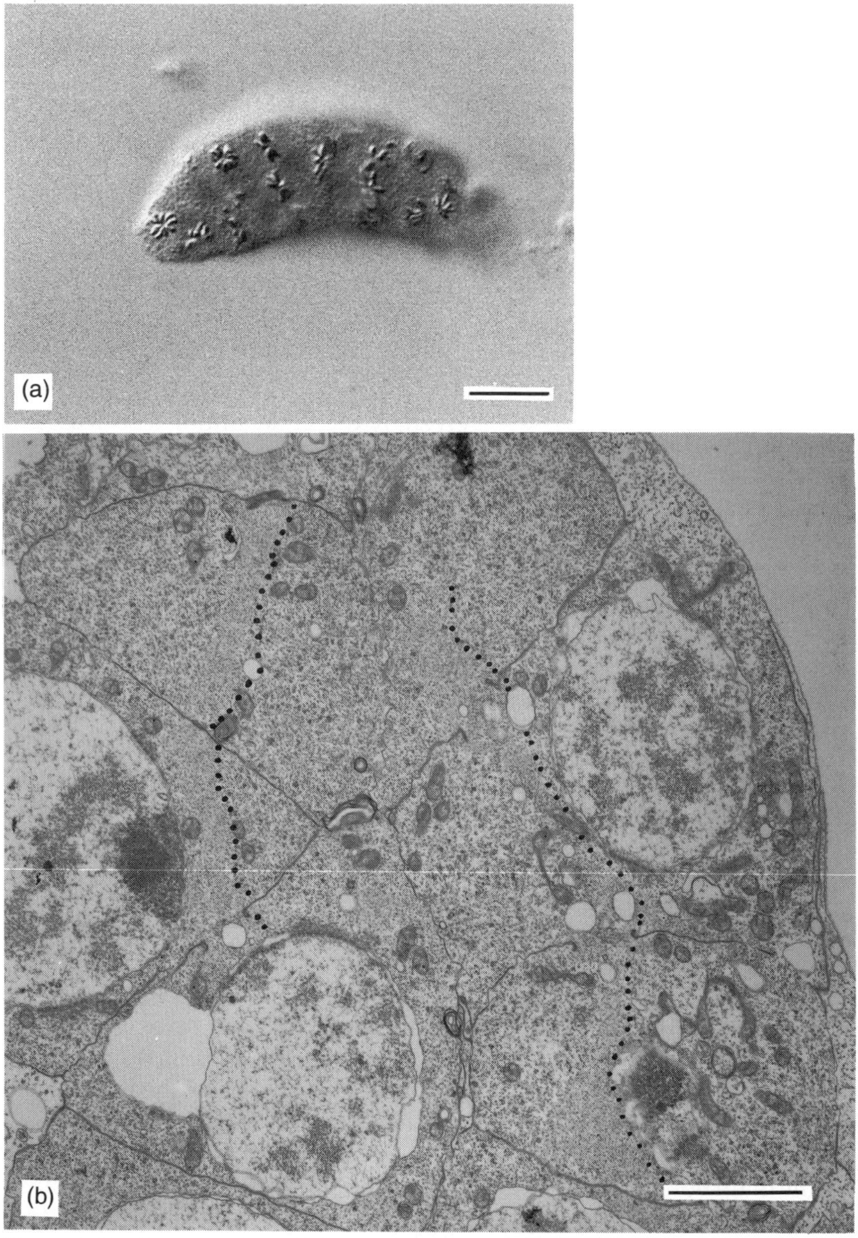

Figure 3.82 Ontogenesis of Ephemeroptera ovariole. (a) Synchronous divisions of germ cell siblings occur during early larval stages, as shown here for *Cloëon* sp., 4 months before final moult. Bar represents 10 μm. (b) Electron micrograph showing part of ovariole, slightly older. Polyfusome (one margin is stippled) stretches through intercellular bridges. No rosette formation is found. Bar represents 2 μm. (From Sohanka and Büning, 1993, with permission of Springer Verlag.)

found in the oocytes of other systems. The macromolecules may be stored in greater aggregates, which become visible ultrastructurally at least in the later phases of growth. In this way, a weak osmotic gradient may be established in this telotrophic ovary. Indications of enhanced synthesis of nurse cells, which would increase this polarity, are not yet available.

The next order in which telotrophic meroistic ovaries are found is the Hemiptera. This order includes scale insects, aphids, white flies, lantern flies and true bugs. As will be shown (section 3.5.3), ovary architecture varies from suborder to suborder, but the highest complexity occurs in true bugs (Heteroptera) and most information about transport phenomena in telotrophic ovarioles comes from this suborder.

The nutritive chamber of heteropterans is a highly polarized structure,

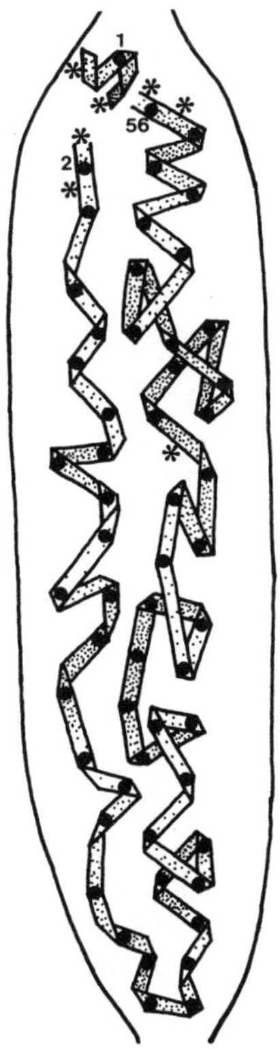

Figure 3.83 Reconstruction of the three-dimensional architecture of siblings in an ovariole of *Cloëon*. Fifty-five siblings form a linear cluster bent in one great curve through the whole region. The centre of the cells is marked by a black dot. Tight stippling indicates that cells are below, rare stippling indicates position above the plane. One cell is isolated in the tip and shows subsequent intercellular bridges without nuclei in between. All regions showing this phenomenon are marked with an asterisk.

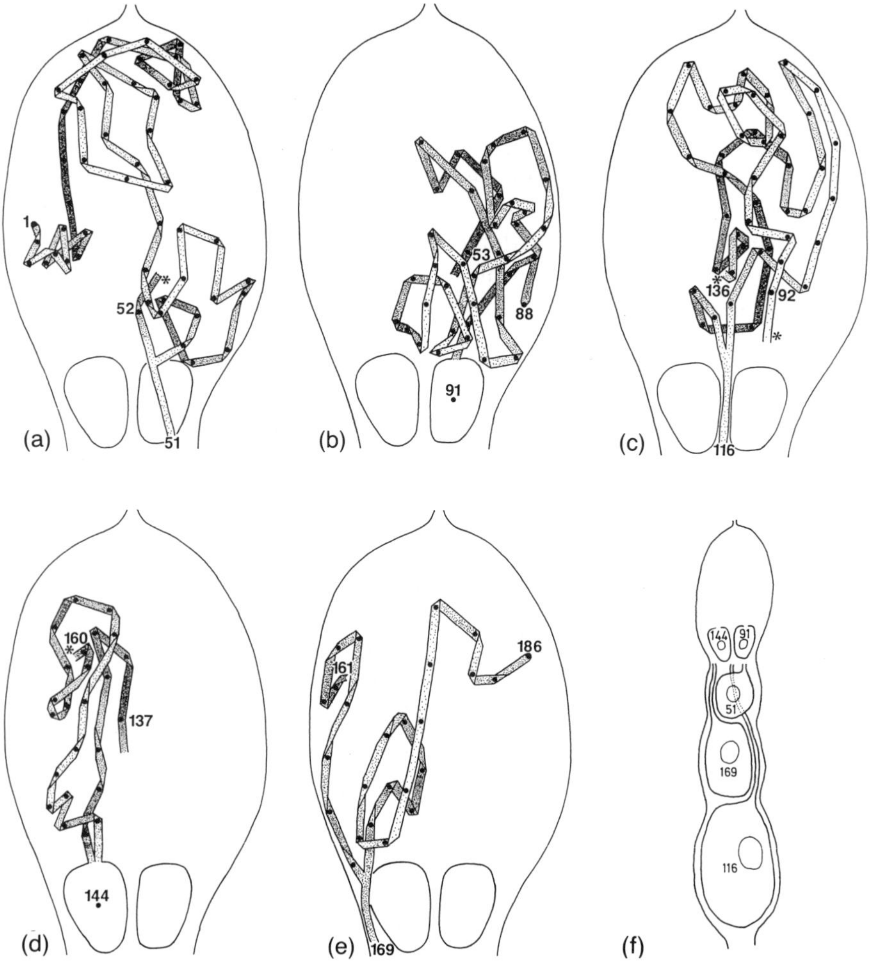

Figure 3.84 Cluster analysis in an ovariole of *Siphlonurus*, analysed by ultrathin serial sectioning. Arrangement of each subcluster (a–e) is shown separately for better understanding. Note that terminal cells (indicated by numbers) of one subcluster are always situated next to terminal cells with a closed intercellular bridge (*) of next subcluster. (f) Overview of the same ovariole, showing the position of the oocytes and nutritive cords.

which houses at its anterior part some permanently dividing nurse cells. These nurse cells are connected primarily via intercellular bridges to the central trophic core region. Later on (i.e. more basally) intercellular bridges lose their typical rims and nurse cells become connected to the central core by elongated, small tubes. At this stage these tubes can twist with each other, forming a membrane labyrinth. In the next zone, the tubes can enlarge to more or less wide openings. Finally, nurse cells can be autolysed at the bottom of the trophic chamber. Oocytes are all basally located and also connected to the central trophic core via very thin nutritive cords. Only those nutritive cords that communicate with growing oocytes enlarge. Nutritive cords, the central core region and the small tubes of nurse cells or the broad openings in basal regions of nurse cells are all filled with masses of microtubules, ordered parallel to the long axis of cords, tubes and trophic core. These structures have been studied as part of the transport system (Hamon and Folliot, 1969; Mays, 1969, 1972; Brunt, 1970; Huebner and Anderson, 1970; Macgregor and Stebbings, 1970). Consequently, contributions to microtubule-mediated transport are at the centre of transport research in heteropterans.

Valdimarsson and Huebner (1989) showed by exact staging of *Rhodnius prolixus* larvae and watching the birefringency of paralleled microtubules in the nutritive cords and core, that the microtubule system grows and enlarges tremendously about 1 week before final moult.

Stebbings and coworkers have tried to elucidate the role of microtubules during the transport of macromolecules from nurse cells to oocytes in heteropteran ovaries (Stebbings, 1988). They introduced *in vitro* systems to analyse explanted ovaries as well as isolated nutritive cords (Stebbings and van Bueren, 1981). In addition, they followed the ontogenetic development of nutritive cords in different species, showing that in *Corixa punctata* mitochondria are excluded from transport via microtubules, whereas in other bugs like *Notonecta glauca* and *Dysdercus fasciatus* mitochondria are abundant in the cross-sections of nutritive cords (Hyams and Stebbings, 1977, 1979a,b). Microtubules keep species-specific and constant distances between themselves in nutritive cords and nutritive cords enlarge in diameter by the addition of microtubules during previtellogenic growth. In isolated nutritive cords the polarity of microtubules was determined by the addition of tubulin monomers. These monomers can associate with stable microtubules at right angles, while further addition of monomers showed characteristic curves, the direction of which depends on the polarity of the stable microtubule. Looking along a microtubule from its plus end to its minus end, the added tubulin monomers assemble to short polymers, bent anticlockwise. This aspect was found when looking at cross-sections of nutritive tubes viewed from the oocyte side. The vast majority of microtubules were decorated with tubulin anticlockwise, indicating that microtubules are orientated with their minus ends to the oocyte, whereas the plus ends are orientated towards the trophic core (Stebbings and Hunt, 1983).

This orientation is to be expected when the microtubules are directly derived from the microtubules of the former spindle during the mitotic events of cluster formation, and the addition of microtubules during the growth of nutritive tubes does not change their orientation. Unfortunately, the authors gave no information about the orientation of microtubules from nurse cells to the trophic core, which is expected to show the same orientation as found in oocyte nutritive cords, when the former cleavage furrow lies in the trophic core. If this interpretation is correct (Figure 3.85), transport via microtubules from nurse cells to oocytes must switch from minus (nurse cell nucleus) to plus (trophic core), to plus (trophic core) to minus (oocyte). Such a switch in the trophic core region divides the microtubule-mediated transport into two parts if we assume transport via microtubule associated

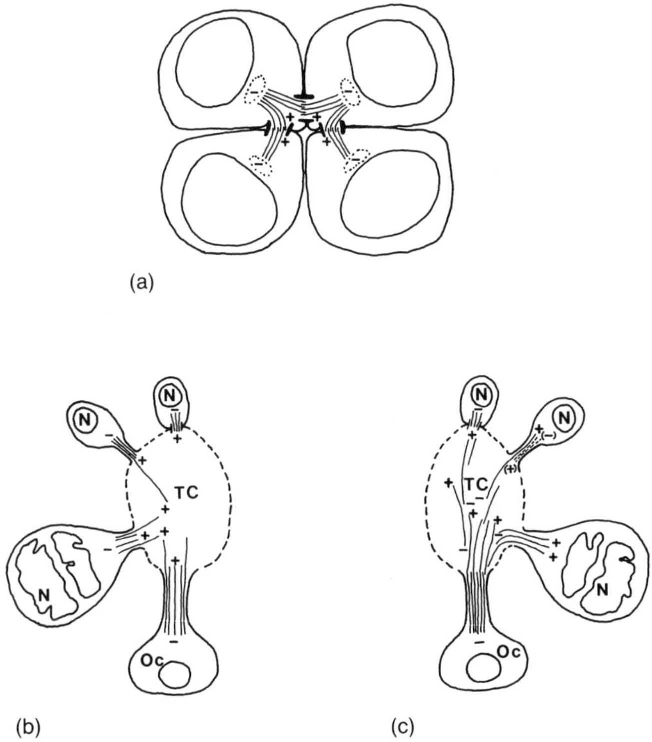

Figure 3.85 Hypothetical models of microtubule arrangement in telotrophic ovaries of Hemiptera. (a) Arrangement of microtubules in dividing clusters as shown here for a rosette-forming four cell cluster. In each cell, microtubules adhere with their negative ends to the centrosome (stippled areas). In intercellular bridges the positive ends meet. In Hemiptera, trophic core (TC) emanates from central arrangement of intercellular bridges. Two hypothetical models are shown. (b) Conservative substitution: orientation of microtubules remains as found during divisions. Additional microtubules are orientated in the same way. This leads to an accumulation of positive ends in the centre. (c) Polar substitution: orientation of microtubules remains only in the nutritive cords of oocytes (Oc). In nurse cells (N) orientation changes and additional microtubules follow the orientation of those of nutritive cords. Degenerating microtubules are stippled, their polarity is shown in parentheses.

proteins (MAPs), which have a unidirectional carrier function as dynein (+ to –) or kinesin (– to +) (Vale, 1987; Vallee *et al.*, 1988). Such difficulties could be avoided if bidirectional moving MAPs are engaged (Euteneuer *et al.*, 1988), or by strong asymmetry of growing microtubules in which the nurse cell half of the former mitotic spindle would be suppressed and substituted by the half-spindle of the oocyte.

Several attempts were made to analyse the character of microtubule bundles in nutritive cords and identify and isolate MAPs from nutritive tubes (Stebbings and Bennet, 1976; Stebbings *et al.*, 1985, 1986; Stebbings and Hunt, 1985, 1987; Stebbings and Sharma, 1989; Anastasi *et al.*, 1990). Evidence for MAPs in nutritive tubes was found indirectly by the cork-screwing behaviour of 'native' bundled microtubules isolated from nutritive tubes, which does not occur after salt pretreatment, during which MAPs are believed to dissociate from microtubules (Stebbings and Scharma, 1989). *In vitro* studies with bundles of microtubules isolated from nutritive tubes and depleted of endogenous MAPs by salt treatment could resume their ability to transport latex beads when substituted with extracts of insect ovarian MAPs (Anastasi *et al.*, 1990). Further biochemical characterization showed that kinesin is responsible for anterograde (– to +) movement. However, kinesin could not be demonstrated to be a native compound of nutritive cords. Retrograde (+ to –) transport could not definitively be ascribed to a particular protein, but dynein is the preferred candidate. However, substitution with purified fractions of insect dynein could not generate bead translocation.

Transport along microtubules in heteropteran ovaries was also studied in whole ovarioles, kept in Ringer's solution (Dittmann *et al.*, 1987; Münz and Dittmann, 1987). Transport of mitochondria was confirmed in trophic core regions by analysing the positions of fluorescein-labelled mitochondria, injected and fixed after different incubation times. The results show that mitochondria are transported within the trophic core towards the oocyte side and not to the nurse cells. However, transport through nutritive cords could not be demonstrated for technical reasons. In a second attempt, Dittmann *et al.* (1987) showed direct transport of mitochondria in nutritive tubes over a period of 90 min, using video-enhanced differential interference contrast. Analysis of 258 video-recorded mitochondria revealed that 10–30% of all mitochondria showed a clear translocation during analysis. Of moving mitochondria, 76% travel in the oocyte direction with a mean velocity of 3.37 μm/min, whereas 24% move towards the tropharium (2.84 μm/min). The character of movement was mainly that of the saltatory type I, as known from axonal transport mediated by microtubules (Figure 3.86). The individual velocity of mitochondria was the same in both directions; however, approximately 75% of all moving mitochondria shifted towards the oocyte side with a slightly higher mean velocity. The overall effect results in a discharge of mitochondria from the trophic core into the previtellogenic oocyte.

Figure 3.86 Model of microstream movements of mitochondria among microtubules in trophic cord of *Dysdercus* ovarioles; schematic cross-sectional view. Microtubules form regular arrays at distances of 300 nm. Each microtubule is surrounded by a rapid primary microstream region of 50 nm radius (crosshatched areas), which is thought to correspond to the circumtubular clear zone found in ultrathin sections around microtubules (Hyams and Stebbings, 1977). Slower floating microstreams in the opposite direction are supposed to exist in spaces between microtubule-bound primary microstreams. A mitochondrion (m_1) evenly spaced between microtubules will be transported slowly in one direction. Another mitochondrion (m_2), which attaches slightly to the fast microstreams will not move, since forces compensate. A third mitochondrion (m_3) which is tightly attached to two microtubules will move quickly in the opposite direction to m_1. (Figure from Dittmann *et al.*, (1987) with permission of Springer Verlag.)

The idea of a two-phase transport mechanism comes from ontogenetic arguments (see above) and some morphological and physiological data. Huebner and Gutzeit (1986) found an F-actin meshwork around the trophic core. Actin was also found in isolated nutritive cords forming discrete bundles orientated parallel to the long axis of nutritive cords (Huebner and Diehl-Jones, 1993). Furthermore, electrically mediated transport of macromolecules was demonstrated in the tropharium supporting the transport of negatively charged molecules from nurse cells into the trophic core in *Rhodnius prolixus* (Figure 3.87) (Telfer *et al.*, 1981, 1982) and *Oncopeltus fasciatus* (Woodruff and Anderson, 1984); however, further transport from the trophic core region via nutritive tubes into previtellogenic growing oocytes could not be demonstrated. In *Dysdercus intermedius*, flow of negatively charged molecules from the trophic core region into the oocytes was shown; however, the authors suggest a number of intercellular transport subsystems may work together in bug ovaries (Dittmann *et al.*, 1981, 1987; Münz and Dittmann, 1987). In the tropharium an electrical gradient is supposed to be the main force for transport into the trophic core. The main forces for transport into the oocyte via nutritive cords are supposed to be microtubule-based mechanisms, as known from axonal transport. Therefore, the microstream concept is preferred in which microtubule-associated force-generating enzymes produce microstreams along the surfaces of microtubules which carry the transported organelles, such as mitochondria and ribosomes. These microstreams exist around

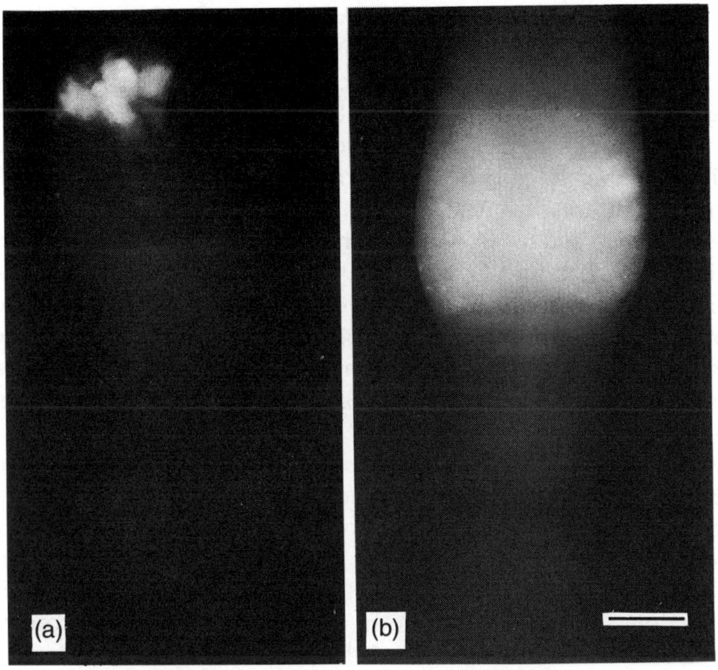

Figure 3.87 When positively charged molecules (here fluorescein-labelled lysozyme, FLY) are injected into trophocytes of *Rhodnius prolixus*, no transport occurs. When the same molecule is negatively charged (methylcarboxylated fluorescein-labelled lysozyme, McFLY) it spreads from the point of injection into all other nurse cells, to the trophic core and to the beginning of the trophic cords, indicating charge-mediated transport within the terminal chamber. Bars indicate 20 μm. (Courtesy of E. Huebner, University of Manitoba, Canada.)

microtubules, covering a region of 40–50 nm. The space between parallel microtubules differs from species to species and varies within the same nutritive tube to a great extent (50–300 nm). If spaces are large, a generally slower microstream in the opposite direction is supposed, since this microstream covers a larger area than the microtubule-associated microstream. This model can explain all kinds of movements of mitochondria watched directly by video-enhanced differential interference contrast microscopy (Figure 3.86) (Dittmann *et al.*, 1987).

Similar experiments on *Rhodnius prolixus* ovarioles were done by Diehl-Jones and Huebner (1992, 1993) and Huebner and Diehl-Jones (1993). They measured extracellular electrical currents along the telotrophic ovariole using a 2-D vibrating probe combined with ion substitutions, channel blockers and inhibitors. Ca^{2+} appears important to the currents around the tropharium and to transient inward currents at the apical area of mid-vitellogenic oocytes. This transient inward current correlates with the closure of trophic cords in *Rhodnius* ovarioles. Na^+ efflux at the apical and basal part of the vitellogenic follicle and Na^+ influx combined with Cl^- efflux on the middle of the vitellogenic follicle serve for current loops along the vitellogenic follicle. Similar currents occur along the tropharium by Na^+ efflux at the base of the tropharium and by influx of Na^+ and Ca^{2+} ions over middle and apical regions of the tropharium. Thus, in *Rhodnius* as in *Dysdercus*, combined transport mechanisms are supposed.

We have analysed the ovaries of a small group of Hemiptera, the psyllids (Psyllidae) (Beyrich and Gottanka, 1992; Büning *et al.*, in preparation; see section 3.5). In these insects nutritive cords are not filled with bundled microtubules; thus, microtubule-associated transport systems do not exist in this hemipteran group. This indicates that other transport mechanisms must substitute in this group and these other mechanisms might also exist in other hemipterans.

Two other telotrophic meroistic ovaries exist among holometabolous insects. Their different origin from each other as well as from Hemiptera is well documented. One is found in polyphage Coleoptera and the other in two closely related groups, the Sialidae (Megaloptera) and the Rhaphidioptera (Büning, 1972, 1978, 1979a,b,c, 1980; King and Büning, 1985; sections 3.5 and 3.6). In both, terminal chambers are connected to growing oocytes via nutritive cords lacking microtubules ordered in parallel bundles. In these types, as in Psyllidae, other mechanisms of transport must exist.

In polyphage Coleoptera one of these transport mechanisms might be based on peristaltic movements of the tunica externa of each ovariole. This can be seen in excised ovarioles of weevils, where the two ovarioles in each ovary are very long, containing up to 50 oocytes queued one after the other in the previtellogenic growth zone (see section 3.2.2; Büning, unpublished). Here, pulsating waves of the ovariole sheath are initiated at the base of the tropharium and run down the whole length of the ovariole. Each wave squeezes the nutritive cords and oocytes in a similar way to the peristaltic movements of the alimentary canal. However, these peristaltic movements were seen in only one out of ten females analysed. Thus, such movements do not occur regularly and can also be interpreted as stress-induced artefacts (Büning, unpublished). Similar observations on tenebrionid and coccinellid beetles were published by Stebbings (1981). The tunica externa of these beetles contains striated muscles over the whole length of the previtellogenic and vitellogenic growing oocytes, which could be used for peristaltic movements.

In addition to this mechanism, one must be aware of osmotic and electrical forces which might develop between the syncytial trophic chamber and the growing oocytes. Extracellular current has been recorded along the terminal chambers and the vitellarium of *Ips perturbatus* (Coleoptera:

Ipidae), indicating that electrical forces may be used for molecular transport (Huebner, 1984b).

In the ovary of *Sialis flavilatera* radioactive precursors for RNA synthesis are incorporated very slowly. The transport of synthesized molecules cannot be demonstrated by a front of silver grains running down the nutritive tubes or by apical–basal gradients in oocytes, as is possible in coleopteran or heteropteran ovarioles (Coleoptera, Polyphaga: Büning, 1972; Ullmann, 1973; Ray and Ramamurty, 1979; Heteroptera: Mays, 1972; Davenport, 1974, 1976). One can only see an accumulation of silver grains over different times, about 10 times slower than in Coleoptera, Polyphaga or Heteroptera (Figure 3.48; Table 3.5; Büning, unpublished).

This unexpected behaviour corresponds well with another unexpected behaviour of *Sialis* ovarioles. When they were treated with a vibrating probe, no extracellular current could be measured (Huebner, 1984b). Both findings clearly show that in the *Sialis* ovarioles, other less successful transport mechanisms occur compared with the telotrophic ovaries of Heteroptera or Coleoptera, Polyphaga.

The architecture of all four telotrophic meroistic ovaries combined with the ultrastructure of nurse cells, nutritive cords and oocytes clearly show that each telotrophic ovary has it own combination of transport systems. The most primitive one is that of Ephemeroptera, in which no membrane reduction occurs and each oocyte is supported by its own separated subcluster. Another transport system which is also not well developed is found among Sialidae and Rhaphidioptera. In polyphage beetles the transport system is more advanced, since here radioactive precursors reach the oocyte within hours. However, the most advanced transport type is that of the hemipterans, which show parallel bundled microtubules in nutritive cords and the nutritive core. Here, several of the above-mentioned mechanisms for transport of macromolecules and organelles are now being studied.

3.5 CLUSTER FORMATION OF GERM CELLS AND ARCHITECTURE OF OVARIOLE TYPES

When discussing details of germ cell cluster formation in females, the following terms will be used, based on definitions by King (1970). Germ cells are all those cells which are members of the germ line. Usually, germ cells become separated from somatic cells early during embryogenesis, before germ layers (i.e. ectoderm, mesoderm and entoderm) are built. Stem cells are defined by their potency to undergo unequal divisions, one sibling will be a stem cell again while its sister cell will behave as a cystoblast. A cystoblast cell is defined by its ability to undergo some species-specific cycles of equal divisions followed by incomplete cytokinesis. The siblings are called cystocytes. Among cystocytes, final transformation can occur which segregates definitive oocytes from abortive oocytes, which transform to nurse cells. All these terms are defined without any reference to the place at which the events will happen. Normally, the place is the germarial region of the ovarioles. However, in a variety of species these events can occur before or while ovarioles are being built. Such ovarioles will be devoid of germarial regions from the very beginning or the germarial regions will contain arrested cystocyte clusters and/or cystoblasts but no stem cells.

When discussing the early events of spermiogenesis (section 3.3.1), it can be shown that stem cells as well as cystoblasts undergo incomplete cytokinesis, which results in germ cell clusters in the male line. The following sequence of basic events (characters) occurs.

(1) Stem cell(s) undergo incomplete cytokinesis and cluster(s) of interconnected cystoblast cells can be generated.
(2) All these cells remain in contact with somatic apical cell(s) and with somatic cells of the inner sheath.
(3) Intercellular bridges of stem cell cluster(s) contain fusomal material.
(4) Cystoblast cells pinch off from the stem cell cluster, lose contact with apical cell(s) and become nested into somatic cells, bordered tightly to the inner sheath.
(5) Each cystoblast undergoes a species-specific programme of synchronized mitotic cycles.
(6) Clusters follow the 2^n-rule, i.e. a limited number of cycles (n) exists and all cystocytes will divide according to this programme.
(7) Cytokinesis of cystocytes is arrested, leaving intercellular bridges.
(8) Mitotic spindles will be degraded to midbodies and replaced by fusomal material.
(9) One spindle pole remains tight to the fusome of the preceding division, the other pole is situated just opposite. Thus, spindle axes are rectangular to preceding fusomes more or less and branching is maximized.
(10) Fusomal material of different mitotic events will fuse to polyfusomes.
(11) Rosette formation occurs during different mitotic cycles; i.e. after incomplete cytokinesis, intercellular bridges shift toward a centre, next to the first intercellular bridge.
(12) Synchrony of development of each cell of the cluster is strictly obeyed.

This strategy is most powerful when sperms are built (section 3.3.1), but needs at least some additional inventions optimizing the potency of syncytial clusters when oocytes should be developed. This section describes the architecture of the ovarioles found in insect orders. Thereby, phenotypical characters become separated by comparative views on ovarioles. These

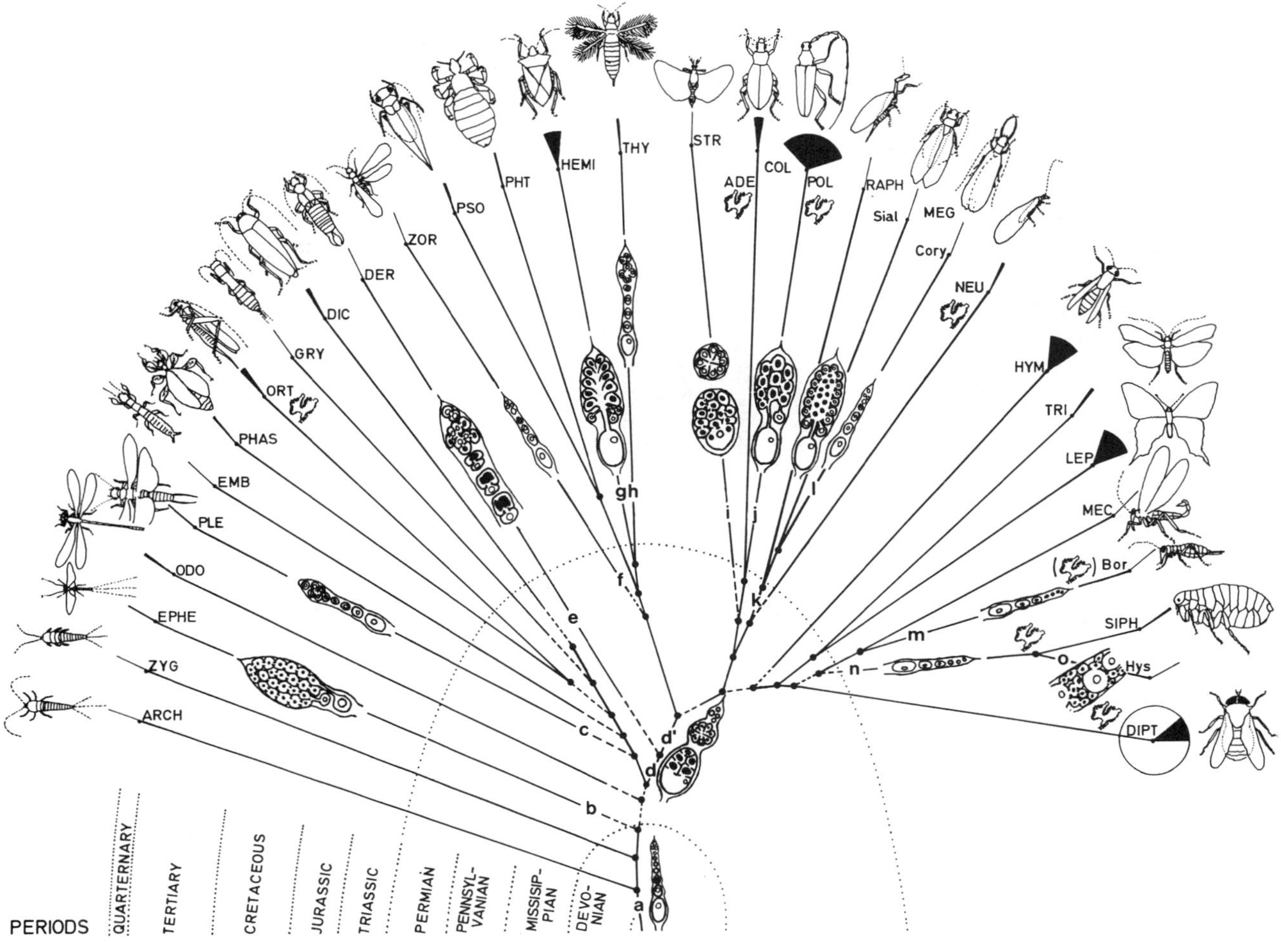
PERIODS
QUARTERNARY
TERTIARY
CRETACEOUS
JURASSIC
TRIASSIC
PERMIAN
PENNSYL-VANIAN
MISSISSIP-PIAN
DEVO-NIAN
ARCH
ZYG
EPHE
ODO
PLE
EMB
PHAS
ORT
GRY
DIC
DER
ZOR
PSO
PHT
HEMI
THY
STR
ADE
COL
POL
RAPH
Sial
MEG
Cory
NEU
HYM
TRI
LEP
MEC
Bor
SIPH
Hys
DIPT
a
b
c
d
d'
e
f
gh
i
j
k
l
m
n
o

characters will be used later on to reconstruct evolutionary pathways (section 3.7).

3.5.1 Panoistic ovaries

(a) Orders without cluster formation

The majority of ancient insect groups built their eggs through the panoistic pathway (Figure 3.88). We will first focus on those orders in which cystoblast mitoses of the female pathway are repressed to zero, i.e. the female cystoblast will differentiate directly into the oocyte. However, in respect to this question only a few of these taxa are investigated and further investigations are needed. Candidates for this type of panoistic ovary are the

---

Figure 3.88 Synopsis of ovariole types among insect orders. Phylogenetic relationships are adopted from Hennig (1981) and Kristensen (1981). Uncertain relationships are stippled. (a) At the base of true insects, a panoistic ovary is found, most probably by blocking cystocyte divisions in females. (b) In mayflies (Ephemeroptera) linear clusters arise, the cells of which are fixed in the germarium. Only oocytes will pass down the ovariole during growth, combined to upper cells by nutritive cords. Thus, a telotrophic ovary is built. (c) In stoneflies (Plecoptera) clusters are built in the germarium, but cells become isolated later on and all develop to oocytes, thus, a panoistic ovary is built, which is the first step to meroism. (d) At this phylogenetic point, the foremothers may have developed an effective differentiation between oocytes and nurse cells, including endomitosis in nurse cells. (d') Except Dermaptera, all other polytrophic meroists have in common one type of cluster development in which only one of the two innermost cells develops as an oocyte, all the others become nurse cells. (e) Earwigs (Dermaptera) show several autapomorphies (stem cell cluster, subcluster out of a cystocyte cluster), which separates this line from other polytrophic meroistic ovaries. (f) Zoraptera are said to be panoistic but ultrastructure and development are unknown. (g) Hemiptera are all telotrophic meroistic, the subgroups of which show a set of synapomorphies. (h) Thrips (Thysanoptera) have clusters in the germarium, which show similar characters to those from stoneflies. However, according to close relatives it is assumed that this panoistic ovary is born by reduction from a meroistic foremother. (i) Strepsiptera are endoparasitic insects, in which the ovaries show great reductions of somatic tissues. Cluster formation is similar to the basic type, thus, this ovary is supposed to be polytrophic, showing many autapomorphies. (j) Coleoptera: Polyphaga have telotrophic meroistic ovaries of unique ontogenesis and differ clearly from all other telotrophic ovaries. (k) Rhaphidioptera and Megaloptera: Sialidae also have a unique telotrophic ovary, which clearly has only one root. (l) Megaloptera: Corydalidae are panoistic, which is supposed to be a reduction from meroistic foremothers. (m) Similar reductions are assumed for the panoistic ovary of Mecoptera: Boreidae, and for (n) fleas (Siphonaptera). In Siphonaptera: Hystrichopsyllinae a polytrophic ovary is found, which has some autapomorphies. The distribution of extrachromosomal rDNA amplification is shown by a ring-symbol underneath the taxon abbreviation. Since the existence of extrachromosomal rDNA in Boreidae is still uncertain the symbol is in parentheses. For further discussion see sections 3.5 and 3.6.

Archeognatha, the Zygentoma, the Odonata, the Embioptera, the Phasmatodea, the Orthoptera, the Grylloblattaria and the Dictyoptera.

Table 3.1 lists some characters of this type of panoistic ovary, found in the different species of the above orders. As one can see, the list is incomplete and the number of investigated species is rather low. Ontogenetic development of ovarioles has not been looked at by ultrathin serial sectioning. Thus, there is no real proof for the total repression of cystoblast mitoses. However, germaria, if they exist, of all these ovarioles are small, cone like, and not separated by a neck region from the vitellarium (Figure 3.55a). The early ontogenetic development of *Thermobia domestica* gonads has been investigated by Klag (1977). At the end of his report, he mentions intercellular bridges which connect germ cells. Although he termed these cells oogonia, he gave no figure or description which allows a clear definition between oogonia and spermatogonia. A reinvestigation is needed. Our own, preliminary investigations on *Lepisma saccharina* ovaries have failed to demonstrate persistent intercellular bridges (Büning, unpublished). A detailed ultrastructural analysis of previtellogenesis in dragonflies (Odonata) has been given by Halkka (1980). Our own repeated investigations of germarial regions during ontogenetic development using ultrathin serial sectioning have failed to demonstrate clusters of cystoblasts. Thus, we believe that cystoblast divisions are repressed to zero in odonate female germaria (Gottanka and Büning, unpublished).

Ultrathin serial sectioning of ovarioles was also done in a Burmesian species of Embioptera and in *Carausius morosus* (Phasmida), with no success in demonstrating clusters (Gottanka and Büning, unpublished). Studies of early ontogenetic stages of ovarioles of *Bacillus rossius* (Phasmida) have been published (Taddei *et al.*, 1992) and will be discussed below.

In Orthoptera, we could not demonstrate clusters during the early development of female germaria in *Locusta migratoria.* The situation is different in *Gryllus bimaculatus* females. In this species, as in other gryllids, many germ cells degenerate during ontogenetic development in the germarial region (Buchner, 1909; Telfer, 1975). This behaviour has caused discussion about the cluster formation of germ cells in females of these species (Telfer, 1975). Ultrathin serial sectioning of ovarioles during ontogenetic growth has not clarified the situation (Gottanka and Büning, unpublished). Permanent intercellular bridges do not exist during adult and the final larval stages, when degeneration of about 50% of germ cells is found. However, during larval stage 2, some intercellular bridges are found connecting two cells each, the roles of which are not clear. The twins might be products of a stem cell division or of a cystoblast division (Figure 3.89).

All the intercellular bridge complexes seen are located in regions of the gonadal tube or at the base of differentiating ovarioles. Interestingly, all those membranes which contact the intercellular bridges directly are in

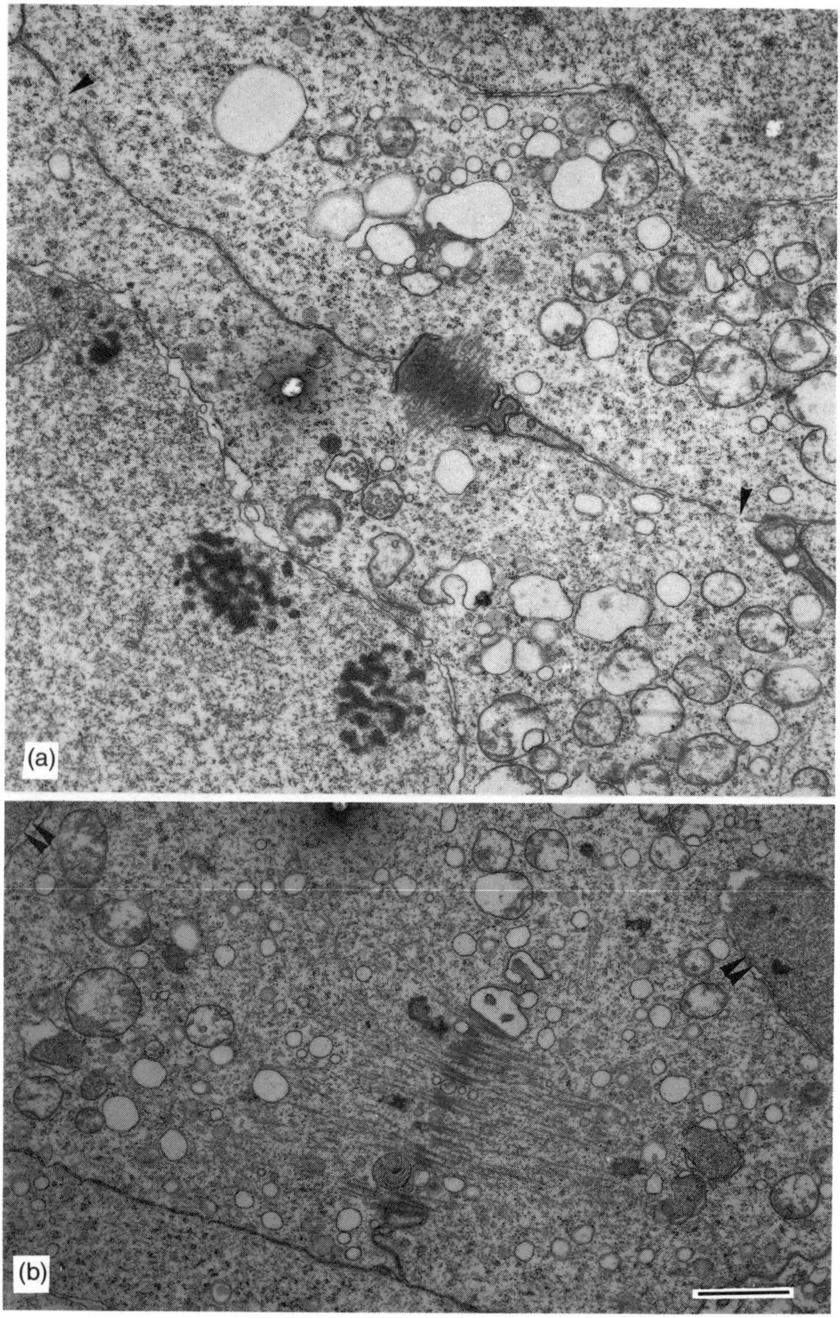

Figure 3.89 (a) Intercellular bridge between germ cell siblings from ovary of *Gryllus bimaculatus*, derived from a second larval instar. Both cells show the same volume and identical morphology. Note midbody in the intercellular bridge and breakdown of cell membrane between siblings (arrowheads). (b) The same ovary shows, in another plane, a more advanced stage of siblings, in which one part of the cell membrane is totally reduced. Arrowheads point to nuclei. (Courtesy of J. Gottanka, University of Erlangen-Nürnberg, Germany.)

different stages of destruction. Thus, we assume that these divisions will produce cells with at least double nuclei. Such cells must degenerate as a whole or one nucleus must degenerate. The first case can be seen in further stages, but we failed to demonstrate the second alternative. Thus, it seems as if cluster formation can be started in some part of cystoblasts, but the species will regulate 'overproduction' of oocytes by degeneration processes (Gottanka and Büning, unpublished).

Another example, which should be looked at for cluster formation, possibly combined with degeneration of one of the hypothetical siblings, is the ovary of *Tachycines asynamores* (Saltatoria: Ensifera, Gryllacridoidea). This ovary was investigated thoroughly at the light-microscopical level (Von Kraft, 1960a,b), showing some details which are important to our discussion. Von Kraft found constancy of germ cell descendants in ovarioles from larval stage 3 onwards. No mitotic figures are found in germ cells older than stage 3. Each ovariole houses about 18 germ cells, of which all but two are in a resting stage (L3). This shifts to equal amounts in L8 females, about nine still resting in the germarium and another nine, which have entered the growing zone, the vitellarium, the oocytes of which are single lined and all in subsequent previtellogenic growth phases. In young adults, the germarium houses only a few (2–4) arrested germ cells. All other germ cell descendants are growing, with the terminal oocyte starting vitellogenesis. In old adults, a germarial zone is missing and the anterior end of the ovariole houses a previtellogenic growing oocyte. The total number of oocytes has reduced to five or seven at the end of the egg laying period. In young larvae especially, the germarium contains pycnotic nuclei, the number and structure of which are similar to those found in *Gryllus*. Germ cells of the germarium are enclosed by somatic inner sheath cells, which are in contact with few interstitial cells. Prefollicular cells emerge continuously from these cells via mitosis. Young oocytes are enveloped by only some, flat follicular cells, the number of which increases during previtellogenic growth until a cuboid follicular epithelium is formed just before vitellogenesis starts. At this stage, the nuclei of follicular cells have become enlarged by endomitosis and finally the nucleus seems to split into two nuclei which keep tightly together like peanut twins in their common seed-case. Possibly, the nucleus is not halved totally as was demonstrated for *Gryllus* follicular cells during vitellogenesis (Gross, 1903; Dennis and Bradley, 1989). In another closely related locust, *Isophya leonorae* (Saltatoria: Ensifera), a germarial region does not exist in ovarioles. Even in ovariole anlagen of the first larval instar we found a constant number of about eight oocytes, the morphology of which resembles that of oocytes during previtellogenic growth (Figure 3.55b) (Demuth and Büning, unpublished). During larval life, the number of oocytes decreases to about six, and one or two degenerating cells are watched in each ovariole. This behaviour shows clearly that stem cell divisions with or without cluster formation have shifted to very early stages of gonad development during organogenesis in embryos. Taddei *et al.* (1992, 1993) have investigated the development of ovarioles of the stick insect *Bacillus rossius* during larval and adult stages. They could not find mitotic figures of germ cells during this period and no cluster formation was found. Germ cells of the germarial region remain arrested in late diplotene meiotic stage. The vitellarium contains some previtellogenic growing oocytes. Summing up the last examples, one must conclude that cystoblast cells transform directly into oocytes without cluster formation and mitotic activity of stem cells becomes shifted to very early stages, which can result in the total loss of a germarium in an ovariole, as has been shown to happen in *Isophya* ovarioles. Such a shift is also known from other species like coccids (Hemiptera) or most midges (Diptera). All panoistic ovaries without cluster formation follow a similar pattern of oogenesis, showing few variations (Table 3.1).

The ovary of Dictyoptera, especially of diverse cockroaches, has been investigated in many laboratories, since these animals can be reared easily and they are ideal organisms for physiological and behavioural study. The histology and ultrastructure of ovaries has been investigated but the germ cell character in germaria has not been studied in detail (Bonhag, 1959; Bonhag and Arnold, 1961; Anderson, 1964, 1969). Thus, proof of cluster formation is still missing.

Besides these 'primitive' orders with panoistic ovaries, there are some taxa which have panoistic ovaries of the same kind as just described, but which belong to 'modern' insect groups the close relatives of which have meroistic ovaries (Figure 3.88). Thus, all these examples are thought to be reduced forms, emerging from meroistic ancestors which lost the character of cystocyte divisions in females (see section 3.6). These ovaries are therefore denoted as secondary panoistic.

The first case is that of dobsonflies (Corydalidae), which are the sister group of Sialidae, having telotrophic meroistic ovaries (Table 3.3; Figure 3.88) (Matsuzaki and Ando, 1977; Büning, 1979c). Matsuzaki and coworkers (1985) investigated *Protohermes grandis* and could not find any signs of cluster formation. Mitosis of germ cells is found only during larval stages in the anterior region of ovarioles. Persisting intercellular bridges were not reported. Thus, one can conclude that stem cells cease during late larval stages and oocytes develop from cystoblasts directly without cystocyte divisions. Follicle cells develop from sparse nests of prefollicular cells, which do not form a coherent prefollicular tissue. The germinal vesicle is highly active, and numerous nucleolus-like aggregates are found in the germinal vesicle indicating extrachromosomal rDNA amplification.

The second example is found in snow fleas (Mecoptera: Boreidae) (Table 3.4; Figures 3.56 and 3.88) (Steiner, 1937, Büning, unpublished). They have panoistic ovaries of the above type, whereas scorpion flies like *Panorpa communis* (Mecoptera) have polytrophic meroistic ovaries (Gross, 1903;

Grell, 1942; Ramamurty, 1964a,b, 1967, 1968). In two female adults of boreids, showing panoistic ovaries, a germarial region was no longer existent and the most anterior oocyte had already started previtellogenic growth. Thus, nobody knows if cluster formation takes part in the germaria of snow flea ovaries. If clusters are built, this phase must occur during earlier stages. However, this example shows clearly, that stem cell divisions cease at least during adult stages in snow fleas. Furthermore, rDNA genes of oocytes seem to be amplified in this species, showing numerous nucleolar bodies (Figure 3.56). Follicular cells rise from small nests of prefollicular cells, the nuclei of which are found concentrated in wedged corners between young oocytes. Later on, high mitotic activity will build the follicular epithelium in which cytokinesis can be arrested and siblings remain connected by small intercellular bridges (Figure 3.20b). This character is shared with some hymenopterans and dipterans. The ooplasm shows some characteristics which are also found in flea ooplasm (see below); one can distinguish between early cytoplasm, poor in free ribosomes, and late cytoplasm, rich in ribosomes, which occurs during previtellogenic growth, and occupies about 90% of the volume when vitellogenesis begins.

Another example is found in the order of fleas (Table 3.4; Figure 3.88) (Siphonaptera) (Büning and Sohst, 1988, 1989). The majority of species have panoistic ovaries, in which clusters have not been detected. Other characters, especially the rise of follicular cells, are fully comparable with those of panoistic ovaries described above, especially that of snow fleas. Here, too, stem cells cease their mitotic activity, oocyte nuclei contain extrachromosomal amplified rDNA (Figure 3.69), and follicular cells do not develop from a coherent prefollicular tissue. The order of fleas is relatively young, the nearest relatives belong to the mecopteran complex and have polytrophic ovaries (Figure 3.88) (see section 3.6 for further discussion). Among fleas, one group has developed a polytrophic meroistic ovary, the Hystrichopsyllinae (this ovary will be described in section 3.5.2).

(b) Orders with cluster formation

The first order described, in which panoistic ovaries are built out of cystoblast clusters, is the thrips (Thysanoptera) (Table 3.2; Figure 3.88) (Pritsch and Büning, 1989). The germarial region is clearly separated from the vitellarium by a constriction (Pritsch and Büning, 1989; their Figure 1). By serial sectioning of an ovariole of a young adult female, the following structures were found. In the tip of the germarium one large germ cell was found, which might be a stem cell (Figure 3.90).

A four-cell cluster follows, the cells of which are all in telophase. The central intercellular bridge is filled with fusomal material, whereas the other two intercellular bridges contain remnants of the mitotic spindle apparatus. The next cluster contains eight cells; the intercellular bridges of three cells each form small rosettes (for definition see section 3.5.2). Fusomal material of the intercellular bridges in rosettes is combined to form polyfusomes. Posteriorly, single germ cells are found, showing no intercellular bridges. However, two isolated complexes could be detected, in which fusomal material, intercellular bridges and cell membranes are mixed. In ovarioles of older females, a germarial region with clusters is no longer evident. All germ cells are found in advanced previtellogenesis.

As in the examples described before, these data clearly show that stem cells stop their dividing programme during the reproductive phase. The

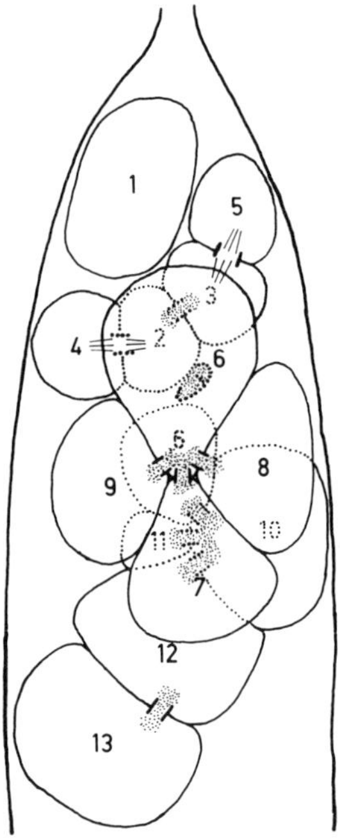

Figure 3.90 The upper half of a germarium of a young adult female of *Parthenothrips dracenae* (Thysanoptera). Cell 1 is proposed to be the stem cell. Cells 2–5 are combined by intercellular bridges to the young four-cell cluster. Cells 6–13 represent the older cluster, in which two rosette configurations are seen, each bearing a polyfusome (fine stippling). Cells below others are indicated by broken lines. Cell 6 is shown in both optical planes. Intercellular bridges are shown by thick lines. Cell 11 contains only 6% of the average volume of its sister cells and has no nucleus. A separated intercellular bridge complex is situated between cells 2 and 6 (lower plane). (From Pritsch and Büning, 1988; with permission of Springer Verlag.)

same behaviour is found in many other species, irrespective of their panoistic or meroistic ovariole character, except the drosophilids and possibly other closely related groups of dipterans (King, 1970). Thus, it is assumed that a restricted number of stem cell mitoses reflect the basic condition of germ stem cell behaviour in insect ovaries. The consequences of this panoistic ovary in Thysanoptera in an environment of orders having meroistic ovaries are discussed in section 3.6.

The second order, in which panoistic ovaries are built with cluster formation (Gottanka and Büning, 1990), is the very ancient order of stoneflies (Plecoptera) (Table 3.1; Figure 3.88).

Each ovary consists of numerous short ovarioles, ordered perpendicularly around two strong lateral oviducts, which fuse anteriorly above the alimentary canal to form one continuous, horseshoe-like ovary (Figure 3.14). The ovary sheath embraces all the ovarioles in young gonads but disrupts later on, and ovarioles are bathed freely in the haemolymph of the body cavity, while remnants of the ovary sheath still embrace the lateral oviduct with trachea and muscle cells. An outer ovariole sheath is not developed and a terminal filament is missing in the ovarioles of adults. Germ cell descendants are enclosed by an inner sheath in the anterior regions of the ovariole and posteriorly by follicular cells. Inner sheath cells are continuous, with few interstitial cells and both continue in small nests of prefollicular cells, emanating from the follicular cells, embracing each oocyte with a monolayer.

In some genetically determined males of plecopterans (Perlidae), the gonads develop ovarioles in anterior regions, which degenerate later on, whereas testicular follicles will function (Brandt, 1878; Schoenemund, 1924).

Cluster analysis was done on the ovarioles of *Nemoura* sp. (Gottanka and Büning, 1990). The analysis of an ovariole, cut in ultrathin serial sections, showed 36 germ cell descendants in the anterior region, distributed among six different syncytia. Figure 3.91 represents the situation schematically.

In the tip (i.e. most anteriorly) a two-cell cluster exists, containing fusomal material in its intercellular bridge. The next three syncytia are separated from the first by protrusions of somatic cells (prefollicular cells) and are situated side by side in the same area. One of these contains six germ cell descendants, the other two contain eight cells each. The three syncytia are followed by one, which contains seven cells but has increased its volume significantly, occupying a similar volume to the three other upper syncytia. The bottom of the germarium is occupied by five cells, one of which is small and pycnotic while the other four are large. One of these larger cells has no intercellular bridge and no remnant of fusomal material. The second has one closed bridge with fusomal material and one isolated area of fusomal material. The other two are combined by an intercellular bridge through which a polyfusome stretches, ending in a closed intercellular bridge. The four big syncytia show rosette formation in part, and have

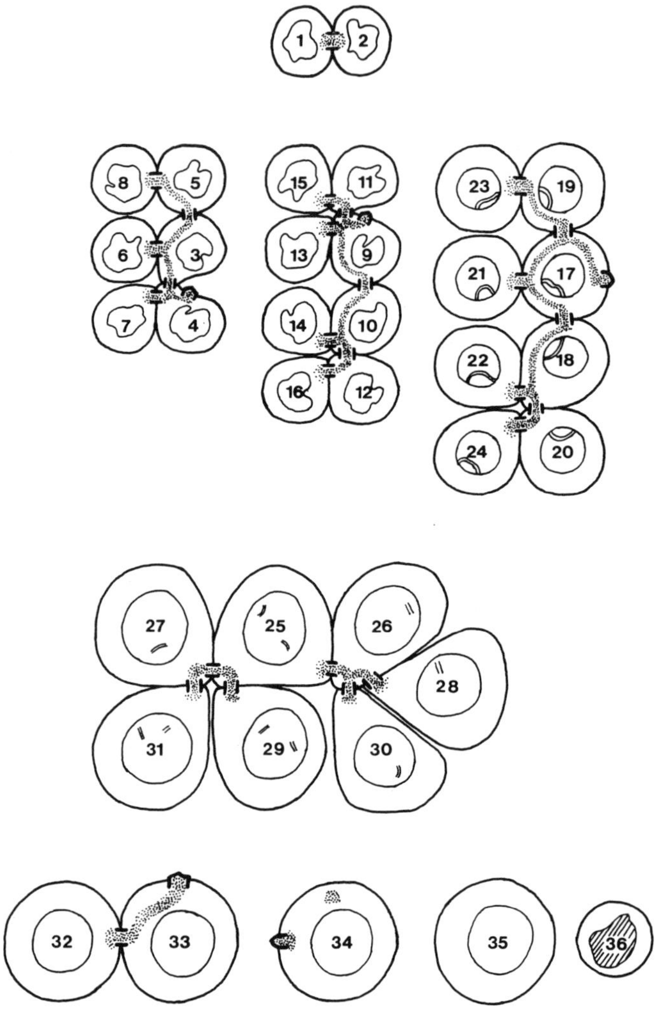

Figure 3.91 All cystocytes of the germarial region of a female last larval instar of *Nemoura* sp. (Plecoptera). Cells are put in one plane and clusters are separated since it is impossible to give the correct three-dimensional arrangement without loss of clarity. Synaptonemal complexes are indicated by double lines in cell nuclei. Fusomes and polyfusomes are stippled, rosettes are indicated by juxtaposition of intercellular bridges. Cell 36 is pycnotic. (From Gottanka and Büning, 1990; with permission of Pergamon Press.)

the polyfusome stretching through them. A polyfusome may be needed for rosette formation but it is not sufficient alone, since other regions show polyfusomes but no rosette formation. Five blind intercellular bridges are found, indicating that the second cytokinetic furrow follows the old, arrested one or is at least very close to it. Isolated rosette aggregates, as in *Parthenotrips dracenae* (Thysanoptera), are not found. The ultrastructure of interconnected cells clearly demonstrates that development of clustered cells is synchronized. No signs indicating differentiating processes between siblings are found. Thus, all cells continue their development as oocytes after splitting.

3.5.2 Polytrophic meroistic ovaries

Meroistic ovaries differ by one important criterion from panoistic ovarioles. A certain amount of interconnected cystocytes will be transformed into nurse cells, which can now differentiate their genomes by polyploidization and/or over- and/or under-replication of special genes (see section 3.3.4). A second event is also necessary for successful functioning, i.e. a transport system must be developed, by which synthesized products of nurse cells will reach the oocyte (see section 3.4).

Polytrophic meroistic ovaries differ from telotrophic ones by close association of nurse cells and oocytes of common origin, both of which become ensheathed by follicular cells and grow as a physiological unit, the follicle, which is separated from other follicles by follicular stalk cells.

(a) Psocodea and the basic type of polytrophic ovaries

When discussing polytrophic ovaries, we will focus first on the ovary of barklice (Psocodea: Psocoptera) in which the basic type of polytrophic meroistic ovary is shown in its clearest form. Thereafter, all orders will be looked at, in which polytrophic ovaries show variations of the basic type (Tables 3.1–3.4; Figure 3.88).

If one constructs a 'basic type' of polytrophic meroistic ovary, one should be aware that this type is a suggested model, derived from a huge mass of investigations of many orders, each showing some specialized developments. The model presented here is the extended continuation of some fundamental work (Figure 3.92) (Giardina, 1901; Knaben, 1934; Hirschler, 1945; King, 1970; Telfer, 1975; King *et al.*, 1982; Huebner, 1984a; King and Büning, 1985; Büning and Sohst, 1990).

When going through the characters of the hypothetical basic type of polytrophic meroistic ovaries of insects, we can follow the fate of germ cells and somatic cells in time (i.e. through ontogenetic development) and in space, since in most ovarioles the ontogenetic development of germ cell descendants is reflected by the polar reorganization of the ovarioles from anterior (early) to posterior (late) regions.

Anterior sheath cells (Figure 3.92a) are somatic cells which are homologous to apical cells in testis follicles during early spermiogenesis. They are thought to serve as 'docking cells' for the germarial stem cells. Since stem

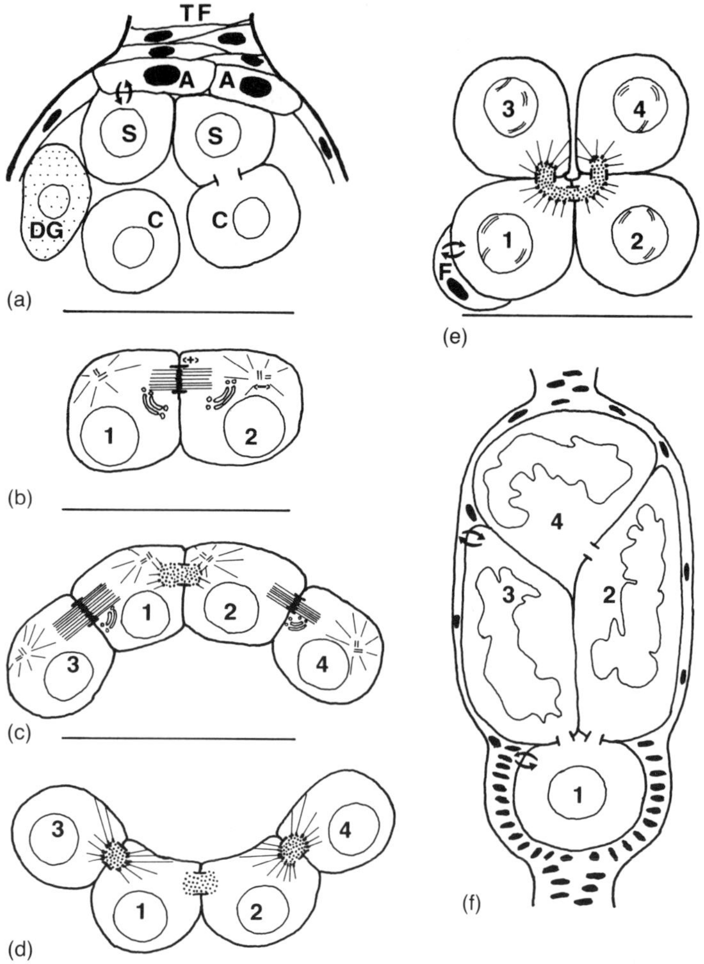

Figure 3.92 Synopsis of early events of cluster formation in female gonads. Terminal filament (TF); anterior sheath cells (apical cells) (A); stem cells (S); cystoblast cell (C); degenerating germ cell (DG); prefollicular cell (F); arrows indicate inductional processes (cell–cell communication). Somatic nuclei are shown in black. (a) Early events during ovariole formation. (b–e) Cluster formation. Cystocytes are numbered. Fusomal areas are stippled; polarity of microtubules is shown by + or - symbol in (b). Meiotic prophase is indicated by parallel lines in nuclei of (e). (f) The final follicle.

cells cease during the reproductive life time (see below), the anterior sheath cells will function only during larval or young imaginal stages. These anterior somatic cells can be morphologically different from lateral sheath cells, as found in spittlebugs (Ksiazkiewicz-Kapralska, 1989a,b), moths (Grünberg, 1903; Sato, 1932; Godula, personal communication), barklice

(Büning, unpublished) or flies (Gregorio *et al.*, 1990a,b). The cells are prismatic, their cytoplasm is often clear and the nuclei are in most cases spherical and heterochromatic areas are rare. Most probably, it is this type of cell from which a small group of cells segregate in Diptera, which contact the oocyte after migration through the nurse cell complex as anterior (terminal or polar) cells in *Drosophila* follicles and other flies like Muscidae (Verhein, 1921; King, 1970; Gutzeit, 1986a; Geysen *et al.*, 1988; De Loof *et al.*, 1990) (Figure 3.82). In all stem cells looked at so far, the number of primary invading germ cells into the ovariole anlagen is low and only some of them will be able to undergo differential mitoses creating cystoblasts and stem cells. Atresia of germ cells found in the young ovariole anlagen of many species is the morphological consequence of this behaviour.

Functional stem cells will remain in the anterior part of the germarium via communication with apical cells as is assumed above, and will undergo a limited number of differential mitotic divisions, i.e. during the time of reproduction, the number of prospective oocytes becomes fixed. However, during the normal life time of insects the number of prospective oocytes is much larger than the capacity of the species to raise all these cells to eggs.

Differential mitosis of stem cells produces siblings of which one behaves as the stem cell and one as the cystoblast (see below). At the end of programmed stem cell divisions both siblings must be converted to cystoblast cells or the stem cell remains as an undeveloped germ cell, which will finally degenerate. This is still an open question and may be regulated differently in different species. In *Hystrichopsylla talpae*, the mole flea, Büning and Sohst (1988) found no isolated stem cells and no cells in pycnosis in the germarial region of adult ovarioles at the end of reproduction, but only small clusters of cystocytes. This indicates that in the end stem cell(s) transform directly into cystoblasts. However, pycnosis of stem cells during pupal stages, which were not investigated, cannot be ruled out.

Each cystoblast undergoes a genetically fixed programme of synchronized division cycles, which results in a clone of interconnected cells by incomplete cytokinesis (Figure 3.92b–e). The final clone or cluster contains 2^n cells, generated by n mitotic cycles. The geometry of final clusters is always the same. Each cell divides in each cycle, whereby one sibling gets all preceding intercellular bridges, the other none. Thus, the cluster shows maximal branching (Coleoptera, Adephaga: Giardina, 1901; Lepidoptera: Knaben, 1934; Lepidoptera and other insect orders: Hirschler, 1945; Diptera: King, 1970; insect orders in general: Telfer, 1975; King *et al.*, 1982; King and Büning, 1985). Measurements of siblings show that all cells are of equal volume during the first mitotic cycles (Brown and King, 1964; Koch and King, 1966, 1969; Büning and Sohst, 1990). If the cluster becomes very large, the end-standing cells will be a bit smaller (*Hystrichopsylla talpae*, Siphonaptera: Büning and Sohst, 1988; Coleoptera, Adephaga: Jaglarz, 1992). However, this behaviour seems to be more an expression of onset of nutritive transport to the innermost cells (i.e. the oocyte) than an expression of qualitative unequal divisions. Later on, volumes of nurse cells can follow an apical–basal gradient by different amounts of endomitotic cycles of nurse cells (see section 3.3.4).

Between subsequent mitotic cycles, all intercellular bridges move during interphase to the centre of the cluster, forming a rosette (Figure 3.92e) (Hirschler, 1945; Telfer, 1975). In each intercellular bridge a transient midbody occurs, which is the remnant of the preceding spindle equator (Figure 3.89). Spindle microtubules decrease, whereas an amorphous substance, the fusome, is increasing and substituting the central area of the spindle and the midbody. Fusomes of preceding divisions fuse to polyfusomes, which can combine all siblings in the end (Hirschler, 1945; Telfer, 1975; King *et al.*, 1982; King and Büning, 1985, Büning and Sohst, 1990). Thereby the polyfusome shortens concomitantly with rosette formation (Figure 3.81). Polyfusomes contain F-actin, at least in some species (section 3.4.1). Polyfusomes and rosettes exist as long as divisions of cystocytes are continuing. The polyfusome can be looked at as a centre of cytoskeletal elements, especially of microtubules as was shown in the barklice ovary (Büning and Sohst, 1990). However, the polyfusome is not a necessary prerequisite for rosette formation, as shown in the aphid ovary, in which rosette formation occurs without developing a fusome. In this group, as in most other hemipteran groups, the microtubules of the spindle persist (Büning, 1985) (see also section 3.5.3). After the mitotic programme of cystocytes has finished, polyfusomes cease and intercellular bridges can move apart into their final positions during previtellogenic growth (Figure 3.92f) (Büning and Sohst, 1990).

During this cluster formation, the definite oocyte will be determined, whereas all other cells will be transdetermined into nurse cells. Transdetermination of nurse cells from a hypothetical predetermination of germ cells to oocytes is deduced from the behaviour of clusters of panoistic ovaries, in which all cells of a cluster become oocytes as shown for stone flies (Plecoptera) (Gottanka and Büning, 1990) and thrips (Thysanoptera) (Pritsch and Büning, 1989) (see section 3.5.1). Although we have no information about the exact time of determination (oocytes) and transdetermination (nurse cells) or of the gene activities behind it, we have information about the amount and position of the cells remaining as oocytes in the basic type: it is always one of the two oldest siblings in each cluster, derived from the first mitotic cycle, or in other words, it is always one of the two innermost cells of the branched cluster.

In a series of papers, King and coworkers pointed out that the polyfusome may be used as a control element for pro-oocyte determination in *Drosophila* ovaries (King, 1970, 1979; King *et al.*, 1982; King and Storto, 1988; Storto and King, 1987, 1988, 1989). In this ovary a cluster of 16 cells comes into being, of which the two innermost cells possess four intercellular bridges each. Both have the largest compartments of the polyfusome and both start with meiotic prophase, whereas the other siblings do not share this process. One of these two cells will go on with meiosis, the other sibling will stop during pachytene and enter the nurse cell pathway as did the other siblings.

This concept cannot be adopted for the basic type of polytrophic ovary for several reasons. There are many examples from a variety of orders in which all siblings enter meiosis up to pachytene (Lepidoptera: Knaben, 1934; Mecoptera, Panorpidae: Büning, unpublished; Coleoptera, Adephaga: Ribbert and Weber, 1970; Psocoptera: Büning and Sohst, 1990; Dermaptera: Yamauchi and Yoshitake, 1982). The same is found in telotrophic meroistic ovaries, which are thought to evolve from polytrophic ancestors (Hemiptera, Coccoidea, Aleyrodoidea: Büning, 1985; Nowak and Büning, unpublished; Coleoptera, Polyphaga: Büning, 1972; Megaloptera, Sialidae: Büning, 1979c). In other dipteran orders more than one sister cell will enter meiosis and follow, at first, the oocyte pathway as was described for some tipulids (Bayreuther, 1952, 1956). Similar distribution is known from *Chrysopa perla* (Neuroptera) (Rousset, 1978a)(see below).

Thus, fundamental events leading to oocyte–nurse cell transdetermination must have occurred once, at the base of meroistic ovaries of true insects. As stated before, we know nothing about the molecular processes leading to oocyte determination and to nurse cell transdetermination. However, some speculation, based on morphological findings, can be given. If we look at all insect groups including the entognathans, we must consider that oocyte determination is most easily explained by gradient theories not bound to the polyfusome. In entognathans we have such situations as linear clusters of which only one cell becomes an oocyte but their nearest neighbours show intermediate morphology between nurse cells and oocytes (see section 2). In the very primitive true insects of mayflies, we found linear clusters too, of which only one cell develops into an oocyte, the position of which is not fixed from the very beginning but depends on the topological position of the cell (see section 3.5.3). A similar aspect is found in the telotrophic ovary of sialids, in which only posterior tapetum cells will develop into oocytes (section 3.5.3). Such gradients may be constructed in different ways. The first and most basal event may be cell-to-cell communication between germ cells and somatic cells. Contact between apical cells and germ cells will decide that germ cells become oocytes, when the female pathway is used (Figure 3.92a). Without apical cell contact, germ cells will not develop further. This mechanism can easily explain all panoistic ovaries regardless of whether they develop clusters or not. If clusters are built, one additional event must be evolved; the cluster of oocytes must split into single cells before growth phases are started.

When evolving meroistic ovaries, another signal must be developed, which can decide between nurse cells and oocytes. One possibility is the contact of cluster siblings to the prefollicular tissue (Figure 3.92e). Only those cells which have a strong contact remain oocytes; all the others will develop into nurse cells, perhaps by signals coming from sheath cells (Figure 3.92f). Other possibilities may use only signals coming from germ cells, perhaps as time counting signals. However, as we know from sex determination (section 3.3.1), oocyte–nurse cell determination is most probably a process in which both cell types, germ cells and somatic cells, are involved. Using all mentioned signals one can construct all the different types of ovarioles found in insects and in other animals and it might be that even the polyfusome may have some function in the oocyte–nurse cell determination process. Thus, one can think of a cascade of gene activities, which is used to establish the different types of ovarioles, comparable with the cascades known from pattern formation in early insect embryogenesis.

The characters for the basic type of polytrophic meroistic ovary, which have just been described, are found nearly unaltered in the ovary of barklice (Büning and Sohst, 1990) and are consistent with the findings of polytrophic ovaries in holometabolans (Giardina, 1901; Hirschler, 1945; Telfer, 1975; King *et al.*, 1982; King and Büning, 1985). Together with the germ cell–somatic cell interactions described above, the following events occur in the basic type (Figure 3.92).

(1) Few germ cells and somatic apical cells of the inner sheath contact each other. Under sex control, germ cells develop into stem cells, which follow the female pathway (a).
(2) Stem cells undergo a limited number of differential mitoses, bearing a limited number of cystoblasts (a).
(3) The first mitosis of a cystoblast is followed by incomplete cytokinesis (b).
(4) The resulting intercellular bridge is arranged asymmetrically: i.e. the bridge and the centrioles are all ordered above the newly formed nuclei (b).
(5) The inner surface of intercellular bridges is decorated with a belt of F-actin molecules, which are remnants of the arrested cytokinesis.
(6) A transient midbody develops from the spindle apparatus, which, at least in part, may consist of MAPs at the plus-end of microtubules (b).
(7) Fusomal material may be transported from Golgi complexes to the centre of the midbody, where it is deposited. Thereby, MAPs and microtubules of the former spindle are moved passively and decorate the rims of the fusome (b).
(8) While preparing for the second cycle of synchronized mitotic divisions, the first fusome will lose most of its microtubules (c).
(9) The second synchronous mitotic cycle gives rise to four cystocytes. As before, transient midbodies are generated during incomplete

cleavages and the resulting intercellular bridges are again arranged asymmetrically (c).

(10) One centriole is anchored near the preceding fusome, but it does not border on the fusomal material directly. The first connection of both systems may arise by aster microtubules (c).

(11) After formation of two new fusomes, the former spindle microtubules of the second division contact in part the cell membrane. The contraction of intercellular bridges (rosette formation) and the fusion of the new emerging fusomes with the old one, generating the polyfusome, seems to be mediated by those microtubules which are remnants of the spindle apparatus of the second division (d).

(12) After the last mitotic cycle and during the final steps of rosette formation, only the polyfusome may serve as an orientation centre for the syncytial cytoskeleton. The centrioles lose their predominant role and shift away from the polyfusome (e).

(13) After rosette formation, most of the microtubules of the former spindle apparatus disappear, whereas the polyfusome persists (e).

(14) Additional mitotic cycles, which embrace all cystocytes, follow by repetition of points (3)–(13), according to the species-specific programme, leading to a maximal branched cluster.

(15) After mitotic divisions, all cystocytes enter into the prophase of meiosis (e).

(16) During pachynema–diplonema, the contraction during rosette formation is greatest: all cells lie in the corners of a polyeder, while the polyfusome and the intercellular bridges are in the centre (e).

(17) At the latest at this point of cluster development, oocyte–nurse cell differentiation will happen, leading to one, centrally located oocyte, which is always one of the two oldest siblings. All other cystocytes will be transformed into nurse cells. This germ cell differentiation is under the control of cell–cell communication of germ cells with somatic, prefollicular cells, which differentiate or maintain the oocyte character, or with lateral sheath cells, differentiating and/or maintaining nurse cell characters (e,f).

(18) Nurse cells polyploidize their genomes. Thereby, special genes may be under- or over-replicated. The oocyte-specific programme of synthesis (RNA and protein) remains unaltered in nurse cells. The oocyte genome goes on with its own synthesis unaltered or oocyte synthesis becomes reduced. This behaviour is species specific and may differ to a great extent between orders and species (f).

(19) When nurse cells start their rapid growth, the polyfusomes vanish and the intercellular bridges joining nurse cells to nurse cells can move away from each other. Oocyte–nurse cell intercellular bridges keep their positions nearly unaltered. This region will build the nutritive appendix later on (f).

(20) An intensive transport from nurse cells to the oocyte is established once the clusters form a single file row in the ovariole.

These findings and hypotheses can easily explain all polytrophic ovaries found in hemi- and holo-metabolans and strengthen the hypothesis of only one basic type once developed by common ancestors (see section 3.6).

All the above characters are found in barklice (Psocoptera) and only one, additional character may be autapomorphic, that is the early growth and tremendous folding of intercellular bridge rims (Figure 3.23a).

The sister group (Phthiraptera), to which the true lice (Anoplura) and the bird lice (Mallophaga) belong, is not as thoroughly investigated regarding characters of cluster genesis. According to older literature (Gross, 1906; Ries, 1932; Ries and van Weel, 1934; Goss, 1954), each follicle contains five nurse cells and one oocyte in true lice (Anoplura). This shows a reduction in cluster cell mitoses in which seven nurse cells are expected when three cycles of mitotic division exist. Similar reductions are also known among holometabolan polytrophic ovaries (see below). However, it is not known which mitoses are reduced as long as cluster analysis in Anoplura is missing. A mallophagan species, *Eomenacanthus stramineus*, has been investigated (Figure 3.23b) (Bilinski and Jankowska, 1987), which followed the 2^n-rule: seven nurse cells support the oocyte in a follicle. As shown in Table 3.2, all those elements, which are typical for polytrophic meroistic ovaries are verified among Phthiraptera. This also supports the idea of a singular path during evolution of the polytrophic meroistic ovary at the base of Paraneoptera and Holometabola (see section 3.7).

Other characters of phthirapteran ovaries like amitotic division of follicle cell nuclei or the accessory nuclei in oocytes are discussed in other contexts (see sections 3.2.5, 3.3.2 and 3.6.2).

(b) Dermaptera

In 1982, Yamauchi and Yoshitake published an ultrastructural investigation of cluster formation in the germarium of *Anisolabis maritima*. They showed that in the anterior region of the larval germarium, stem cells exist which are mitotically active (Table 3.1; Figure 3.88). However, in contrast to normal unequal divisions, siblings do not separate immediately but remain connected by intercellular bridges. Thus, a clone of interconnected stem cells arises, the members of which emerge from subsequent mitotic events (Figure 3.93).

'Old' siblings will detach later on and become cystoblasts. Cystoblast cells undergo three cycles of synchronized mitoses with incomplete cytokinesis, which results in eight interconnected cystocytes. Thereby, mitotic spindles of the second and third mitotic cycle are more or less at right angles to the spindle axes of the preceding cycle and to axes of their neighbours.

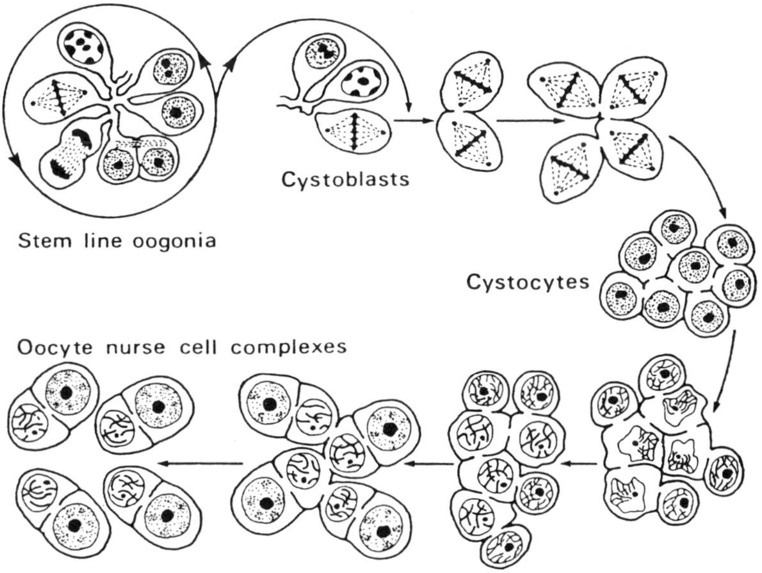

Figure 3.93 Ontogenetic events leading to the polytrophic meroistic ovary of *Anisolabis maritima* (Dermaptera). Note that two characters are unique (autapomorphic) and not found during ontogenetic events leading to the polytrophic meroistic ovary of the basic type: the existence of a cluster of stem line oogonia in the germarium, and the closure of the innermost intercellular bridges together with the splitting of the eight-cell cluster into four two-cell clusters. For further information see text. (From Yamauchi and Yoshitake, 1982; with permission of Pergamon Press.)

This gives rise to two innermost cells containing three intercellular bridges: two cells with two intercellular bridges and four cells with only one intercellular bridge. All siblings will go into meiotic prophase, as is shown by synaptonemal complexes. However, cells with only one intercellular bridge will stop meiotic prophase and will be transformed later on into nurse cells; the other four cells will continue meiosis and remain oocytes. All intercellular bridges connecting oocytes with each other will be closed later on, i.e. the eight-cell cluster will split into four two-cell clusters, which will all enter the vitellarium as separated follicles.

The ultrastructure of the oogonial cluster shows midbodies within each intercellular bridge (Yamauchi and Yoshitake, 1982; their Figure 3). Furthermore, the authors mention that 'the bridge (of syncytial oogonia) always contains vesiculate material and a midbody'. 'Vesiculate material' is a synonym for fusome as one can see from their Figures 10 and 11, which show a polyfusome, combining at least three intercellular bridges of cystocyte clusters. In both clusters intercellular bridges are found next to each other, indicating rosette formation (see next section). When final follicles enter the vitellarium, the single nurse cell undergoes endomitotic replication cycles by which nurse cell genomes become multiplied.

This mode of cluster formation is found in *Anisolabis maritima*, a species which is believed to be near to the basis of earwigs (Sakai, 1984). Investigations of other dermapteran species are rare. Brauns (1912) investigated the ovary of *Forficula auricularia*. He found only one mitotic event in the germarial region in larval stages. This was the mitotic event which cystoblast cells undergo to produce one oocyte and one nurse cell, connected by a persisting intercellular bridge. Other mitoses were never found. Our investigations on the same species support this finding. Thus, in advanced Dermaptera, stem cell cluster formation is reduced to zero or is shifted to very early stages before ovariole development is induced (Büning, unpublished). However, more investigations are needed to get a correct view on the whole order (see section 3.6 for phylogenetic discussion).

(c) Strepsiptera

Among holometabolan orders, we will focus first on an order which shows many adaptations to endoparasitic life, the order of Strepsiptera (Table 3.3; Figure 3.88). The adult females remain larviform and in most species they do not leave their hosts, which are small cicadas (Delphacidae) or Hymenoptera or apterygote insects. The adult larviform female is filled with developing embryos, which finally leave the mother organism via the brood canals, which open outside the host. This is the same method used previously for the fertilization of eggs. The free-living adult male copulates with this opening, which itself disrupts the intersegmental tissue of the host sternites.

The oocytes develop from germ cells, which are first positioned at ventro-lateral sides of the alimentary canal of the young, immigrating larva, the L1 (Baumert, 1957). It is not known if the somatic tissues of ovaries are reduced from the very beginning, or if they will be reduced during early larval stages. This question can be answered best by looking at embryos or L1 of *Mengenilla* species, in which females show lesser reductions and leave their hosts as wingless adults (Silvestri, 1943). In older larvae, germ cell descendants spread through the body cavity, mixed with fatbody cells (Figure 3.79).

Up to the third larval stage (L3), male and female larvae are indistinguishable. We have analysed, by ultrathin serial sectioning, the fate of the gonads in larvae (L2–L4) and female adults of *Elenchus tenuicornis*, a parasite of delphacids (Hemiptera) (Büning and Sohst, unpublished). In the second larval stage, we found many globular aggregates, which are composed of two cell types. Uniform siblings are found centrally, connected to each other by intercellular bridges which themselves are filled with polyfusomes (Figures 3.79 and 3.94).

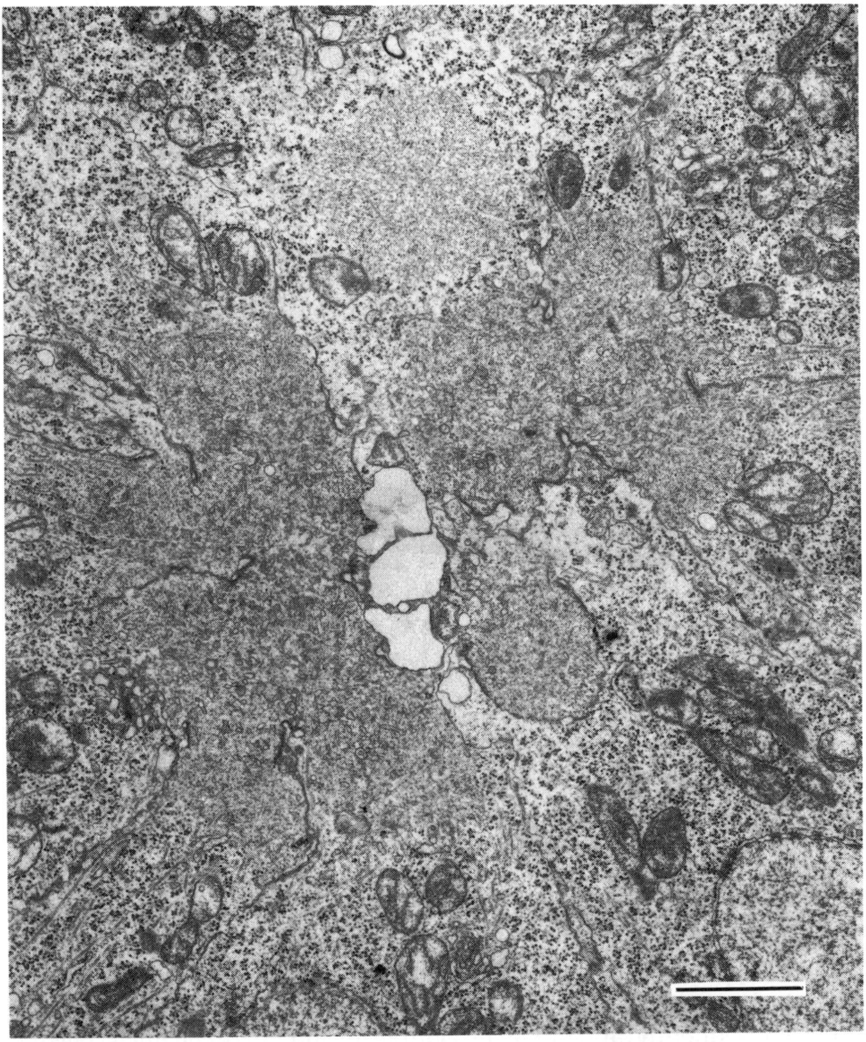

Figure 3.94 *Elenchus* ovary. Germ cell clusters are connected via intercellular bridges and form a rosette. A polyfusome stretches through all bridges. Bar represents 1 μm.

Thus, these cells form one, huge germ cell cluster. Each cluster is ensheathed by some flat somatic cells, which transform later on into the follicular cells. These somatic cells produce a very thin basal lamina towards the body cavity.

All clusters show rosette formation, i.e. all intercellular bridges assemble in the centre and all these intercellular bridges are connected by coherent fusomal material. New mitotic cycles are induced simultaneously in each cluster (Figure 3.79b). One of the spindle poles is always situated next to the polyfusome, while the other pole is just opposite. After each mitotic cycle, the new intercellular bridge moves to the centre and the spindle apparatus transforms into a new fusome, which contacts with the old polyfusome. Most probably, seven mitotic cycles occur, as is deduced from volume measurements which count about 130 cells in the biggest aggregates of L3. In older female larvae (L4), many more of these aggregates occur and spread through the whole body. They are the products of aggregate fissions (Figure 3.79b). This is concluded from two facts. Firstly, the development of all aggregates in a larva of *Elenchus* is strongly synchronized, i.e. in a given larva one will find only one stage of aggregate development. Secondly, in late L3 stages, aggregates occur, which are preparing for fission. Aggregates elongate and show in the middle equatorial furrows. In others, only half of the aggregate goes through mitosis, i.e. synchrony has split into two equal parts before fission is visible. According to Baumert (1957), each aggregate gives rise to about 3–5 final oocytes. Thus, there are 3–5 additional mitotic cycles to maintain about 130 cells in each aggregate. The outer somatic cells follow the furrow and separate follicles after some time (Figure 3.79b). During these fissions, sometimes several follicles remain connected for a longer time, producing ramified rows.

In the last larval instars, transformation of aggregates occurs (Figures 3.79c and 3.95). As a first step, rosette formation becomes destroyed. The polyfusome vanishes and cells move with persisting intercellular bridges to their final positions. In such transformed aggregates no preferential axes could be determined and cells seem to be equal in volume. The first morphological sign of oocyte differentiation is the rise of some yolk platelets, occurring in only one peripheral clonal cell of the aggregate. All other cells function as nurse cells. However, nurse cells do not undergo any transformation like polyploidization of genomes or volume enhancement during the phase after mitotic activity. There is no special previtellogenic growth phase, but cell membranes between the oocyte and nurse cells break down subsequently, and nurse cell cytoplasm and nurse cell nuclei become part of the final egg. Nucleus envelopes of integrated nurse cells break down after some time and only the oocyte nucleus survives. Thus, there is no need for an active transport system from nurse cells to the oocyte, since the whole cluster becomes integrated into the egg.

Proteinaceous yolk platelets arise by endocytosis. However, clathrin decorated vesicles are rare and the oocyte membrane remains smooth and does not develop microvilli. From these basal parts endocytosis takes part normally. Follicle cells grow slightly only when they are in contact with the oocyte. Those which ensheath nurse cells remain flat. Interfollicular spaces are small and not restricted to oocyte–follicle cell boundaries but also occur

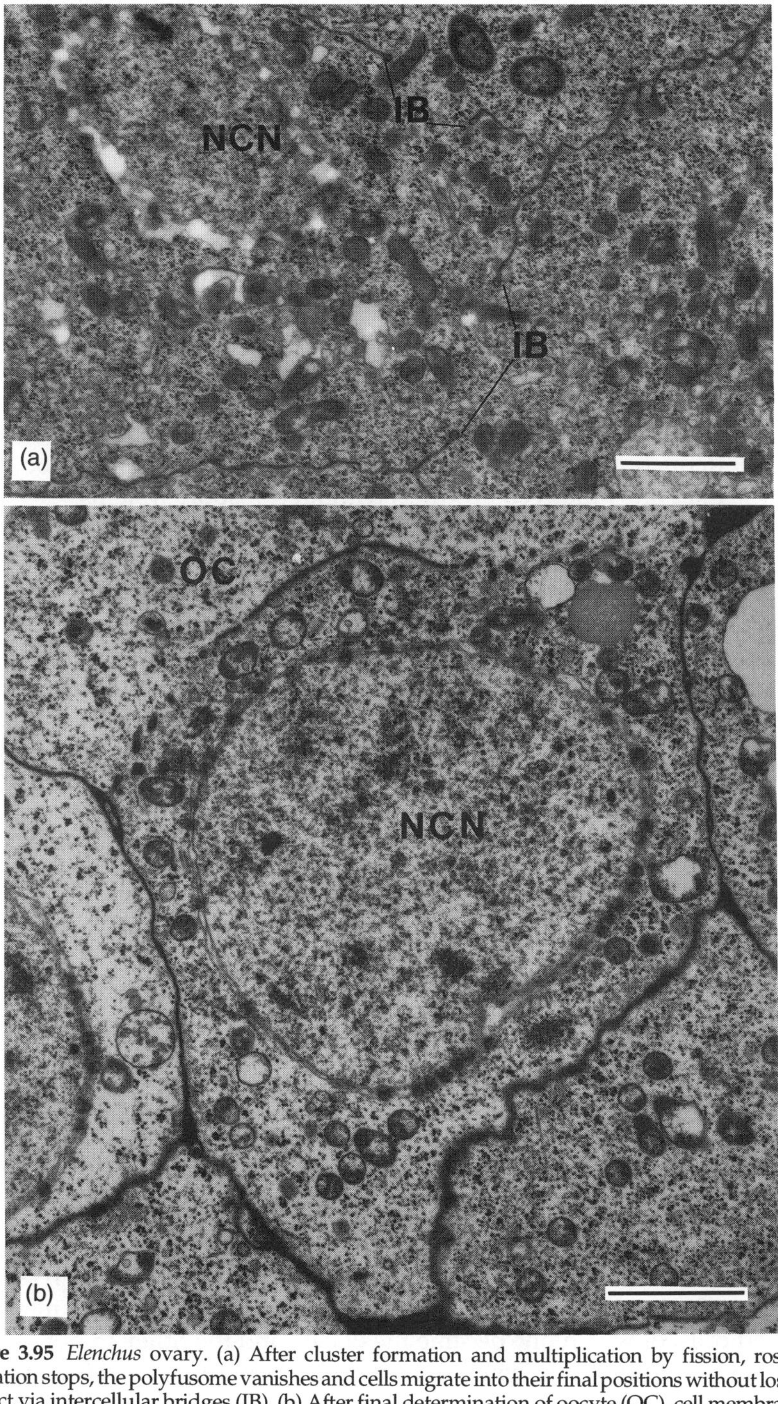

Figure 3.95 *Elenchus* ovary. (a) After cluster formation and multiplication by fission, rosette formation stops, the polyfusome vanishes and cells migrate into their final positions without losing contact via intercellular bridges (IB). (b) After final determination of oocyte (OC), cell membranes between siblings break down, and most cytoplasm of nurse cell siblings is integrated into the oocyte. Nurse cell nuclei (NCN) degenerate. Bars represent 1 μm.

between degenerating nurse cells and the oocyte. Finally, a thin vitelline layer is deposited by the follicular cells.

As shown, germ cell cluster formation in *Elenchus tenuicornis* females shows all the early characters necessary for the construction of a polytrophic ovary, including the determination of only one cell as oocyte. Autapomorphic characters are:

(1) The fission of clusters during the phase of mitotic divisions, which leads to an amplification of follicles and finally of oocytes.
(2) Reduction of somatic tissues necessary for the construction of an ovariole, as outer sheaths, terminal filament, apical cells, prefollicular tissue, pedicel and lateral oviduct. Only some somatic sheath cells survive.
(3) Total reduction of previtellogenic growth phase. Euplasm is generated only during mitotic cycles and during vitellogenesis.
(4) Reduction of nurse cell membranes, which enables the oocyte to integrate nurse cell products without transport systems.
(5) Protection of the germinal vesicle, which survives, whereas nurse cell nuclei degenerate in the egg.
(6) Reduction in vitellogenesis, leading to small eggs, rich in euplasm. Nevertheless, egg axes form during vitellogenesis. The area of the degenerating nurse cell cap becomes the anterior pole, whereas the position of the very early egg cell marks the posterior pole.
(7) Reduction in chorionogenesis. Embryogenesis proceeds in the maternal body cavity, thus a strong chorion is not needed. Furthermore, there is some evidence that the embryo will be nourished in part by the maternal haemolymph as known from parthenogenetic growing aphid embryos.

These data have been confirmed by Gu *et al.* (1992), investigating *Elenchinus japonicus* with the light microscope. In this species 256 cells were counted for each cluster forming a rosette.

(d) Coleoptera: Adephaga

The next insect order to which the strepsipterans are related is the coleopterans (Table 3.3; Figure 3.88). Beetles are subdivided into two infraorders, the Coleoptera: Adephaga and the Coleoptera: Polyphaga. All adephagous beetles have polytrophic meroistic ovaries. Among Adephaga, some families belong to water beetles (Hydradephaga), whereas most belong to ground beetles (Geadephaga). Giardina (1901) was the first to unravel the relationship of nurse cells and oocytes when he analysed the ovary of *Dytiscus marginalis* (Dytiscidae). He found clusters of 16 interconnected cells, of which one became the oocyte and the other 15 nurse cells. Thereby, he found four synchronized cycles of mitotic divisions, starting

Table 3.7 Cluster characters in follicles of ground beetle (Carabidae) (Data from Jaglarz, 1992; with permission of M. Jaglarz and Longman Group UK Ltd.)

Subfamily	Tribe	Species	Number of nurse cells[a]	Number of bridges[b]
Cicindelinae		*Cicindela campestris*	(31, 31), 31	5
		Cicindela hybrida	(31, 31, 31), 31	5
Carabinae	Carabini	*Calosoma inquisitor*	(120, 181)	8
		Carabus coriaceus	(119, 76)	7
		Carabus violaceus	(121, 120), 123	7
		Carabus obsoletus	(113, 17, 84), 52,	7
			75, (102, 128), 120	
		Carabus linnaei	(51, 57)	7
	Cychrini	*Chychrus caraboides*	(56, 61)	6
	Nebriini	*Nebria brevicollis*	(15, 15, 15), 15	4
	Nothiophilini	*Nothiophilus biguttatus*	(15, 15), 15	4
	Elaphrini	*Elaphrus riparius*	15, 15	4
		Elaphrus ullrichii	(15, 15), 15	4
Harpalinae	Bembidiini	*Bembidion properans*	(15, 15, 15)	4
		Bembidion semipunctatum	15, 15	4
		Bembidion lunatum	(15, 15), 15	4
		Bembidion rupestre	15, 15	4
		Bembidion articulatum	15, 15, 15	4
	Trechini	*Trechus pulchellus*	(15, 15), 15	4
		Trechus pulpani	15, 15	4
	Amarini	*Amara familiaris*	(15, 15, 15, 15), 15	4
	Pterostichini	*Pterostichus vernalis*	15, 15, 15	4
		Pterostichus oblongopunctatus	(15, 15), 15, 15	
		Pterostichus vulgaris	(31, 52), 31	4
		Pterostichus madidus	15, 15	5
		Pterostichus foveolatus	(31, 47, 31), 31,	4
			31, 31	5
		Pterostichus pilosus	(31, 31), 31	
		Abax parallelepipedus	(15, 15), 15, 15	5
		Agonum sexpunctatum	(15, 15, 15), 15, 15	4
		Agonum assimile	(15, 11, 15, 15),	4
			(15, 15, 15), 15, 15, 15	4
	Licinini	*Badister bipustulatus*	(15, 15), 15, 15, 15	4
	Harpalini	*Harpalus rufipes*	15, 15, 15	4

[a]Numbers in parentheses are the numbers of nurse cells in consecutive egg chambers.
[b]Number of intercellular bridges directly connecting nurse cells with an oocyte.

from one germ cell. Cytokinesis of each mitosis remains incomplete; thus, siblings remain connected primarily via intercellular bridges. One cell was always distinguishable by an extrachromosomal DNA body, which was inherited by only one cell, the later oocyte or its direct progenitor. This body has been attributed by Giardina as an oocyte determining factor. This suggestive interpretation was accepted by many authors and further research in other species by other scientists followed, but failed in most cases. However, due to this search, some clusters of germ cells in other species were analysed. Günthert (1910) repeated Giardina's work and confirmed his findings. Debaiseaux (1909) did the same but interpreted the extrachromosomal body as nucleolus like. Although he could not know at the time, his feeling was correct (see section 3.3.4). Hegner and Russel (1916) investigated the whirligig beetle *Dineutes nigrior*. They found a cluster of eight cells, of which one developed as an oocyte, the others as nurse cells. All other characters are the same as in *Dytiscus*.

Whereas the number of siblings is low but constant in water beetles, this number is much more variable in ground beetles (Table 3.7) (Kern, 1912; Jaglarz, 1988, 1992). In follicles of the common tiger beetle (*Cicindela campestris*) 31 nurse cells support each oocyte (Bilinski and Jaglarz, 1987; Jaglarz, 1988). The oocyte is connected via five intercellular bridges to the branched subclusters of nurse cells. These oocyte–nurse cell intercellular bridges are kept tightly together and the growing oocyte forms in its anterior part a so-called nutritive appendix, a chimney-like protrusion, which becomes necessary to remain connected with the growing nurse cells (Bilinski and Jaglarz, 1987). Such nutritive appendices are found in all those polytrophic meroistic ovaries in which the number of supporting nurse cells is high and the nurse cells undergo a tremendous increase in their volume (see below). Among *Carabus* species, variations in the numbers of nurse cells are much more pronounced and in many cases the numbers vary even in one ovariole (Table 3.7; Figure 3.57) (Jaglarz, 1992). Here, it comes in later division cycles to intercalar divisions, i.e. intercellular bridges remain at opposite sides of the cell and rosette formation happens only at the first mitotic cycles. Moreover, sometimes two nuclei are found in one nurse cell (Kern, 1912); this character is also found in the telotrophic ovary of some polyphagous beetles (Büning, 1979a) (see section 3.5.3). A second character concerns somatic tissues. Each follicle has at its anterior end a distinct nurse cell chamber. In other orders, this chamber is surrounded by inner sheath cells, which form a flat epithelium. In addition, in adephagous beetles, somatic interstitial cells of the trophic chamber intermingle with nurse cells forming a three-dimensional network. This character is also found in the sister group, the polyphagous beetles, which have telotrophic ovaries (section 3.5.3).

(e) Neuroptera

In the lacewing, *Chrysopa perla* (Neuroptera), 12–14 cystocytes develop in each cluster rather than 16 (Table 3.3; Figure 3.88) (Rousset, 1978a; King *et al.*, 1982). In this species, the two pro-oocytes and their siblings of the third mitotic cycle do not share in the last round of cell division, which results in 12 cystocytes (Figure 3.96).

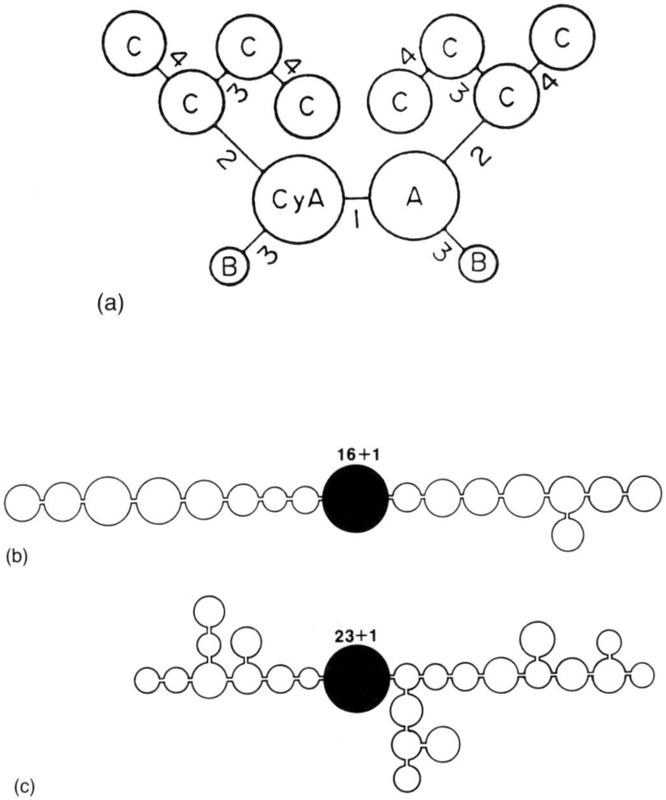

Figure 3.96 (a) Cluster formation in *Chrysopa perla* (Neuroptera: Chrysopidae). Numbers represent cycles of cystocyte divisions. Three types of nurse cells are found, according to their volume and ultrastructural architecture. CyA develops into the oocyte. (Redrawn after Rousset, 1978a, modified.) (b,c) Cluster formation in *Osmylus fulvicephalus* (Neuroptera: Osmylidae). Note the irregular bifurcation, interpreted as repression of mitotic activity. (Courtesy of Janusz Kubrakiewicz, University of Wroclaw, Poland.)

Sometimes, the pro-oocytes share in the fourth mitotic cycle, generating clones with 13 or 14 cells each. Possibly, there are different strains in *Chrysopa perla*, since there are some older reports counting seven or even five nurse cells in each cluster (Gross, 1903; Gruzova *et al.*, 1972). Investigation of other species is rare. The ultrastructure of *Chrysopa septempunctata* ovaries was investigated by Matsuzaki (1978). He counted seven nurse cells accompanying each oocyte and described the rod-like polyfusome, as did Rousset in the same year. In later papers, Matsuzaki's group investigated ovarioles of other neuropterans, such as spongillaflies *Sisyra nikkoana*, Sisyridae (Matsuzaki and Enomoto, 1990), *Ascalaphus ramburi* and *Protidricerus japonicus*, Ascalaphidae (Matsuzaki, 1987), *Eumantispa harmandi*, Mantispidae (Matsuzaki and Satoh, 1986). In all these species, the number of nurse cells accompanying the oocyte is not fixed and varies between 11 and 14. Thus, cluster formation in neuropterans seems to be uniform and represented at best by the analysis of Rousset (1977, 1978a,b).

Besides its variation in nurse cell number, *Chrysopa perla* shows some other characters (Rousset, 1978a,b), which should be kept in mind when looking for characters of the whole group. Cystocytes differentiate into three groups, A, B and C, which emerge during four division cycles (Figure 3.96a). Lateral cystocytes, type C, develop into nurse cells without entering the prophase of meiosis. Type A cystocytes, which are the two innermost cells, and B cystocytes, which are siblings of type A cells of the third mitotic cycle, start with meiotic prophase up to pachynema. Type B cells are small and stop their mitotic programme definitively, whereas type A cells can enter the fourth mitotic cycle (31% only one A cell, 8% both A cells), as do type C cells. One of the A cells becomes the oocyte, all other cells become nurse cells. Thus, clones of 12–14 cells arise in which 11–13 nurse cells support one oocyte.

Type A cells develop extrachromosomal DNA, which is most probably rDNA (Gruzova *et al.*, 1972; Rousset, 1977; Cave, 1982). The oocytes show high activity of extrachromosomal DNA, as is seen by numerous electron-dense spots and by autoradiography (Gruzova, 1966; Gruzova *et al.*, 1972; Matsuzaki, 1978). The oocyte chromosomes form a karyosome, in which RNA synthesis is reduced, whereas extrachromosomal DNA is highly active as shown by ^{3}H-uridine incorporation studies (Gruzova *et al.*, 1972). During previtellogenic growth, large endobodies, which consist mainly of proteins, accumulate in the oocyte nucleus. During vitellogenesis, the karyosome becomes encapsulated by unknown rod-like structures (Gruzova *et al.*, 1972).

The polyfusome appears not to be concentrated around one point, near to the first intercellular bridge, as in 'classical' rosettes (Psocoptera, Hemiptera, Lepidoptera, Strepsiptera) but the polyfusome is stretched and looks like a short cylinder to which all cells open with their intercellular bridges. Similar rodlike structures are also known from hymenopteran polyfusomes (see below and Figure 3.97).

Follicular cells of the germarial region are poorly developed, enclosing germ cell clusters at the beginning by some flat cells, the nuclei of which are found in small nests in wedged corners between subsequent follicles, as is known from most panoistic ovaries (Matsuzaki, 1978). A reinvestigation of some other species of neuropterans by Kubrakiewicz shows different results (personal communication; Kubrakiewicz, 1993). Cluster formation in several species of Neuroptera (*Osmylus fulvicephalus*, Osmylidae; *Euroleon nostras*, Myrmeleonidae) was analysed by serial semi-thin sections and showed the following results. The constant and

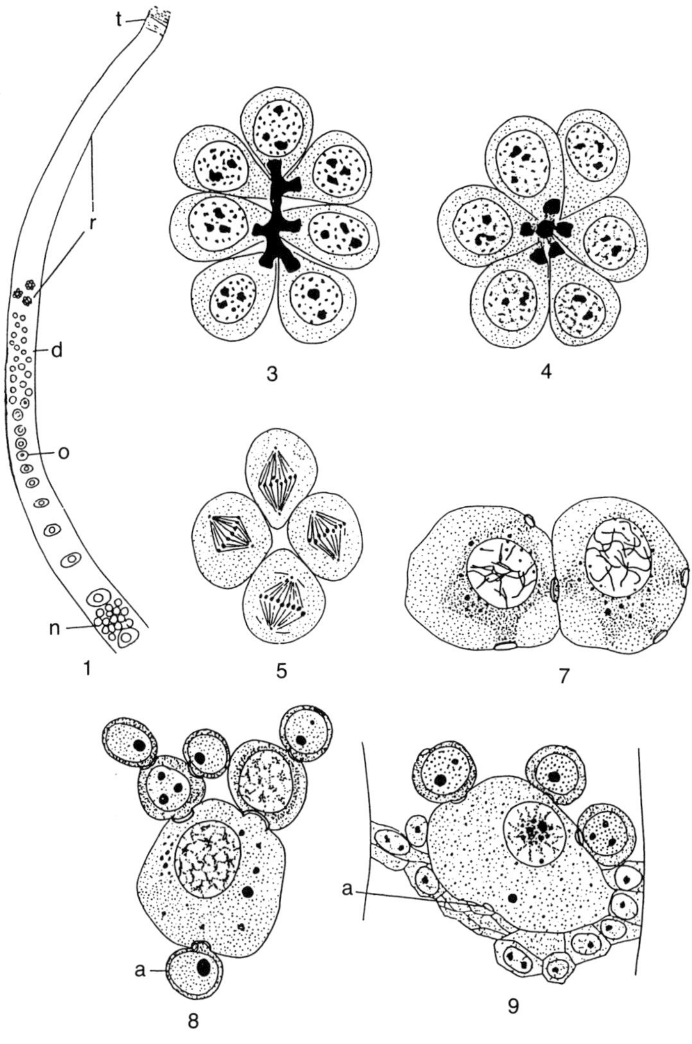

Figure 3.97 Cluster formation in Hymenoptera (*Apis*) after Hegner (1914). His Figure 1 shows the apical part of an ovariole. Zone of rosette formation (r). Zone of oocyte–nurse cell differentiation (d). Oocyte (o); nurse cell (n). Subfigures (3) and (4) show longitudinal sections and transverse sections through rosettes. Note the polyfusome (dark area). Subfigure (5) shows synchrony of divisions in clusters. In the differentiation zone only the oocyte and one of its sister cells shows pachytene configurations of chromosomes (7). Subfigures (8) and (9) show the three-dimensional distribution and contact via intercellular bridges of the next siblings of the oocytes before (8) and after (9) growth of prefollicular cells between adjacent follicles.

regular number of intercellular bridges of oocytes is only two and not four. This character has also been stated for *Chrysopa*. The 2^n pattern with regular bifurcations as known from *Drosophila* and other representatives of the basic type is strongly deviated from. Clusters are much more linear but keep the oocyte in a middle position. Most clusters of *Osmylus* contain 16 nurse cells; however, 19 or even 23 nurse cells can occur. Bifurcations are more common when the number of nurse cells is high. Nurse cells differ in their volume. Those next to the oocytes and some elsewhere remain small, whereas others enlarge probably by additional rounds of replications (Figure 3.96b,c). In *Euroleon*, 13 nurse cells occur regularly with no or only one bifurcation; however, this number can be reduced to 11 (no bifurcation was found) or can be raised to 16 (one bifurcation on each linear row of nurse cells). These clusters are interpreted as more or less linear clusters caused by repressed mitotic divisions at bifurcation points.

(f) Hymenoptera

Among bees and wasps (Hymenoptera) modern analysis of germ cell cluster formation is rare (Table 3.4; Figure 3.88). Thus, we must mainly refer to some older reports and some of our own preliminary and unpublished observations, which can give only a hint as to cluster formation characters (Blochmann, 1886; Paulcke, 1900; Marshall, 1907a; Hegner, 1914, 1915; Buchner, 1918; Henschen, 1928; Hagan, 1954a,b,c; Bobrova and Gruzova, 1967; Engels, 1968; Cassidy and King, 1969, 1972; King and Fordy, 1970; King *et al.*, 1982; Hatakeyama *et al.*, 1990; Bilinski, 1991a,b). Generally, all the characters of the basic type of polytrophic meroistic ovary are found in this group; however, there are modifications.

The first character, which varies to a great extent, is the number of ovarioles and, with an increase in the number, the different regions of each ovariole also elongate. Each ovary of *Habrobracon* contains only two ovarioles (Henschen, 1928; Cassidy and King, 1972). In parasitic bees and wasps there are three ovarioles per ovary (Halictidae and Megalichidae). The same number is found in non-parasitic Colletidae, Andrenidae and Melittidae (Alexander and Rozen, 1987). In other solitary groups there are four or five ovarioles per ovary. Within the Apidae the number of ovarioles varies to a greater extent. In *Bombus*, as in the stingless bee *Scaptotrigona postica*, there are only four ovarioles per ovary, whereas in the presumed sister group of *Psithyrus* the number of ovarioles is variable, ranging from six to 18 per ovary (Staurengo da Cunha, 1979; Alexander and Rozen, 1987). Ant queens like *Formica* sp. possess about 30 ovarioles per ovary (Billen, 1985). In *Apis* each ovary houses about 160–180 ovarioles (Snodgrass, 1956; Engels, 1968). In army ants (*Eciton hamatum*) about 1300 ovarioles are counted per ovary (Hagan, 1954b).

The germarial region is still an enigma. According to Cassidy and King

(1972) the germarial region begins with a zone of solitary germ cell descendants, which consist of 50–100 cells in *Habrobracon*. However, mitotic activity is reported only from the next zone, in which cystocyte divisions take place. Thus, it is unknown if the germ cells of the most anterior zone are stem cells or cystoblasts or a mixture of both. In physogastric queens (*Apis mellifera*) each germarial region may house some hundred resting germ cells; however, their morphology and division behaviour is unknown (Figure 3.97). Recently, we analysed several *Apis* queens, freshly hatched from pupae, but were unable to find single mitotic figures, either by differential interference contrast of living ovarioles, or by Hoechst 33258 stained whole mounts. Thus, it remains unknown if in the adult stage, hymenopterans have the ability for stem cell mitoses or not. Nothing is known about diurnal, annual or other periodic rhythms, or their possible influence on stem cell mitoses. Such rhythms are known to regulate the physogastric cycles of army ants (Hagan, 1954a–c).

In eusocial hymenopterans only the queens can produce females by providing eggs with sperms. Workers, which are genetically females, are prevented from egg-laying normally by social interactions, which are under the register of the queen. In bumblebees, each worker has four ovarioles per ovary as has the queen; thus, only social interaction, pheromone and quality of feeding regulate workers' egg-laying behaviour (Duchateau and Velthuis, 1989). In honey bees like *Apis* and *Scaptotrigona*, as well as in ants, the number of ovarioles is reduced in workers. This shows that feeding of larvae has an enormous impact on ovary structure. Furthermore, workers are able to produce nutritive eggs as well as normal functional eggs which will develop into males since they cannot be fertilized. The morphology of these two egg types is altered. Functional eggs of *Scaptotrigona postica* have normal yolk spheres, whereas vitellogenesis in nutritive eggs is disturbed. In addition, nutritive eggs contain more accessory nuclei than functional eggs, which have more staples of annulate lamellae (Staurengo da Cunha, 1988).

As the germarial regions in ovarioles of queens are elongated, all the other parts of an ovariole are also enlarged, i.e. the zone of rosette formation (Figure 3.97) (Hegner, 1914), the zone of previtellogenic growth and the zone of vitellogenic growth. Consequently, these ovarioles are bent several times in the insect abdomen and the ovariole measures about 1 mm in length in virgin queens of *Apis*, whereas ovarioles enlarge to more than 6 mm in length during vitellogenic stages (Bier, 1954). In ovarioles of all hymenopterans, the germarial region is not separated by a neck region from the previtellogenic growing clusters, but opens continuously. The long terminal filament is not separated by a transverse septum (Figures 3.58 and 3.97).

Rosette formation is well known from bees, wasps and ants (Marshall, 1907a; Maziarski, 1913; Hegner, 1915; Billen, 1985) (Figure 3.97). As in

Table 3.8 Variation in number of nurse cells supporting oocytes in *Athalia rosae ruficornis* (Hymenoptera)

	No. of nurse cells per egg chamber					
	Egg chamber position					
Ovariole*	1	2	3	4	5	6
A–I	61	60	58	52	19	–
A–II	59	64	79	56	47	–
A–III	63	60	51	54	43	–
A–IV	60	62	58	49	–	–
A–V	61	31	56	27	–	–
A–VI	83	43	62	52	–	–
B–I	40	63	58	59	53	31
B–II	40	61	63	32	21	–
C–I	63	48	50	63	30	30
C–II	58	58	47	58	36	45

*A, B and C denote different ovaries and Roman numerals individual ovarioles. Data from Hatakeyama *et al.*, (1990), with permission of K. Oishi and Balaban, Philadelphia.

Chrysopa (Neuroptera), the polyfusome has a rod-like appearance, to which the sister cells are orientated in rectangular directions. Each rosette is built by synchronous mitotic cycles.

In *Habrobracon juglandis* five cycles of synchronous divisions happen, resulting in 32 cells, of which one of the two innermost cells, i.e. one of the two cells with five intercellular bridges, becomes the oocyte (Cassidy and King, 1972; King *et al.*, 1982). In *Apis mellifera*, the same number of synchronous divisions is assumed, since five intercellular bridges are found in the oocyte (Engels, 1968). However, there are some additional mitoses, which raise the number of nurse cells supporting each oocyte to between 50 and 60 (own measurements). Paulke (1900) assumed 47 nurse cells by measuring volumes of nurse cells and volumes of the nurse cell chamber. Unfortunately, he was misunderstood and this number has been treated as a fact in numerous papers. It is unknown if these supernumerical mitoses follow the same rules, i.e. that each mitotic division produces a dichotome ramification, or if intercalar mitoses occur, leading to single line rows in these late divisions.

In solitary hymenopterans, especially in those which care for their eggs and which therefore produce low numbers, the germarial region of the adult ovariole is normal or even reduced (Figure 3.58). In the sawfly, *Athalia rosae ruficornis* (Hymenoptera, Symphyta), about 50–60 nurse cells were counted as was found in *Apis*, but a much larger variety is noticed with the lowest number being 19 and the highest 83 (Table 3.8) (Hatakeyama *et al.*, 1990). Irrespective of real cluster formation in this species, it is obvious that mitotic cycles are not performed by all cystoblast descendants. In *Solenius vagus* (Hymenoptera, Sphecoidea), Buchner (1918) found only 15 nurse cells supporting one oocyte.

After rosette formation, each cluster goes through a series of morphological changes, which result finally in a special and fixed topology during growth phases. At first, the polyfusome vanishes. The two cells which have the highest number of intercellular bridges are orientated in the centre, while all other cells are stretched irregularly in all directions. Thus, the two pro-oocytes have intercellular bridge openings to all sides (Figure 3.97) (Hegner, 1915). During further oocyte differentiation, the prospective oocyte will enter a central and basal position. Thus, all nurse cell connections will move to the anterior side and finally, when growth phases have started, all intercellular bridges are combined in a conus-like area of the oocyte. Such an oocyte nutritive appendix (Cassidy and King, 1972) is also known from the polytrophic ovaries of tiger beetles and ground beetles (Bilinski and Jaglarz, 1987; Jaglarz, 1988, 1992), which also have large numbers of nurse cells. While the oocyte will enter the prophase of meiosis after cluster formation, nurse cells give no morphological signs for entering meiotic prophase, except the other pro-oocyte; thus, only the two pro-oocytes will reach the pachytene stage as can be deduced from the figures of Hegner (1915) (Figure 3.97, subfigure 7), Buchner (1918), and Cassidy and King (1972). One of the two becomes the definitive oocyte; the other will transform into a nurse cell as all others have done before (Figure 3.97, subfigure 8).

Nurse cells during previtellogenic growth are ordered above the oocyte more or less in several layers, with four or more nurse cells in the same plane. The volumes of nurse cells and their nuclei seem to be enlarged slightly towards the oocyte; thus, a slight gradient will be formed in each follicle after rosette formation (e.g. *Formica rufa* queen ovary; Bier, 1952). Nurse cell nuclei undergo endomitotic polyploidization. It is not known whether the DNA content of nurse cells is the same in all nurse cells. Direct cytophotometric measurements are needed.

The chromosomes of the oocyte nucleus will concentrate into a karyosome after pachynema (Buchner, 1918; Bobrova and Gruzova, 1967; Engels, 1968). According to ^{3}H-uridine autoradiography, the karyosome has a strongly reduced activity, whereas in young oocyte nuclei, which are in early pachynema, measurable amounts of RNA are synthesized (Engels, 1968).

Germinal vesicles of Hymenoptera produce so-called accessory nuclei. This character might be in some aspects autapomorphic, but further investigations are needed. These structures were first noticed by Blochmann (1886) in the oocytes of ants and wasps. The name was given by Buchner (1918) in his very detailed investigation of these structures, which he found to exist in all hymenopteran families, except perhaps the Chalcididae which have very small eggs and hence, possibly, any accessory nuclei may have been overlooked. A detailed description of these structures and their possible functions is given in section 3.3.2.

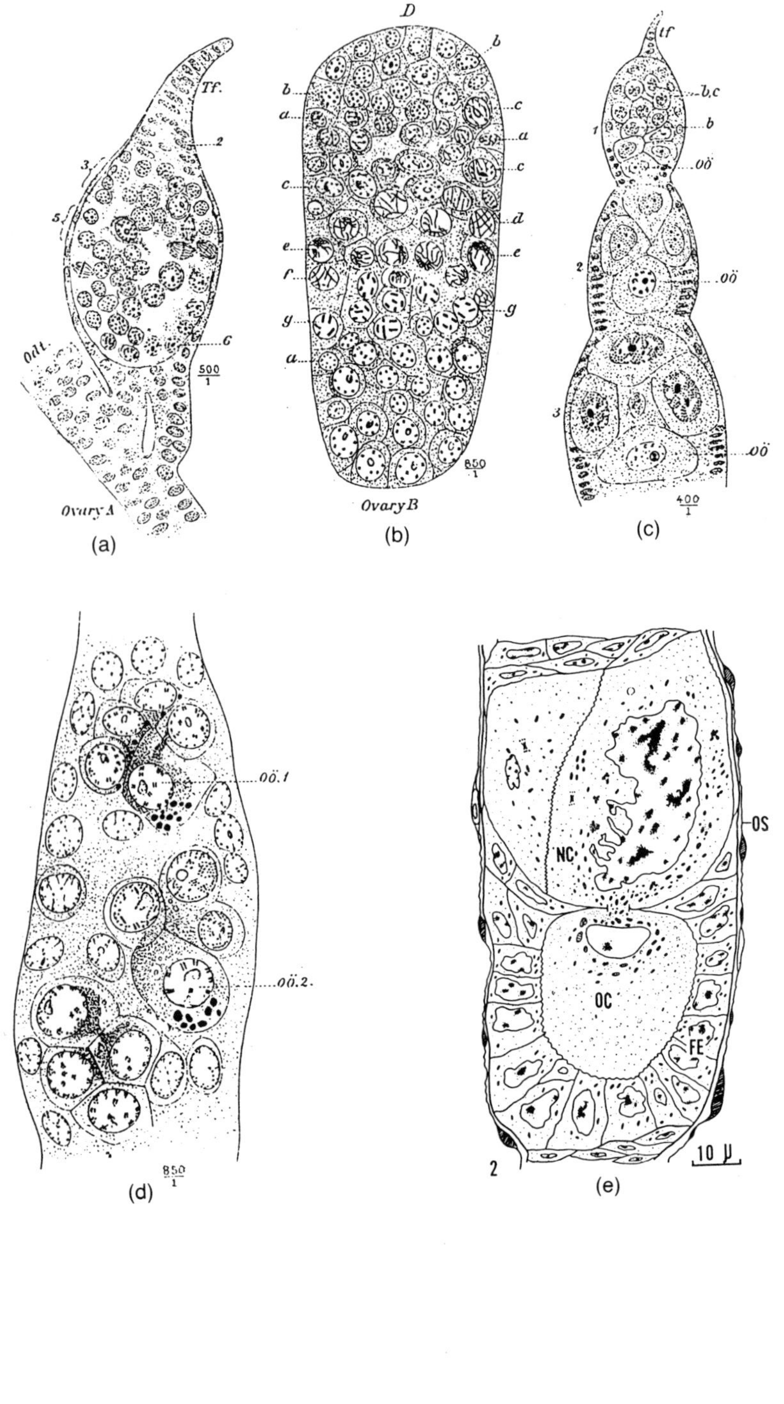
Tf.
Ovary A
500
1
(a)
D
Ovary B
850
1
(b)
tf
b,c
oö
400
1
(c)
oö.1
oö.2.
850
1
(d)
NC
OC
FE
OS
10 µ
2
(e)

Differentiation of intercellular bridge rims may produce, in some species, additional structures, as found in *Habrobracon juglandis* where eight tubular lamellae encircle the bridge opening. The inner ring of circumferentially arranged microtubules and microfilaments encompassing each bridge in *Apis mellifera* (Cassidy and King, 1969, 1972; Ramamurty and Engels, 1977; King *et al.*, 1982) is thought to be part of the normal equipment of the intercellular bridge, since these structures were also found in intercellular bridges of other orders (Gutzeit, 1986a; Gutzeit and Huebner, 1986; Geysen *et al.*, 1988). Further investigations of other species of Hymenoptera are needed to interpret these findings.

Finally, one character of the follicular cells has to be mentioned which is also shared with the follicular cells of Mecoptera: Boreidae and Diptera. When dividing, cytokinesis is incomplete thus, intercellular bridges occur which have a constant diameter of about 0.45 μm. These bridges have been found in *Apis* queens and *Formica* worker ovaries (Ramamurty and Engels, 1977; Billen, 1985).

(g) Trichoptera

Very little is known about the ontogenetic development of polytrophic meroistic ovaries of caddis flies (Trichoptera) (Table 3.4; Figure 3.88). This order is the sister group of Lepidoptera, and all investigators stated strong similarities between both ovaries. Ontogenetic development was first investigated by Marshall (1907b), who analysed larval and pupal ovaries of *Platyphylax designatus*, and for a second time by Matsuzaki (1972), who looked at the ultrastructure of ovaries from late pupae and early adults of the net-spinning caddis fly *Parasthenopsyche sauteri*.

Figure 3.98 Cluster formation in *Platyphylax designatus* (Trichoptera). (a) Figure 1 of Marshall's work (1907b) shows a larval ovary, in which cluster formation occurs, according to the direction of mitotic spindle axes on the left side in the middle. (b) Shows an older larval ovary in which all germ cells undergo prophase of meiosis (cells c–g) in an anterior–posterior gradient. Mitotic activity is no longer found. (c) Represents a stage after oocyte nurse cell differentiation. A flow of cytoplasm from nurse cells into the oocyte (oö.1) and from nurse cell to nurse cell (at the bottom) is shown. In the light of today's knowledge, this cytoplasmic material seems to be a polyfusome stretching through intercellular bridges. (d) Represents an ovariole of a young imago. Note that no germarium exists. The terminal filament borders directly on to a young follicle. This picture shows the structures of cytoplasm floating through channels, which are now known to be polyfusomal materials and intercellular bridges. (e) Follicle of *Parastenopsyche sauteri* (Trichoptera: Stenopsychidae). In contrast to the regular number of seven nurse cells found in follicles of Trichoptera, this species only has three. The figure shows one of three big nurse cells, connected to the oocyte via an intercellular bridge. (With permission of Fukushima University.)

In *Platyphylax* the entire ontogenetic development will occur during larval and pupal stages. In late pupal stages no germarium exists; the terminal filament borders directly to the youngest follicle, which is in previtellogenic growth phase. Thus, stem cells cease their mitotic activity as known from most insect orders (Figure 3.98d).

The youngest ovary Marshall investigated shows a terminal filament, which borders anteriorly a germarial region. At its posterior end a pedicel is found, which opens into the lateral oviduct. In the germarial region, germ cell descendants are to be seen, which in part show mitotic figures (Figure 3.98a). Most of them, if not all, will represent cystocyte divisions in modern terms, as can be deduced from their clustered appearance and their spindle orientation seen in the section. In Figure 3.98b, which shows the germarial region in an advanced stage without somatic tissues, one can see a zone showing leptotene or pachytene stages of meiosis, which embraces all germ cells visible in that horizon. This clearly shows that all germ cell descendants will go through the early stages of meiotic prophase, irrespective of their later destination as oocytes or nurse cells. Oocyte differentiation is to be seen morphologically prior to the polyploidization of nurse cell genomes, by basal orientation of the oocyte in the cluster and by accumulation of some granules in the posterior cytoplasm (Figure 3.98c). The same picture shows some strands, connecting the oocyte with nurse cells, which still show distinct chromosomes. These strands are interpreted as parts of a polyfusome. Marshall did not count the number of nurse cells belonging to one follicle. However, his pictures from longitudinal sections of previtellogenic growing follicles run through three to five nurse cells. This is exactly the same number as can be seen in longitudinal sections of lepidopteran follicles of the same stage. Most lepidopterans contain seven nurse cells in a follicle thus, it seems justified to estimate the same number of nurse cells in *Platyphylax*.

However, as in other orders, the number of nurse cells can be reduced as is reported from *Parastenopsyche* by Matsuzaki (1972). He counted only two nurse cells and one oocyte in a cluster, but sometimes the cluster was further reduced to one nurse cell connected to the oocyte. The cluster arises by two divisions. The first bears the oocyte and the first nurse cell, the second divides the nurse cell into two. Therefore, the oocyte has only one intercellular bridge which connects to the nurse cells (Figure 3.98e). Thus, according to the 2^n-rule, the second division of the oocyte is repressed. Nurse cell nuclei enlarge enormously and show the same morphology as known from Lepidoptera (King and Aggarwal, 1965; Miya *et al.*, 1969). Therefore, it seems justified to assume a high level of endomitotic polyploidization in these nuclei.

The above characteristics are exactly the same as those known from studying the basic type. No autapomorphic characters could be found.

(h) Lepidoptera

The order of moths and butterflies (Lepidoptera) is one in which polytrophic meroistic ovaries have been investigated most successfully (Table 3.4; Figure 3.88). The main findings concerning cluster formation were made by Knaben (1934), Hirschler (1942, 1945) and Mandelbaum (1980; see also Telfer, 1975). Investigations of early development of gonad anlagen were known from Grünberg (1903) (*Bombyx mori, Phalera bucephala, Gastropacha rubi, Pieris brassicae, Vanessa io*); Zick (1911) (*Pieris brassicae, Vanessa io, V. polychloros*); Machida (1926), Kawaguchi (1928) (*Bombyx mori*); Sato (1932) (*Lymantria dispar*); and Miya (1959, 1975) (*Bombyx mori*). Because of the similar life styles among lepidopterans, the structure and genesis of ovarioles differs only slightly among species.

The first events of gonad development differ in some respect from 'normal' development as was described before (section 3.2.1). Miya (1959, 1975) has given a very detailed description of these processes, which will have some impact on the phylogenetic discussion (section 3.6).

During embryogenesis, about 15 germ cells distribute more or less equally in small nests in the abdomen prior to segmentation during mesoderm formation. After abdominal segmentation, nests of germ cells concentrate in segments, most of them in segments 6–8. When the neural groove is formed, the mass of germ cells in each segment divides into two aggregates, one to each lateral side. Each aggregate becomes enclosed by a special mesodermal tissue, the genital ridge, which develops from the splanchnocoel. The genital ridges of segments 6–9 connect with each other, forming cord-like tissues on both sides of the body. These cords concentrate and the gonad anlagen are formed as short and, finally, closed tubes, stretching from the end of the seventh segment to the beginning of the ninth. Not all germ cells enter these gonad anlagen. Some of them are lost during the formation of gonads. Germ cell mitoses are still going on and the gonadal tube houses 100–200 germ cells, the number of which depends on strain differences and on sex. Male gonads house more germ cells than female ones.

In an ultrastructural study, Miya (1975) describes a large branched cytoplasmic process found regularly in the germ cells of the gonad anlagen, which also contains intercellular bridges of the foregoing mitotic events. His observations, made on single ultrastructural sections, fit exactly with the modern idea of germ cell clusters forming a rosette of interconnected siblings. Consequently, we must accept that germ cells can form clusters of interconnected siblings during their migration phase, i.e. long before ovarioles are built (see also section 3.6).

In males and females, each gonad differentiates four tubes: follicles in males, ovarioles in females. In both sexes, the early events of cluster formation are similar. Each tube differentiates an apical cell, which is believed to be somatic by most authors. In females, this cell persists in its apical position and its role is unclear. In males, this cell will move a bit more posteriorly and sometimes it will lose its connection to the other sheath cells surrounding the follicle (Grünberg, 1903). For its possible function in males see section 3.3.1.

The further development of ovarioles shows some group specificities. All four ovarioles are kept tightly together in germarial regions by the ovary sheath, which widens posteriorly in the region of vitellogenic growing follicles. Each ovary is anchored by the suspensory ligament. This enables the lepidopteran ovary to reduce the terminal filaments of ovarioles. Thus, each ovariole ends blindly and is closed by the tunica propria. Inside each ovariole, the apical cell terminates anteriorly (Grünberg, 1903; Gross, 1903; Zimowska *et al.*, 1991). Ovarioles of Lepidoptera are very long tubes which can reach 50 mm in length (*Bombyx, Hyalophora*), and where more than 50% is occupied by the elongated pedicel, which serves in Lepidoptera as a single row shelter for ripe eggs. However, the vitellarium can have dimensions of about 9 mm in length, housing many follicles which proceed in growth posteriorly. The ovariole sheath and the tunica propria are similar to the general scheme. A prefollicular tissue at the posterior pole of the germarium is not developed. Prefollicular cells rise from small nests of somatic cells in the germarium, which are connected to the inner sheath cells, which embrace all germ cell descendants.

Differential stem cell mitoses in ovaries of Lepidoptera have not been investigated. However, if there are some, the mitotic activity of stem cells must occur during larval or early pupal stages, since in adults germarial regions are no longer active. Recently, Zimowska *et al.* (1991) have marked the ovarioles of the Indian meal moth with immunofluorescent dye coupled to a polyclonal antiserum against an embryonic protein, which is found in pupae only in cystocytes and in nuclei of the ovary sheath and not in follicular tissue. In the youngest ovary of early pupae (their Figure 1a), about eight pronounced cells in the very tip of the ovarioles are found, ordered as a cap, which might be stem cells.

Cluster formation of follicles has been investigated at the light microscopical level by Knaben (1934) and Hirschler (1942, 1945) and its ultrastructure was elucidated by King and Aggarwal (1965) and by Mandelbaum (1980). The pioneering work of Knaben showed that the lepidopteran cluster contains eight cells (seven nurse cells and one oocyte) always arranged in the same manner. The oocyte and one of its nurse cells contain three intercellular bridges, two nurse cells contain two intercellular bridges and four nurse cells are connected by only one intercellular bridge to their sister cells (Figure 3.81a).

As shown, he gave the correct interpretation of three cycles of mitoses, whereby each cell divides perpendicular to the axis of the preceding division, which results in maximal branching of the cluster. He used the symbol 2^n for the first time for the cluster with n=3 in the case of lepidopterans. Such an amount is typical for Lepidoptera, although Knaben found one example (*Orneodes hexadactyla*) in which the number was reduced to $4=2^2$. In one species (*Macrothylacia rubi*), he found, besides clusters of eight cells, some with 16 (2^4–1 nurse cells and one oocyte). This shows a great constancy among lepidopterans. However, in *Galleria melonella* Gruzova *et al.* (1990) found highly variable numbers of nurse cells between three and seven. These nurse cells also differ in their amount of endomitosis. Only those three nurse cells which are in direct contact with the oocyte show high amounts of DNA. Those far away remain small and their ploidy is low. In the Indian meal moth *Ploida interpunctella*, only one nurse cell does not follow the endomitotic cycles of its six sister nurse cells and remains small (Zimowska *et al.*, 1991).

Hirschler (1942, 1945) introduced the term 'fusom' for the cytoplasmic material found inside the intercellular bridges during cluster formation. He was the first to draw attention to the fact that intercellular bridges, concomitantly with the shortening of their fusomes, concentrate during cluster formation (Figure 3.81b), a reaction which had been well known and described as 'rosettes' from the cluster development in males and females of other species. The term 'rosette formation' for this process was given by Telfer in his review in 1975, when discussing the main points of cluster genesis in *Hyalophora cecropia*, as described by Mandelbaum who was working in Telfer's laboratory. Most of the ultrastructural details of cluster formation of the basic type were first elaborated by Mandelbaum (1980) (section 3.5.2). Thus, one can see that Lepidoptera have the full set of symplesiomorphic characters. Nurse cells undergo endomitotic polyploidization after cluster formation (for details see section 3.3.4).

Apomorphic characters, shared only by species of Lepidoptera, are not known concerning cluster formation processes.

(i) Diptera

Among the midges and flies (Diptera), the ultrastructure and cluster formation of the ovary of *Drosophila* is best known, especially from the work of King and his coworkers during the last three decades. He introduced terms like 'stem cell' for those oogonia which undergo differential mitoses; 'cystoblast' for the stem cell sibling which passes to cluster formation; and 'cystocyte' for sibling cells during and after synchronous mitotic divisions of the cystoblast. Most of the characters outlined for *Drosophila* can be adopted for other dipteran families, but not all (Table 3.4; Figure 3.88).

Before going into the details of germ cell characters, we will focus on somatic cell characters typical for dipterans. One character, shared with hymenopteran follicular cells as well as with the follicular cells of snow fleas (Mecoptera: Boreidae), is the development of intercellular bridges arising during mitoses of follicular cells (see section 3.2.5). Clusters of these cells have been analysed in part by Fiil (1978) (Figure 3.20c). In the beginning, most cells divide intercalarilly (i.e. all cells are in a line). Later on, branchings are found, without showing any clear pattern. Some apical sheath cells of the nurse cell chamber have a special fate. During the late stages of previtellogenic growth (stage 9 in *Drosophila*, King, 1970), so-called border cells move down centrally from the apex of the nurse cell chamber to the apical centre of the oocyte, which they reach at stage 10A. This process was documented first in *Calliphora* (*Musca*) (Figure 3.82a) by Verhein (1921). Of these border cells, the two innermost cells are differentiated further in *Sarcophaga bullata* (Geysen *et al.*, 1988) (Figure 3.82b), since only these cells and adjacent intercellular spaces between these cells and the oocyte can be marked with larval serum protein antibodies. A corresponding pair of follicle cells exists at the posterior pole, showing the same reaction. Both pairs of cells exhibit a constant topology to each other and to the final position of the germinal vesicle. The nuclei of anterior polar cells and the germinal vesicle share the same plane, but the posterior polar cells are orientated perpendicular. In *Drosophila*, these anterior and posterior polar cells have been seen via P-element mediated transformation (O'Kane and Gehring, 1987; Fasano and Kerridge, 1988). These cells are the only follicular cells that permanently express the membrane protein fasciclin III, known to be expressed during neurogenesis. Expression of fasciclin III is necessary during ovariole differentiation for stalk cell development and for

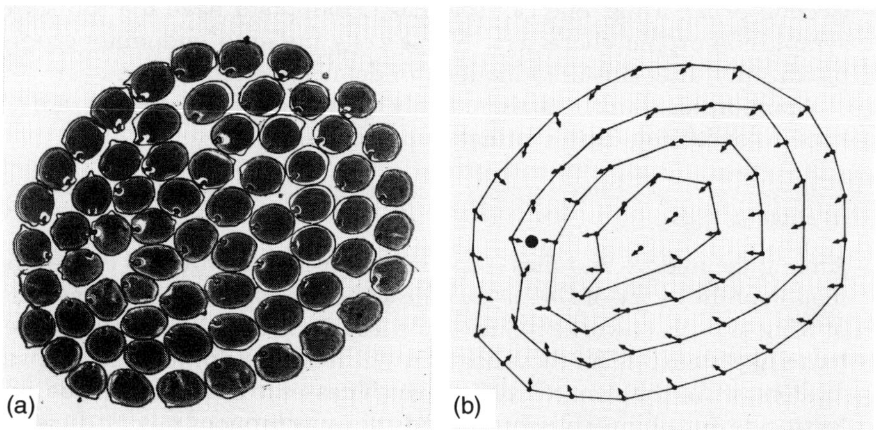

Figure 3.99 (a) Cross-section cut through *Musca domestica* (Diptera: Muscidae) ovary during chorionogenesis, showing dorsal ventral polarity of oocytes by raphe structure and orientation of ovarioles to an imaginal centre inside the ovary. (b) Each bar shows polarity of ovarioles (centre towards raphe) and their orientation to the imaginal centre of the ovary (large dot). (From Kleine-Schonnefeld and Engels, 1982; with permission of H. Kleine-Schonnefeld and Springer Verlag.)

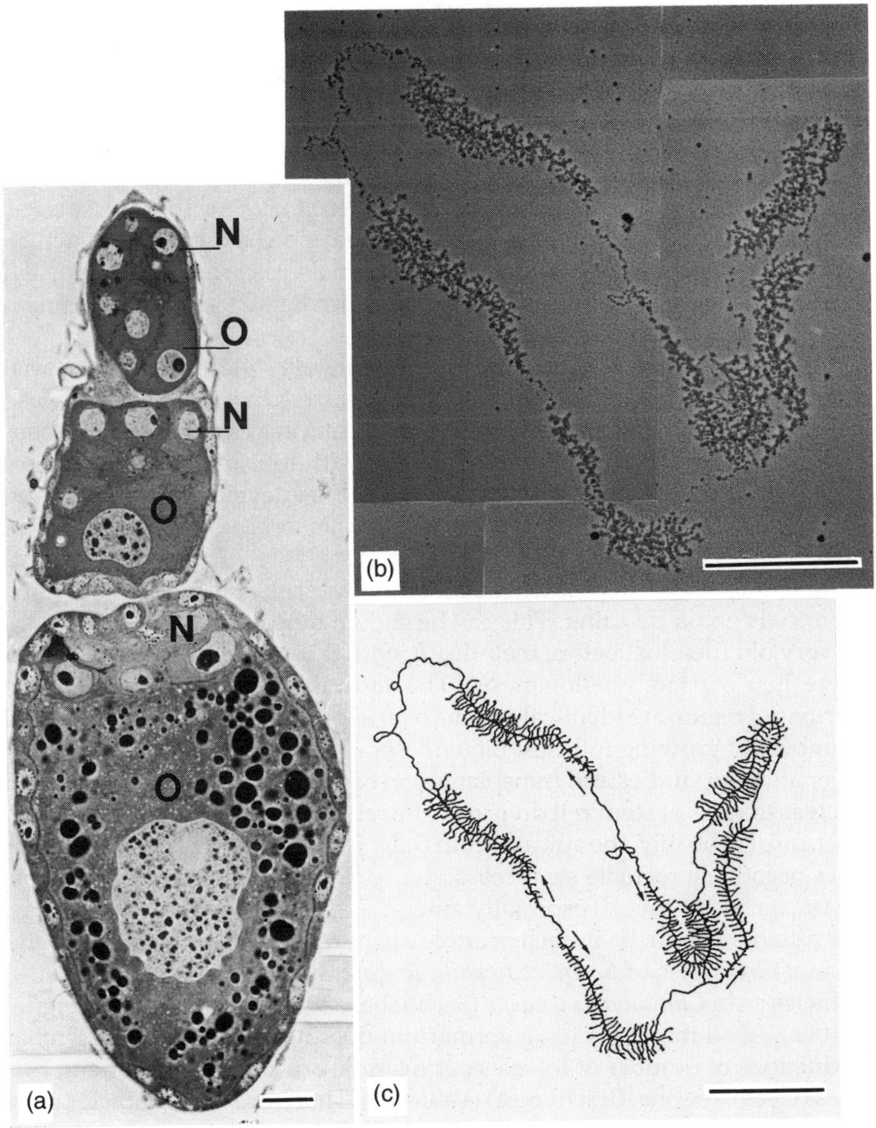

Figure 3.100 (a) The ovariole of *Tipula* sp. (Diptera: Tipulidae) is built by three follicles, enclosed by somatic tissues, the outer epithelial sheath and a tunica propria. A germarium does not exist. Note that nurse cells (N) do not grow, whereas the oocyte nucleus (O) grows tremendously, showing numerous spots of extrachromosomal rDNA rings, shown in (b). Bar represents 10 µm. (b, c) Typical for extrachromosomal rDNA in *Tipula* is the change in direction of transcription (arrows). Bars represent 1 µm.

the separation of the first follicle from the germarial region (Patel *et al.*, 1987; Ruohola *et al.*, 1991). However, further reports on other species are needed before this character can be adopted for the whole group of Diptera, including the suggested implications for dorso-ventral axis determination in ovarioles.

The orientation of ovarioles and their developing follicles around a centre in each ovary has been reported for *Musca* (Kleine-Schonnefeld and Engels, 1981) and for *Sarcophaga* (Geysen *et al.*, 1988). In both species, follicles are orientated with their future dorsal side next to a centre, which is first indicated by the position of the germinal vesicle and later on by the position of raphe building follicular cells, which lie adjacent to the germinal vesicle (Figure 3.99).

As shown in Figure 3.34, cluster formation takes place in the germarial region of the ovary, by means of four synchronized divisions of a cystoblast, followed by incomplete cytokinesis. This results in a cluster of 16 interconnected cells, of which one of the two cells with four intercellular bridges develop into the oocyte, whereas the other cells develop into nurse cells. All characters of the basic type are verified in *Drosophila* except the following two.

In *Drosophila*, stem cells divide indefinitely; even at the end of their life, stem cells go on dividing. This can be shown by comparing the ovarioles of very old flies, just before their death, with those of young adults, which have just started vitellogenesis. The terminal chambers housing the germarial region are identical and the ovarioles of both have about the same numbers of growing follicles. Lin and Spradling (1993) confirmed this by laser ablation studies and transplantation of germaria to hosts. They found increased rates of stem cell divisions after ablation of the terminal filament and, most probably, the apical sheath cells. This suggests that apical somatic cells negatively regulate stem cell divisions. In other dipteran families this character is not found, especially among midges (Nematocera), in which the number of follicles in each ovariole can be reduced to only one, as found in most ovarioles of *Sciara coprophila* (Figure 3.10), or it is reduced to 3–4 follicles as in *Chironomus tendens* (Sachtleben, 1918) or in *Tipula* sp. (Figure 3.100a). In all three species, a germarium does not exist in adults. Similar reductions of number of follicles per ovariole are also known among true flies (Cyclorrhapha, Brachycera) (Table 3.4). Therefore, this character must be proved even in these groups.

Not all siblings of a cluster enter the prophase of meiosis. In *Drosophila* as well as in most other dipterans, only the two pro-oocytes enter this pathway, all the others go directly into the nurse cell line, the nuclei of which polyploidize.

As in other orders, within each follicle an apical–basal gradient of nurse cell volumes develops. The smallest cells are situated at the anterior end and the largest nurse cells are connected directly to the oocyte (Verhein,

1921; Brown and King, 1964; King, 1970). This behaviour is based on different amounts of endomitotic polyploidization of nurse cells. The highest amount of polyploidization occurs in those cells which are connected directly to the oocyte. In *Drosophila* there is one exception, the former pro-oocyte, which remains smaller than the others (Brown and King, 1964). During polyploidization, nurse cell nuclei can develop polytene chromosomes. This has been shown for *Drosophila* (Painter and Reindorp, 1939; King *et al.*, 1981), *Calliphora* (*Musca*) *vomitoria* (Verhein, 1921, the so-called 'Raupenchromosomen'), *C. erythrocephala* (Bier, 1957, 1958, 1959; Ribbert and Bier, 1969; Ribbert, 1979), *Anopheles* (Mariani and Bruno-Smiraglia, 1957; Coluzzi, 1968; Coluzzi *et al.*, 1970; Saifuddin *et al.*, 1978; Redfern, 1981; Graziosi *et al.*, 1990). Whereas in *Anopheles* the pattern of bands and interbands of polytene chromosomes of nurse cells is directly comparable to those of somatic cells (Redfern, 1981), the pattern of *Calliphora* nurse cell polytene chromosomes differs completely from that of somatic cells (Ribbert, 1979). In *Drosophila*, King *et al.* (1981) found giant chromosomes the patterns of which are comparable only in part. These differences may reflect the physiological status of the cell, when it undergoes polyploidization. Nuclear proteins, thought to be necessary for banding patterns, may be influenced by the physiological conditions of the cell during polyploidization (Ashburner, 1981).

Comparing the number of cystocytes that build a female cluster in dipterans, one can assume that the ground plan consists of a 16-cell cluster, as found in *Drosophila, Musca, Calliphora, Nycteribosca, Ascodipteron, Psychoda* and *Tipula* (Table 3.4). However, there is a tendency to reduce the number of nurse cells. In *Melophagus ovinus*, as in other Hippoboscidae, the cluster contains seven nurse cells and one oocyte (Ulrich, 1963). In this species, reduction of nurse cells and reduction of ovarioles to only two in each ovary is probably due to extreme brood care, in which only one larva can be fed by milk glands in the uterus. The cluster is generated by three synchronous divisions, the intercellular bridges of which are arranged in the same manner as described for lepidopteran ovaries (Ulrich, 1963) and which is identical to *Drosophila* if one omits the last division cycle. In *Aedes* and *Culex* each cluster also contains seven nurse cells and one oocyte (Roth and Porter, 1964; Fiil, 1978; Ramadan *et al.*, 1986a,b). In *Sciara coprophila* or *S. ocellaris* only one nurse cell supports the oocyte (Gutzeit, 1985; own investigations) (Figure 3.10), and the same number is found in *Chironomus* as well as in the cecidomyiid family. The last three examples belong to the midges (Nematocera), in which large and synchronous developing batches of eggs are laid. This becomes possible by raising the number of ovarioles to more than 100, while the number of follicles per ovariole is reduced (Table 3.4).

In tipulids, Bayreuther (1952, 1956) found extrachromosomal DNA bodies in pro-oocytes generally, and in some species additional extrachromosomal DNA bodies in one or two other siblings of the later

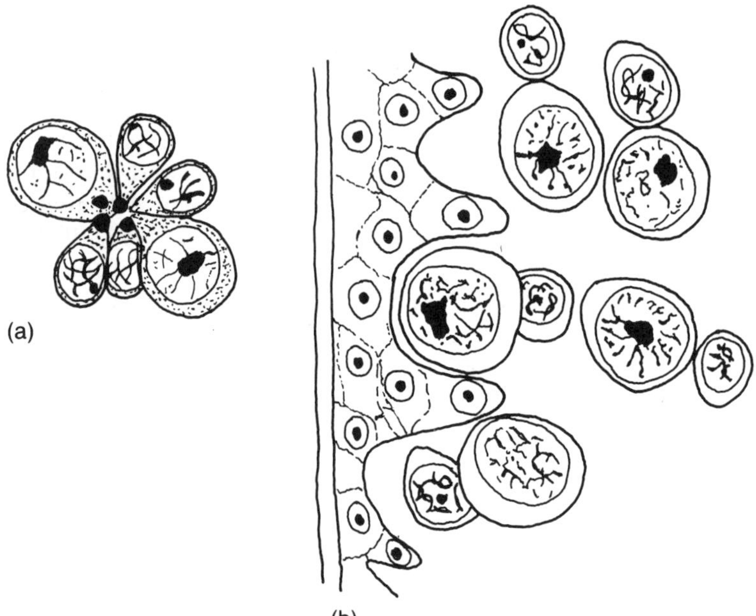

Figure 3.101 (a) This figure from Sachtleben's thesis (1918) shows a rosette of six germ cells of *Chironomus* early larval female gonad before the ovarioles are built. Two cells are enlarged, indicating nurse cell character. (b) Several pairs of two connected cells (one oocyte and one nurse cell) will be enclosed by protrusions of somatic cells in young larva of *Chironomus*, indicating the first steps in ovariole formation. Compare Figure 3.10b showing the same process in *Sciara* ovary.

oocyte. Lima-de-Faria and Moses (1966) analysed these bodies on the ultrastructural level. They noticed different behaviour between the oocyte and nurse cell bodies. Whereas the oocyte bodies change their morphology during growth phases and contain hundreds of 'nucleoli' inside, nurse cell DNA bodies do not change their compact appearance and remain small. Thus, one can conclude there is a different regulation mechanism in oocyte and nurse cell nuclei. The DNA body contains extrachromosomal rDNA, as is confirmed by the morphology of extrachromosomal rings, containing some active rRNA genes (Domazer-Schmidl and Büning, unpublished). These rings show a high percentage of inverse orientated genes (Figure 3.100b,c). Another interesting finding is the very small nurse cell nuclei, which seem not to grow, suggesting no or a low grade of polyploidy (Figure 3.100a). Comparing the nurse cell volumes of the three follicles, it becomes clear that the contribution of nurse cells to oocyte growth is low (Bauer, 1933).

The highest reduction of nurse cell numbers occurs in chironomids (Grimm, 1871; Sachtleben, 1918; Wülker and Götz, 1968; Wülker and Winter, 1970), and in fungus gnats (Sciaridae) (Berry, 1941; Gutzeit, 1985; own investigations, unpublished) (Table 3.4; Figures 3.10 and 3.101).

Ontogenesis of ovarioles is shifted completely to larval and pupal stages. Figure 3.10a shows a compact female gonad anlagen from a final larval instar of *Sciara coprophila*, housing about 550 germ cell descendants which all look alike. During the pupal stages, morphological differentiation of oocytes and nurse cells becomes apparent (Figure 3.10b). Oocyte chromosomes remain more or less compact, although differences between S- and E-chromosomes are easy to recognize since they occupy different compartments of the nucleus and show a different degree of higher order chromosome structure (Figure 3.10b,c) (for E-chromosomes see section 3.3.4). The nurse cell nucleus grows tremendously, indicating endomitotic polyploidization. The morphologically differentiated two-cell clusters become ensheathed by some somatic cells of the gonadal tube. The gonadal tube itself differentiates into a dorsal and a ventral mass of somatic cells. Two-cell clusters become orientated to these tissues so that nurse cells border the dorsal mass of somatic cells of the ovary, while its oocytes share the ventral mass which differentiates later on into the lateral oviduct (Figure 3.10b). The dorsal somatic tissue differentiates into a very short terminal filament, inner sheath cells, the extracellular basal lamina, and the outer ovariole sheath cells during pupal stages. The beginning of this differentiation is seen in the posterior part of the elongating ovary (Figure 3.10b). Compartments arise, enclosing one to four two-cell clusters. These compartments develop by elongation and growth of follicular cells into the ovarioles, in which the two-cell clusters become ordered later on in a single row. In each ovariole, posterior follicles grow earlier than the second or third, which degenerate in most ovarioles before maturation (Figure 3.10c). In the gonads of adults, each ovary is composed of 70–100 ovarioles tightly packed in four rows, more or less, along a lateral oviduct. Anteriorly located ovarioles house only one follicle, whereas posterior ovarioles can shelter up to three follicles. However, the total number of follicles is limited to 100–150 follicles in each ovary and a germarium is not built in any stage. From these morphological data it can be concluded that the 550 germ cell descendants seen in the ovary of the last larva are the two-cell clusters prior to morphological differentiation. This example shows that differentiation of ovarioles can occur even after cluster formation of germ cells. Consequently, a germarium is not built from the very beginning.

This deviated behaviour in which germ cell differentiation precedes differentiation of somatic tissues during gonadal development is possibly shared by all midges (Nematocera). Similar development is reported from chironomids (Balbiani, 1885; Sachtleben, 1918; Wülker and Winter, 1970), *Psychoda* (Koch, 1929), and *Tipula* (Bauer, 1933). Another example, in which ovarioles arise after cluster formation of germ cells, is known from telotrophic ovaries of scale insects (section 3.5.3).

Before moving to the most deviated ovary among diptera, one should discuss the findings of Sachtleben (1918) who investigated the larval gonads of *Chironomus tendens* (Fabr.). His investigation is based on the work of Balbiani (1885), who clearly showed rosettes of germ cells in both sexes in gonad anlagen before ovarioles have been formed.

Ontogenesis of the ovary of *Chironomus tendens* is similar to that found in other midges, i.e. cluster genesis occurs before ovariole induction, as has been shown above. Sachtleben describes rosettes of six or eight cells in the cavity of the gonad anlage (see his figures 17, 18, here Figure 3.101a,b), which clearly undergo oocyte–nurse cell differentiation in such a way that four cells develop as oocytes and four cells develop as nurse cells. After differentiation is initiated, the cluster becomes enclosed by finger-like projections of somatic cells. This is the beginning of ovariole development. These findings were confirmed by Wülker and Götz (1968) and by Wülker and Winter (1970). Concomitantly with the rise of ovarioles, the primary cluster must split into four subclusters, which become ordered into a single file row of which only the most basally located subcluster will undergo maturation, while the others will degenerate. If this interpretation is correct (and it must be confirmed by stage-specific ultrathin serial sectioning) this finding is comparable to the cluster formation shown in primitive dermapterans (see section 3.5.2). Its influence on phylogeny is discussed in section 3.6.3.

The most deviated ovary among dipterans occurs in gall midges (Cecidomyiidae). Many gall midges were known to be facultative paedogenetic, i.e. larvae can reproduce parthenogenetically (Metschnikow, 1866). It has been known for decades that the number of nurse cells nourishing the oocyte varies in these ovaries to a great extent. In *Miastor*, 5–26 nurse cell nuclei were found (Kahle, 1908; Ivanova-Kasas, 1965), whereas in *Heteropeza* lower numbers (3–7) were counted (Reitberger, 1940; Ivanova-Kasas, 1965; Camenzind, 1966; Panelius, 1968). Matuszewski (1968) found 6–20 nurse cells in egg chambers of *Mikiola fagi*, and 3–12 in follicles of *Boucheella artemisiae*. These numbers contradict the 2^n-rule and a somatic origin of nurse cells was suggested very early on (Kahle, 1908; Hegner, 1914; Panelius, 1968). On the other hand, Reitberger (1940) described morphological differences between nurse cell nuclei. He found one large nurse cell nucleus in the nurse chambers of *Oligarces (Heteropeza) paradoxus*, accompanied by different numbers of small nuclei. The same hypothesis was investigated by Hauschteck (1962), Camenzind (1966), and by Madhavan (1973), who all stated these differences. The last author was the first to give the correct interpretation of the dual origin of the nurse chamber in gall midges; however, his investigation was on the light microscopical level and needed further support. Mahowald and Stoiber (1974) gave the first data at the electron microscopical level showing that one nurse cell nucleus is indeed the sister of the oocyte. They showed morphological characters in the large nurse cell nucleus of *Miastor* which are found only in germ line

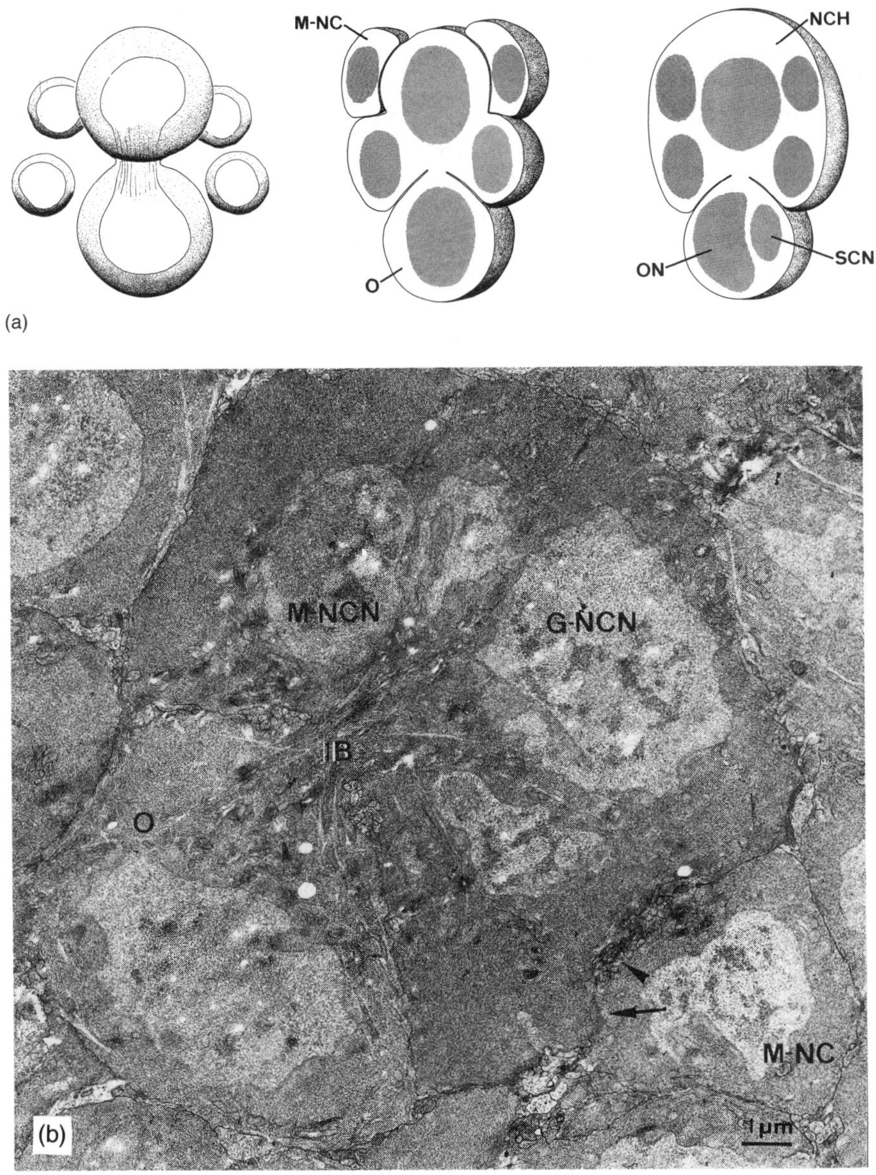

Figure 3.102 (a) Scheme of somatic cell integration into the two germ cell descendants, the germarial nurse cell and the oocyte, in *Mycophila speyeri* (Diptera: Cecidomyiidae). This process is unique among insects. Mesodermally derived prospective nurse cells (M-NC); nurse chamber (NCH); oocyte (O); oocyte nucleus (ON); somatic cell nucleus (SCN).(b) Electron micrograph, demonstrating integration of somatic cells by cell fusion. Mesodermal nurse cell nucleus (M-NCN); germ cell derived nurse cell nucleus (G-NCN); intercellular bridge (IB); oocyte (O). Asterisk indicates another somatic cell which has just fused. Arrowheads point to crystalline membrane stacks in the fusion area; higher magnification is shown in the inset. Arrows point to the fusion zone. (From Schüpbach and Went, 1983; with permission of D.F. Went and Springer Verlag.)

cells like pole cells, and, furthermore, they found only one intercellular bridge, connecting the oocyte with the syncytial nurse chamber.

The whole story about germ cell lineage from pole cells to oocytes, as well as the fusion of somatic cells with the nurse chamber and with the oocyte, was analysed by Schüpbach and Wendt (1983) and by Schüpbach and Camenzind (1983) using differential interference contrast microscopy on living embryos and isolated follicles of *Mycophila speyeri*. Follicles of the different stages of somatic cell association to the two-cell cluster were fixed for electron microscopy. During blastoderm formation, the embryo develops at its posterior four pole cells which migrate during organogenesis into the gonad anlagen. Each gonad houses two oogonia which are the former pole cells. Thus, mitotic activity during the migration phase is absent. The two oogonia are embedded in the posterior part of the ovary, whereas in the anterior part about 30–40 mesodermal cells are found. In the gonads each germ cell undergoes three mitotic divisions without growth phases, which terminate in 16 smaller final oogonia embedded in somatic cells. The somatic cells also divide; thus about 500 mesodermal cells are found finally. The whole ovary elongates during these divisions; however, ovarioles and germ chambers will not develop. In the final mitotic division of the 16 final oogonia, the definitive oocyte and the definitive nurse cell are built, connected to each other by an intercellular bridge, derived from incomplete cytokinesis of this last division. These early follicles are surrounded by somatic cells, which can fuse with the germ cell cluster. In a first fusion reaction, about four somatic cells approach the nurse cell compartment (Figure 3.102).

Their nuclei shift toward the annealing membranes, which develop a membrane labyrinth before fusion. In a second fusion reaction, additional somatic cells can fuse with the syncytial nurse chamber. Thus, a nurse chamber is established in which various amounts of nurse cell nuclei serve the oocyte. Somatically derived nurse cell nuclei may come under the register of the germ line nurse cell nucleus, as one can assume from the dense bodies (nuage material) associated tightly to both types of nurse cell nuclei (Mahowald and Stoiber, 1974; Mahowald, 1975). In a detailed study, Matuszewski (1968) compared nurse chamber volumes with nurse cell nuclei volumes, which correspond directly with their endomitotic activity. He found the total volume of nurse cell nuclei remained constant, irrespective of the number of nurse nuclei in a chamber. This means that the total number of genome equivalents serving each oocyte is measured and kept constant by the organism in an unknown way. After nurse cell fusion, two somatic cells fuse with the oocyte one after another, leading to an oocyte with two small somatic nuclei in addition to the oocyte nucleus. These 'small nuclei' had been known to exist for some time (Hauschteck, 1962; Camenzind, 1966; Matuszewski, 1968), and must be clearly distinguished from accessory nuclei which derive from the germinal vesicle during oogenesis (see sections on Hymenoptera and Mallophaga). Small nuclei degenerate during embryogenesis, but in some cases they contribute to early embryonic development (Camenzind, 1966, 1971). The final follicles are surrounded by additional mesodermal cells, which differentiate as follicle cells. The follicles are now completed and the ovary sheath is ruptured; thus, the follicles develop freely in the body cavity of the mother. Growth of follicles and early embryogenesis follow each other without arrest. This indicates that final meiotic arrest is absent (Went, 1982).

(j) Mecoptera

Our knowledge about ovary architecture among Mecoptera is fragmentary and based on only a few investigations of the scorpion fly ovary, *Panorpa communis* (Table 3.4; Figure 3.88) (Gross, 1903; Grell, 1942; Gruzova, 1962, 1967; Ramamurty, 1963, 1964a,b, 1967, 1968; Bier and Ramamurty, 1964; Rottmar, 1966; Büning, unpublished). The panoistic ovary of *Boreus hiemalis* (Mecoptera: Boreidae) is described in section 3.5.1. In *Panorpa*, each ovary consists of 10–12 ovarioles, arranged in a comb-like fashion. Ovariole sheaths are combined anteriorly to a suspensory ligament. Each ovariole ends in a terminal filament, which degenerates in adults. Ovariole differentiation takes place during the larval and pupal stages (Grell, 1942; Rottmar, 1966). In the ovarioles of adults, a short anterior region is found, which houses siblings of cystoblasts only, all arrested in the prophase of meiosis. Thus, stem cells cease their dividing programme. Possibly, they degenerate, as is suggested by the pycnotic cell found in the anterior region (Figure 3.47a). Four cystocytes are combined by intercellular bridges; one of the inner cells develops as an oocyte, all the others as nurse cells. Siblings are arranged in rosettes and a polyfusome stretches through all bridges (Figure 3.47b) (Ramamurty, 1963, 1964a; Büning, unpublished).

Inner sheath cells surround all clusters of the germarium, between which some somatic interstitial cells are found. Interstitial cells and inner sheath cells are continuous and not separated from prefollicular cells, which are highly mitotically active. At the base of the germarial region, which opens continuously into the growing zone of the vitellarium, nurse cells become morphologically differentiated from oocytes by enormous growth of nuclei, indicating high synthesis. In this way, the nucleus becomes heavily lobed. Numerous nucleolar aggregates are scattered within chromatin, beside faintly staining globular compartments which may be endobodies (see section 3.3.2). Concomitantly with nurse cell nucleus growth, the cytoplasm becomes enriched with numerous free ribosomes. Mitochondria are abundant and show elongated and often Y-shaped forms, indicating high division rates. The nurse cell cytoplasm is not homogenous but differentiates into certain morphologically distinct compartments (Figure 3.47) (Gross, 1903; Grell, 1942; Ramamurty, 1964a; Büning, unpublished).

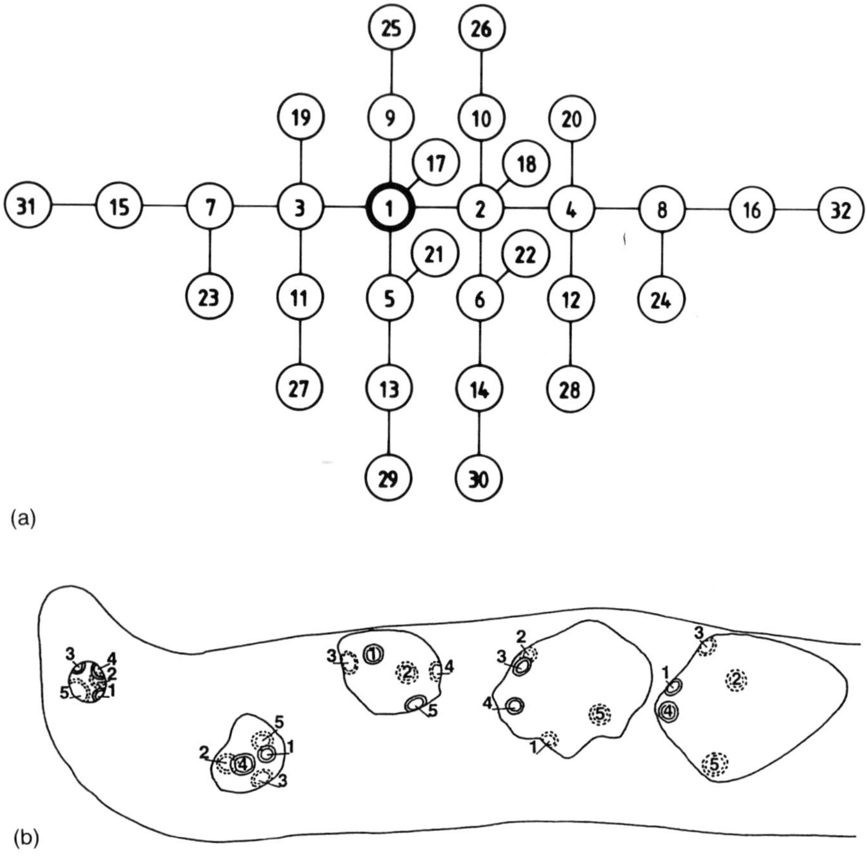

Figure 3.103 (a) Cluster analysis in the polytophic meroistic ovary of *Hystrichopsylla talpae* after ultrathin serial sectioning. Each cluster consists of 32 siblings, of which one of the innermost cells develops as an oocyte. All the others serve as nurse cells. Intercellular bridge distribution among siblings follows the basic scheme exactly. Lettering of siblings follows the scheme of Brown and King (1964) developed for *Drosophila*. (From Büning and Sohst; with permission of Longman Group.) (b) Distribution of intercellular bridges on the surface of oocytes as found in the analysed ovariole. Bridges which open to the lower surface are shown with broken lines, those bridges which can be directly seen are as shown with closed lines. Numbers indicate division cycles: 1 = intercellular bridge connecting the oocyte (1) with nurse cell (2); 2 = (1) with (3); 3 = (1) with (5); 4 = (1) with (9); 5 = (1) with (17). No preferential orientation of subclusters exists. Membrane growth of oocytes of posterior parts is positive allometric.

Oocytes of young clusters may synthesize to some extent, since small nucleolar aggregates are found. During later stages of previtellogenesis, chromosomes lose their contact with the germinal vesicle envelope and form a karyosome (Gruzova, 1962). According to Gruzova's figures, one or some big endobodies grow, together with some dark staining nucleolar bodies.

From the results, one can conclude that cluster formation follows the 2^n-rule and is of the basic type.

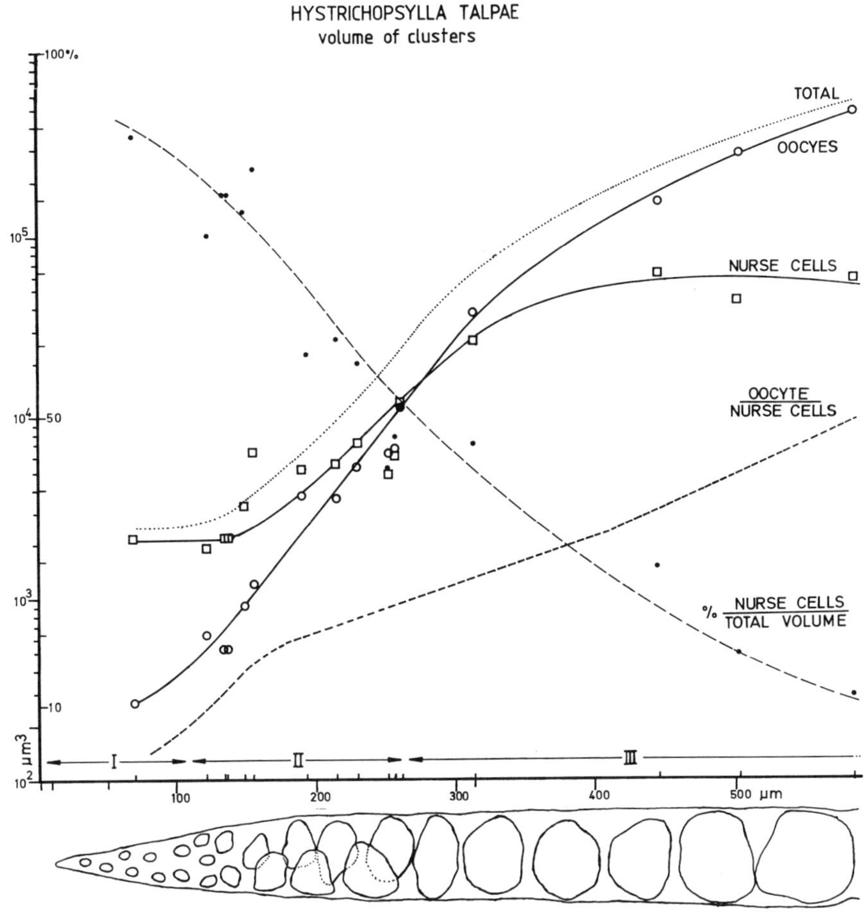

Figure 3.104 Volume measurements in several clusters of an ovariole of *Hystrichopsylla talpae*. Position of clusters is indicated by bars on the *x*-axis. Open circles represent volumes of oocytes; open squares represent volumes of total amount of nurse cells of each cluster at this position; full circles represent the quotient of nurse cell volume to total volume, the rapid decline shows that nurse cells have little influence on oocyte growth. The ovariole below shows the positions of the oocytes, nurse cells are omitted. Compare with Figures 3.21 and 3.30.

(k) Siphonaptera: Hystrichopsylla and Stenoponia

Most species of fleas (Siphonaptera) house panoistic ovarioles (Table 3.4; Figure 3.88). However, some light microscopical studies suggested some species were polytrophic meroistic. First reports on flea ovaries were given by Landois (1867), Gross (1903) and Lass (1905). All authors investigated one or two species, which were all found to have panoistic ovarioles. The first report in which flea ovaries were investigated comparatively is that of Kunitskaya (1960). She compared 22 species belonging to both superfamilies of fleas (Pulicoidea, Ceratophylloidea) and found panoistic ovaries in 18 species, belonging to both superfamilies, whereas strong evidence for a strange polytrophic meroistic ovary was found in four species: *Hystrichopsylla talpae, H. (Hystoceras) satunini, Stenoponia ivanovi* and *S. insperata,* which belong to the subfamilies of Hystrichopsyllinae and Stenoponiinae (Ctenophthalmidae, Ceratophylloidea). A second report of such a polytrophic meroistic ovary was given by King and Teasley (1980) for *S. tripectinata.* In *S. insperata* nine, 11 and 13 nurse cells were counted providing an oocyte (Kunitskaya, 1960), whereas in *S. tripectinata* 8–13 nurse cells serve each oocyte (King and Teasley, 1980). Cluster analysis of *H. talpae* was made by ultrathin serial sectioning by Büning and Sohst (1988, 1989). Each cluster contains 32 cells, 31 nurse cells and one oocyte. As shown (Figure 3.103a), the position of the oocyte and intercellular bridge distribution among cystocytes follows exactly the 2^n-rule. Since cluster genesis in *Stenoponia* is not elucidated, we have no information on the architecture of the clusters and mitotic cycle reduction in Stenoponiinae.

Nothing is known about rosette formation and polyfusomes in meroistic fleas, since those processes occur, if they exist, at the late larval or pupal stages, which were not under investigation. However, stem cells cease their dividing programme. This is concluded from two observations. Mitotic activity was never found among germ cell descendants, and in old females the anterior end of the ovariole contains only some clusters with oocytes at mid-previtellogenesis. No isolated germ cells were identified.

The position of subclusters to the oocyte and to the long axis of the ovariole is random, thus, a chaotic appearance of each cluster results (Figure 3.103b). Nurse cells grow only at the beginning of previtellogenesis (Figure 3.104). In this way, those nurse cells which are in direct contact with the oocyte via intercellular bridges grow faster than those far apart. This is the case especially for the former pro-oocyte, which connects the oocyte with the largest subcluster of 16 nurse cells. During the early stages of growth it is impossible to decide which of the two pro-oocytes will develop as the final oocyte. No morphological character is given except growth. Thus, in this ovary it seems possible that growth alone will finally decide oocyte determination.

The five intercellular bridges of oocytes do not move together in a special apical appendix as do most polytrophic ovaries (Coleoptera, Adephaga; Hymenoptera; Lepidoptera), but their relative positions remain unaltered, i.e. the distances between the bridges grow concomitantly with early growth. Later on, the posterior region of the oocyte shows positive allometric growth (Figure 3.103b).

Cytophotometric measurements of cystocytes at the beginning of previtellogenic growth indicate that all cystocytes have values of DNA larger than 4C, but no preference for 8C or 16C could be detected (Figure 3.30b). This indicates low endomitotic activity, if any, and extrachromosomal DNA existing in all cystocytes, which is unique among all those species that have extrachromosomal DNA. Active rDNA genes however, are found predominantly in oocytes as is concluded by ultrastructure. Ribosomes accumulate only in oocytes, not in nurse cells. On the other hand, nurse cell nuclei show the same changes in ultrastructure as do oocyte nuclei during previtellogenic growth, but to a lesser extent. Miller spreads of extrachromosomal rDNA rings, clearly derived from nurse cells, or ultrastructural and autoradiographic data on ^{3}H-uridine incorporation are needed to clarify the amount of nurse cell rRNA synthesis.

Follicle cells are produced by the high mitotic activity of small nests of prefollicular cells, scattered among germ cell clusters before growth phases have started (Figure 3.30a). This mimics exactly the situation known from the panoistic ovaries of fleas and of most other panoistic ovaries except those of Thysanoptera and Plecoptera, in which prefollicular tissues are found at the base of the germarial region.

Nurse cells do not form a compact trophic chamber, as known from other polytrophic meroistic ovaries. Instead, they are integrated into the growing follicular epithelium and vanish at the end of previtellogenic growth (Figure 3.21). Thus, a special interaction of nurse cells with somatic cells (as found in all other meroistic ovarioles), which leads to inner sheath cells, is missing.

3.5.3 Telotrophic meroistic ovaries

This ovary type also develops from germ cell clusters, which differentiate later on into definitive nurse cells besides oocytes. However, telotrophic ovaries are clearly cut from polytrophic ones by one anterior trophic chamber to which all oocytes are connected via nutritive cords. As we will see in the forthcoming sections, this type evolved independently four times. Common to all types, each nutritive cord is part of the oocyte, connected primarily via an intercellular bridge to other siblings of the germ cell cluster. There is no example of secondary connections of oocytes to nutritive tissues via outgrowing processes, which develop into nutritive cords. Secondary connections play a role in the nurse cell chamber, in which cell membranes between nurse cells can be reduced, as found in Coleoptera,

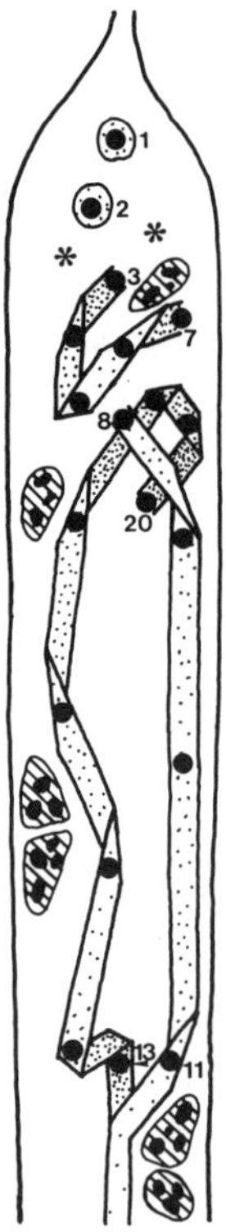

Figure 3.105 In the late stages of vitellogenesis, the trophic chamber degenerates, as shown in the ovariole of *Ephemerella*, close to subimaginal moult. Two isolated cells in the anterior tip are stippled. Terminal cells of subclusters are numbered. Two isolated intercellular bridges are marked (*). Cell aggregates in pycnosis are striped.

Polyphaga, in Sialidae and in Hemiptera. In all four telotrophic meroistic ovaries, germarial regions do not exist in functioning ovarioles; thus stem cells cease early in development.

(a) Ephemeroptera

It has been shown that mayflies have ovaries of the telotrophic type (Table 3.1; Figure 3.88) (Gottanka, 1992; Gottanka and Büning, 1993). Three species were investigated and all showed the same pattern of cluster formation (*Cloëon* sp., *Siphlonurus armatus* and *Ephemerella ignita*). Mayflies have comb-like ovaries, each with 60–80 ovarioles (Soldan 1979, 1981). During larval stages, about 4 months before the final moult, metachrone cluster mitoses occur in ovariole anlagen (Figure 3.83a). A linear arrangement of siblings is apparent. Ultrastructural analysis shows that siblings remain connected by intercellular bridges, while a polyfusome stretches through them (Figure 3.83b). No indication for rosette formation is found. After mitosis, the whole terminal chamber is filled with long and coiled rows of siblings. The three-dimensional arrangement of clusters was analysed by ultrathin serial sections through three ovarioles representing three different stages of development. The first ovariole came from an idividuum just after mitotic activity of siblings (*Cloëon*) (Figures 3.83 and 3.84). The second was derived from an older larvae just at the beginning of vitellogenesis (*Siphlonurus*) (Figure 3.85). The third came from an old larva at the end of vitellogenic growth, which already had one to two ripe eggs per ovariole (*Ephemerella*) (Figure 3.105).

The first example shows only one linear cluster of 57 cells, and one cell, isolated in the very anterior part of the ovariole (Figure 3.84). All cluster cells are combined by intercellular bridges through which one continuous polyfusome stretches. The cluster begins near the anterior end of the germarial region, turns down to the posterior end, bends up and finally reaches the anterior end again. Prospective oocytes and nurse cells cannot be distinguished at that stage. The isolated germ cell at the anterior end contains one nucleus and a polyfusome, which extends in two ventricles. Each ventricle has constrictions by intercellular bridges, ordered linearly. This strange appearance of the isolated cell can most easily be interpreted by a decoupling of incomplete cytokinesis from S-phases, which are repressed in this apical germ cell, while the cytoskeleton follows some periodic signals, which may come from the dividing cluster or from somewhere else.

The terminal chamber of *Siphlonurus* is enlarged and clearly separated from the vitellarium. It houses 183 germ cells, combined to five linear clusters, coiled with each other (Figure 3.85a–e). At the posterior end of the terminal chamber each cluster has developed one oocyte. When this oocyte has developed from a middle position, as is found in most cases, it is joined to its neighbouring siblings of the cluster through two intercellular bridges. However, when the oocyte has developed in a terminal position only one bridge is seen. Comparing the positions of the terminal germ cell descendants from one cluster to the next, as shown by terminal cells 52 and 53, 91 and 92, 136 and 137, 160 and 161, it becomes obvious that the close neighbourhood is not random but strongly suggests the former connections of the clusters to a single, huge, linear cluster. A second argument comes from intercellular bridge distribution. There are only four cells which have one, additional intercellular bridge with a blind end; all of these belong to the end-to-end pairs mentioned above. Within all linearly ordered clusters only one branch exists. It is cell 88 which is not in line with the others. This is the only exception found among 186 germ cell descendants. A polyfusome and isolated fusomal material is no longer found at the beginning of vitellogenesis. The cytoplasm of bridge regions is the same as found in adjacent cells. Growing oocytes are connected via nutritive cords to the siblings which remain in the terminal chamber (Figure 3.85f). In this way, intercellular bridges persist.

The physiologically oldest terminal chamber, analysed by ultrathin serial sectioning, has become slender and the number of germ cell descendants is low (Figure 3.105). Two larger linear clusters were found, one in the apical region, housing five cells, the other containing seven cells and opening in a nutritive cord. Besides these clusters, three isolated cells were found, two in the anterior tip, the third in a middle position. Furthermore, two isolated intercellular bridge complexes and several degenerating aggregates were found, the contours of which resemble those of germ cells rather than those of somatic cells.

Comparing all three terminal chambers, which represent not only three different species but three different physiological stages, the ontogenetic path of germ cells in mayflies can be reconstructed. During early larval stages, the two elongated gonadal tubes house about 100 germ cell descendants, which function as cystoblasts. Stem cell function in ovarioles does not exist. Each cystoblast divides synchronously, at least in the beginning. During early divisions, ovariole induction occurs, enclosing only one developing cluster. During later divisions, synchrony is lost and the long, linear cluster is cut into pieces of more or less equal length. The most basally positioned cell develops as an oocyte; all the other cells remain undifferentiated or they are definitively transformed to nurse cells. However, there are no morphological characters by which one can distinguish between the above categories. After oocyte determination, the development of additional oocytes in the same cluster must be prevented, since we have never found more than one oocyte developing in a subcluster. This may be done by production of nurse cell determinants by the oocyte, which opens special gene sets in siblings, and/or by production of oocyte inhibitors, which keep the oocyte pathway silent in siblings, which would have the same effect. At the end of vitellogenesis, the subcluster breaks into pieces and isolated nurse cells degenerate (Figure 3.105). Thus, the terminal chamber becomes reduced at the end of vitellogenesis.

Somatic cells of the terminal chamber are small and form a flat inner sheath, from which some protrusions reach between germ cell clusters. These cells are clearly separated from follicular cells, which from the very beginning are more or less cuboid and are always attached to oocytes. This behaviour shows that communication between oocytes and follicle cells is established at once. A continuous prefollicular tissue, as found in all other telotrophic ovaries, does not exist. Follicular cells contain from the very beginning many ribosomes, which contrasts to inner sheath cells which have only a few. Oocyte nuclei and nurse cell nuclei show the same elements during growth phases: both contain diffuse chromatin and a well-developed nucleolus. The cytoplasm of nurse cells, nutritive cords and oocytes is of similar aspect, housing masses of free ribosomes and mitochondria, whereas ER compartments, Golgi complexes, and microtubules are rare. A preferential orientation of mitochondria and/or microtubules in nutritive cords is not observed.

(b) Hemiptera

Among hemimetabolous insects, hemipterans are the most important group since they have the highest number of species, about 56 000, and show a great radiation into eight subgroups of equal rank (Hennig, 1981). The bugs (Heteroptera) are the largest subgroup with about 25 000 species, living in all climate zones and with species which can adapt to the water (*Gerris, Notonecta*), to special plants, or to life as ectoparasites (*Cimex*). A small group with only 19 species are the Coleorrhyncha, restricted to South America and Australia. The ovary has not been investigated, as far as I know, and therefore this ovary is not included in the forthcoming pages. The next taxon, the Auchenorrhyncha, includes about 23 000 species, and consists of two sister groups, lantern-flies (Fulgoriformes) and cicadas (Cicadiformes), most of them living in subtropical and tropical regions. The last four groups are combined in the taxon of Sternorrhyncha which includes moth lice or white flies (Aleyrodoidea) with about 200 species, the psyllids (Psylloidea) with 1500 species, the aphids (Aphidoidea) with 2000 species, and the scale insects (Coccoidea) with about 3000 species. All of them have developed telotrophic ovaries, generally of the same kind, and we will focus on the characters they have in common and on the characters shared by members of one or some groups, when discussing cluster formation and ontogenesis of this hemipteran ovary.

For a better understanding of this type of telotrophic ovary, we will start with Sternorrhyncha and, within this group, with the Coccoidea, in which clusters are small (Table 3.2; Figure 3.88). Scale insect ovaries have been analysed, besides others, by Emeis (1915), Hughes-Schrader (1925, 1927), Weglarska (1961, 1962), Ksiazkiewicz (1980), King and Büning (1985) and Foldi (1990a,b).

During the larval stages of *Quadraspidiotus ostreaeformis*, each ovary consists of an elongated, closed tube, housing germ cells which divide mitotically until diapause, where a resting stage is reached (Weglarska, 1961). After diapause, mitotic activity occurs, leading to groups of 6–12 germ cell descendants, developing synchronously. These groups are encompassed by some epithelial cells, derived from the peritoneal sheath surrounding the whole ovary. During this differentiation, a gradient is established from posterior to anterior in each ovary, in which groups of cells of the posterior part precede those of the anterior parts. Out of each group, subgroups of four cells become enclosed by some epithelial cells, from which one is always found in the tip of the just emerging ovariole (Figure 3.59g). Each ovary produces about 50 ovarioles during this first cycle. Later on, additional ovarioles occur, intermingling with the older ones (Figure 3.11a).

This description is similar to that given by Hughes-Schrader (1925, 1927) describing the gonadal development of *Iceria purchasi*. This scale insect develops hermaphrodites in which sperms and oocytes are built in the same gonads. Gonadal development starts during embryogenesis before katatrepsis. A small mass of germ cells becomes clearly distinguishable, situated between the cells of the symbiontic organ and the mesoderm germ band, which developed further. After katatrepsis, the gonad splits into two halves, situated on each side of the alimentary canal which has been built meanwhile. During both stages, mitoses are found in germ cells and gonads elongate. Development of male germ cells precedes the development of female determined germ cells. In both cases, three mitoses give rise to eight cells, which all start with meiosis. In male cysts, meiosis is completed and 32 sperm develop; in females the eight cells differentiate into one oocyte and seven nurse cells since they stop their meiotic programme and the genomes undergo endomitotic replications. Female determined clusters will be accompanied first by one somatic cell, which quickly divides, and the cells of which ensheath and protrude from the whole cluster to the periphery and give rise to the small ovarioles, housing only one cluster. Male determined cysts are ensheathed also with some somatic cells but remain inside the gonad, the lumen of which serves as the vas deferens and the lateral oviduct in later stages. Seven nurse cells and one oocyte are also found in *Pseudococcus maritimus* (Figure 3.59f).

As in Nematocera (Diptera) germ cell differentiation precedes the differentiation of somatic parts of the gonad. Stem cell mitoses cease long before ovarioles are built; therefore, no germarial region exists in coccid ovarioles and ovarioles seem to emerge by 'budding' processes from the tubular gonad. The circumferential peritoneal sheath degenerates to some extent during the rise of ovarioles; thus, ovarioles seem to be free in the body cavity.

Since meiosis occurs in all members of a cluster simultaneously, it is concluded that the preceding mitoses also occur in synchronized cycles. The 2^n-rule seems to be obeyed in scale insects with four or eight cystocytes respectively, of which only one develops as an oocyte finally, whereas the others are transformed to nurse cells. As we will see in forthcoming parts, this character seems to be unique among hemipterans. In all other taxa more than one oocyte will be found in each cluster. Ultrastructure of developing clusters is missing, thus we can only assume that processes found in the sister group of aphids can be adopted for coccids also. This implies that cytokinesis is arrested during mitotic cycles and intercellular bridges arise. No direct evidence is given for rosette formation; however, the structure of differentiated ovarioles of scale insects mimics exactly that of aphids. Nurse cells and the nutritive cord open into a central trophic core region, only the number of nurse cells and oocytes is reduced (Figure 3.59f,g) (Ksiazkiewicz, 1980; King and Büning, 1985; Büning, 1985). Thus one can assume rosette formation in coccids, but it must be verified. Another open question is the existence of a temporal polyfusome. In aphids, a polyfusome could not be found. A character shared with most hemipteran telotrophic ovaries is the existence of massive bundles of microtubules in the nutritive cord and in the trophic core (Ksiazkiewicz, 1980, King and Büning, 1985; Foldi, 1990a,b).

The sister group of scale insects are the aphids (Aphidoidea), the ovaries of which have been investigated from the very beginning of oogenetic research in insects (e.g. Leydig, 1850; Huxley, 1859; Lubbock, 1859; Witlaczil, 1884; Stevens, 1905; Tannreuther, 1907; von Baehr, 1909; Hirschler, 1912; Toth, 1933). All authors noticed the large nutritive cord, connecting the growing oocyte with the trophic chamber. There was also no doubt that nurse cells open to a central core. However, ontogenetic development and cluster formation remained unclear. The question of whether parthenogenetic ovarioles and those of egg laying females are comparable with each other remained open for decades. The situation was clarified by Blackman (1978), who compared the development of parthenogenetic ovarioles of several species with the light microscope, and by Büning (1985), who studied cluster formation in different morphs and different species with the light and electron microscope.

Ovariole number in aphids is species specific but changes can occur between different morphs (Leather and Wellings, 1981; Büning, 1985). Many species have five ovarioles in each ovary (Figure 3.59a,b), but fundatrix generations (i.e. spring generations, which develop from overwintering eggs) of *Drepanosiphum platanoides* regularly contain six. Generally six are found in *Macrosiphum rosae* and *Brevicoryne brassicae* (Couchman and King, 1979). However, in some species higher and more variable numbers occur, as found in *Megoura viciae*, in which 6–11 ovarioles are found in each ovary (Brough and Dixon, 1989). Development of ovarioles is asynchronous but polar in each ovary. The most lateral inserted ovarioles antecede those in the central parts (Figure 3.59a,b).

No general differences are found between the ovarioles of virginoparous and oviparous females except of growth. Each ovariole houses only one germ cell cluster, ensheathed by somatic cells (Figure 3.59a–e). A terminal filament extends from the anterior part of the ovariole and connects with other terminal filaments to form a short rope, which anneals to the body wall dorsolaterally. Next to the terminal filament, inner sheath cells surround the nurse cells of the terminal chamber forming a flat epithelium. Between the terminal filament and the inner sheath cells of the anterior part of the terminal chamber, a transversal septum is found, which continues into the tunica propria. Outer epithelia are reduced in functioning ovaries (Büning, 1985; Brough and Dixon, 1989). The posterior part of the terminal chamber houses the arrested oocytes which are in contact with the

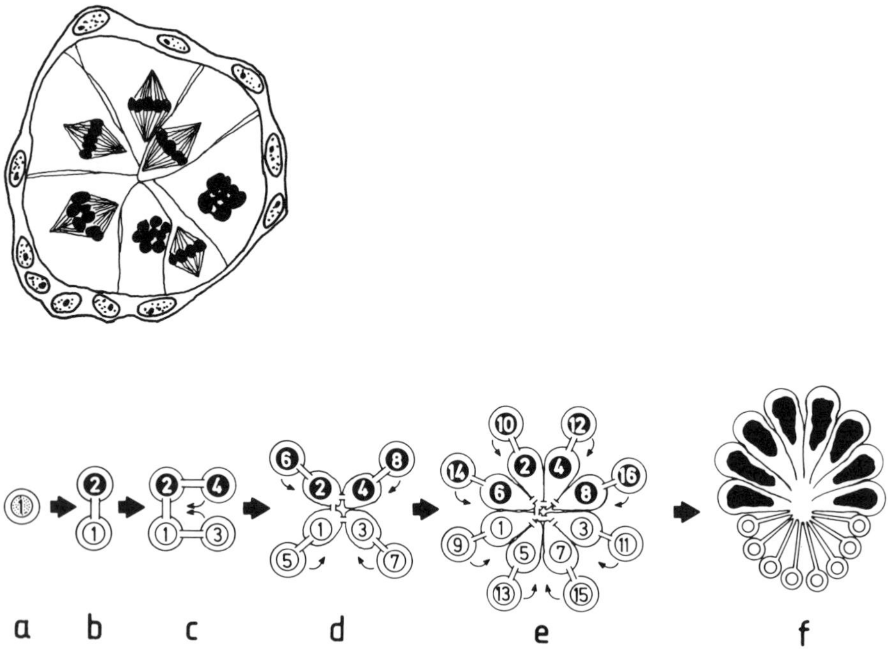

Figure 3.106 (a) Synchronous divisions occur during cluster formation. Note that spindle axes point to a centre, indicating rosette formation (redrawn from Toth, 1933). (b) Model of cluster formation in aphids. Five consecutive, synchronized divisions (thick arrows) produce a cluster of 32 cells (2^5). In *Drepanosiphum,* half of the cells develop as oocytes, the other half as nurse cells. Movement of cells during rosette formation is indicated by small arrows. Lettering is the same as in diagrams of polytrophic meroistic ovaries of *Drosophila* (King, 1970). Closed circles = nurse cells; open circles = oocytes. (From Büning, 1985; with permission of Wiley and Sons.)

Figure 3.107 (a,b) Two consecutive, longitudinal semi-thin sections through an ovariole of *Trialeurodes vaporariorum* (Hemiptera: Aleyrodina). Nurse cells (N) open to a central core (C) from which a nutritive cord (NC) connects the previtellogenic growing oocyte (O). Encompassing follicle cells (F) are mononuclear. (c) Cut through an ovariole of *Trialeurodes* during early larval stage 4. All cystocytes show pachytene stages. Note the polyfusome area (F) and intercellular bridges. Somatic sheath cells (S). (d) Another part of the central area at higher magnification in which the polyfusome (F) stretches through two intercellular bridges (B), which are ordered nearly rectangular to each other. (Courtesy of T. Nowak, University of Erlangen, Germany.)

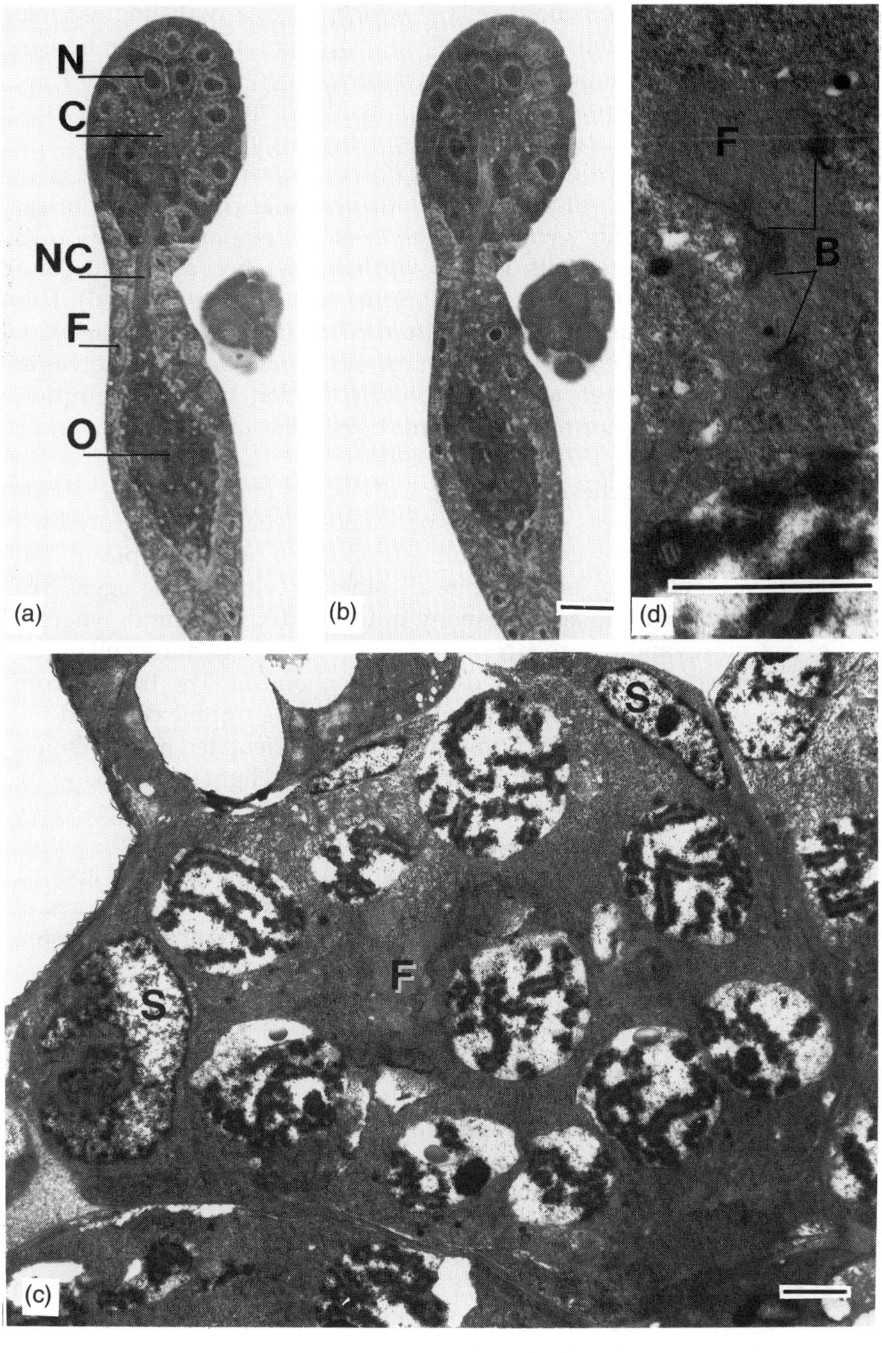
N
C
NC
F
O
(a)
(b)
F
B
(d)
S
F
S
(c)

prefollicular tissue, the cuboid cells of which are clearly distinguishable from flat inner sheath cells (Figure 3.59c,d). Each terminal chamber houses a constant, species-specific number of nurse cells and oocytes (Toth, 1933; Blackman, 1978; Büning, 1985). Nurse cells differ in the amount of endomitotic polyploidization when comparing virginopare and ovipare females. In ovipare generations, 512 (2^9) copies can exist, whereas in virginoparae up to 16C values are found in nurse cells. However, in the last virginopare generation, which gives birth to the ovipare generation, C values are rising in nurse cells. This shows that differences are gradual and not regulated by 'all or nothing' mechanisms (Büning, unpublished). This finding agrees with rare examples of intermediate (ambiphasic) females, in which normally parthenogenetic eggs are built in one ovariole, whereas in others full grown eggs are produced. However, in rare examples, parthenogenetic and normal vitellogenic eggs occur in the same ovariole (Stevens, 1905; Crema, 1973; Hardie, 1984).

During previtellogenesis, the volume of oocytes in ovipare generations is raised to 10^5, whereas oocytes in parthenogenetic females grow by a factor of about 50. However, the ultrastructure of nurse cell plasm and ooplasm is identical in both. As in all other previtellogenic eggs, free ribosomes are predominant and many mitochondria are found, whereas Golgi complexes and ER are rare. Nuage material around nurse cell nuclei is very common, often associated with mitochondria. The trophic core region, as well as the openings of nurse cells to the trophic core and the nutritive cords, are filled with numerous, parallel orientated microtubules, the masses of which allow birefringency of polarized light (Figure 3.59a,b). Nutritive cords, connecting the trophic chamber with growing eggs, have diameters of 15–20 μm in oviparae and 2–3 μm in virginoparae. Trophic cords to resting oocytes are extremely thin and have diameters of about 1 μm or less. In oocytes, chromosomes are visible throughout all stages of oogenesis. They remain attached to the nuclear envelope and are arrested during previtellogenic growth in a post diplotene stage.

During embryogenesis, ovarioles are formed in a compact ovary anlage, in which very early ovarioles differentiate. Each ovariole houses one cluster of siblings, generated by synchronous divisions (Toth, 1933) (Figure 3.106a).

Thereby, a rosette is formed from the very beginning. Intercellular bridges assemble in a central area, filled with microtubules which are thought to be remnants of the former spindle apparatus (Figure 3.46). A polyfusome is not found. After five mitotic cycles a cluster of 32 cells is formed. Posterior cells are destined to become oocytes, whereas anterior cells will develop into nurse cells. This differentiation is not only triggered by position, since constant numbers of oocytes and nurse cells exist as can be easily explained by the 2^n-rule: in *Drepanosiphum platanoides*, 16 nurse cells serve for 16 oocytes; in *Metopolophium dirhodum*, 24 nurse cells nourish eight oocytes; and in *Macrosiphum rosae*, eight or 11 oocytes are connected

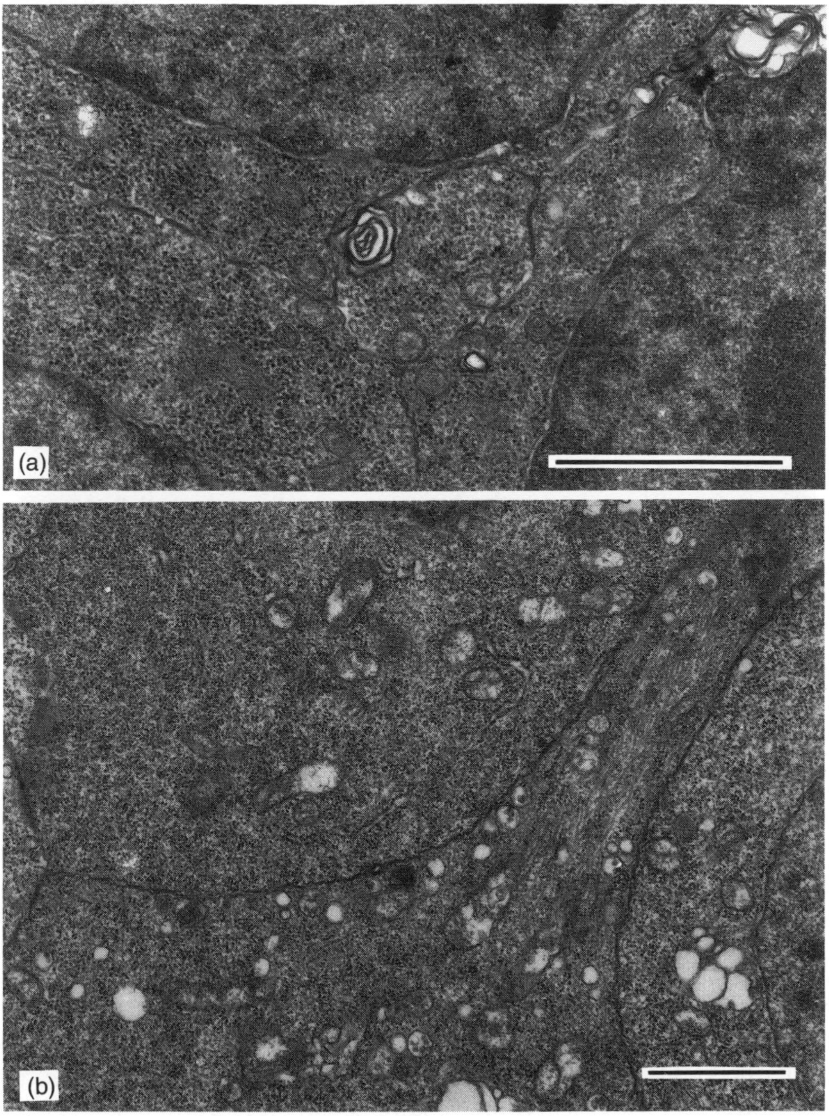

Figure 3.108 Cross-section through the nutritive cord connecting the oocyte to the central core at the beginning of the previtellogenic growth of *Trialeurodes* ovariole. Note tightly packed microtubules. (b) Same stage as before. Longitudinal section through nutritive cord opening into oocyte. Bars represent 1 μm. (Courtesy of T. Nowak, University of Erlangen, Germany.)

to 21 nurse cells. If we assume cluster formation in aphids follows the pattern known from the basic type, a pattern of nurse cell and oocyte differentiation may happen within the cluster as shown in Figure 3.106b.

Oogenesis of white flies (Aleyrodoidea) has not been studied, except for some remarks on ovary structure (Weber, 1934, 1935). Recently, we have analysed the ovary of *Trialeurodes vaporariorum* by means of light and transmission electron microscopy, elucidating the three-dimensional organization of the ovariole by ultrathin serial sectioning (Nowak and Büning, in preparation).

Each ovary houses about 15 ovarioles, which show different developmental stages, ordered in the same polarized manner as in aphids. The ovariole consists of a terminal chamber, enclosed by a flat inner sheath, the apical cell of which is pronounced in young ovariole anlagen (Figure 3.107a).

A rudimentary terminal filament is separated from the terminal chamber by a thin transverse septum. In young ovarioles, nurse cells and oocytes cannot be distinguished by morphological criteria. Cytoplasm and nuclei are identical. All siblings enter prophase of meiosis synchronously up to pachynema, as can be shown by synaptonemal complexes (Figure 3.107c,d). Even during growth phases, nurse cells and oocytes seem to be identical. Nurse cell nuclei do not enlarge significantly; thus their DNA content seems to be arrested at 4C and no endomitosis will occur. Oocytes become distinguishable from nurse cells only by the very narrow and long (4–10 μm in resting oocytes) nutritive cords, the diameters of which are about 0.5 μm. Nurse cells have conus-like, short and broad openings (about 2 μm in diameter). A second character by which oocytes can be separated from nurse cells is the bundles of microtubules, which are found in Aleyrodinae ovarioles only in nutritive cords and not in the central core region nor in the mouth region of nurse cells (Figure 3.108).

During previtellogenic growth, nutritive cords enlarge to diameters of about 4 μm and a total length of about 80 μm. Although they are filled with microtubules orientated in parallel to the long axis, birefringency cannot be seen because the total amount of microtubules is too low (about 1000), although the density/μm^2 with about 105 microtubules is comparable to the density known from other hemipteran nutritive cords (100–125/μm^2 in *Rhodnius* (Huebner, 1981b); 175/μm^2 in *Notonecta* (Stebbings, 1988)).

During synchronous divisions of cystocytes, intercellular bridges are built, filled with fusomal material (Figure 3.107c,d). Fusomes concentrate in a central region of the terminal chamber and build a polyfusome, which persists even after synaptonemal complexes are lost (Figure 3.109). At this stage, individual intercellular bridges cannot be analysed, since intercellular bridge walls intermingle with each other and with cell membranes, which develop the membrane labyrinth. When the polyfusome vanishes and the membrane labyrinth reduces, the trophic core region is finished.

As shown, each ovariole consists of one germ cell cluster. By ultrathin

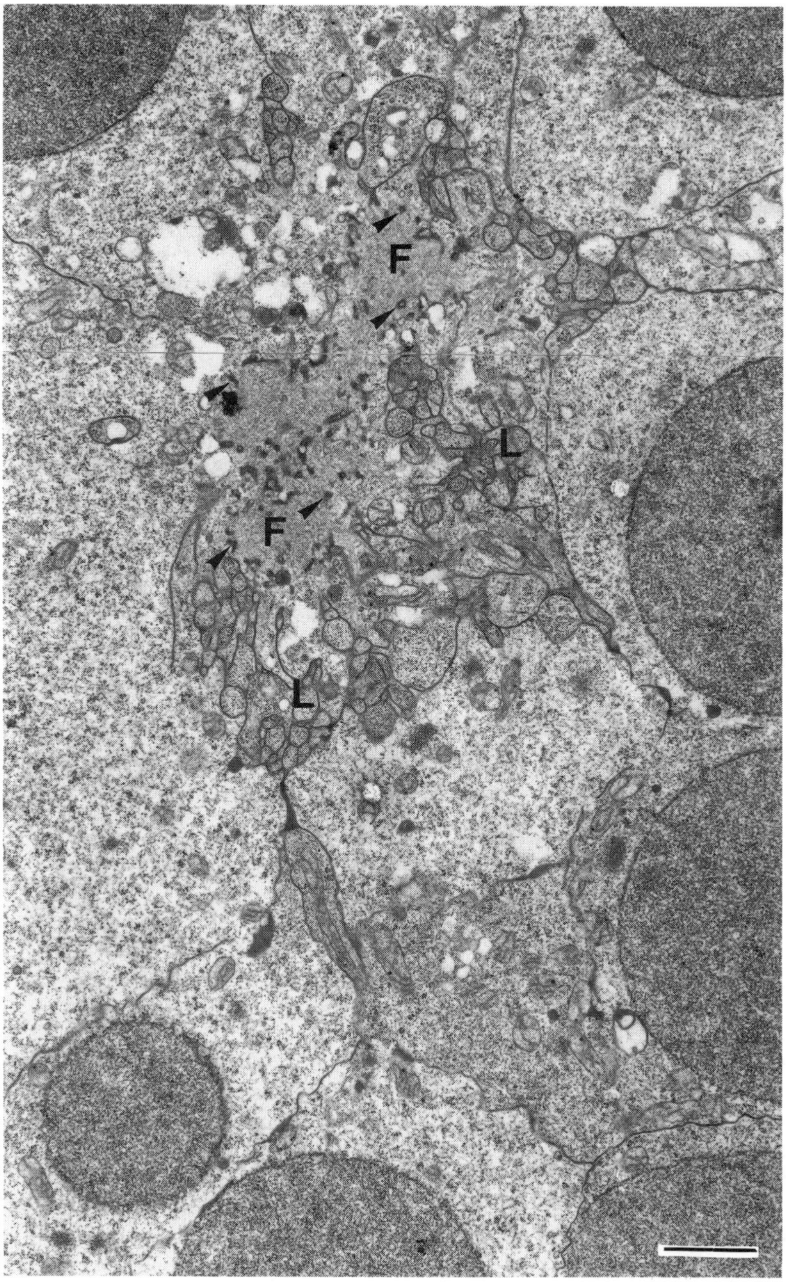

Figure 3.109 Central part of an ovariole of *Trialeurodes vaporariorum* (Hemiptera: Aleyrodinae) during late larval stage 4. Cystocyte nuclei look absolutely uniform. The rising central core houses remnants of polyfusome (F) and intercellular bridges (arrowheads) next to membrane labyrinth (L) which continues into nurse cell membranes. Bar indicates 1 μm. (Courtesy of T. Nowak, University of Erlangen, Germany.)

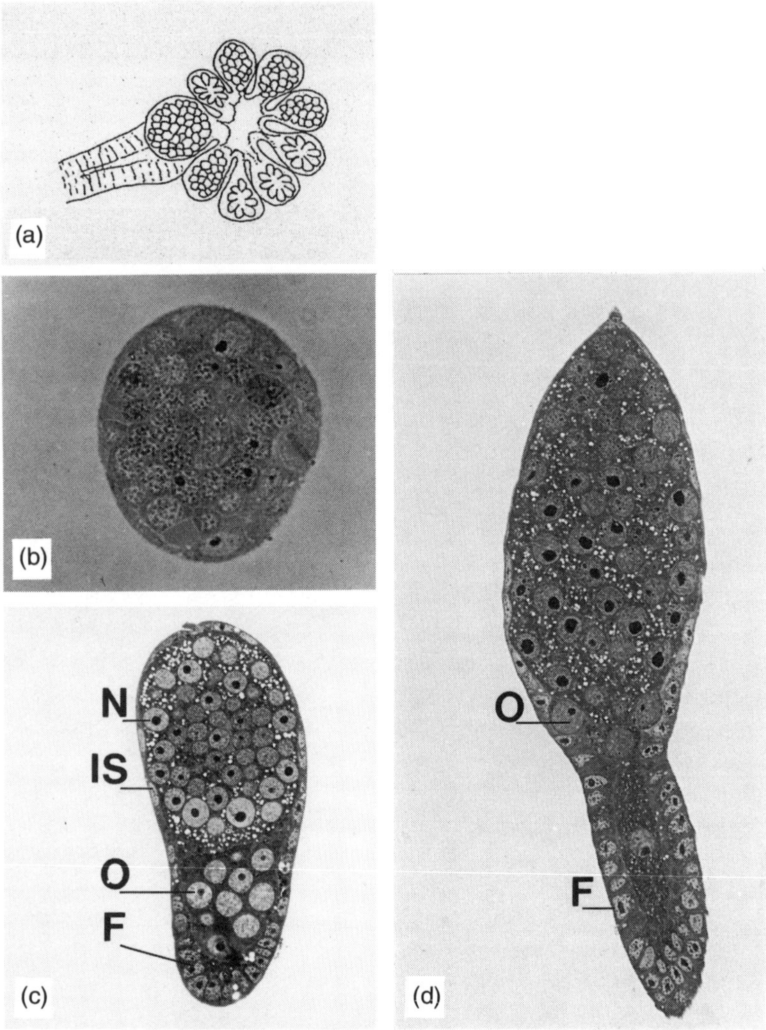

Figure 3.110 (a) The ovary of *Rhinocola speciosa* (Psylloidea). Four ovarioles are in rosette configuration, the others are further developed, comparable to (b). (Redrawn from Witlaczil, 1885.) (b–d) Ovary of *Psylla alni* (Hemiptera, Psylloidea: Psyllidae). (b) In late larval stages, siblings undergo synchronized mitotic divisions as shown here by prophase configuration of chromosomes. Cell in metaphase is a somatic inner sheath cell. (c) In young adults, cell membranes of nurse cells (N) reduce totally, whereas those of presumptive oocytes (O) remain unaltered. Differentiation between inner sheath cells (IS) embracing the nurse cell syncytium and follicle cells (F) embracing the oocyte layer, becomes apparent. (d) During growth phases of the oocyte, nurse cell syncytium fills the whole terminal chamber, except some arrested oocytes (O) at the bottom. Follicular cells (F) are mononucleate. Bars represent 20 μm. (Courtesy of C. Beyrich, University of Erlangen, Germany.)

serial sectioning, the number of nurse cells was determined as 56 and the number of oocytes was eight; thus, the cluster contains 64 cells, which obeys the 2^n-rule. We could not analyse the geometry of the cluster during each division cycle, which happens during late embryonal and early larval stages. However, distribution of intercellular bridges reveals that cluster formation is of a branching type with strong rosette formation; thus, as a theoretical model we use the branching type as known from polytrophic meroistic ovaries to describe early events.

The sister group of white flies are the psyllids (Psylloidea). Ovaries of psyllids have been investigated by Witlaczil (1885). Each ovary consists of a lateral oviduct, the anterior end of which opens into a small bulbus which shows numerous small ovarioles. In the last larval instars and in young adult females each ovariole consists of a terminal chamber, which houses germ cell siblings in a rosette formation (Figure 3.110a).

In older ovarioles, the rosette configuration is lost but differentiation into oocytes and nurse cells becomes apparent. Five to ten oocytes are found at the bottom of the terminal chamber, clearly separated from the smaller nurse cells which are numerous. The growing oocyte has been found attached to the terminal chamber by a nutritive cord.

From this report one can conclude that in psyllids, as in aphids or white flies, each ovariole consists of only one germ cell cluster, which is constructed after the same principles (i.e. during divisions of siblings a rosette is formed). However, later on this concept seems to change and all parts of the terminal chamber are filled with nurse cell nuclei and the trophic core region seems to be highly reduced. Nurse cell nuclei do not enlarge, therefore one can conclude that no endomitosis happens, which has also been concluded for the sister group of white flies. The number of nurse cells is enhanced compared with whitefly ovarioles, whereas the number of oocytes remains low. Thus, we might expect a cluster of 128 or 256 cells, when the 2^n-rule is obeyed strictly, or even more cells if it is not.

We analysed the ovary of *Psylla alni* during different adult stages, using semi-thin and ultrathin serial sectioning (Figure 3.110b–d) (Beyrich and Büning, unpublished). Just after the final moult, the ovariole houses cystocytes in the pachytene stage. Nurse cells and oocytes show similar morphology. All cells belong to one cluster, which forms a rosette. Intercellular bridges are filled with fusomal material, which combines to a polyfusome. However, shortly thereafter, nurse cells lose their synaptonemal complexes and cell membranes including intercellular bridges of nurse cells vanish totally. Thus, all nurse cell nuclei are found in a common cytoplasm and no cell boundaries can be observed (Figure 3.110c,d). This behaviour parallels the development of the central syncytium of *Sialis* telotrophic ovaries. A clear portion of basally located germ cells do not lose their membranes but enter a resting stage. These cells are arrested oocytes and their nuclei remain in pachynema for a longer period. Nurse cell nuclei do not grow during the phase of synthesis, indicating that these cells do not raise their DNA content by endomitotic replication rounds but stay at their 4C level. Only one out of the 35–40 presumptive oocytes of each ovariole starts with previtellogenic and later on with vitellogenic growth. Most of the others will be integrated later on into the nurse cell syncytium (Figure 3.110d). About 280 nurse cell nuclei support this growth, i.e. each cluster has more than 300 interconnected cells, which is clearly more than $256 = 2^8$; thus, the 2^n-rule is not obeyed strictly.

The second largest group of Hemiptera are the Auchenorrhyncha, including Cicadiformes (cicadas, spittle bugs, leafhoppers) and Fulgoriformes (lantern flies). However, our knowledge on terminal chambers of ovarioles is very fragmentary and ultrastructural information comes from very few species: *Cercopis sanguinolenta, Aphrophora alni, Philaeus spumarius* of the spittle bug family Cercopidae (Ksiazkiewicz-Kapralska, 1985, 1989a,b), and *Bothrogonia japonica (Cicadella ferruginea)*, a member of the leaf-hopper family Tettigellidae (Matsuzaki, 1975).

Somatic tissues are identical to those of bugs (Heteroptera) and will be described there. In *A. alni*, apical cells of the inner sheath of the tropharium are shown to be morphologically different forms of lateral sheath cells. Whereas lateral cells are extremely flat, apical cells form a short cuboidal tissue, directly beneath the terminal filament, from which they are separated by a strong transverse septum. During the fifth larval instar, these apical cells have lobed nuclei and their cytoplasm contains Golgi complexes, rER cisternae, many ribosomes and mitochondria. Furthermore, they contain vacuoles filled with aggregates of electron dense material. The author suggests that this material may be used for the generation of the transversal septum, since this septum cannot be built by outer somatic epithelia or by blood cells, as is thought for other basal lamina (see section 3.2.3)(Ksiazkiewicz-Kapralska, 1989a).

Lateral inner sheath cells border, at the posterior end of the terminal chamber, a plug of prefollicular tissue, with which arrested oocytes are in contact. Prefollicular cells are spindle shaped and mitotically active. When oocytes start with previtellogenic growth they are enclosed by prefollicular cells and, while growing, transported into the longitudinal axis of the ovariole, forming a single file row. Meanwhile, prefollicular cells undergo transformation into follicular cells. Lateral cells form a single layer of follicular cells, bordering anteriorly to the oocyte and basally to the basal lamina. The nuclei enlarge, indicating endomitotic polyploidization although proof of this is lacking. The last step of greater morphological events is the amitotic division of nuclei into two nuclei of equal volume, a character shared with the follicular cells of heteropteran ovaries.

As in heteropteran species (see below), the terminal chamber is enlarged and houses many nurse cells, which show an anterior–posterior gradient of development. Anterior nurse cells have small nuclei and one nucleus is found in each cell. In two species (*C. sanguinolenta, P. spumarius*), mitotic divisions occur in anterior nurse cells, followed by incomplete cytokinesis, as revealed by intercellular bridges, whereas in *A. alni* mitosis of apical nurse cells seems to be restricted to larval stages. The same may be found in *B. japonica*, in which gradients of nurse cells are described but mitotic activity is not reported. The further fate of nurse cells is exactly that of heteropteran nurse cells and will be described in detail. After a resting stage, nurse cell nuclei grow by endomitotic polyploidization. The nuclei become heavily lobed and amitotic division of nuclei is suggested. In the middle and posterior part of the terminal chamber, nurse cell nuclei become pycnotic in part and basophilic vacuoles are found in the trophic core and in nutritive cords, suggesting transport of these substances into the oocytes. As in all hemipterans, oocytes occupy the posterior part of the terminal chamber. Only oocytes enter the prophase of meiosis during the last larval instar (*A. alni*, Ksiazkiewicz-Kapralska, 1989b). Trophic core and nutritive cords are filled with parallel orientated microtubules. Nurse cell cytoplasm is very rich in free ribosomes and mitochondria are abundant, whereas rER and Golgi complexes are rare. Early events of cluster formation, which will happen during embryonal and early larval stages, are not known at present. However, morphology of adult ovarioles and the last larval stage is identical to heteropteran species, thus there is no doubt that the ontogenetic rise of ovarioles follows the same fundamental events in cicadas and bugs.

The bug (Heteroptera) ovary is investigated best among all Hemiptera; however, it is the most difficult to understand. In the following text a short description is given of the main parts of the adult ovary and of ontogenetic events, as far as they are clearly elucidated. Thereafter, the two main models of ontogenetic development are given, and finally, they are compared with ovary development in other taxa.

We will discuss gross morphology and somatic tissues, then germ cell development. The bug ovary consists of several ovarioles, which insert at the anterior part of each lateral oviduct. The number of ovarioles per ovary is species specific and in many species seven are found, but in some eight or five is the regular amount. Each ovariole is in the same developmental stage; thus, no developmental gradient among the ovarioles is found. During the development of the larval stages, the ovary is found as a compact somatic organ, in which ovarioles are embedded into well-developed somatic sheaths as described in section 3.2. Early events occurring during embryogenesis are reported by Seidel (1924) describing the rise of gonads in *Pyrrhocoris apterus*. He found seven well-separated ovariole anlagen distributed in equal distances on both sides of the alimentary canal, and only one pair of ovariole anlagen was found in any given abdominal segment. Seidel stated that this finding reflects an ancestral status in which gonads are found in each segment of Articulata, and he adopted segmentally arranged gonads as part of the ground plan of insects. We will discuss the validity of this in section 3.6.1.

Fundamental work on heteropteran ovaries was described by Gross (1901, 1903). He introduced the term 'telotrophic' for all those meroistic ovaries in which only oocytes pass down the ovariole during growth phases, whereas nurse cells remain in an endstanding chamber, the terminal (trophic) chamber or the tropharium. Gross compared the ovaries of 13 species, all of them belonging to pentatomorph families, most of them to Pentatomidae, Coreidae and Pyrrhocoridae. No fundamental differences are found in the architecture of the terminal chamber. But there are differences in the number of oocytes in vitellogenic growth. In *Pyrrhocoris* several oocytes (5–6) develop synchronously in each ovariole during vitellogenesis, whereas in Coreidae and Pentatomidae only 1–3 oocytes are in vitellogenesis, but all oocytes are in sequential developmental stages down the ovariole. Synchronous development is also found in *Dysdercus intermedius* (Pyrrhocoridae) (Dittmann *et al.*, 1984, 1990) and may be a family-specific character.

In the terminal chamber, Gross found enlarged somatic apical cells, which may have a role in determination processes. Gross assumed a close connection between these apical cells and the lateral sheath cells to the closely adhered tunica propria. He found three regions in the terminal chamber and described the fate of nurse cell nuclei, their mitotic activity in the anterior zone, their enlargement from anterior to posterior regions followed by amitotic divisions, and finally their pycnosis. Furthermore, he describes the amitotic division of follicular cell nuclei, leading to double nuclei of equal volume. According to his investigation, nurse cell nuclei become more or less lobed; thus the nucleus can split into two or even more smaller nuclei, which are not always equal in volume. However, he also noticed that inside the nucleus chromatin assembles in median areas prior to division. Thus, he assumed an unknown but regular mechanism of amitosis. Today, we are far from understanding amitosis, especially of nurse cells or follicular cells. However, cytophotometric measurements of follicular cells and nurse cells indicate that amitosis split nuclei in compartments, the C values of which are roughly multiples of the 2C value (Dittmann and Maier, 1987; Dittmann *et al.*, 1984). Gross also noticed differences between the ovarioles of young and old adults, in which the first zone of mitotic active nurse cells is shortened, indicating that mitotic activity will cease during adulthood.

In a series of papers Bonhag and coworkers investigated the ovary of another pentatomorph species, the milkweed bug, *Oncopeltus fasciatus* (Lygaeidae) (Bonhag and Wick, 1953; Bonhag, 1955a,b; Wick and Bonhag, 1955). They were the first to give a complete description of the ontogenesis of ovaries during all five larval stages by light microscopy (Wick and Bonhag, 1955). Germ cells of the first instar assemble in a globular mass, enclosed by somatic tissues. All these germ cells show mitotic activity. The aspect remains similar up to the fourth larval stage, except for an increase in the number of germ cells by mitotic activity. At the end of the fourth stage, differentiation of germ cells can be noticed. Posteriorly, oocytes are clearly separated from anterior nurse cells. Only oocytes enter the prophase of meiosis, whereas nurse cells increase their volume. Several nuclei are found next to each other in a common cytoplasm, which assembles around an acellular area, the trophic core. More anteriorly, nurse cells occupy a region in which they are in an arrested stage: they are mononucleate and chromatin is found in coarse chromatin granules. At the anterior part of the terminal chamber, small nurse cell nuclei are found; some of them show mitotic figures. The authors interprete the rise of multinucleate compartments in lower parts by nurse cell fusion, not by amitotic divisions as Gross did.

Ultrastructural analysis of terminal chambers of adults and of larval stages was done in another species, *Rhodnius prolixus* (Reduviidae), which belongs to the Cimicomorpha (Stys and Kerzhner, 1975; Schuh, 1986). In a series of papers, Huebner and coworkers showed that this species follows the general line elucidated previously at the light microscopical level by Gross and Bonhag (Huebner and Anderson, 1972a,b,c; Huebner, 1981a,b, 1984a,b; Lutz and Huebner, 1980, 1981; Huebner and Injeyan, 1980, 1981). The mean facts are:

(1) Morphological differentiation among germ cell descendants happens during the fifth instar.
(2) Early in larval stage 5, all germ cell descendants are equal in size, can multiply by mitosis, and siblings seem to be orientated to a centre (Lutz and Huebner, 1980, their Figure 6).
(3) In anterior areas, nurse cells are connected to each other by intercellular bridges.
(4) Midbodies and a surrounding mass of small vacuoles fill the space of intercellular bridges, which seem to be equivalent to the fusome.
(5) Microtubules of the former spindle degrade, and before the massive microtubules of the adult stages arise, nutritive tubes and connections of nurse cells house only a few microtubules.
(6) Oocytes are connected with each other by intercellular bridges.
(7) Oocytes develop synaptonemal complexes, which clearly show the entrance of oocytes into the prophase of meiosis.
(8) During the fifth instar, a membrane labyrinth arises in the central area, which transforms later on into the trophic core.
(9) The rise of multinucleate areas of nurse cell tissue is interpreted by membrane reduction processes as proposed by Bonhag.
(10) In the posterior region of the trophic chamber, just above the prefollicular tissue, a 'germarial lumen' is found, in which germ cell

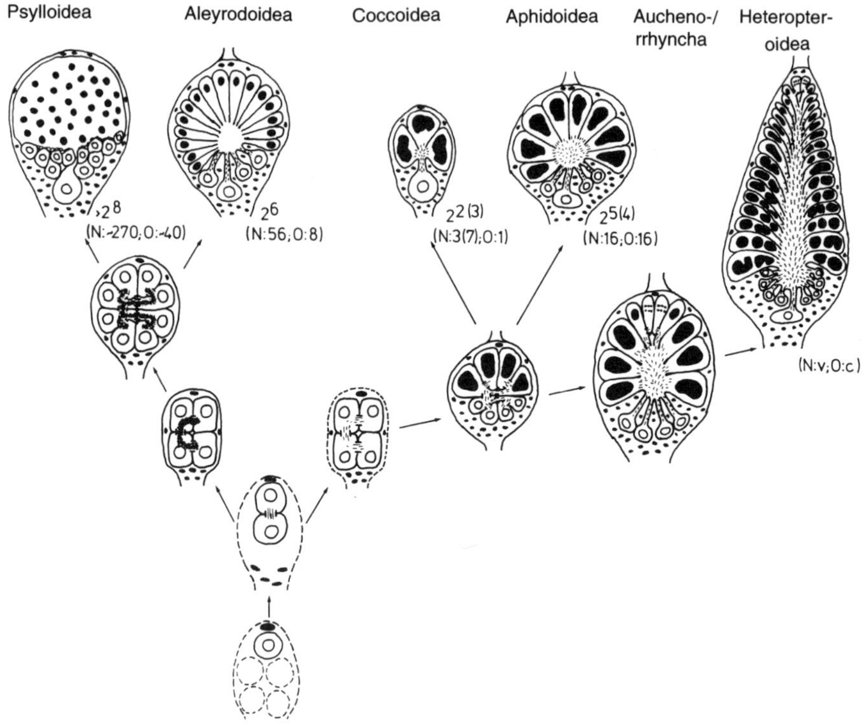

Figure 3.111 Rise of cluster of germ cell descendants in Hemiptera. Development starts with only one cell, attached to apical somatic cells of the ovariole anlagen. Other germ cells might start with development, but will degenerate later on. The germ cell in favour undergoes synchronous mitotic divisions followed by incomplete cytokinesis. There is a strong tendency to reduce the polyfusome, except in Aleyrodoidea and Psylloidea. The amount of synchronous divisions is at least family specific. In scale insects there are two to three cycles (four or eight cells); in aphids there are four to five cycles (16 or 32 cells); in white flies six cycles (64 cells); in psyllids about eight cycles, but asynchrony may happen in late cycles; in Cicadiformes and Fulgoriformes the oocyte number is constant, but the nurse cell number is variable as in most heteropterans, except mirids and bed bugs, in which the oocyte number and nurse cell number are constant. In the last groups and as a new invention, mitotic activity of anterior nurse cells is long lasting.

descendants become lysed. This area may be consistent with the displaced oocyte batch while Wick and Bonhag (1955) mentioned in *Oncopeltus* ovaries.

(11) Apical nurse cells are shut off from the syncytial area of nurse cells and oocytes, according to distribution of injected fluorescent dyes.
(12) In *Rhodnius* only one oocyte starts vitellogenesis in an ovariole.

The above findings have focused on a model of cluster formation in *Rhodnius*, which was published first by Huebner and Anderson (1972c):

some large syncytial clusters arise by mitotic activity of germ cells. These clusters fuse with their central areas, which develop into the trophic core. Each cluster contains future nurse cells and future oocytes. Some anterior germ cells remain isolated, divide mitotically and produce later clusters, which are pure nurse cell clusters. These cells fuse with the pre-existing clusters. Thus, fusion of anterior isolated germ cells and fusion of nurse cells during later differentiation is the predominant feature of this model besides cluster formation.

In 1956, Davis published his investigation on male and female reproductive organs of *Cimex lectularius* (Cimicidae). He was unable to demonstrate mitotic divisions in the anterior nurse cells in ovarioles of adults. In *Lygocoris pabulinus* and in *Lygus lineolaris* (Miridae) as in other mirids, the terminal chamber of adult females consists of only one type of nurse cell. Their nuclei are heavily lobed and it is not possible to analyse their true three-dimensional character by light microscopy (Wightman, 1973; Ma and Ramaswamy, 1987; Büning, unpublished) (Figure 3.24). No developmental gradient exists among nurse cells. Thus, the number of nurse cells and oocytes is constant and both germ cell descendants may belong to the same cluster. Although the ontogenetic development of ovarioles of mirids is not known, the fundamental similarities with aphid ovaries are striking.

Based on these investigations, together with investigations on fire bugs (*Pyrrhocoris apterus*) and cotton bugs (*Dysdercus intermedius*) (Figure 3.35) a second model of cluster formation of Hemiptera was elaborated, published first by King and Büning (1985) (Figures 3.6 and 3.111).

Each ovariole houses only one cluster. Development starts with one cell, attached to the apical somatic cells of the ovariole anlagen. Communication of this germ cell with apical somatic cells will determine this cell as cystoblast. Other germ cells may start with development, but will degenerate later on. The reasons may be insufficient communication with apical cells and/or retardation by the successful cystoblast or its siblings. The germ cell in favour undergoes synchronous mitotic divisions followed by incomplete cytokinesis. Rosette formation takes place during division cycles. Differentiation of nurse cells and oocytes happens early during cluster formation, which follows the 2^n-rule and subsets differentiate into oocytes or nurse cells. Polyfusome becomes reduced in most taxa, except in white flies (Aleyrodoidea) and psyllids (Psylloidea). The amount of synchronous division is at least family specific: in scale insects (Coccoidea) there are two to three cycles (four or eight cells); in aphids (Aphiodoidea) there are four to five cycles (16 or 32 cells); in white flies six cycles (64 cells); in psyllids more than six cycles, and asynchrony may happen; in Cicadiformes and Fulgoriformes the oocyte number is constant, but the nurse cell number is variable as in most heteropterans, except mirids (Miridae) and bed bugs (Cimicidae), in which oocyte number and nurse cell number are constant. In Heteroptera and Cicadiformes/Fulgoriformes the mitotic activity of anterior nurse cells is long lasting. This character is a new invention and may be a synapomorphic character of the above taxa. This model avoids fusion events and generates the whole structure of the terminal chamber by cluster formation processes and interaction of germ cells with somatic cells. Furthermore, the model explains the existence of degenerating cells in the trophic chamber, which are found in each ovariole. In larval ovarioles of *Pyrrhocoris apterus*, some cells regularly degenerate in the central position; whether they are somatic or germ line cells is not known (Figures 3.35 and 3.36). In the milkweed bug, Wick and Bonhag (1955) found a tight batch of 'displaced germ cells' in the prefollicular tissue, which degenerate (their Figure 6). Lutz and Huebner (1980) found similar areas of degenerating germ cells in *Rhodnius* ovarioles, which were termed 'germarial lumen'. Some degenerating cells at the base of the germarium, probably germ cells, are also found sometimes in aphid ovaries, in which they were supernumerary to the 32 cells of the cluster (Büning, unpublished).

Using the rules of parsimony, the last model should be preferred; however, proof is still missing. Ultrastructural serial sectioning can resolve the problems at least in part. Are apical nurse cells isolated or combined to the central syncytium by intercellular bridges? Do nurse cells fuse after endomitotic replication or split their nuclei via amitosis? The question of very early fusion processes of germ cells can be resolved only by ultrastructural analysis of embryonal gonads and/or genetically with the help of germ cell markers.

Reports on the ovariole structure of taxa other than Cimicomorpha and Pentatomorpha are rare. The ovary of *Gerris remigis* (Gerromorpha) has been analysed ultrastructurally by Choi and Nagl (1976; 1977a,b,c) during the late larval and adult stages. They found the general organization of heteropteran ovaries was as described by earlier authors. In addition, they found ultrastructural details, indicating amitotic multiplication of nurse cell nuclei after endomitosis. In mononucleate trophocytes of zone II, intranuclear membranes occur like collapsed pockets of the nuclear envelope. Furthermore, they found bundles of microtubule-like fibrils inside nurse cell nuclei of zone III, which were interpreted by the authors as transport vehicles for RNPs. However, these structures can also be instruments for amitotic divisions. In this zone, zone III, mono-, bi-, and trinucleate trophocytes occur.

In *Sphaerodema rusticum* (Nepomorpha: Belostomatidae) Jawale and Ranade (1990) found nurse cells, which are mostly mononucleate, but sometimes binucleate, and which open with slender projections into the trophic core. In the same species, Sahai (1975) was not able to find mitotic activity in the anterior regions of the tropharium, although a developmental gradient of nurse cells exists.

The ovary of *Saldula saltatoria* (Leptopodomorpha: Saldidae) was investigated recently by Bilinski *et al.* (1990). Here, only mononucleate nurse cells occur, each attached to the trophic chamber by slender projections called 'trophic processes'. Remnants of intercellular bridges clearly demonstrate the primary nature of trophic processes. A mitotic active zone exists in the apex of the tropharium and nurse cell nuclei enlarge in the posterior parts of the tropharium, indicating endomitotic polyploidization.

The last two species belong to infraorders, which are not as 'advanced' as Cimicomorpha or Pentatomorpha (Schuh, 1986). However, it seems to be too early to use the character 'mononucleate' as representative for 'primitive' Heteroptera. Investigation of species belonging to Enicocephalomorpha and Dipsocoromorpha, which are the most 'primitive' infraorders, is badly needed.

(c) Megaloptera (Sialidae) and Raphidioptera

Although Gross (1903) termed the ovary of *Sialis* as telotrophic, he could not demonstrate nutritive cords. The architecture and ultrastructure of the ovaries of adults (*Sialis mitsuhashii*) was elucidated by Matsuzaki and Ando (1977). Ontogenetic development of *Sialis flavilatera* ovaries was investigated by Büning (1979c). Ovarioles of *Raphidia* species were investigated at the light microscopical level by Achtelig (1978), and the ultrastructure of *Raphidia flavipes* was published by Büning (1980).

Since all investigations revealed very similar architecture of the ovarioles, it is of no doubt that both telotrophic meroistic ovarioles have a common origin (Table 3.3; Figure 3.88). Each ovary consists of numerous ovarioles, which are organized in *Sialis* in a comb-like fashion. About 100 ovarioles insert along four rows to a lateral oviduct. In *Raphidia* about 40–60 ovarioles assemble tightly around a calyx of the lateral oviduct. The anterior end of each ovariole is occupied by a terminal filament, which is not separated from the inner sheath cells by a transverse septum. Morphologically pronounced somatic apical cells are not found. A flat inner sheath encloses the germ cell descendants of the terminal chamber. In young adults just after final moult, a central syncytium is found, housing numerous nuclei within a common cytoplasm (Figure 3.37a). This area is ensheathed by cuboid tapetum cells, of which each cell opens to the central syncytium with an intercellular bridge. No bridges are found connecting tapetum cells with each other. Nuclei of tapetum cells are indistinguishable from nuclei of the central syncytium. They are all of the same magnitude and each has a well-developed nucleolus. RNA synthesis proceeds equally in all germ cell descendants (Figure 3.54). During ontogenetic development, all cells pass the leptotene stage before meiotic arrest. Thus, endomitosis is absent and all germ cell descendants are believed to have 4C values. At the posterior region of the tropharium, some of the tapetum cells, which are in contact with a very small prefollicular tissue, enlarge. These tapetum cells, now clearly destined to be oocytes, grow down the ovariole. Each oocyte remains connected to the central syncytium via a persisting intercellular bridge, which opens to a growing nutritive cord. Masses of microtubules are not found in *Sialis*, but many microtubules occur in *Raphidia* nutritive cords. However, contrasting with microtubules in the nutritive cords of Hemiptera, they are not aligned in a parallel fashion and have no preferential direction within nutritive cords.

In which way does this telotrophic ovary come into being? Periods of oogenesis during the life cycle are shown in Figure 3.9. During the early larval stages, a gonadal tube exists, from which ovarioles begin to branch (Figure 3.7). Germ cells emigrate from the gonadal tube into the ovariole anlagen, which consists in the beginning of a well-differentiated terminal filament to which the young ovariole adheres. It is composed of a tunica propria, inner sheath cells and some large somatic cells, which become replaced by germ cells. Germ cells multiply by mitosis. Thereby, an anterior–posterior gradient is established. Anteriorly, germ cells continue with mitosis. Thereby, clusters are formed which show rosette formation. The exact number of cystocytes is still not known, because this investigation was done using glass knives which does not allow serial sectioning in a range needed for such analysis (Büning, 1979c). Intercellular bridges of clusters are filled with microtubules, which are interpreted as remnants of the former spindle apparatus. A polyfusome was not found in any stage of development investigated. The central area of the ovariole is filled first with a membrane labyrinth, which opens to the intercellular bridges. Later on, or more basally, this labyrinth vanishes, and nurse cell nuclei float freely in a common cytoplasm. Cell membranes of those germ cell descendants, which are in contact with somatic inner sheath cells, can resist. These cells are the tapetum cells. In late larval stages, mitosis of germ cells is no longer found. In the anterior part the membrane labyrinth still exists, but ceases during the pupal stages.

Previtellogenic growth starts during the larval stages and the vitellarium consists of about five growing oocytes. Vitellogenesis begins in the most basally located oocyte at pupal moult. At the imaginal moult three to four ripe eggs are produced and the tropharium begins to degenerate.

A preliminary model of cluster formation is given in Figure 3.37b. Several stem cells are fixed to the anterior part of the young ovariole by germ cell–somatic cell interaction. Germ cells multiply with a limited programme of differential mitotic divisions. The resulting cystoblasts also divide with a limited programme of division cycles. By incomplete cytokinesis a cluster arises, which shows maximal branching of siblings. In the beginning, rosettes are formed. After the last mitotic cycle, siblings, which are in tight contact with inner sheath cells, keep their position and transform to tapetum cells, which are prospective oocytes. All centrally located siblings will reduce their cell membranes totally. During this transformation, a transient membrane labyrinth develops.

(d) Coleoptera, Polyphaga

All polyphage Coleoptera have terminal chambers in their ovarioles to which growing oocytes adhere via nutritive cords (Büning, 1972, 1979a,b). The first to see nutritive cords clearly in coleopteran ovarioles was Gross (1903), who also introduced the terms polytrophic and telotrophic. A good histological description of the ovariole of *Tenebrio molitor* was given by Schlottman and Bonhag (1956).

Before discussing the architecture of the ovarioles and especially of the terminal chambers, some remarks are given on the gross morphology of beetle ovaries. The number of ovarioles varies to a great extent. Whereas in blister beetles like *Meloe proscarabaeus* about 1000 ovarioles are found in each ovary, in some scarabeid species, one ovary is reduced totally and in the other only one ovariole develops functional oocytes. It is always the left ovary which can develop: the right ovary disintegrates during larval stages, as do five ovarioles from the left side. Only one ovariole out of six can mature (Heymons, 1930; Willimzik, 1930; Robertson, 1961; Matsuda, 1976; Halffter and Lopez, 1977; Büning, 1979a; Pluot, 1979; Crowson, 1981). Ovariole number is family-specific when the number is low, but varies to a greater extent when the number of ovarioles per ovary is rising (Büning, 1979a). The arrangement of ovarioles within the ovary varies among different families. Comb-like or pectinate ovaries occur as well as ramified ones (Stein, 1847; Matsuda, 1976). The problem of ovary architecture in view of phylogenetic interpretation has been discussed earlier (section 3.2.6).

Early development of coleopteran gonads is similar to the general description (section 2.3.1) in *Tenebrio* (Saling, 1907), *Donacia* (Hirschler, 1909) and *Leptinotarsa* (Richard-Mercier, 1977; Richard-Mercier *et al.*, 1986, 1988). There is one, compact gonad anlage on each side of the alimentary canal, situated around the seventh segment. Out of this gonad, ovarioles rise, of which the terminal filament differentiates first. Contrary, transitory segmental arrangement of germ cells is found in *Lytta viridana* (Church and Rempel, 1971), in which small clusters of germ cells split from a former compact mass, which itself is derived from pole cells. The small clusters settle in segments 2–7 next to their anterior border and symmetrically on each side of the alimentary canal, closely associated with the splanchnic mesoderm. At this time, gonads are not yet closed. The phylogenetic relevance will be discussed in section 3.6.1.

The somatic tissues of coleopteran ovarioles are similarly organized to those of other taxa with one exception; the somatic cells of the terminal chamber are not restricted to the inner sheath, but intermingle with nurse cells, forming a three-dimensional net of interstitial cells (Table 3.3; Figures 3.17, 3.18, 3.53 and 3.78) (Schlottman and Bonhag, 1956; Büning, 1972). In functioning ovarioles, interstitial cells are clearly separated from prefollicular cell tissue which they border in the posterior part of the terminal chamber. During ontogenesis on prepupal and pupal stages, such a clear distinction which indicates a common origin, is not possible (Büning, 1972, 1978). The anterior tip of the trophic chamber is separated from the terminal filament by a transverse septum. Oocytes have close connections to the prefollicular cells, whereas nurse cells border a basal layer of interstitial cells through which only the nutritive cords of oocytes can pass.

Follicular cells develop from prefollicular tissue. Each oocyte will be bordered by mononucleate follicular cells, which synthesize the vitelline membrane during the later stages of vitellogenesis. Finally, they synthesize the different layers of the chorion (see section 3.2.5). The ontogenetic development of germ cell descendants of polyphage Coleoptera has been approached several times by Büning (1972, 1978, 1979a,b) and King and Büning (1985) investigating *Bruchidius obtectus* (Bruchidae) and over 30 other coleopteran species; by the group of Matuszewski (Kloc, 1976; Kloc and Matuszewski, 1977; Matuszewski *et al.*, 1985), who investigated *Creophilus maxillosus* (Staphylinidae) and three other species; and Kozhanova and Pasicnik (1979) who looked at the ontogenesis of the ovarioles of *Coccinella septempunctata* (Coccinellidae).

The very early rise of ovarioles is not known. Thus, one can only assume similar processes as known from other genera like *Drosophila*, in which some anterior germ cells communicate with somatic cells, which differentiate the cells of the terminal filament first.

Mitotic activity of germ cells is found during larval stages. The exact termination of the divisional programme is species specific, and differs slightly among species (Matuszewski *et al.*, 1985). Preferentially, siblings are found in long rows, parallel to the long axis of the terminal chamber. Divisions of cystocytes are synchronous or metachronous (Kloc and Matuszewski, 1977; Kozhanova and Pasichnik, 1979; King and Büning, 1985; Matuszewski *et al.*, 1985). However, next to the posterior part of the terminal chamber, the architecture becomes more complicated. Matuszewski *et al.* (1985) showed basally located connections between rows of siblings, which were interpreted as intercellular bridges. The same authors showed a nurse cell of *Dermestes*, which has three intercellular bridges, indicating clearly the ramification of this cluster. King and Büning (1985) have shown in *Bruchidius* prepupal ovarioles that spindle axes of mitotic active germ cell clusters are not orientated in single-line rows, but suggest ramification inside a cluster. In *Asclera coerulea*, intercellular bridges persist in functioning ovaries. Distribution of intercellular bridges at the base of the tropharium clearly indicates ramification (Figure 3.17) (Büning, 1978; and unpublished). In an ultrathin serial section of *Asclera* ovarioles, clear evidence is given that clusters are ramified, especially in the posterior part of the ovarioles. Siblings with at least five intercellular bridges are found, connected to other siblings, by which the classical ramification is apparently not followed. Further analysis will clarify

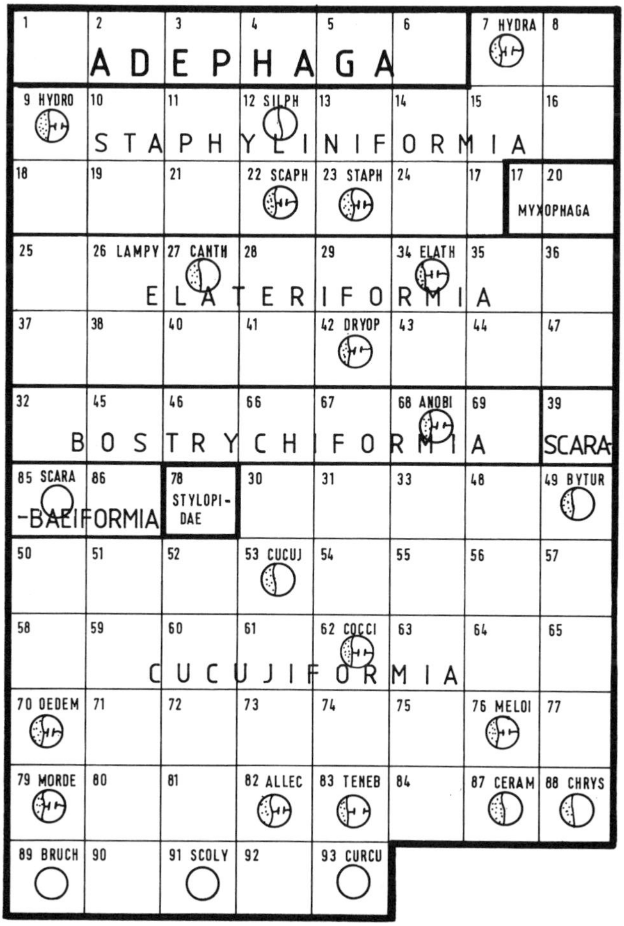

Figure 3.112 Distribution of subtypes of telotrophic ovaries of polyphage Coleoptera, according to membrane reduction. Coleopteran families and higher ranked groups according to Crowson (1981). Each circle symbolizes the situation found in representatives of this group. A circle without any other symbol shows full reduction of nurse cell membranes. Only those membranes between nurse cells and somatic cells persist. Circles with one curved line symbolize those groups in which nurse cell membranes between subclusters persist, whereas those membranes between direct siblings degenerate together with their intercellular bridges. Circles with curved lines and an intercellular bridge symbolize all those groups which show no membrane reduction in functioning ovarioles. Stippling indicates differences in concentration of ribosomes found in subclusters.

whether the whole terminal chamber is housed by one, huge and basically ramified cluster (Figure 3.18c), from which long rows of unbranched or poorly branched subclusters arise anteriorly, or if there are several clusters, as was assumed before (Figure 3.18b) (Büning, 1979a, unpublished; King and Büning, 1985; Matuszewski *et al.*, 1985). Irrespective of this question, one has to assume more than one oocyte per cluster. Oocytes are situated primarily not in a basal and terminal position in each cluster as was previously assumed (Kloc and Matuszewski, 1977; King and Büning, 1985), but in a central position in the whole cluster. The oocytes are finally located in the basal part of the terminal chamber, since the middle and upper parts are filled by later nurse cells born by additional and mostly intercalar and metachrone divisions. These late divisions give rise to the final aspect of linear subclusters (Figure 3.18b,c) (Matuszewski *et al.*, 1985; Büning, unpublished).

Later development is characterized by species-specific degradation of cell membranes between nurse cells. There are species in which intercellular bridges persist in functioning ovarioles, and species in which intercellular bridges and adjacent membranes vanish. In a third group of species, all nurse cell membranes become reduced and nurse cell nuclei are stabilized in their position only by the three-dimensional net of interstitial cells (Figure 3.18a) (Büning, 1979a). Comparing the three types among families and suborders of polyphage Coleoptera, the tendency of membrane reduction in more advanced species is obvious (Figure 3.112). However, the material is not large enough to settle clear cladistical help for family series specificity.

Nurse cell nuclei undergo few endomitotic replication cycles as is revealed by ^{3}H-thymidine autoradiography and volume measurements (Büning, 1972; 1979a; Ullmann, 1973; King and Büning, 1985). Thus, only minimal anterior–posterior gradients are found, in which anterior and sometimes also posterior nurse cell nuclei are smaller than those of the middle regions (Gerber *et al.*, 1978; Büning, 1979a; Ray and Ramamurty, 1979). In addition to endomitotic replication of nurse cells, in some species double nuclei, which may be products of amitotic division, occur (Figure 3.53). This is suggested since double nuclei occur also in those species in which membranes are not reduced and intercellular bridges are well preserved (Büning, 1979a).

Oocytes become distinguishable from nurse cells late as in *Bruchidius* (Büning, 1972) or early as in *Creophilus* (Kloc and Matuszewski, 1977). In *Creophilus*, only the oocyte shows synaptonemal complexes, whereas in *Bruchidius* oocytes and prospective nurse cells start with meiosis. In *Creophilus* the oocyte nucleus has an extrachromosomal DNA body, whereas nurse cells are lacking (Kloc, 1976; Matuszewski and Kloc, 1976; Kloc and Matuszewski, 1977).

Chromosomes of oocyte nuclei are decondensed during early previtellogenesis and are believed to be in a lampbrush stage (Gruzova, 1979a,b; Gruzova and Batalova, 1979). In later stages a karyosome is formed. However, RNA synthesis is going on at least in part, as has been shown by ^{3}H-uridine incorporation studies (Büning, 1972; Ullmann, 1973;

Chaminade and Laverdure, 1976; Ray and Ramamurty, 1979). In all germinal vesicles looked at so far, nuclear bodies occur during karyosome formation, the ultrastructure of which resembles that of nucleoli, besides others, which are mainly proteinaceous and are denoted as endobodies (see section 3.3.2) (Figures 3.50 and 3.52). It is unknown whether these nuclear bodies contain extrachromosomal genes which are associated with their gene products and proteins, or if they are only gene products associated with proteins. Candidates are of course rDNA genes; however other genes may also contribute to these nuclear bodies (see section 3.3.2).

3.6 PHYLOGENY OF INSECT OVARIOLES

The preceding section has highlighted many characters, which have to be compared to evaluate their phylogenetic significance. The first part will deal with the very early events during formation of the gonad anlagen and the rise of ovarioles. Thereafter, the somatic tissues of female gonads will be discussed. Finally, characters of germ cells and their descendants will be analysed.

This book is not the place to go deeply into fundamental discussions of phylogenetic analysis. However, some definitions will be described briefly.

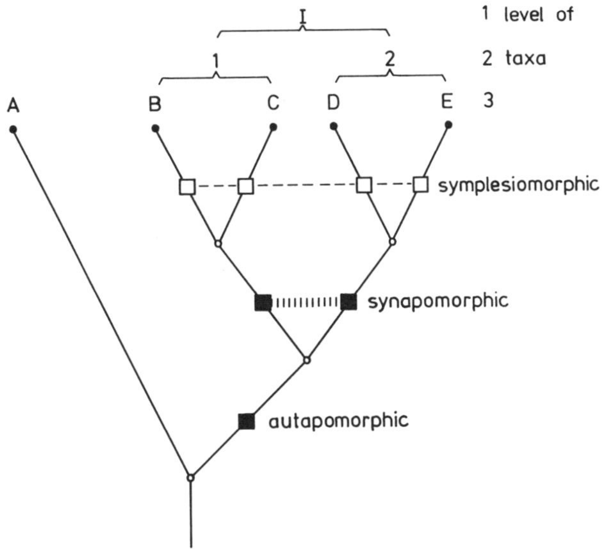

Figure 3.113 The value of a given character changes with levels of taxa. An autapomorphic (phylogenetically new invention) character (black square) of taxon I changes to a synapomorphic character when the subtaxa 1 and 2 are looked at (black squares combined by vertical stripes). At level 3, i.e. within taxon B or C or D or E the same character can be valued only as symplesiomorphic (blank squares, combined by broken line), because this character cannot be used to root sister groups at that level.

The first and most crucial point is the definition of the term 'character'. There is no natural unit of character, which is defined by physical constants, but its definition depends on persons and their view of things (in biology mostly on individuals). 'Ein Merk-Mal ist ein Mal oder Zeichen, das *man* bemerkt, *sich* merkt und auf das man *andere* aufwerksam macht. Ein Merkmal ist mit anderen Worten eine Einheit, die *man* beobachtet, festlegt und mitteilt.' ('A character is a mark or sign, *one* will notice, which *one* will bear in mind, and which will be shown to *others*. In other words, a character is a unit, noticed, described and communicated by *someone*.') (Werner, 1970; italics by J. Büning). Thus, the basic instrument, the character, is not a natural unit and cannot be described by physical means. The character is subjective, *per se*. The corrective is the acceptance of arguments for a character by other scientists, which of course is also subjective.

Characters must be separable from each other, which depends of course on the methods available at the time. Keeping in mind these fundamentals, Hennig has elaborated a method, by which phylogeny of taxa can be determined, first judging homologue characters as plesiomorphic or apomorphic and second by valuing the mosaic set of characters. An autapomorphic character is a phylogenetically new invention and is found only in members (individuals) of a given taxon and not outside. This character will change to a synapomorphic character when it is looked at at the next higher level of taxa, shared by only two subtaxa, which can be rooted by this character. Plesiomorphic characters are shared by members of a given taxon as well as by members of other taxa, which cannot be rooted directly (Figure 3.113) (Hennig, 1969, 1981, 1982; Ax, 1984).

In which way is the term 'character' used in this section? At first, morphological criteria are used which enable decisions like 'yes' or 'no' to be made, for example do intercellular bridges between follicle cells exist or not? As shown in preceding sections, all characters which participate in the construction of germ cell clusters are found in the male line and furthermore, in other higher ranked taxa. Does this mean that occurrence of cystocyte clusters in insect females cannot be used for Hennigian analysis? Of course not. Although all these characters are very old inventions and can be looked at as plesiomorphic, one has to look at the developmental process in which the character is embedded. In other words, the expression or the retention of possible characters of the female pathway will value a character as autapomorphic, synapomorphic or symplesiomorphic. Thus, we value not only the morphology but also the genetic background.

This book deals only with the characters of gonads, mainly ovaries. Therefore, it cannot give sufficient phylogenetic diagrams, which have to embrace other characters. The diagram of phylogeny of ovaries is therefore superimposed mostly on the classical phylogenetic analyses of Hennig in its revised form (1981) and Kristensen (1975, 1981). However, the characters and views on mosaics of ovariole characters elaborated here, can be used for phylogenetic analysis of hexapodan orders, as has been done recently (Stys and Bilinski, 1990; Bilinski, 1993; Büning, 1993 and in preparation).

3.6.1 Formation of female gonad anlagen

There is no doubt today that pole cells are the progenitors of germ cells in insects. Furthermore, it is commonly accepted that germ cells, once determined, move anteriorly as singles, only loosely connected to each other during germ band formation. They are guided by the extracellular matrix of the underlying mesodermal cells. If germ cell connection with each other is weak, as shown for instance in *Periplaneta* (Heymons, 1895), germ cells can assemble preliminarily opposite the intersegmental grooves during germ cell movement. In other cases, as shown here for *Drosophila*, the germ cell mass moves more or less as a unit and splits later on into two, symmetrically arranged parts to the left and right side of the alimentary canal (Figure 3.5; Lasko and Ashburner, 1990). There, germ cells nest into the ovary anlagen as a whole. Such compact ovary anlagen can be found in most insects, like in *Drosophila* (Diptera), *Tenebrio*, *Leptinotarsa* or *Donacia* (Coleoptera: Polyphaga), or they can be slightly elongated, as in *Gryllus* (Orthoptera) or more elongated as in *Sialis* (Megaloptera) (Figures 3.1, 3.7, 3.8).

The preliminary segmental arrangement of ovary anlagen in *Pyrrhocoris* and the rise of only one ovariole in each anlagen has convinced Seidel (1924) that this behaviour is ancestral and not only a remnant of the old arthropodal body plan in which each segment has its pair of gonads, and that segmental gonad anlagen also belong to the body plan of insects. This assumption has been corroborated recently by the finding of one pair of gonads in seven abdominal segments in *Machilis helleri* (Archeognatha) during larval and adult stages (Weyda, 1989). Development of insect gonads from primary segmental anlagen was accepted by Weber (1933) and Weber and Weidner (1974) and was followed by many authors, like Gustavson (1950), although Snodgrass (1935) strongly opposed it with many arguments. He believed that ovarioles are raised by differentiation processes inside a primarily compact ovary anlagen, as is found in most insects and has been shown here in detail for *Drosophila*. Such inner and secondary differentiation of ovarioles is found not only in 'advanced' insects but also occurs in most primitive insects in which cells of the terminal filament are the first cells which differentiate inside a pre-existing mesodermal gonadal tube, as was described first by Heymons (1891) in the *Blatta (Blatella)* ovary, and later on in a comparative study for *Forficula* (Dermaptera), *Gryllus*, *Gryllotalpa* (Orthoptera), *Blatta*, *Ectobia*, and *Periplaneta* (Dictyoptera) (1895). A similar rise of ovaries in *Lepisma* (Zygentoma) was published in 1897: during embryogenesis, on each side of the alimentary canal, a continuous, elongated ovary anlagen is built, each of which develops five ovarioles. In newly hatched larvae, one pair of ovarioles is found in abdominal segments

2–6. In males, double pairs of follicles develop in segments 4, 5 and 6. During larval development, gonoducts shorten and the segmental arrangement is disturbed. This shows that association of ovarioles and male follicles to special abdominal segments is variable even in the same species. Arguments against primarily segmental gonad anlagen in insects also comes from the group of Archeognatha. In *Pretobius lohmanderi*, Birket-Smith (1974) found eight pairs of ovarioles, of which six are housed in abdominal segments 1–5, and a double pair is housed in the third thoracic segment. This shows that ovarioles are distributed among abdominal segments, not in close associations but randomly, even in this old insect order.

A separated development of ovarioles, as described for *Pyrrhocoris* (Seidel, 1924), is assumed to occur also in a coleopteran species, *Lytta viridana*, in which germ cells assemble in intersegmental grooves, as described above (Church and Rempel, 1971). However, one has to decide between preliminary assembly of germ cells during movement, and rise of the mesodermal gonad anlagen which occurs later on. As pointed out (section 3.5.2), Miya (1959) showed an intermediate development of ovaries in *Bombyx mori*. There, germ cells move isolated from each other and are found first distributed at random along a mesodermal sheath. Later on, the number of germ cells is raised by mitosis and germ cells will group together, following more or less segmental boundaries. Germ cells nest finally in special mesodermal sheaths, the genital ridges, which shorten later on to a compact gonad.

This interpretation is supported by a detailed analysis of *Drosophila* gynandromorphs (Szabad and Nöthiger, 1992). These flies were raised by special crosses in which one X-chromosome is lost during the early stages of embryogenesis. All cells with XX follow the female pathway, whereas all cells with single X follow the male pathway. Analysing fertile and sterile crosses, they found a clear correlation between female and male contributions of somatic gonadal tissue, which can be interpreted most easily by only one genital primordium on each side of the alimentary canal, located at segment 4. This primordium will be derived from about 10 progenitor cells located within a single segment.

However, to make the story complicated, Brookman *et al*. (1992) followed the development of gonadal mesoderm by analysing the expression of the 412 retrotransposon which can be seen in whole mounts by dioxigenin-labelled hybridization probes. This retrotransposon is identical to parts of isolated genomic clones, which are target sequences for *Ultrabithorax*. *Ultrabithorax* is a homeotic gene, also engaged in the formation of the mesoderm. The pattern of the 412 retrotransposon expression is highly regulated and shows a broad stripe of cells in each segment in the early stages. Later on, the signal is restricted to both sides of the alimentary canal and concentrates sometime later small stripes with which germ cells associate. Shortly thereafter, germ cells assemble in the two round gonadal primordia found at segment 5 on each side of the alimentary canal. The authors interpret their findings as a strong indication for primary segmental gonad anlagen. However, what they have shown is the activity of the retrotransposon 412 which might be one step somewhere in the cascade of gene activities which determines the gonadal mesoderm.

If the descriptions and interpretations are correct, we will again have a cascade of differentiation processes, in which each single event can be shifted in time and, consequently, also in place, but in which the essentials remain constant. Apical cells of the inner sheath may play a role in the further differentiation processes of germ cells within the ovariole, but the fate of germ cells is not engaged in primary differentiation processes of ovarioles (see section 3.2.1).

The 'normal' development may happen in insects following the scenario outlined here.

(1) Determination of germ cells early in blastoderm formation by interaction of posterior nuclei with pole plasm.
(2) Interaction of germ cells with early somatic tissues, i.e. mesodermal cells.
(3) Transport of germ cells to final positions along a path of extracellular matrix of mesodermal cells.
(4) During movement of germ cells, preliminary assembly of small groups of germ cells in intersegmental grooves can happen.
(5) Anterior movement and lateral movement assemble germ cells finally into two symmetrically arranged gonad anlagen on both sides of the alimentary canal.
(6) Closing of the genital ridge to a tube-like ovary anlagen.
(7) During stages 2–6 germ cells can multiply by mitosis.
(8) Germ cells assemble only in the middle regions of the primary gonad anlagen. The anterior part, thoracic segments and the first abdominal segments, differentiate into the suspensory ligament, which anchors the ovary anteriorly. The posterior part, including the posterior abdominal segments up to segment 7 or 8, are also free of germ cells finally, and degenerate or transform to the lateral oviduct.
(9) The posterior somatic cells of that part of the genital ridge in which germ cells are nested, differentiate into the anterior part of the lateral oviduct, often called the calyx.
(10) Interaction of some germ cells with anterior somatic cells. Only these germ cells will transform to the later stem cells. The anterior somatic cells transform into the apical cells of the later germarial region of the ovariole. These anterior cells are in close contact with the terminal filament, which has been built before.
(11) The differentiation of lateral inner sheath cells follows, by which growth and differentiation the ovariole anlage is raised.

(12) Lateral inner sheath cells of ovarioles are in contact with lateral and posterior (basal) somatic cells of the ovary (gonadal tube), from which the prefollicular tissue and the pedicel differentiate.

These developmental events can vary in time and, therefore, in place, which results in the various forms of gonad development outlined before. Thus, from today's point of view, variations can occur independently of the phylogenetic position of the species. Only one or two phenotypes must vary to produce an elongated ovarium, in which only one pair of ovarioles is formed per segment, roughly following segmentation boundaries, as found preliminarily in *Phyrrhocoris*, or permanently in Archaeognatha (Seidel, 1924; Snodgrass, 1935; Birket-Smith, 1974; Weyda, 1989). Archaeognathan species should be analysed in greater detail to elucidate the cascade of events during gonad formation.

3.6.2 Somatic tissues of ovaries

Some of the somatic characters have been looked at in the preceding section (3.6) and are listed in the appropriate tables. By comparing these data one will see only a few characters of which one can say that this special character is group specific (i.e. autapomorphic for a given taxon). Most characters are not analysed by modern ultrastructural methods to enable firm decisions and in many orders the number of species analysed is too low to evaluate group specificity. However, some characters can be used.

First of all, one should evaluate the ground plan of the insect ovary; thereafter the reductions should be looked at, and finally, the new inventions of appropriate taxa should be valued. The early events during ontogenesis of insect ovaries have shown that the first tissue which differentiates is the outer ovary sheath and its basal lamina. This tissue belongs to the ground plan and exists, at least during early periods of development, in nearly all orders. The only exception is the order of Strepsiptera, in which this tissue could not be found. However, as shown in section 3.5, the very early events of gonad development in *Elenchus* have not been analysed and the more primitive family of Mengeidae, in which females leave their hosts and show less reductions, is not looked at.

The second somatic tissue of ovaries which differentiates, and concomitantly the first tissue of ovarioles, is the terminal filament. The terminal filament is composed of a row of cells aligned laterally by a basal lamina. To its posterior end it can be bordered directly by apical inner sheath cells, as found in some insect orders, such as Diptera and Hymenoptera. In most orders, however, the basal lamina continues in a transverse septum, separating the terminal filament from the apical inner sheath cells. Such a clear decision is not always possible. In young ovarioles of fleas, just after pupal moult, the transverse septum is not clearly established. In old ovarioles a transverse septum is apparent. This shows firstly that this character (transverse septum) is dependent on the physiological status of the ovariole; and secondly, terminal filament and apical inner sheath cells have the same progenitors. The transverse septum is therefore not a well-separated character and its occurrence may have happened several times during phylogeny.

In functional ovarioles, apical cells of the inner sheath are not clearly distinguishable from lateral cells. However, during ontogenesis apical cells are prominent constituents in all those ovaries which are looked at more closely, as in dipterans, lepidopterans, hemipterans and psocopterans. These cells may play a key role in early ovariole development, since they keep contact with germ cells, especially the stem cells (Lin and Spradling, 1993). Following this hypothesis, in panoistic and polytrophic meroistic ovaries only stem cells will be kept in the apical position by interaction with apical sheath cells. In telotrophic meroistic ovaries, this communication is extended to all cystocytes and lateral sheath cells, except those cystocytes which become definitively determined as oocytes and may get early contact to prefollicular cells. If this interpretation of the role of apical sheath cells is correct, these cells belong to the fundamental equipment of insect ovarioles. This character would be apomorphic on the level of true insects and plesiomorphic to all lower ranked taxa, like orders.

Interstitial cells as persisting elements with a special function are found in the trophic chambers of adephagous and polyphagous Coleoptera, as well as in the trophic chambers of Hymenoptera. In lepidopteran trophic chambers some cells are found in young trophic chambers, and in nurse cell chambers of Diptera a specialized group of somatic cells, the border cells, pass the trophic chamber during early phases of previtellogenic growth (see below).

Persisting interstitial cells are continuous with lateral sheath cells. Ultrastructure and function are similar; they fix the three-dimensional arrangement of nurse cell-determined cystocytes. Interstitial cells develop from some somatic cells intermingling with germ cells in early ovary anlagen. Therefore, in early ovariole anlagen, some interstitial cells are also found in the germaria of all other orders. The rise of persisting interstitial cells with special functions in the architecture of trophic chambers in Coleoptera and in Hymenoptera is coupled with great numbers of nurse cells. Comparative and intensive ultrastructural investigation may be helpful to decide between parallel development and synapomorphic character.

In all polytrophic meroistic ovaries, single mesodermal cells of early ovariole anlagen intermingle with germ cell clusters and develop into small nests of prefollicular cells, giving rise to follicular cells and stalk cells between follicles.

These prefollicular cells, which are scattered in small nests between germ cell clusters, are not always easily distinguished from coherent prefollicular tissues, which occur at the posterior part of the germarium. Most coherent prefollicular tissues cause a constriction between the germarium and the prefollicular growth zone, often called the neck region. Enlarged prefollicular tissues are found only in the telotrophic ovaries of Heteroptera, Cicadiformes and polyphage Coleoptera. The number of cells in prefollicular tissues reduces in psyllids, aleyronids, aphids and coccids, compared with the above taxa. A small prefollicular tissue is also found in the panoistic ovariole of Thysanoptera. Psocoptera have no prefollicular tissue. Thus, prefollicular tissue of Hemiptera and Thysanoptera can be synapomorphic.

However, outside the Paraneoptera, a well-developed prefollicular tissue is found in Coleoptera, Polyphaga as mentioned above. In addition, small prefollicular tissues are also found in Raphidioptera and Sialidae (Megaloptera), which both have telotrophic ovaries (section 3.6.3). These examples are believed to be convergent developments.

In all other orders, irrespective of their ovariole type, prefollicular cells are organized in nests between growing germ cell descendants. In lepidopterans and in some dipterans these nests enlarge abruptly in the posterior parts of the germaria, as known from prefollicular tissue. However, there is always a continuity between these nests and some smaller nests or single somatic cells inside the germarial region, which is never found in prefollicular tissues. Since most holometabolous orders and most hemimetabolous orders have small nests of prefollicular cells, which are not always clearly distinguishable from interstitial cells and somatic inner sheath cells, it is assumed that this organization is the oldest and belongs to the ground plan of insect ovarioles.

Apical border cells of dipteran follicles are a typical example of apomorphic characters of orders, here of Diptera. In this group alone, a clone of apical sheath cells of follicles (not of germaria) will pass through trophocytes during previtellogenic growth, keeping a median line until they reach the oocyte. In the final position, at least two inner cells develop further to border cells, which mark the apex of the oocyte (anterior–posterior axis of the egg) and may also be involved in dorso-ventral axis determination (for further discussion see section 3.5.2).

Another character of follicular cells is their syncytial arrangement via intercellular bridges, which was found in Diptera, Hymenoptera and just recently in Mecoptera: Boreidae (Büning, unpublished). Our own investigations on hemipteran, psocopteran, strepsipteran, coleopteran and siphonapteran ovarioles failed to demonstrate intercellular bridges among follicular cells. Reports of other orders, which describe the ultrastructure of follicular cells, gave no hints of intercellular bridges between follicular cells. However, one should be careful with negative results. In all three cases, intercellular bridges exist during early vitellogenesis and it may be that not all late divisions of all subgroups of follicular cells will build intercellular bridges. Thus, in other orders they may be overlooked. All three cases belong to orders of the mecopteroidean complex. Further investigations are needed before one can value this character.

Doubled nuclei in differentiated follicular cells are found in Heteroptera and Cicadiformes. They develop from uninuclear cells by one amitotic division during endomitotic cycles. Among the next relatives, embracing all other taxa of Hemiptera, follicular cells remain unicellular as found in most other orders. However, in bark lice, which are members of Paraneoptera and close to the stem, follicular nuclei become heavily lobed during their high synthetic phase at the beginning of vitellogenesis and in many follicular cells the nuclei become nearly halved by deep furrows. This may be a prerequisite for amitotic divisions. In the sister group of Psocoptera, in the Phthiraptera, which embraces true lice and mallophagans, follicular cells develop doubled nuclei (Figure 3.23b) (Ries, 1932; Ries and van Weel, 1934; Bilinski and Jaworska, 1987). Thus, this form of doubled nuclei in follicular cells may be synapomorphic for the above taxa. Similar furrows, which seem to halve the follicular nuclei, are reported in *Acheta* (Gross, 1903; Dennis and Bradley, 1989). However, these nuclei remain mononucleate and their lobes remain connected by small nucleoplasmic bridges. Among other taxa, double nuclei in follicular cells are reported for *Aeschna* and *Gomphus* (Odonata), for *Lepisma* (Zygentoma) (Gross, 1903) and for *Bacillus* (Phasmida) (Daiber, 1905). These examples need re-examination and are looked at preliminarily as parallel developments.

3.6.3 Characters of germ cell descendants and their influence on ovariole types

The interaction of some germ cells with special somatic cells, the apical cells of gonads, is the first induction on the way to male or female differentiation of germ cells to sperms and oocytes.

The surprising identification of intercellular bridges among germ cells nested around apical cells can be interpreted as stem cell clusters in *Locusta* follicles (section 3.3.1), and has important consequences for the phylogenetic debate. Using the rules of parsimony, it seems obvious to value this character as a basic one which is expressed in the male pathway but is kept silent in the female pathway except in primitive earwigs (Dermaptera) and in which all characters of cluster formation of stem cells and cystoblast cells are used to establish a polytrophic meroistic ovary. However, at least three other insect groups should be looked at more closely for possible stem cell clusters: the coccids (Hemiptera), the Strepsiptera and the midges (Diptera). In all these groups, cluster formation of cystoblasts is shifted to stages prior to ovariole development, or development of ovarioles is reduced totally (Strepsiptera).

As shown in the preceding section (3.5), in most species only a few stem cells are found which do not create stem cell clusters. Therefore it seems justified to believe that in all orders above the level of Dermaptera, this character has been kept silent in the female pathway, but may have been used again in coccids, Strepsiptera and midges.

The few stem cells have a limited programme of differential mitoses, which results in a fixed number of oocytes per ovariole. This character is believed to be a basic invention of the female pathway and belongs to the ground plan of insect ovarioles. The few exceptions, as for instance the unlimited stem cell divisions found in *Drosophila* ovaries (King, 1970; Wieschaus and Szabad, 1979; Lin and Spradling, 1993), are new inventions and therefore autapomorphic. Social insects may also have this character of unlimited stem cell divisions, or they may have huge amounts of stem cells with a limited programme of differential mitoses. For a clear decision, additional information on the ontogenesis of ovarioles is needed.

Stem cells differentiate cystoblasts. These cells divide equally into cystocytes by a limited programme of synchronous mitotic cycles. Thereby, cytokinesis is blocked, generating intercellular bridges, which persist. Each cystocyte undergoes the next cycle of division so that one sibling gets all preceding intercellular bridges, the other none (i.e. maximal branching occurs in the cluster). During cluster formation, spindle remnants form a fusome which combines to a polyfusome. Intercellular bridges of all division cycles move together, generating a rosette configuration of the cluster. Thereby, the polyfusome shortens. These characters are traded as a unit in most species during cluster formation. Therefore, this unit belongs to the ground plan. Moreover, when looking into sperm development, one can see the same unit of characters in the male line; thus, this combination is independent of sex differentiation and may be a very old invention of germ cell development. Similar cluster formation, following the 2^n-rule, is known to occur in nearly all large taxonomic units, for instance in Crustacea (Anteunis *et al.*, 1966), Mammalia (Franchi and Mandel, 1962; Ruby *et al.*, 1969; Fawcett, 1971; Moens and Hugenholtz, 1975) or even in Ctenophora (Dunlap Pianka, 1974) in both pathways in the male, and sometimes reduced in the female pathway (King and Büning, 1985). Only one exception is known today, the Nematoda (Wirth, 1984). This indicates symplesiomorphy of this fundamental character combination.

Each deviation, occurring in taxonomic units, must be apomorphic at its time of emergence. In this sense, the whole unit of cluster formation characters in females in so-called primitive insects is a symplesiomorphic character, in which apomorphy occurred at the base of metazoa when oogamy had been developed. These types of panoistic ovaries are called primary panoistic here, in contrast to those secondary panoistic ovaries which can be derived from meroistic ancestors. Primary panoistic ovaries are found in the following orders: Archaeognatha (?), Zygentoma (?), Odonata, Embioptera, Phasmatodea/Orthoptera, Grylloblattaria (?) and Dictyoptera (Isoptera ?), although the taxa indicated by question marks have not been investigated sufficiently.

In one ancient order, the Plecoptera, cluster formation takes place, which initially follows the 2^n-rule with the above character combination. However, one crucial invention has not been developed in this order, the differentiation of nurse cells. Thus, all siblings split later on and follow the oocyte pathway.

In mayflies (Ephemeroptera), cluster formation also occurs but two characters of the basic combination of cluster formation have changed. Here, only one linear cluster occurs, which fills the germarium by bending back and forth. This strongly suggests that only one germ cell transforms directly into a successful cystoblast. All other germ cells, if they exist, must degenerate. A stem cell population is not necessary. Permanent oocytes develop only when appropriate siblings are in permanent contact with prefollicular cells. All other siblings will not differentiate to definitive oocytes. Sometime later, the whole cluster breaks into smaller units, in which only one oocyte exists. This indicates that siblings come under the register of the oocyte and serve as nurse cells. Endomitotic cycles in nurse cells are absent. After previtellogenesis, the rows split into smaller units and nurse cells degenerate. The above characters are considered apomorphic, since this behaviour of germ cell descendants is unique among insects, and since no polytrophic ovaries have been developed in other ancient insect orders, this telotrophic ovary must have been born directly from panoistic ancestors.

Cluster formation in Dermaptera is also unique among insects (Figure 3.93). This order is the only one in which stem cell clusters are built in the female pathway. Siblings of subsequent stem cell divisions remain combined. Once cystoblasts lose contact with the germ cell cluster they undergo three cycles of synchronized divisions. Polyfusome and rosette formation occur as known from other insects. However, all four cells of the centre develop into oocytes, whereas the four other siblings of the last division cycle do not share in rosette formation and develop as nurse cells. The whole cluster splits into four subclusters, containing one oocyte and one nurse cell each. Stem cell cluster and splitting of cystocytes into subclusters are autapomorphic characters. No other polytrophic meroistic ovary of an insect shows these characters; thus, this type has evolved independently, using the same background of cluster formation, but has converted with autapomorphic inventions to a functional polytrophic meroistic ovary. The phylogeny of Dermaptera is still unsettled (Boudreaux, 1979; Hennig, 1981; Kristensen, 1981; Stys and Bilinski, 1990). Thus, it remains open if the ovary of Dermaptera is a convergent development to the basic type, which is correct if the Dermaptera are not closely related to Paraneoptera (Hennig, 1981; Kristensen, 1981; Pritsch and Büning, 1989; Büning and Sohst, 1990). On the other hand, if the systematic position of Dermaptera as proposed by Boudreaux (1979) and Stys and Bilinski (1990) is correct and can be confirmed by other characters, the basic type shifts more basally in the stem of true insects, including the Dermaptera. Characters of the basic type remain the same and autapomorphic characters of the basic type are: (1) transdetermination of nurse cells; (2) communication of nurse cells with somatic follicular cells to construct a trophic chamber; and (3) endomitotic polyploidization of nurse cells. In a cladistic argument, one might say that the whole set of information which is entangled in cluster formation can be looked at as autapomorphic, since these characters are absent in ovaries of close relatives like Dictyoptera. However, in the males of panoistic insects these characters exist. Thus, these characters are plesiomorphic and the polytrophic meroistic ovary of the basic type is rooted on this plesiomorphic complex. The characters of the basic type of polytrophic meroistic ovary have been shown above and the characters of cluster formation and the determination pattern of oocytes and nurse cells are constant throughout this type. Deviations are coupled with the rise of new ovariole types, which are telotrophic (Hemiptera, Megaloptera: Sialidae/Rhaphidioptera, and Coleoptera, Polyphaga); or they are reduced to panoistic ovaries, which are therefore secondary panoistic (Thysanoptera, Megaloptera: Corydalidae, Siphonaptera, Mecoptera: Boreidae).

Before going into the details of telotrophic ovaries, those inventions which do not change the type of ovarioles will be looked at, i.e. basically polytrophic meroistic, but with significant impact on physiology and structure. The first is the order of Strepsiptera, in which all somatic tissues are reduced, except for some follicular cells. Cluster formation seems to be classic but nurse cells do not polyploidize and two autapomorphic characters occur, showing a separated pathway. Clusters can split into equal halves during cluster formation. Thereby, the final number of eggs is raised. The oocyte nucleus is prevented from degradation processes, which attack all nurse cell nuclei housed in the same ooplasm after degradation of nurse and cell membranes. Degradation of nurse cell membranes is found also in telotrophic meroistic ovaries of Coleoptera and Polyphaga. Following the line of argument of Crowson (1981), who puts the Strepsiptera as part of the Coleoptera, Polyphaga, this character would be autapomorphic for the Coleoptera, Polyphaga. Following Kinzelbach (1971) and his arguments for the early separation of both groups (Strepsiptera and Coleoptera) the above character would be a convergent development, since Coleoptera: Adephaga do not have this character and the telotrophic ovary of polyphage Coleoptera is derived from ancestors with polytrophic ovaries, which are similar to that of adephage Coleoptera (see below). Investigation of Mengeidae is needed, since only in this family of Strepsiptera do we find an ovary which is not as strongly reduced as that of *Elenchus* (Silvestri, 1943). Thus, classification of this ovary as a highly reduced form of a polytrophic ovary is preliminary and must be settled by further investigation.

The second group, in which nurse cells do not polyploidize but cluster formation follows the 2^n-rule, is that of *Hystrichopsylla* (Hystrichopsyllinae, Siphonaptera). In addition, nurse cells have lost specific communication with somatic sheath cells; thus a specific trophic chamber, which isolates nurse cells from the oocyte chamber, as found in follicles of all other polytrophic meroistic ovaries will not be built. Each subcluster of nurse cells becomes integrated into the follicular sheath surrounding each oocyte. This character is apomorphic for Hystrichopsyllinae. Characters like the 2^n-rule (branched cluster), oocyte–nurse cell determination (2^n–1 nurse cells, one central oocyte) and a limited programme of stem cell mitoses, are valued as symplesiomorphic, since they are shared with the basic type. Oocytes of all fleas, panoistic as well as polytrophic meroistic, show extra-chromosomal amplification of rDNA. This character is shared by many, not closely related insect species and occurs among panoistic and polytrophic ovaries. Therefore, this character is valued today as demonstrating convergent development. Further investigation of structure and biochemical analysis of rDNA will give additional information to decide which amplifications are convergent and which might be apomorphic.

Looking only at ovary characters, two methods of phylogenetic interpretation are open: the polytrophic meroistic flea ovary is primary and the panoistic ovary is of secondary origin, or the polytrophic meroistic ovary is secondary and derived from the secondary panoistic ovary. A clear decision is not possible; thus, other arguments must be added to get a conclusive result. However, the systematics of fleas as well as of the mecopteran complex are still in debate and data from ovaries fit with both lines of argument. If one follows the arguments of Hinton (1958), Rothschild (1976) and Schlein (1980), the genus Hystrichopsylla is a young and modern flea genus. Since all other flea species, except the Hystrichopsyllinae (*Hystrichopsylla* and *Stenoponia*) have panoistic ovaries, the phylogenetic line becomes clear if one obeys the rule of parsimony: panoistic species came first and the polytrophic ovary of Hystrichopsyllinae is a new invention. If this line of argument is not accepted, then Hystrichopsyllinae must be an ancient group of fleas, or, when accepting other characters as 'modern', this small group has retained ancient characters especially in its ovaries. Another point of debate concerns the close relatives of fleas. The above authors have given some arguments for the close relationship of fleas and snowfleas (Boreidae, Mecoptera). Following Schlein's arguments (1980), fleas have evolved from snowfleas. Snowfleas have, as have the majority of fleas, panoistic ovaries. Our own preliminary investigations on boreid ovaries show that this ovary is panoistic and shows the same elements of panoistic ovaries found in fleas. These include a limited programme of stem cells, small nests of prefollicular tissue, extrachromosomal DNA, and two types of cytoplasm, one poor in ribosomes, which is early and another one rich in ribosomes, which occurs later. This congruence is suggestive but not sufficient since all these data are symplesiomorphic or can be convergent. Thus, further investigation is needed to clarify the relationship between both groups. Recently, this interpretation was opposed by Stys and Bilinski (1990). They accept the phylogenetic interpretation of Willmann (1987), who values the Boreidae as a family among Opisthogonophora (2.2), including also Meropeidae and Eomeropidae, whereas Bittacidae (2.1) and Nannochoristidae (1) are ranked as two suborders. Thus, he interpreted the Mecoptera as a true monophyletic entity and accepting his arguments, Siphonaptera can be only the sister group. Although only Panorpidae are looked at more closely (section 3.5), Mecopterans as a whole are believed to have polytrophic meroistic ovaries of the basic type. Thus, panoistic ovaries of both taxa have evolved independently. Büning and Sohst (1988) have interpreted the polytrophic meroistic ovary of *Hystrichopsylla* as secondary polytrophic, derived from the secondary panoistic ovary which occurs in most other fleas. This was opposed also by Stys and Bilinski (1990), who said: 'A reappearance of trophocytes once lost is hardly credible'. They believe that *Hystrichopsylla* has retained the ancestral polytrophy, whereas other characters are more advanced. An objective decision between both hypotheses is not possible today. However, the following arguments may be helpful for better understanding the hypothesis of Büning and Sohst. Cluster formation occurs generally in the male germ line with all the basic characters, such as synchrony of divisions, 2^n-rule (dichotomy), polyfusome and rosette formation. Thus, in females these characters are not lost, but they are repressed by female-specific regulatory gene(s). In the simplest mode, there is only one regulatory gene responsible for activation of the whole set of characters found in germ line clusters. A second event is the determination process, which makes it possible to decide between permanent oocytes and 'nurse cells'. A third event is the endomitotic polyploidization of nurse cells, followed by special regulation for under- and/or over-replication of special genes, which is correlated species-specifically according to the way of life of these animals. A fourth event is the formation of a tropharium, i.e. special interaction of nurse cells with anterior follicular cells by which these cells will convert into a flat inner sheath. This interaction keeps nurse cells together, forming one physiological unit. In *Hystrichopsylla* only the first two events are realized. Following the interpretation of Stys and Bilinski, this ovary is on the way to reduce its polytrophic character by losing the last characters in favour of a panoistic ovary. This is possible but contradicts all the other events by which *Hystrichopsylla* is denoted as 'advanced'. In the interpretation of Büning and Sohst, only the first two characters are switched on again in the female line of *Hystrichopsylla* and its next relatives. The other characters, polyploidization and building of a trophic chamber, have been decoupled from female-specific regulation during the silent phase, in which the foremothers of *Hystrichopsylla* have been panoistic, as have all other fleas. Thus, *Hystrichopsylla* is on the way to re-establishing a polytrophic meroistic ovary.

Another ovary, in which nurse cells do not have the full equipment of characters, is that of tipulids. Here, nurse cells do not multiply their genes by endomitosis but remain small. All other characters for nurse cell development are established. Similar to fleas, extrachromosomal rDNA is transcribed in oocytes during the late stages of previtellogenesis and early vitellogenesis. In some nurse cells rDNA is also amplified, but is not transcribed (Bayreuther, 1952, 1956; Lima-de-Faria and Moses, 1966; Lima-de-Faria *et al.*, 1968; Lima-de-Faria, 1974). Contrasting with all other extrachromosomal rDNA rings, in *Tipula* the amount of antiparallel ordered genes is high (Figure 3.100b) (Büning and Domazer-Schmidl, unpublished). This autapomorphic character indicates a separated line in rDNA development.

In all Nematocera, differentiation between oocytes and nurse cells is shifted to larval periods and antecedes the development of somatic tissues of ovarioles. If the interpretation of Sachtleben's (1918) findings is correct, we will have a second example for cluster segregation after determination of oocytes and nurse cells. This would mimic exactly the cluster behaviour of earwigs (Dermaptera), except for one point. In the dermapteran ovary, the cluster splits in the germarial region of the ovariole, whereas in the ovary of chironomids these possible events will happen much earlier in the gonadal tube. Since earwigs and midges are not closely related, both events are looked at as parallel developments.

In Strepsiptera only cluster formation processes are preserved and all other characters are reduced. Autapomorphic characters are the fission of clusters and the resorption of nurse cell genomes, when they become integrated into the ooplasm by membrane reduction. The characters found in strepsipteran ovaries do not answer the question from which type of ovary, telotrophic or polytrophic, this ovary is derived. It retains all plesiomorphic characters of cluster formation and the basic type of oocyte determination. Therefore, this ovary is typed as polytrophic.

Membrane reduction is also found in the ovarioles of Cecidomiidae (Diptera), in which somatic nurse cells fuse with one germ cell-derived nurse cell to a membrane-free syncytial nurse chamber. This is an autapomorphic character of Cecidomiidae. Fusion of other somatic cells with the oocyte is also autapomorphic for the same taxon.

In Coleoptera, Polyphaga, as well as in Sialidae (Mecoptera), and perhaps some Heteroptera, which all have telotrophic meroistic ovarioles, nurse cell reduction and nurse cell fusion are also characters, which have great impact on the morphology of ovarioles. According to other characters, all three types have evolved independently (see below), therefore, membrane reduction in all three telotrophic types is attributed as convergent.

The number of mitotic cycles during cluster formation is a character, which is species-specific, sometimes genus- or family-specific and only in some higher taxa, suborders or orders, constant. Thus, the majority of Diptera, Brachycera and Cyclorrhapha have clusters of 16 cells, which can be looked at as typical for these taxa; however in some species of these suborders the number of siblings is reduced. In midges (Diptera: Nematocera), only in some species as in tipulids does the cluster contains 16 (n=4) cells, whereas in most others the amount of division cycles is reduced to n=1 (two siblings) or n=2 (four siblings). High constancy of division cycles is found in Lepidoptera with n=3 (eight siblings).

Aberrant numbers of cluster cells can be established in two ways: (1) repression of mitotic activity in some siblings; and (2) supernumerary mitoses in other siblings. Furthermore, the character of maximal branching can be altered in favour of linear clusters in which both siblings will get old bridges. All these characters are found among Neuroptera which have developed, therefore, linear clusters showing only few branches. This character seems to be unique and is valued as autapomorphic of this order. In ground beetles of the genus *Carabus* (Coleoptera: Adephaga), the disorder is regulated in other ways; each oocyte undergoes regularly seven mitotic cycles, which results in seven intercellular bridges found at the nutritive appendix of each oocyte. However, the number of nurse cells differs tremendously, thus, nurse cell divisions are not regulated as precisely and may be independent of oocyte divisions. Sometimes the number of mitoses is reduced, i.e. there are less than 128 siblings, or nurse cell replications are supernumerary, which results in more than 128 siblings. Such irregular mitotic activity of nurse cell siblings is tremendously enhanced in the telotrophic ovaries of polyphage Coleoptera, in which only one or perhaps some huge clusters arise (see section 3.5.3 and below).

Among telotrophic ovaries, the hemipteran ovary has been best elucidated and all ovariole characters are consistent with the commonly accepted cladogram of Paraneoptera (Hennig, 1981; Kristensen, 1981) except for the order of Zoraptera, the ovary of which is supposed to be panoistic (Zaffagnini, 1969; Matsuda, 1976) and its phylogenetic position among Paraneoptera is doubtful (Stys and Bilinski, 1990). Paraneopterans without Zoraptera are also known as Acercaria (Boudreaux, 1979; Stys and Bilinski, 1990). At the base of Condylognatha (Thysanoptera/Hemiptera) one can assume a reduction of stem cells to only one functional cell, which communicates with somatic apical cells. Furthermore, a prefollicular tissue might have evolved at this point. In the thysanopteran line, nurse cell transdetermination is lost, thus clusters arise, the siblings of which are all oocytes. Oocytes become isolated from each other by a secondary detachment process. In the hemipteran line, five autapomorphic changes occurred. Stem cells have lost their capacity to undergo differential mitosis, i.e. they transform directly into a cystoblast. Consequently only one cluster can arise. Furthermore, the stabilized oocyte–nurse cell gradient (2^n–1 =

nurse cells, one oocyte) of polytrophic meroistic ovaries of the basic type becomes variable and shifts in favour of oocytes. Rosette formation becomes stabilized by the rise of a trophic core. Microtubules become stabilized in all those cytoplasmic areas which are used for the transport of nurse cell cytoplasm into the oocyte, i.e. projections of nurse cells to the trophic core, the trophic core itself, and the nutritive cords, connecting oocytes to the trophic core. All these elements are shared by nearly all hemipteran families but not by the sister group of Thysanoptera and not by Psocodea. Thus, these elements are synapomorphic among suborders of Hemiptera. All these characters are found unaltered in aphid ovaries, as well as in ovaries of Miridae and Cimicidae (Heteroptera) and perhaps in some Cicadomorpha. At the base of Heteroptera/Cicadomorpha one autapomorphy might have evolved. Nurse cell mitosis of apical nurse cells is prolonged into the functional ovary. This has special effects in the trophic chamber. A developmental gradient becomes apparent by which the trophic chamber can be divided into three or four zones: (1) apical zone of dividing nurse cells; (2) resting zone, in which nurse cells do not undergo morphological changes; (3) zone of endomitosis and high synthetic activity of nurse cells; and (4) zone of resting oocytes. In zone 3, nurse cells grow tremendously, especially their nuclei. During these syntheses, nurse cell nuclei can divide amitotically in some groups of Heteroptera, and it is this zone in which nurse cell fusion can be found. Both characters must be looked at more carefully before one can attribute them as apomorphic or convergent among subtaxa of heteropterans.

Today, it is not possible to decide at which point the character of prolonged nurse cell divisions should be positioned, at the base of, above taxa or within the taxa, since constant numbers of nurse cells are also found in some species (see above). Further investigation, especially of more 'primitive' groups of Heteroptera and Cicadiformes, is needed. The character of double nuclei, which occur in follicle cells, can be looked at as synapomorphic for Heteroptera/Cicadomorpha, when the occurrence of the same character in lice and mallophagans is looked at as parallel development. However, this needs further investigation to give a definitive answer. Psylloidea/Aleyrodoidea have two characters in common, not found in the sister group of Coccoidea/Aphidoidea: the nurse cells have reduced endomitosis in favour of mitosis (Figure 3.111). Thus, their terminal chambers have many, small nurse cells. In aleyrodids and psyllids the polyfusome persists for a long period and the trophic core does not fill with bundles of persisting microtubules. In *Psylla* the tropharium reduces all membranes among nurse calls, thus, no trophic core can be seen (Figure 3.110c,d) (Büning and Beyrich, unpublished). Among coccids, the cluster reduces to two or three cycles of mitotic divisions and only one cell develops as an oocyte, all others as nurse cells. Thus, formally, the original determination scheme between oocytes and nurse cells becomes restored: 2^n–1 nurse cells and one oocyte. However, it is unknown today which cell becomes the oocyte, one of the two innermost cells or any other cell of the cluster.

In the sister group of Paraneoptera (Acercaria), the Holometabola, a telotrophic ovary is found twice. The first taxon is the suborder of Coleoptera, Polyphaga. Its sister group, the Coleoptera, Adephaga, has polytrophic meroistic ovaries (see above). Comparing the characters of both ovaries, one can see some characters in common (Table 3.3): interstitial cells are found in both, serving as a three-dimensional net to keep nurse cells in position. Clusters of cystocytes become more and more disordered among ground-beetle families, whereby the pattern of cystocyte division remains constant by the innermost cells, especially the oocyte, although nurse cells undergo great variations in their mitotic programme. This is correlated with endomitotic activity. Thus, clusters with few nurse cells have bigger nuclei than those with many nurse cells. Both characters are found in telotrophic meroistic ovaries of Coleoptera, Polyphaga, too. Furthermore, cluster development seems to be divided into two phases. In the first phase, the cluster develops normally, i.e. like the clusters of the basic type, with one crucial exception: oocyte determination has changed in favour of more than one oocyte per cluster. Those cells which will be transdetermined to nurse cells, will prefer a division programme in most species, leading to more or less linear subclusters. These changes are sufficient to explain the rise of a telotrophic ovary, when it is restricted to only one huge cluster. If there is more than one cluster, two other characters must be predicted. Clusters must have the ability to fuse, at least at their posterior regions. This can be followed from membrane reductions found in many terminal chambers of polyphage Coleoptera (Büning, 1972, 1979a,b). Secondly, clusters must be kept in the terminal chambers side by side, i.e. there must be some cooperation between cystocytes and somatic cells of the terminal chamber to keep clusters in their original position, except the oocytes, which run down the ovariole when they become ensheathed with follicular cells.

The second taxon of holometabolans in which telotrophic ovaries are found, are the Megaloptera/Raphidioptera. Together with Neuropterida, there is no doubt that all three groups are closely related and of common origin. Neuropterida have polytrophic meroistic ovaries showing in all subgroups a reduction of bifurcation of clusters during cystocyte mitosis, which results in variable amounts of nurse cells ordered more or less linearly. All other characters are common to the basic type. Therefore, one can assume that progenitors of Neuroptera have evolved this type only once. However, within the subgroups of Megaloptera/ Raphidioptera: Sialidae, Corydalidae, Chauliodidae as well as Raphidiidae and Inocelliidae, the phylogeny has not been clearly unravelled. Stys and Bilinski (1990) have given six alternative views on cladogenesis of all the above groups, proposed by different authors (Figure 3.114).

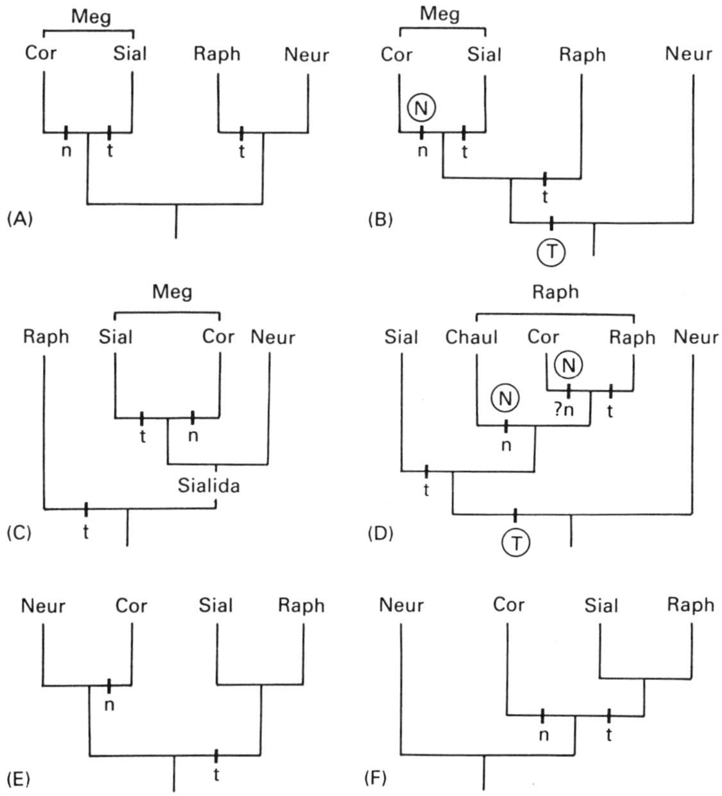

Figure 3.114 Cladogram of Stys and Bilinski shows several possibilities of relationships between Corydalidae, Sialidae, Raphidioptera and Neuroptera. Changes in ovariole types are indicated by n = neopanoism; t = telotrophy. Encircled capitals symbolize autapomorphies. For further discussion see text.

Concerning ovariole structure, we know that Corydalidae are panoistic (Matsuzaki *et al.*, 1985) while Raphidiidae and Sialidae are telotrophic (Matsuzaki and Ando, 1977; Achtelig, 1978; Büning, 1979c, 1980) and undoubtedly of the same origin (Table 3.3; Figure 3.88). Thus, one can propose that telotrophy has been raised only once in this taxon. Stys and Bilinski (1990) made general objections against reductional processes leading to secondary panoism (neopanoism) out of a telotrophic meroistic background. Their arguments are based on the assumption that it would be easier to reduce a meroistic ovary from the polytrophic subtype than from the telotrophic subtype, since one has to do one step less in evolution. This is not conclusive because one has to look only at reductionary steps which are needed to gain a panoistic ovary from a given and higher evolved status. Looking at the characters of telotrophy of Sialidae/ Raphidiidae, one can see that panoism can be obtained by only one reduction, the reduction of cystoblast divisions to zero. This is exactly what one has to do to reduce a polytrophic ovary to panoism. Today, no information is available about the ovaries of Chaulioidae and Inocellidae. Furthermore, no information on possible cluster development in the panoistic ovary of Corydalidae is available. Thus, cladograms based only on ovary structures are too speculative and one should use those cladograms which are most accepted, in this case cladogram B with only two inventions, one at the base of Raphidioptera/ Megaloptera which enables the unique telotrophy in both taxa, and another which reduces the ovary to secondary panoism in the Corydalidae (shown by encircled symbols).

3.6.4 Final remarks

At the end of this monograph on insect ovarioles, some remarks should be added. All that has been shown in the preceding sections is of course not finished, but work is going on in many laboratories, to elucidate the anatomical, histological, cellular and molecular aspects of oogenesis as well as to elaborate further models of anagenesis of insect ovarioles, and finally, of phylogeny of insects.

Today, most cellular and molecular work on oogenesis is done with model objects like *Drosophila*. No doubt, investment into such organisms speeds our knowledge and is fully accepted among most scientists. However, one should always keep in mind that *Drosophila* is a single, and in many ways a highly specialized, insect from a variety of about 1 million insect species. Thus, the *Drosophila* ovary is not 'the typical' insect ovary and not even 'the typical polytrophic meroistic' ovary, and represents only one, very well explored example. Nature is one of the best experimenters and nearly all the varieties one can think of on cluster formation have been probed by nature. This clearly shows that science must use the supply that nature offers to get a more complete model of evolutionary processes.

A monograph, such as this book, will give an overview of the main topics of research in ovary development at the time of publication. This includes personal decisions by authors and in that sense this book is not objective. However, the reader should use this 'fault' as a chance to learn, since controversial discussions are the best way to forward science.

References

Achtelig, M. (1978) Entwicklung und Morphologie der inneren und äusseren weiblichen Genitalorgane der Kamelhalsfliegen (Neuropteroidea, Raphidioptera). *Ent. Germ.*, **4**, 140–163.

Adijodi, K.G. and Adiyodi, R.G. (1975) Morphology and cytology of the accessory sex glands of invertebrates. *Int. Rev. Cytol.*, **43**, 353–398.

Alberts, B., Bray, D., Lewis, J. *et al.* (1988) *Molecular Biology of the Cell*, 2nd edn, Garland Publishing, New York.

Alexander, B. and Rozen, J.G. (1987) Ovaries, ovarioles, and oocytes in parasitic bees (Hymenoptera, Apoidea). *Pan-Pacific Entomol.*, **63**, 155–164.

Al-Atia, G.R., Fruscolini, P. and Jacobs-Lorena, M. (1985) Translational regulation of mRNAs for ribosomal proteins during early *Drosophila* development. *Biochem.*, **24**, 5798–5803.

Allen, E.R. and Cave, M.D. (1969) Cytochemical and ultrastructural studies of ribonucleoprotein containing structures in oocytes of *Acheta domesticus*. *Z. Zellforsch.*, **101**, 63–71.

Allen, E.R. and Cave, M.D. (1972) Nucleolar organization in oocytes of Gryllid crickets, subfamilies Gryllidae and Nemobiidae. *J. Morphol.*, **137**, 433–447.

Al-Khalifa, M.S. (1981) The spermathecal structure of the flour beetle *Tribolium castaneum* (Herbst.) (Col., Tenebrionidae). *Ind. J. Zool.*, **9**, 1–6.

Amaldi, F., Bozzoni, I., Beccari, E. and Pierandrei-Amaldi, P. (1989) Expression of ribosomal protein genes and regulation of ribosome biosynthesis in *Xenopus* development. *Trends Biochem. Sci.*, **14**, 175–178.

Amrein, H., Maniatis, T. and Nöthiger, R. (1990) Alternatively spliced transcripts of the sex-determining gene *tra-2* of *Drosophila* encode functional proteins of different size. *EMBO J.*, **9**, 3619–3629.

Anastasi, A., Hunt. C. and Stebbings, H. (1990) Isolation of microtubule motors from an insect ovarian system, characterization using a novel motility substratum. *J. Cell Sci.*, **96**, 63–69.

Anderson, E. (1964) Oocyte differentiation and vitellogenesis in the roach *Periplaneta americana*. *J. Cell Biol.*, **20**, 131–155.

Anderson, E. (1969) Oogenesis in the cockroach, *Periplaneta americana*, with special

references to the specialization of the oolemma and the fate of coated vesicles. *J. Microsc. (Oxford)*, **8**, 721–738.
Anderson, K.V. and Lengyel, J.A. (1979) Rates of synthesis of major classes of RNA in *Drosophila* embryos. *Dev. Biol.*, **70**, 217–231.
Anderson, L.M and Telfer, W.H. (1969) A follicle cell contribution to the yolk spheres of moth oocytes. *Tissue Cell*, **1**, 633–644.
Anderson, L.M. and Telfer, W.H. (1970) Trypan blue inhibition of yolk deposition – a clue to follicle cell function in the *Cecropia* moth. *J. Embryol. Exp. Morphol.*, **23**, 35–52.
Angelier, A. and Lacroix, J.C. (1975) Complexes de transcription d'origines nucleolaire et chromosomique d'ovocytes de *Pleurodeles waltlii* et *P. poireti* (Amphibiens, Urodeles). *Chromosoma (Berl.)*, **51**, 323–335.
Angelier, N., Paintrand, M., Lavaud, A. and Lechaire, J.P. (1984) Scanning electron microscopy of amphibian lampbrush chromosomes. *Chromosoma (Berl.)*, **89**, 243–253.
Ansari, M.S. and Murad, H. (1981) Histomorphology of the female reproductive organ of the cattle fly, *Hippobosca magulata* LCH (Diptera, Hippoboscidae). *Netherlands J. Zool.*, **31**, 466–471.
Anteunis, A., Fautrez-Firlefyn, N. and Fautrez, J. (1966) La structure de ponts intercellulaires 'obturés' et 'ouverts' entre oogonies et oocytes dans l'ovaire d'*Artemia salina*. *Arch. Biol.*, **77**, 645–654.
Artavanis-Tsakonas, S. and Simpson, P. (1991) Choosing a cell fate: a view from the *Notch* locus. *Trends Genet.*, **7**, 403–408.
Asaba, H. and Ando, H. (1978) Ovarian structures and oogenesis in *Lepidocampa weberi* Oudemans (Diplura, Campodeidae). *Int. J. Insect Morphol. Embryol.*, **7**, 405–414.
Ashburner, M. (1981) Some aspects of the structure and function of the polythene chromosomes of the Diptera. *Symp. R. Entomol. Soc. London*, no. 10, 65–84.
Ashhurst, D.E. (1985) Connective tissues, in *Comprehensive Insect Physiology, Biochemistry and Pharmacology*, (eds G.A. Kerkut and L.I. Gilbert), Pergamon Press, Oxford, Vol. 3, pp. 249–287.
Ashhurst, D.E. and Bailey, A.J. (1980) Insect collagen, morphological and biochemical characterization. *Eur. J. Biochem.*, **103**, 75–83.
Ashhurst, D.E. and Constin, N.M. (1974) The development of a collagenous tissue in the locust, *Locusta migratoria*. *Tissue Cell*, **6**, 279–300.
Avancini, R.M.P. and Pires do Prado, A. (1986) Oogenesis in *Chrysomya putoria* (Wiedemann) (Diptera, Calliphoridae). *Int. J. Insect Morphol. Embryol.*, **15**, 375–384.
Ax, P. (1984) *Das Phylogenetische System*, Gustav Fischer, Stuttgart.
Baker, B.S. (1989) Sex in flies, the splice of life. *Nature*, **340**, 521–524.
Baker, B.S., Nagoshi, R.N. and Burtis, K.C. (1987) Molecular genetic aspects of sex determination in *Drosophila*. *BioEssays*, **6**, 66–70.
Balbiani, E.G. (1885) Contribution à l'étude de la formation des organes sexuels chez les insectes. *Recueil Zool. Suisse*, **2**, 527–588.
Baldarelli, R.M., Mahoney, P.A., Salas, F. *et al.* (1988) Transcripts of the *Drosophila* blastoderm-specific locus, *terminus*, are concentrated posteriorly and encode a potential DNA-binding finger. *Dev. Biol.*, **125**, 85–95.
Banerjee, B. (1964) Cytogenesis in the ovarioles of *Odontotermes redemanni* (Wasmann). *Cellule*, **2**, 223–233.
Barbier, R. (1975) Différenciation de structures ciliaires et mise en place des canaux au cours de l'organogenèse des glandes collétériques de *Galleria melonella* L. (Lépidoptère, Pyralidae). *J. Microsc. Biol. Cell.*, **24**, 315–326.
Bardsley, A., McDonald, K. and Boswell, R.E. (1993) Distribution of tudor protein in the *Drosophila* embryo suggests segregation of functions based in the site of localization. *Development*, **119**, 207–219.

Bareth, C. (1972) L'ovogenèse chez un Arthropoe du sol, *Campodea remyi* (Diploures). Structure et évolution de l'ovaire. *C. R. Acad. Sci. Paris*, **275**, 1887–1890.

Bast, R.E. and Telfer, W.H. (1976) Follicle cell protein synthesis and its contribution to the yolk of the *Cecropia* moth oocyte. *Dev. Biol.*, **52**, 83–97.

Batalova, F.M. and Gruzova, M.N. (1987) Chromosomes of the raptorial gall midge *Aphidoletes aphidimyza* (Diptera, Cecidomyiidae) in the late oogenesis and early embryogenesis. *Citologija (Moscow)*, **29**, 642–649.

Batalova, F.M., Gruzova, M.N. and Troshkov, V.A. (1990) Oogenesis of the raptorial gall midge *Aphidoletes aphidimyza* Rond. V. Peculiarities of growth and formation of the egg follicle in ovarioles of the raptorial gall midge. *Citologija (Moscow)*, **32**, 12–20.

Bauer, H. (1933) Die Histologie des Ovars von *Tipula paludosa* Meig. *Z. Wiss. Zool.*, **143**, 53–76.

Baumeister, H.G. (1973) Lampbrush chromosomes and RNA-synthesis during early oogenesis of *Brachydanio rerio*. *Z. Zellforsch.*, **145**, 145–150.

Baumert, D. (1957) Mehrjährige Zuchten einheimischer Strepsipteren an Homopteren. 1. Hälfte. Larven und Puppen von *Elenchus tenuicornis* Kirby. *Zool. Beitr. NF*, **3**, 365–421.

Bayreuther, K. (1952) Extrachromosomale feulgenpositive Körper (Nukleinkörper) in der Oogenese der Tipuliden. *Naturwiss.*, **39**, 71.

Bayreuther, K. (1956) Die Oogenese der Tipuliden. *Chromosoma (Berl.)*, **7**, 508–557.

Bayreuther, K. (1957) Extrachromosomales DNS-haltiges Material in der Oogenese der Flöhe. *Z. Naturforsch.*, **12b**, 458–461.

Beams, H.W. and Kessel, R.G. (1969) Synthesis and deposition of oocyte envelopes (vitelline membrane, chorion) and the uptake of yolk on the dragonfly (Odonata, Aeschnidae). *J. Cell. Sci.*, **4**, 241–264.

Bedo, D.G. (1987) Polytene chromosomes mapping in *Ceratitis capitata* (Diptera, Tephritidae). *Genome*, **29**, 598–611.

Belote, J.M., Handler, A.M., Wolfner, M.F. *et al.* (1985) Sex-specific regulation of yolk protein gene expression in *Drosophila*. *Cell*, **40**, 339–348.

Benes, H. and Cave, M.D. (1985) Distribution of 5S ribosomal RNA genes in somatic and germ cells of the house cricket, *Acheta domesticus*. *Chromosoma (Berl.)*, **93**, 31–37.

Berleth, T., Burri, M. Thoma, G. *et al.* (1988) The role of localization of *bicoid* RNA in organizing the anterior pattern of the *Drosophila* embryo. *EMBO J.*, **7**, 1749–1756.

Berlot, J. and Goodman, C.S. (1984) Guidance of peripheral pioneer neurons in the grasshopper: Adhesive hierarchy of epithelial and neuronal surfaces. *Science*, **223**, 493–684.

Bernard, E.C. (1979) Egg types and hatching of *Eosentomon* Berlese (Protura, Eosentomidae). *Trans. Am. Microsc. Soc.*, **98**, 123–126.

Berry, S.J. (1982) Maternal direction of oogenesis and early embryogenesis in insects. *Ann. Rev. Entomol.*, **27**, 205–227.

Berry, S.J. and Johnson, E. (1975) Formation of temporary flagellar structures during insect organogenesis. *J. Cell Biol.*, **65**, 489–492.

Berry, R.O. (1941) Chromosome behavior in the germ cells and development of the gonads in *Sciara ocellaris*. *J. Morphol.*, **68**, 547–583.

Beyrich, C. and Gottanka, J. (1992) 2 Telotrophe Ovarien unterschiedlicher Herkunft, Ephemeroptera und Psylliden (Hemiptera). *Verh. Dtsch. Zool. Ges.*, **85**(1), 251.

Bielanska-Osuchowska, Z. (1957) Kwasy nukleinowe w jajach ryjkowca *Phyllobius urticae* D. *Folia Morphol.*, **8**, 257–269.

Bielanska-Osuchowska, Z. (1960) Histochemical studies in insect telotrophic ovaries. *Zool. Polon.*, **10**, 131– 163.

Bier, K. (1952) Beziehungen zwischen Nährzellkerngrösse und Ausbildung ribonukleinsäurehaltiger Strukturen in den Oocyten von *Formica rufa rufo-pratensis minor* Gösswald. *Verh. Dtsch. Zool. Ges.*, **40**, 369–374.

Bier, K. (1954) Über Phasen gesteigerter Protein-und Kohlehydrateinlagerungen und die Fettverteilung im Hymenopterenovar. *Zool. Anz. Suppl.*, **18**, 422–429.

Bier, K. (1957) Endomitose und Polytänie in den Nährzellkernen von *Calliphora erythrocephala* Meigen. *Chromosoma (Berl.)*, **8**, 493–522.

Bier, K. (1958) Beziehungen zwischen Wachstumsgeschwindigkeit, endometaphasischer Kontraktion und der Bildung von Riesenchromosomen in den Nährzellen von *Calliphora. Z. Naturforsch.*, **13b**, 85–93.

Bier, K. (1959) Quantitative Untersuchungen über die Variabilität der Nährzellkernstruktur und ihre Beeinflussung durch die Temperatur. *Chromosoma (Berl.)*,**10**, 619–643.

Bier, K. (1963a) Synthese, interzellulärer Transport und Abbau von Ribonukleinsäure im Ovar der Stubenfliege *Musca domestica. J. Cell Biol.*, **16**, 436–440.

Bier, K. (1963b) Autoradiographische Untersuchungen über die Leistungen des Follikelepithels und der Nährzellen bei der Dotterbildung und Eiweissynthese im Fliegenovar. *Roux's Arch. Entwicklungsmech.*, **154**, 552–575.

Bier, K. (1965a) Über den Transport zelleigener Makromoleküle durch die Kernmembran. I. RNS-Synthese und RNS-Transport unter Sauerstoffmangel und bei herabgesetzter Temperatur. *Chromosoma (Berl.)*, **16**, 58–69.

Bier, K. (1965b) Zur Funktion der Nährzellen im meroistischen Insektenovar unter besonderer Berücksichtigung der Oogenese adephager Coleopteren. *Zool. Jb. Physiol.*, **71**, 371–384.

Bier, K. (1967a) Lampenbürstenchromosomen und RNS-Versorgung in Insekteneizellen. *Umschau Wiss. Techn.*, **14**, 453.

Bier, K. (1967b) Oogenese, das Wachstum von Riesenzellen, *Naturwiss.*, **8**, 189–195.

Bier, K. (1968) Insect oogenesis with special reference to nuclear structure and function. *Accademia Nationale dei Lincei*, **104**, 61–72.

Bier, K. (1970) Oogenesetypen bei Insekten und Vertebraten, Ihre Bedeutung für die Embryogenese und Phylogenese. *Verh. Dtsch. Zool. Ges. 1969, Zool Anz.* Suppl. **33**, 7–29.

Bier, K., Kunz, W. and Ribbert, D. (1967) Struktur und Funktion der Oocytenchromosomen und Nukleolen sowie der Extra-DNS während der Oogenese panoistischer und meroistischer Insekten. *Chromosoma (Berl.)*, **23**, 214–254.

Bier, K., Kunz, W. and Ribbert, D. (1969) Insect oogenesis with and without lampbrush chromosomes, in *Chromosomes Today*, (eds C.D. Darlington and K.R. Lewis), Oliver and Boyd, Edinburgh, Vol. 2, pp. 107–115.

Bier, K. and Ramamurty, P.S. (1964) Elektronenoptische Untersuchungen zur Einlagerung der Dotterproteine in die Oocyte. *Naturwiss.*, **9**, 223–224.

Bier, K. and Ribbert, D. (1966) Struktur und genetische Aktivität des DNS-Keimbahnkörpers von *Dytiscus. Naturwiss.*, **53**, 115.

Bilinski, S. (1974) Ultrastructure of annulate lamellae in the oocytes and trophocytes of *Tetrodontophora bielanensis* (Waga) (Collembola). *Acta Biol. Crac., Ser. Zool.*, **17**, 177–180.

Bilinski, S. (1975) Origin of dictyosomes in the oocytes of *Tetrodontophora bielanensis* (Waga) (Collembola). *Acta Biol. Crac., Ser. Zool.*, **18**, 125–129.

Bilinski, S. (1976) Ultrastructure studies on the vitellogenesis of *Tetrodontophora bielanensis* (Waga) (Collembola). *Cell Tissue Res.*, **168**, 399–410.

Bilinski, S. (1977) Oogenesis in *Acerentomon gallicum* Jonescu (Protura). *Cell Tissue Res.*, **179**, 401– 412.

Bilinski, S. (1979) Oogenesis in *Campodea* sp. (Diplura), The ultrastructure of the egg chamber during vitellogenesis. *Cell Tissue Res.*, **202**, 133–143.

Bilinski, S. (1983a) Differentiation of the oocyte and nurse cells in an apterygote insect (Campodea). *Tissue Cell*, **15**, 965–973.

Bilinski, S. (1983b) Oogenesis in *Campodea* sp. (Insecta, Diplura): Chorion formation and the ultrastructure of follicle cells. *Cell Tissue Res.*, **228**, 165–170.

Bilinski, S. (1986) The formation of intercellular bridge coating in the hairy whirligig beetle, *Orectochilus villosus*. *Cell Tissue Res.*, **243**, 673–675.

Bilinski, S. (1987) Heterocellular gap junctions in vitellogenic follicles of *Campodea* sp. (Diplura), in *Recent Advances in Insect Embryology in Japan and Poland*, (eds H. Ando and C. Jura), ISEBU, Tsukuba.

Bilinski, S. (1989) Formation and function of accessory nuclei in the oocytes of the bird louse, *Eomenacanthus stamineus* (Insecta, Mallophaga) I. Ultrastructural and histochemical studies. *Chromosoma*, **97**, 321–326.

Bilinski, S. (1991a) Are accessory nuclei involved in the establishment of developmental gradients in hymenopteran oocytes? *Wilhelm Roux Arch. Dev. Biol.*, **199**, 423–426.

Bilinski, S. (1991b) Morphological markers of anteroposterior and dorsoventral polarity in developing oocytes of the hymenopteran, *Cosmoconus meridionator* (Ichneumonidae).*Wilhelm Roux Arch. Dev. Biol.*, **200**, 330–335.

Bilinski, S. (1993) Structure of ovaries and oogenesis in enognathans (Apterygota). *Int. J. Insect Morphol. Embryol.*, **22**, 255–269.

Bilinski, S., Hage, W.J. and Bluemink, J.G. (1985) Gap junctions between the follicle cells and the oocyte during oogenesis in an insect, *Tribolium destructor*. *Wilhelm Roux Arch. Dev. Biol.*, **194**, 296–300.

Bilinski, S.M. and Jaglarz, M. (1987) Oogenesis in the common tiger beetle, *Cicindela campestris* (Coleoptera, Adephaga) II. Unusual structure ensuring the contact between the oocyte and accompanying nurse cells. *Zool. Jb. Anat.*, **116**, 353–359.

Bilinski, S.M. and Jankowska, W. (1987) Oogenesis in the bird louse *Eomenacanthus stramineus* (Insecta, Mallophaga) I. General description and structure of the egg capsule. *Zool. Jb. Anat.*, **116**, 1–12.

Bilinski, S. and Klag, J. (1977) The oogenesis in *Acerentomon gallicum* Jonescu (Protura). An ultrastructural analysis of chorion formation. *Acta Biol. Crac., Ser. Zool.*, **20**, 101–106.

Bilinski, S. and Klag, J. (1982) Gap juctions between oocyte and follicle cells in *Acerentomon* sp. (Insecta, Protura). *Int. J. Invert. Reprod.*, **5**, 331–335.

Bilinski, S. and Klag, J. (1985) Formation of intranuclear annulate lamellae in oocytes of *Acerentomon* species (Protura). *Cytobios*, **43**, 97–100.

Bilinski, S.M., Ksiazkiewicz-Kapralska, M. and Stys, P. (1990) Unusual organization of the tropharium in the telotrophic ovarioles of an insect, *Saldula saltatoria*. *Tissue Cell*, **22**, 213–219.

Bilinski, S. and Larink, O. (1989) Fine structure of the egg envelope and the supporting stalk in the dipluran *Campdea* (Apterygota, Campodeidae). *Int. J. Insect Morphol. Embryol.*, **18**, 199–204.

Bilinski, S. and Petryszak, B. (1978) The ultrastructure and function of follicle cells in *Foucartia squamulata* (Herbst) (Curculionidae). *Cell Tissue Res.*, **189**, 347–353.

Bilinski, S. and Szklarzewicz, T. (1987) Ultrastructural modifications of the follicular epithelium accompanying the onset of vitellogenesis in the whirligig beetle, *Gyrinus natator*. *Cell Tissue Res.*, **249**, 209–214.

Bilinski, S. and Szklarzewicz, T. (1992) The ovary of *Catajapyx aquilonarius* (Insecta, Entognatha): Ultrastructure of germarium and terminal filament. *Zoomorphology*, **112**, 247–251.

Bilinski, S. and Tylek, W. (1987) Intermediate nurse cells in *Campodea* sp. (Diplura). Differentiation and possible role during oogenesis, in *Recent Advances in Insect Embryology in Japan and Poland*, (eds H. Ando and C. Jura), ISEBU, Tsukuba.

Billen J. (1985) Ultrastructure of the worker ovarioles in *Formica* ants. *Int. J. Insect Morphol. Embryol.*, **14**, 21–32.

Birket-Smith, S.J.R. (1974) On the abdominal morphology of Thysanura. *Entomol. Scand. Suppl.*, **6**, 1–67.

Bitsch, J. (1981) Ultrastructural modifications of the glandular epithelium of the receptaculum seminis in *Thermobia domestica* (Insecta Thysanura) during the molting period. *Cell Tissue Res.*, **220**, 99–113.

Bitsch, J. and Palevody, C. (1980) La vitellogenése des insectes aptérygotes. *Bull. Soc. Zool. France*, **105**, 419–425.

Blackman, R.L. (1978) Early development of the parthenogenetic egg in three species of aphids. *Int. J. Insect Morphol. Embryol.*, **7**, 33–44.

Blass, S. and Ruthmann, A. (1989) Fine structure of the accessory glands of the female genital tract of the ichneumonid *Pimpla turionellae* (Insecta, Hymenoptera). *Zoomorph.*, **108**, 367–377.

Blochmann, F. (1886) Ueber die Eireifung bei Insekten. *Biol. Zbl.* **3**, 554–559.

Blumberg, B., MacKrell, A.J., Olson, P.F. *et al.* (1987) Basement membrane procollagen IV and its specialized carboxyl domain are conserved in *Drosophila*, mouse and man. *J. Biol. Chem.*, **262**, 5947–5950.

Blumberg, B., MacKrell, A.J. and Fessler, J.H. (1988) *Drosophila* basement membrane procollagen (alpha)1 (IV) II. Complete cDNA sequence, genomic structure, and general implications for supramolecular assemblies. *J. Biochem. Chem.*, **236**, 18328-18337.

Bobrova, I.F. and Gruzova, M.N. (1967) Morphological and cytochemical studies of the oocyte nuclei in the saw fly in relation to the problem of karyosphere. *Citologija (Moskow)*, **9**, 27–33.

Boggs, R.T., Gregor, P., Idriss, S., Belote, J.M. and McKeown, M. (1987) Regulation of sexual differentiation in *D. melanogaster* via alternative splicing of RNA from the *transformer* gene. *Cell*, **50**, 739–747.

Bohrmann, J. (1991a) *In vitro* culture of *Drosophila* ovarian follicles, the influence of different media on development, RNA synthesis, protein synthesis and potassium uptake. *Wilhelm Roux Arch. Dev. Biol.*, **199**, 315–326.

Bohrmann, J. (1991b) Potassium uptake into *Drosophila* ovarian follicles, relevance to physiological and developmental processes. *J. Insect Physiol.*, **37**, 937–946.

Bohrmann, J., Dorn, A., Sander, K. and Gutzeit, H.O. (1986a) The extra-cellular electrical current and its variability in vitellogenic *Drosophila* follicles. *J. Cell Sci.*, **81**, 189–206.

Bohrmann, J. and Gutzeit, H.O. (1987) Evidence against electrophoresis as the principal mode of protein transport in vitellogenic follicles of *Drosophila. Development*, **101**, 279–288.

Bohrmann, J., Heinrich, U.R., Dorn A., Sander, K. and Gutzeit, H.O. (1984) Electrical phenomena and their possible significance in vitellogenic follicles of *Drosophila melanogaster. J. Embryol. Exp. Morph. Suppl.*, **8**, 151.

Bohrmann, J., Huebner, E., Sander, K. and Gutzeit, H.O. (1986b) Intercellular electrical potential measurements in *Drosophila* follicles. *J. Cell Sci.*, **81**, 207–221.

Bohrmann, J., Kiefer, G. and Sander, K. (1986c) Inverse correlation between mean nuclear DNA content and cell number in nurse cell clusters of *Drosophila. Chromosoma (Berl.)*, **94**, 36–44.

Bohrmann, J. and Sander, K. (1987) Aberrant oogenesis in the patterning mutant *dicephalic* of *Drosophila melanogaster*, time-lapse recordings and volumetry *in vitro. Wilhelm Roux Arch. Dev. Biol.*, **196**, 279–285.

Bonhag, P.F. (1955a) Histochemical studies of the ovarian nurse tissues and oocytes of the milkweed bug, *Oncopeltus fasciatus* (Dallas) I. Cytology, nucleic acids, and carbohydrates. *J. Morphol.*, **96**, 381–439.

Bonhag, P.F. (1955b) Histochemical studies of the ovarian nurse tissues and oocytes of the milkweed bug, *Oncopeltus fasciatus* (Dallas) II. Sudanophilia, phospholipids, and cholesterol. *J. Morphol.*, **97**, 283–312.

Bonhag, P.F. (1956) The origin and distribution of periodic acid- Schiff-positive substances in the oocyte of the earwig, *Anisolabis maritima* (Géné). *J. Morphol.*, **99**, 433–463.

Bonhag, P.F. (1958) Ovarian structure and vitellogenesis in insects. *Ann. Rev. Entomol.*, **3**, 137–160.

Bonhang, P.F. (1959) Histological and histochemical studies on the ovary of the American cockroach *Periplaneta americana* (L.). *Univ. Calif. Publ. Entomol.*, **16**, 81–124.

Bonhag, P.F. and Arnold, W.J. (1961) Histology, histochemistry and tracheation of the ovariole sheaths in the american cockroach *Periplaneta americana. J. Morphol.*, **108**, 107–129.

Bonhag, P.F. and Wick, J.R. (1953) The functional anatomy of the male and female reproductive systems of the milkweed bug, *Oncopeltus fasciatus* (Dallas) (Heteroptera, Lygaeidae). *J. Morphol.*, **93**, 177–230.

Bopp, D., Bell, R.L., Cline, T.W. and Schedl, P. (1991) Developmental distribution of female-specific *Sex-lethal* proteins in *Drosophila melanogaster. Genes Dev.*, **5**, 403–415.

Boswell, R.E. and Mahowald, A.P. (1985) Cytoplasmic determinants in embryogenesis, in *Comprehensive Insect Physiology, Biochemistry and Pharmacology*, (eds G.A. Kerkut and L.I. Gilbert), Pergamon Press, Oxford, Vol. I, pp. 387–405.

Bouderaux, H.B. (1979) *Arthropod Phylogeny with Special Reference to Insects*, John Wiley, New York.

Bouligand, Y. (1972) Twisted fibrous arrangements in biological materials and cholesteric mesophases. *Tissue Cell*, **4**, 189–217.

Bownes, M. (1979) Three genes for three yolk proteins in *Drosophila melanogaster. FEBS Lett.*, **100**, 95–98.

Bownes, M. (1980) The use of yolk protein variations in *Drosophila* species to analyse the control of vitellogenesis. *Differentiation*, **16**, 109–116.

Bownes, M. (1982) Ovarian yolk protein synthesis in *Drosophila melanogaster. J. Insect Physiol.*, **28**, 953–960.

Bownes, M., Dempster, M. and Blair, M. (1983) The regulation of yolk protein gene expression in *Drosophila melanogaster*, in *Molecular Biology of Egg Maturation*, (eds R. Porter and J. Whelan), *Ciba Foundation Symposium*, **98**, pp. 63–79.

Bownes, M., Steinmann-Zwicky, M. and Nöthiger, R. (1990) Differential control of yolk protein gene expression in fat bodies and gonads by the sex-determining gene *tra-2* for *Drosophila. EMBO J.*, **9**, 3975–3980.

Bownes, M., Ronaldson, E., Mauchline, D. and Martinez, A. (1993) Regulation of vitellogenesis in *Drosophila. Int. J. Insect Morphol. Embryol.*, **22**, 349–367.

Brandt, A. (1874) Über die Eiröhren der *Blatta orientalis* (*Periplaneta*). *Mem. Acad. Imp. Sci., VIIme Ser.*, **21**(12), 1–30.

Brandt, A. (1878) *Über das Ei und seine Bildungsstätte*, Wilhelm Engelmann Verlag, Leipzig.

Brauns, F. (1912) Die Entstehung der Nährzelle und die Bedeutung derselben für das wachsende Ei bei *Forficula auricularia* L., *Sitzungsber. Abhandl. Naturfr. Gess. Rostock* N.F. **4**, 99–141.

Bray, D., Heath, J. and Moss, D. (1986) The membrane associated 'cortex' of animal cells, its structure and mechanical properties. *J. Cell Sci., Suppl.*, **4**, 71–88.

Brennan, M.D., Weiner, A.J., Goralski, T.J. and Mahowald, A.P. (1982) The follicle cells are the major site of vitellin synthesis in *Drosophila melanogaster*. *Dev. Biol.*, **89**, 225–236.

Bridges, C.B. (1925) Sex in relation to chromosomes and genes. *Am. Nat.*, **59**, 127–137.

Bridges,C.B. and Brehme, K.S. (1944) The mutants of *Drosophila melanogaster*. Carnegie Institution of Washington, Washington DC, No. 552.

Brookman, J.J., Toosy, A.T., Shashidhara, L.S. and White, R.A.H. (1992) The 4512 retrotransposon and the development of gonadal mesoderm in *Drosophila*. *Development*, **116**, 1185–1195.

Brough, C.N. and Dixon, A.F.G. (1989) Follicular sheath (ovarian sheath) structure in virginoparae of the vetch aphid, *Megoura viciae* Buckton (Homoptera, Aphididae). *Int. J. Insect Morphol. Embryol.*, **18**, 217–226.

Brown, D.D. and David, I.B. (1968) Specific gene amplification in oocytes. *Science*, **160**, 272–280.

Brown, E.H. and King, R.C. (1964) Studies on the events resulting in the formation of an egg chamber in *Drosophila melanogaster*. *Growth*, **28**, 41–81.

Brues, C.T. (1903) A contribution to our knowledge of the Stylopidae. *Zool Jb. Anat. Ontog. Tiere*, **18**, 241– 270.

Brunt, A.M. (1970) Extensive system of microtubules in the ovariole of *Dysdercus fasciatus* Signoret (Heteroptera, Pyrrhocoridae). *Nature*, **228**, 80–81.

Brunt, A.M. (1971) The histology of the first batch of eggs and their associated tissues in the ovariole of *Dysdercus fasciatus* Signoret (Heteroptera, Pyrrhocoridae) as seen with the light microscope. *J. Morphol.*, **134**, 105–130.

Buchner, P. (1909) Das accessorische Chromosom in Spermatogenese und Ovogenese der Orthopteren, zugleich ein Beitrag zur Kenntnis der Reduktion. *Arch. Zellforsch.*, **3**, 335–430.

Buchner, P. (1918) Vergleichende Eistudien 1. Die akzessorischen Kerne des Hymenoptereneies. *Arch. Mikr. Anat.*, **91**, 1–202.

Buchner, P. (1965) *Endosymbiosis of Animals with Plant Microorganisms*. Wiley, New York.

Bunch, T.A. and Brower, D.L. (1993) *Drosophila* cell adhesion molecules. *Curr. Topics Dev. Biol.*, **28**, 81–123.

Büning, J. (1972) Untersuchungen am Ovar von *Bruchidius obtectus* Say. (Coleoptera-Polyphaga) zur Klärung des Oocytenwachstums in der Prävitellogenese. *Z. Zellforsch.*, **128**, 241–282.

Büning, J. (1978) Development of telotrophic-meroistic ovarioles of polyphage beetles with special reference to the formation of nutritive cords. *J. Morphol.*, **156**, 237–256.

Büning, J. (1979a) The trophic tissue of telotrophic ovarioles in polyphage Coleoptera. *Zoomorph.*, **93**, 33–50.

Büning, J. (1979b) The telotrophic nature of ovarioles of polyphage Coleoptera. *Zoomorph.*, **93**, 51–57.

Büning, J. (1979c) The telotrophic-meroistic ovary of Megaloptera I. The ontogenetic development. *J. Morphol.*, **162**, 37–66.

Büning, J. (1980) The ovary of *Rhaphidia flavipes* is telotrophic and of the *Sialis* type. *Zoomorph.*, **95**, 127–131.

Büning, J. (1983) Changes of RNA precursor pools (ATP, UTP) during embryogenesis of *Drosophila melanogaster* (Diptera) and *Bruchidius obtectus* (Coleoptera). *Wilhelm Roux Arch. Dev. Biol.*, **192**, 159–163.

Büning, J. (1985) Morphology, ultrastructure, and germ cell cluster formation in ovarioles of Aphids. *J. Morphol.*, **186**, 209–221.

Büning, J. (1990) Extra-chromosomale rRNA-Synthese und Binnenkörper in Oocyten der Siphonapteren. *Ver. Dtsch. Zool. Ges.*, **83**, 447–448.

Büning, J. (1993) Germ cell cluster formation in insect ovaries. *Int. J. Insect Morphol. Embryol.*, **22**, 237– 253.

Büning, J. and Sohst, S. (1988) The flea ovary: Ultrastructure and analysis of cell clusters. *Tissue Cell*, **20**, 783–795.

Büning, J. and Sohst, S. (1989) The ovary of *Hystrichopsylla*: Structure and pre-vitellogenic growth, in *Regulation of Insect Reproduction IV 1987*, (eds M. Tonner, T. Soldan, B. Bennettova), Academia Praha, pp. 113–124.

Büning, J. and Sohst, S. (1990) Ultrastructure and cluster formation in ovaries of bark lice, *Peripsocus phaeopterus* (Stephens) and *Stenopsocus stigmaticus* (Imhof and Labram) (Insecta, Psocoptera). *Int. J. Insect Morphol. Embryol.*, **19**, 227–241.

Burke, T., Waring, G.L., Popodi, E. and Minoo, P. (1987) Characterization and sequence of follicle cell genes selectively expressed during vitelline membrane formation in *Drosophila*. *Dev. Biol.*, **124**, 441–450.

Busson, D., Limbourg-Bouchon, B., Mariol, M.C., Preat, T. and Lamour-Isnard, C. (1988) Genetic analysis of viable and lethal *fused* mutants of *Drosophila melanogaster*. *Wilhelm Roux Arch. Dev. Biol.*, **197**, 221–230.

Cagan, R.L. and Ready, D.F. (1989) *Notch* is required for successive cell divisions in the developing *Drosophila* pupal retina. *Genes Dev.*, **3**, 1099–1112.

Callahan, P.S. and Cascio, T. (1963) Histology of the reproductive tract and transmission of the sperm in the corn earworm, *Heliothis zea*. *Ann. Ent. Soc. Am.*, **56**, 535–556.

Callan, H.G. (1986) *Lampbrush Chromosomes*. Springer Verlag, Berlin.

Callan, H.G. and Gall, J.G. (1991) Association of RNA with the B and C snurposomes of *Xenopus* oocyte nuclei. *Chromosoma (Berl.)*, **101**, 69–82.

Callan, H.G., Gall, J.G. and Murphy, C. (1991) Histone genes are located at the sphere loci of *Xenopus* lampbrush chromosomes. *Chromosoma (Berl.)*, **101**, 245–251.

Camenzind, R. (1966) Die Zytologie der bisexuellen und parthenogenetischen Fortpflanzung von *Heteropeza pygmaea* Winnerz, einer Gallmücke mit paedogenetischer Vermehrung. *Chromosoma*, **18**, 123–152.

Camenzind, R. (1971) The cytology of paedogenesis in the gall midge *Mycophila speyeri*. *Chromosoma*, **35**, 393– 402.

Campbell, A.G., Fessler, L.I., Salo, T. and Fessler, H.J. (1987) Papilin, A *Drosophila* proteoclycan-like sulfated glycoprotein from basement membranes. *J. Biol. Chem.*, **262**, 17605–17612.

Capco, D.G. and Jeffery, W.R. (1979) Origin and spatial distribution of maternal messenger RNA during oogenesis of an insect, *Oncopeltus fasciatus*. *J. Cell Sci.*, **39**, 63–76.

Cardoen, J., Schoofs, L., Broekaert, D. *et al*. (1986) Polyploidization and localisation of poly(A)+ RNA in the different cell types of the vitellogenic meroistic ovary of the fleshfly, *Sarcophaga bullata*. *Histochem.*, **85**, 305–312.

Cardoen, J., Watson, C., DeLoof, A. and Berry, S.J. (1990) Polyploidy in the nuclei of ovarian nurse and follicle cells of the silk moth, *Hyalophora cecropia*. *Arch. Insect Biochem. Physiol.*, **15**, 93–100.

Carlson, J.G. and Handel, M.A. (1988) Intercellular bridges and factors determining their patterns in the grasshopper testis. *J. Morphol.*, **196**, 173–185.

Carpenter, A.T.C. (1975) Electron microscopy of meiosis in *Drosophila melanogaster*. II. The recombination nodule – a recombination associated structure at pachytene? *Proc. Natl. Acad. Sci. USA*, **72**, 3186–3189.

Carpenter, A.T.C. (1979a) Synaptonemal complex and recombination nodules in wild type *Drosophila melanogaster* females. *Genetics*, **92**, 511–541.

Carpenter, A.T.C. (1979b) Synaptonemal complex and recombination nodules in recombination-defective mutants of *Drosophila melanogaster*. *Chromosoma (Berl.)*, **75**, 259–292.

Carpenter, A.T.C. (1981) EM autoradiographic evidence that DNA synthesis occurs at recombination nodules during meiosis in *Drosophila melanogaster* females. *Chromosoma*, **83**, 59–80.

Carpenter, A.T.C. (1982) Mismatch repair, gene conversion, and crossing-over in two recombination-defective mutants of *Drosophila melanogaster*. *Proc. Natl. Acad. Sci. USA*, **79**, 5961–5965.

Carpenter, A.T.C. (1989) Thought on recombination nodules, meiotic recombination, and chiasmata, in *Genetic Recombination*, (eds R. Kucherlapati and G.R. Smith), Am. Soc. Microbiol., Washington, DC.

Carpucino, M., Yin, C.M., Stoffolano, J.G. Jr., *et al.* (1992) F-actin distribution in the ovaries of previtellogenic and vitellogenic black blowflies, *Phormia regina* (Meigen) (Diptera: Calliphoridae). *Int. J. Insect Morphol. Embryol.*, **21**, 77–86.

Cassidy, J.D. and King, R.C. (1969) The dilatable ring canals of the ovarian cystocyte of *Habrobracon juglandis*. *Biol. Bull.*, **137**, 429–437.

Cassidy, J.D. and King, R.C. (1972) Ovarian development in *Habrobracon juglandis* (Ashmead) (Hymenoptera, Braconidae). I. The origin and differentiation of the oocyte–nurse cell complex. *Biol. Bull.*, **143**, 483–505.

Cave, M.D. (1975) Absence of ribosomal DNA amplification in the meroistic (telotrophic) ovary of the large milkweed bug *Oncopeltus fasciatus* (Dallas) (Hemiptera, Lygaeidae). *J. Cell Biol.*, **66**, 461–469.

Cave, M.D. (1976) Absence of rDNA amplification in the uninucleate oocyte of the cockroach *Blatella germanica* (Orthoptera, Blattidae). *J. Cell. Biol.*, **71**, 49–58.

Cave, M.D. (1982) Morphological manifestations of ribosomal DNA amplification during insect oogenesis, in *Insect Ultrastructure*, (eds R.C. King and H. Akai), Plenum Press, New York, Vol. 1, pp. 86–117.

Cecchini, J.P., Knibiehler, B., Mirre, C. and Le Parco, Y. (1987) Evidence for a type-IV-related collagen in *Drosophila melanogaster*. Evolutionary constancy of the carbocylterminal noncollagenous domain. *Eur. J. Biochem.*, **165**, 587–593.

Chaminade, M. and Laverdure, A. (1976) Étude quantitative des synthèses d'ARN nucléaires dans l'ovaire de *Tenebrio molitor* adulte. *C. R. Acad. Sci. Paris, Ser. D*, **282**, 1297–1300.

Champe, M.A. and Laird, C.D. (1989) Nucleotide sequence of a cDNA from the putative *ovarian tumor* locus of *Drosophila melanogaster*. *Nucl. Acid Res.*, **17** (8), 3304.

Chasan, R., Jin, Y. and Anderson, K.V. (1992) Activation of the *easter* zymogen is regulated by five other genes to define dorsal-ventral polarity in the *Drosophila* embryo. *Development*, **115**, 607–617.

Cheung, H.K., Serano, T.L. and Cohen, D.C. (1992) Evidence for a highly selective RNA transport system and its role in establishing the dorsoventral axis of the *Drosophila* egg. *Development*, **114**, 653–661.

Chi, H.C. and Hui, C.F. (1989) Primary structure of the *Drosophila* laminin B2 chain and comparison with human, mouse and *Drosophila* laminin B1 and B2 chains. *J. Biol. Chem.*, **264**, 1543–1550.

Choi, W.C. and Nagl, W. (1976) Electron microscopic study of the differentiation and development of trophocytes and oocytes in *Gerris najas* (Heteroptera). *Cytobios*, **17**, 47–62.

Choi, W.C. and Nagl, W. (1977a) Electron microscopic study on the development and functional morphology of the ovarian nutritive tissue in *Gerris najas* (Heteroptera). *Biol. Zbl.*, **96**, 513–522.

Choi, W.C. and Nagl, W. (1977b) Ribosome crystals in the oocyte of *Gerris najas* (Heteroptera). *Z. Naturforsch.*, **32c**, 137.

Choi, W.C. and Nagl, W. (1977c) Patterns of DNA and RNA synthesis during the development of ovarian nurse cells in *Gerris najas* (Heteroptera). *Dev. Biol.*, **61**, 262–272.

Chooi, W.Y. (1976) RNA transcription and ribosomal protein assembly in *Drosophila melanogaster*, in *Handbook of Genetics*, Vol. 5, (ed R.C. King), Plenum Press, New York, pp. 219–265.

Church, N.S. and Rempel, J.G. (1971) The embryology of *Lytta viridana* LeConte (Coleoptera, Meloidae). VI. The appendiculate, 72-h embryo. *Can. J. Zool.*, **49**, 1563–1570.

Claus, C. (1864) Beobachtungen über die Bildung des Insecteneies. *Z. Wiss. Zool.*, **19**, 42–54.

Clements, A.N. and Potter, S.A. (1967) The fine structure of the spermathecae and their ducts in the mosquito *Aedes aegypti*. *J. Insect Physiol.*, **13**, 1825–1836.

Cline, T.W. (1978) Two closely linked mutations in *Drosophila melanogaster* that are lethal to opposite sexes and interact with *daughterless*. *Genetics*, **90**, 683–698.

Cline, T.W. (1988) Evidence that *sisterless-a* and *sisterless-b* are two of several discrete 'numerator elements' of the X/A sex determination signal in *Drosophila* that switch *Sxl* between two alternative stable expression states. *Genetics*, **119**, 829–862.

Cline, T.W. (1989) The affairs of *daughterless* and the promiscuity of developmental regulators. *Cell*, **59**, 231–234.

Cmelic, S.H.W., Bursell, E. and Slack, E. (1969) Composition of the gut contents of the third instar tsetse larvae (*Glossina mortisans* Westwood). *Comp. Biochem. Physiol.*, **29**, 447–453.

Coluzzi, M. (1968) Cromosomi politenici delle cellule nutrici ovariche nel complesso *gambiae* del genere *Anopheles*. *Parasitologia*, **10**, 179–184.

Coluzzi, M., Cancrini, G. and DiDeco, M. (1970) The polytene chromosomes of *Anopheles superpictus* and the relationship with *Anopheles stephensi*. *Parasitologia*, **12**, 101–112.

Conti, L., Ciofi-Luzzatto, A. and Autuori, F. (1972) Ultrastructural and histochemical observations on spermathecal gland of *Dytiscus marginalis* L. (Coleoptera). *Z. Zellforsch.*, **131**, 85–96.

Cook, B.J. and Peterson, T. (1989) Ovarian muscularis of the stable fly *Stomoxys calcitrans*: Its structural, motile, and pharmacological properties. *Arch. Insect Biochem. Physiol.*, **12**, 15–30.

Cooley, L., Verheyen, E. and Ayers, K. (1992) *Chickadee* encodes a profilin required for intercellular cytoplasm transport during *Drosophila* oogenesis. *Cell*, **69**, 173–184.

Couchman, J.R. and King, P.E. (1979) Germarial structure and oogenesis in *Brevicoryne brassicae* (L.) (Hemiptera, Aphididae). *Int. J. Insect Morphol. Embryol.*, **8**, 1–10.

Counce, S.J. (1961) The analysis of insect embryogenesis. *Ann. Rev. Entomol.*, **6**, 295–312.

Counce, S.J. (1972) The causal analysis of insect embryogenesis, in *Developmental systems: insects*, Vol. 2, (eds S.J. Counce and C.H. Waddington), Academic Press, London, pp. 1–156.

Courtot, C., Fankhauser, C., Simanis, V. and Lehner, C.F. (1992) The *Drosophila* cdc25 homolog *twine* is required for meiosis. *Development*, **116**, 405–416.

Coyle-Thompson, C.A. and Banerjee, U. (1993) The *strawberry notch* gene functions with *Notch* in common developmental pathways. *Development*, **119**, 377–395.

Crema, R. (1973) Structure and determination of the ambiphasic ovary of *Acyrthosiphon pisum* (Homoptera, Aphididae). *Ent. Exp. Appl.*, **16**, 427–432.

Crippa, M. and Telfer, W.H. (1971) Hybridization analysis of DNA replication in nurse and follicle cells in *Cecropia* moth. *Biol. Bull. Mar. Biol. Lab., Woods Hole*, **141**, 384.

Cronmiller, C., Schedl, P. and Cline, T.W. (1988) Molecular characterization of

daughterless, a *Drosophila* sex determination gene with multiple roles in development. *Genes Dev.*, **2**, 1666–1676.
Crowson, R.A. (1981) *The Biology of Coleoptera*, Academic Press, New York.
Cruickshank, W.J. (1971) Follicle cell protein synthesis in moth oocytes. *J. Insect Physiol.*, **17**, 217–232.
Cummings, M.R. (1972) Formation of the vitelline membrane and chorion in developing oocytes of *Ephestia kühniella*. *Z. Zellforsch.*, **127**, 175–188.
Cummings, M.R. and King, R.C. (1969) The cytology of the vitellogenic stages of oogenesis in *Drosophila melanogaster*. I. General staging characteristics. *J. Morphol.*, **128**, 427–442.
Curtin, T.J. and Jones, J.C. (1961) The mechanism of ovulation and oviposition in *Aedes aegypti*. *Ann. Entomol. Soc. Am.*, **54**, 298–313.
Dabbs, C.K. and King, R.C. (1980) The differentiation of pseudonurse cells in the ovaries of *fs231* females of *Drosophila melanogaster* Meigen. *Int. J. Insect Morphol. Emryol.*, **9**, 215–229.
Daiber, M. (1905) Beiträge zur Kenntnis der Ovarien von *Bacillus rossii* Fabr. nebst einigen biologischen Bemerkungen. *Jenaische Zeitschr. Naturwiss.*, **39**, 176–202.
Dalby, B. and Glover, D.M. (1992) 3′ non-translated sequences in *Drosophila* cyclin B transcripts direct posterior pole accumulation late in oogenesis and peri-nuclear association in syncytial embryos. *Development*, **115**, 989–997.
Dallai, R. (1975) Fine structure of the spermatheca of *Apis mellifera*. *J. Insect Physiol.*, **21**, 89–109.
Dapples, C.C. and King, R.C. (1970) The development of the nucleolus of the ovarian nurse cell of *Drosophila melanogaster*. *Z. Zellforsch.*, **103**, 34–47.
Darlington, C.D. (1931) Meiosis in diploid and tetraploid *Primula sinensis*. *J. Genet.*, **24**, 65–96.
Das, N.K. (1976) Cytochemical and biochemical analysis of development of *Urechis* oocytes. *Am. Zool.*, **16**, 345–362.
Davenport, R. (1974) Synthesis and intercellular transport of ribosomal RNA in the ovary of the milkweed bug *Oncopeltus fasiatus*. *J. Insect Physiol.*, **20**, 1949–1956.
Davenport, R. (1976) Transport of ribosomal RNA into the oocytes of the milkweed bug, *Oncopeltus fasciatus*. *J. Insect Physiol.*, **22**, 925–926.
Davey, K.G. (1985) The female reproductive tract, in *Comprehensive Insect Physiology, Biochemistry and Pharmacology*, (eds G.A. Kerkut and L.I. Gilbert), Pergamon Press, Oxford, Vol. I, pp. 37–82.
Davey, K.G. and Webster, G.F. (1967) The structure and secretion of the spermatheca of *Rhodnius prolixus* Stal: a histochemical study. *Can. J. Zool.*, **45**, 653–657.
Davidson, E.H. (1986) *Gene Activity in Early Development*, Academic Press, New York.
Davidson, E.H. (1990) How embryos work: a comparative view of diverse modes of cell fate specification. *Development*, **108**, 365–389.
Davis, N.T. (1956) The morphology and functional anatomy of the male and female reproductive systems of *Cimex lectulatius* (Heteroptera, Cimicidae). *Ann. Ent. Soc. Am.*, **49**, 466–493.
Dawid, I.B. and Brown, D.D. (1970) The mitochondrial and ribosomal DNA components of oocytes of *Urechis caupo*. *Dev. Biol.*, **22**, 1–14.
Debaisieux, P. (1909) Les débuts de l'ovogenèse dans le *Dytiscus marginalis*. *La Cellule*, **25**, 207–229.
De Cicco, D.V., Wakimoto, B., Kalfayan, L., Levine, J. and Spradling, A.C. (1984) *Drosophila* chorion gene amplification, a model system for the study of chromosome replication. *Phil. Trans. R. Soc. Lond. B*, **307**, 239–245.
Deegener, P. (1928) Geschlechtsorgane, in *Handbuch der Entomologie*, (ed. C. Schröder), Fischer Verlag, Jena, Vol. I. pp. 466–533.

Degrugillier, M.E. (1977) Sperm–egg interaction in the house fly, *Musca domestica* L., release of micropylar cap substance by female accessory gland secretion. *Proc. A. Meet. North Central Br. Ent. Soc. Am.*, **32**, 41.

Degrugillier, M.E. (1985) *In vitro* release of house fly, *Musca domestica* L. (Diptera: Muscidae), acrosomal material after treatment with secretion of female accessory gland and micropyle cap substance. *Int. J. Insect Morphol. Embryol.*, **14**, 381–391.

Degrugillier, M.E. and Leopold, R.A. (1972) Abdominal peripheral nervous system of the adult female housefly and its role in mating behaviour and insemination. *Ann. Entomol. Soc. Am.*, **65**, 689–695.

Degrugillier, M.E. and Leopold, R.A. (1976) Ultrastructure of sperm penetration of house fly eggs. *J. Ultrastr. Res.*, **56**, 312–325.

De Loof, A., Geysen, J., Cardoen, J. and Verachtert, B. (1990) Comparative developmental physiology and molecular cytology of the polytrophic ovarian follicles of the blowfly *Sarcophaga bullata* and the fruitfly *Drosophila melanogaster*. *Comp. Biochem. Physiol.*, **96A**, 309–321.

De Loos, F., Dijkhof, R., Grond, C.J. and Hennig, W. (1984) Lampbrush loop specificity of transcript morphology in spermatocyte nuclei of *Drosophila hydei*. *EMBO J.*, **3**, 2845–2849.

Demina, N.I. (1989) Development of accessory nuclei in the oocytes of some Hymenoptera. *Doclady Akademii Nauk SSSR*, **304**, 1237–1240 (English translation).

Dennis, J.C. and Bradley, J.T. (1989) Ovarian follicle development during vitellogenesis in the house cricket *Acheta domesticus*. *J. Morphol.*, **200**, 185–198.

Diehl-Jones, W. and Huebner, E. (1992) Spatial and transcellular current patterns during oogenesis. *Dev. Biol.*, **153**, 302–311.

Diehl-Jones, W. and Huebner, E. (1993) Ionic basis of bioelectric currents during oogenesis in an insect. *Dev. Biol.*, **158**, 301–316.

Dittmann, F., Ehni, R. and Engels, W. (1981) Bioelectric aspects of the hemipteran telotrophic ovariole (*Dysdercus intermedius*). *Wilhelm Roux Arch.*, **190**, 221–225.

Dittmann, F., Hörner, R. and Engels, W. (1984) Endopolyploidization of tropharium nuclei during larval development and the first gonocycle in *Dysdercus intermedius* (Heteroptera). *Int. J. Invert. Reprod. Dev.* **7**, 279–290.

Dittmann, F., Trenczek, T. and Kleemann-Stumpf, I. (1985) Juvenile hormone-controlled vitellogenin cycles in *Dysdercus intermedius* (Heteroptera). *J. Insect Physiol.*, **31**, 729–739.

Dittmann, F. and Maier, E. (1987) Developmental patterns of DNA-accumulation and nuclear division in the follicle epithelium of the red cotton bug, *Dysdercus intermedius*. *J. Insect Physiol.*, **33**, 191–200.

Dittmann, F., Steinbrück, G. and Münz, A. (1990) Amplification of tropharium rDNA in the telotrophic ovariole of the bug, *Dysdercus intermedius*. *Invertebr. Reprod. Dev.*, **17**, 9–18.

Dittmann, F., Weiss, D.G. and Münz, A. (1987) Movement of mitochondria in the ovarian trophic cord of *Dysdercus intermedius* (Heteroptera) resembles nerve axonal transport. *Wilhelm Roux Arch. Dev. Biol.*, **196**, 407–413.

Dixon, L.K. and Ford, J.P. (1982a) Persistence of nonribosome bound 5S RNA in full grown oocytes of *Xenopus laevis*. *Dev. Biol.*, **93**, 474–477.

Dixon, L.K. and Ford, P.J. (1982b) Regulation of protein synthesis and accumulation during oogenesis in *Xenopus laevis*. *Dev. Biol.*, **93**, 478–497.

Dreyer, C., Wang, Y.H. and Hausen, P. (1985) Immunological relationship between oocyte nuclear proteins of *Xenopus laevis* and *X. borealis*. *Dev. Biol.*, **108**, 210–219.

Driever, W. and Nüsslein-Volhard, C. (1988) The *bicoid* protein determines position in the *Drosophila* embryo in a concentration-dependent manner. *Cell*, **54**, 95–104.

Driever, W. and Nüsslein-Volhard, C. (1989) The *bicoid* protein is a positive regulator of *hunchback* transcription in the early *Drosophila* embryo. *Nature,* **337**, 138–143.

Driever, W., Siegel, V. and Nüsslein-Volhard, C. (1990) Autonomous determination of anterior structures in the early *Drosophila* embryo by the maternal *bicoid* morphogen. *Development,* **109**, 811–820.

Duchateau, M.J. and Veltuis, H.H.W. (1989) Ovarian development and egg laying in workers of *Bombus terrestris*. *Entomol. Exp. Appl.,* **51**, 199–213.

Dufour, L. (1831) Recherches anatomiques et physiologiques sur les Hémiptères. *Mem. Prés. Div. Sav. Acad. R. Sci. Inst. France,* **4**, 129–461.

Dufour, L. (1841) Recherches anatomiques et physiologiques sur les orthoptères, les hymenoptères et les néuroptères. *Mem. Prés. Div. Sav. Acad. R. Sci. Inst. France,* **7**, 265–674.

Dunlap Pianka, H. (1974) Ctenophora, in *Reproduction in Marine Invertebrates,* (eds A.C. Giese and J.S. Pearse), Academic Press, New York, Vol. 1, pp. 201–265.

Duveau-Hagege, J., Lender, T. and Sauzin, M.J. (1963) Données sur l'ovogenèse de *Galleria mellonella*. *Bull. Soc. Zool. F.,* **88**, 422–431.

Elkins, T., Zinn, K., McAllister, L., Hoffmann, M.F. and Goodman, C.S. (1990) Genetic analysis of a *Drosophila* neural cell adhesion molecule: Interactions of fasciclin I and abelson tyrosine kinase mutations. *Cell,* **60**, 565–575.

Emeis, W. (1915) Über Eientwicklung bei den Cocciden. *Zool. Jb. Anat.* **39**, 27–78.

Engelmann, F. (1979) Insect vitellogenin: Identification, biosynthesis, and role in vitellogenesis. *Adv. Insect Physiol.,* **14**, 49–108.

Engels, W. (1966) Der zeitliche Ablauf von Protein-und Kohlenhydratsynthesen in der Oogenese bei *Apis mellifica* L. *Verh. Dtsch. Zool. Ges. Jena,* pp. 243–251.

Engels, W. (1968) Extraoocytäre Komponenten des Eiwachstums bei *Apis mellifica*. *Insectes Sociaux,* **3**, 271–288.

Engels, W. (1969) Geschwindigkeit des RNS-Transports im Einährverband der Dermapteren im Vergleich mit anderen Insekten meroistischen Ovartyps. *Verh. Dtsch. Zool. Ges., Zool. Anz.,* Suppl., **33**, 30–39.

Engels, W. (1971) Verteilungsmuster und Ultrastruktur des Glykogens in der Oocyte von *Musca domestica*. *Wilhelm Roux Arch.,* **167**, 294–298.

Engels, W. (1973) Das zeitliche und räumliche Muster der Dottereinlagerung in die Oocyte von *Apis mellifica*. *Z. Zellforsch.,* **142**, 409–430.

Ephrussi, A. and Lehmann, R. (1992) Induction of germ cell formation by *oscar*. *Nature,* **358**, 387–392.

Erickson, R.P. (1990) Post meiotic gene expression. *Trends Gen.,* **6**, 264–269.

Eschenberg, K.M. and Dunlap, H.L. (1966) The histology and histochemistry of oogenesis in the water strider, *Gerris remigis* Say. *J. Morphol.,* **118**, 297–316.

Euteneuer, J., Koonce, M.P., Pfister, K.K. and Schliwa, M. (1988) An ATPase with properties expected for the organelle motor of the giant amoeba, *Reticulomyxa*. *Nature,* **332**, 176–178.

Fargnoli, J. and Waring, G.L. (1982) Identification and cytogenetic localization of vitelline membrane proteins in *Drosophila melanogaster*. *Dev. Biol.,* **92**, 306–314.

Fargnoli, J. and Waring, G.L. (1984) Identification and genetic localization of vitelline membrane mRNAs of *Drosophila melanogaster*. *Dev. Biol.,* **105**, 41– 47.

Fasano, L. and Kerridge, S. (1988) Monitoring positional information during oogenesis in adult *Drosophila*. *Development,* **104**, 245–253.

Favard-Séréno, D. (1971) Cycles secrécrétoires successifs au cours de l'élaboration des enveloppes de l'ovocyte chez le Grillon (Insecte, Ortoptére). Rôle de l'appareil de Golgi. *J. Microscopie,* **11**, 401–424.

Favard-Séréno, D. and Durand, M. (1963) L'utilisation de nucléosides dans l'ovaire du grillon et ses variations au cours de l'ovogenèse I. Incorporation dans l'ARN. *Dev. Biol.,* **6**, 184–205.

Fawcett, D.W. (1971) Observations on cell differentiation and organelle continuity in spermatogenesis, in *Edinburgh Symposium on the Genetics of the Spermatozoon*, (eds E.A. Beatty and S. Gluecksohn-Waelsch), Bogtrykkeriet Forum, Copenhagen, pp. 37–68.

Feddersen I., Sander, K. and Schmidt, O. (1986) Virus-like particles with host-protein-like antigenic determinants protect an insect parasitoid from encapsulation. *Experientia*, **42**, 1278–1281.

Ferenz, H.J. (1978) Uptake of vitellogenin into developing oocytes of *Locusta migratoria*. *J. Insect Physiol.*, **24**, 273–278.

Ferenz, H.J. (1993) Yolk protein accumulation in *Locusta migratoria* (R and F) (Orthoptera: Acrididae) oocytes. *Int. J. Insect Morphol. Embryol.*, **22**, 295–314.

Fessler, L.I., Cambell, A.G., Duncan, K.G. and Fessler, J.H. (1987) *Drosophila* laminin, characterization and localization. *J. Cell Biol.*, **105**, 2383–2391.

Fiil, A. (1976) Oogenesis in the malaria mosquito *Anopheles gambiae*. *Cell Tissue Res.*, **167**, 23–35.

Fiil, A. (1978) Follicle cell bridges in the mosquito ovary: Syncytia formation and bridge morphology. *J. Cell Sci.*, **31**, 137–143.

Fiil, A. and Moens, P.B. (1972) The development, structure and function of modified synaptonemal complexes in mosquito oocytes. *Chromosoma (Berl.)*, **36**, 119–130.

Fleig, R., Gutzeit, H.O. and Engels, W. (1991) Structural organization of ovarian follicle cells in the cotton bug (*Dysdercus intermedius*) and the honeybee (*Apis mellifera*). *Cell Tissue Res.*, **265**, 297–305.

Fleig, R. and Sander, K. (1988) Honeybee morphogenesis, embryonic cell movements that shape the larval body. *Development*, **103**, 525–534.

Fleischmann, G. (1975) Anlage und embryonale Entwicklung fertiler Gonaden mit und ohne Polzellen bei *Pimpla turionellae* L. (Hymenoptera, Ichneumonidae). *Zool. Jb. Anat.*, **94**, 375–412.

Foldi, I. (1990a) Internal anatomy, in *Armoured Scale Insects, Their Biology, Natural Enemies and Control, Vol. A*, (ed. D. Rosen), Elsevier, Amsterdam, pp. 65–80.

Foldi, I. (1990b) Gametogenesis, in *Armoured Scale Insects, Their Biology, Natural Enemies and Control, Vol. A*, (ed. D. Rosen), Elsevier, Amsterdam, pp. 199–204.

Forlani, S., Ferrandon, D., Saget, O. and Mohier, E. (1993) A regulatory function for K10 in the establishment of dorsoventral polarity in the *Drosophila* egg and embryo. *Mech. Dev.*, **41**, 109–120.

Franchi, L.L. and Mandel, A.M. (1962) The ultrastructure of oogonia and oocytes in the foetal and neonatal rat. *Proc. R. Soc. Lond. B (Biol. Sci.)*, **157**, 99–114.

Francois, J. (1973) Sur la presence de fibroblastes caracteristiques chez le Thysanoure, *Termobia domestica*. *C.R. Acad. Sci. Paris*, **277**, 2505–2507.

Francois, J. (1978) The ultrastructure and histochemistry of the mesenteric connective tissue of the cockroach *Periplaneta americana* L. (Insecta Dictyoptera). *Cell Tissue Res.*, **189**, 91–107.

Franke, W.W., Scheer, U., Trendelenburg, M.F., Spring, H. and Zentgraf, H. (1978) Morphology of transcriptionally active chromatin. *Cold Spring Harbor Symp. Quant. Biol.*, **42**, 755–772.

Frey, A. and Gutzeit, H. (1986) Follicle cells and germ line cells both affect polarity in dicephalic chimeric follicles of *Drosophila*. *Wilhelm Roux Arch. Dev. Biol.*, **195**, 527–532.

Frey, A., Sander, K. and Gutzeit, H. (1984) The spatial arrangement of germ line cells in ovarian follicles of the mutant *dicephalic* in *Drosophila melanogaster*. *Wilhelm Roux Arch. Dev. Biol.*, **193**, 388–393.

Frohnhöfer, H.G. and Nüsslein-Volhard, C. (1986) Organization of anterior pattern in the *Drosophila* embryo by the maternal gene *bicoid*. *Nature*, **324**, 120–125.

Gaino, E. and Mazzini, M. (1989) Chorionic adhesive material of the egg of the mayfly *Habrophlebia eldae* (Ephemeroptera, Leptophlebiidae); morphology and synthesis. *Boll. Zool.*, **56**, 291–298.
Gaino, E. and Mazzini, M. (1990) Follicle cell activity in the ovarioles of *Habrophlebia eldae* (Ephemeroptera, Leptophlebiidae). *Trans. Am. Microsc. Soc.*, **109**, 300–310.
Gall, J.G. (1968) Differential synthesis of the genes for ribosomal RNA during amphibian oögenesis. *Proc. Natl. Acad. Sci. USA*, **60**, 553–560.
Gall, J.G. (1969) The genes for ribosomal RNA during oögenesis. *Genetics*, **61 (1)**, 121–132.
Gall, J.G. and Callan, H.G. (1989) The sphere organelle contains small nuclear ribonucleoproteins. *Proc. Natl. Acad. Sci. USA*, **86**, 6635–6639.
Gall, J.G., Macgregor, H.C. and Kidston, M.E. (1969) Gene amplification in the oocyte of Dytiscid water beetles. *Chromosoma (Berl.)*, **26**, 169–187.
Gall, J.G. and Rochaix, J.D. (1974) The amplified ribosomal DNA of Dytiscid beetles. *Proc. Natl. Acad. Sci. USA*, **71**, 1819–1823.
Gans, M., Audit, C. and Masson, M. (1975) Isolation and characterization of sex-linked female sterile mutants in *Drosophila melanogaster*. *Genetics*, **81**, 683–704.
Garzino, V. (1989) Expression of laminin and of laminin-related antigen during early development of *Drosophila melanogaster*. *Development*, **106**, 17–27.
Gateff, E. (1982) Gonial cell neoplasm of genetic origin affecting both sexes of *Drosophila melanogaster*, in *Embryonic Development*, Part B. *Cellular Aspects*, (eds M. Burger and R. Weber), Alan R. Liss, New York, pp. 621–632.
Gehring, W.J. (1976) Determination of primordial disc cells and the hypothesis of stepwise determination, in *Insect Development*, Vol. 8, (ed. P.A. Lawrence), Symp. R. Ent. Soc. London, pp. 99–108.
Gehring, W.J. (1987) Homeoboxes in the study of development. *Science*, **236**, 1245–1252.
Gerber, G.H., Church, N.S. and Rempel, J.G. (1971) The structure, formation, histochemistry, fate and functions of the spermatophore of *Lytta nuttalli* Say (Coleoptera, Meloidae). *Can. J. Zool.*, **49**, 1595–1610.
Gerber, G.H., Neill, G.B. and Westdal, P.H. (1978) The anatomy and histology of the internal reproductive organs of the sunflower beetle, *Zygogramma exclamationis* (Coleoptera, Chrysomelidae). *Can. J. Zool.*, **56**, 2542–2553.
Geysen, J., Cardoen, J., van Eynde, S., Geens, C. and de Loof, A. (1988) Cellular and molecular markers of anteroposterior and dorsoventral organisation in the vitellogenic follicles of adult *Sarcophaga bullata* (Diptera) and dorso-ventral orientation of follicles in the ovary. *Wilhelm Roux Arch. Dev. Biol.*, **197**, 101–109.
Giardina, A. (1901) Origine dell'oocite e delle cellule nutrici nei *Dytiscus*. *Int. Monatsschr. Anat. Phys.*, **18**, 417–479.
Gilbert, S.F. (1988) *Developmental Biology*, 2nd edn, Sinauer Associates, Inc., Sunderland, MA, USA.
Gilbert, W. and Dressler, D. (1968) DNA replication: The rolling circle model. *Cold Spring Harbour Symp.*, **33**, 473–484.
Gillott, C. (1988) Arthropoda – Insecta, in *Reproductive Biology of Invertebrates, Vol. III, Accessory Sex Glands*, (eds K.G. Adijodi and K. Adijodi), Wiley, Chichester, pp. 319–471.
Giorgi, F. (1978) Intercellular bridges in ovarian follicle cells of *Drosophila melanogaster*. *Cell Tissue Res.*, **186**, 413–422.
Giorgi, F., Bradley, J.T., Vignali, R. and Mazzini, M. (1989) An autoradiographic analysis of vitellogenin synthesis and secretion in the fat body of the stick insect *Bacillus rossius*. *Tissue Cell*, **21**, 517–524.
Giorgi, F., Yin, C.M. and Stoffolano, J.G. Jr. (1991) Permeability barriers and anionic

sites of the ovarian basal laminae in the black blowfly *Phormia regina* (Meigen) (Diptera, Calliphoridae). *Invert. Reprod. Dev.*, **19**, 37–44.

Giorgi, F., Cecchettini, A., Lucchesi, P. and Mazzini, M. (1993) Oocyte growth, follicle cell differentiation and vitellin processing in the stick insect, *Carausius morosus* Br. (Phasmatodea). *Int. J. Insect Morphol. Embryol.*, **22**, 271–293.

Giroux, C.N. (1992) Meiosis: components and process in nuclear differentiation. *Dev. Genet.*, **13**, 387–391.

Glätzer, K.H. (1979) Lengths of transcribed rDNA repeating units in spermatocytes of *Drosophila hydei*. Only genes without an intervening sequence are expressed. *Chromosoma (Berl.)*, **75**, 161–175.

Glätzer, K.H. (1980) Regular substructures within homologous transcripts of spread *Drosophila* germ cells. *Exp. Cell Res.*, **125**, 519–523.

Glätzer, K.H. and Meyer, G.F. (1981) Morphological aspect of the genetic activity in primary spermatocyte nuclei of *Drosophila hydei*. *Biol. Cell*, **41**, 165–172.

Gollin, S.M. and King, R.C. (1981) Studies of *fs (1621)*, a mutation producing ovarian tumours in *Drosophila melanogaster*. *Dev. Genet.*, **2**, 203–218.

Gönczy, P., Viswanathan, S. and DiNardo, S. (1992) Probing spermatogenesis in *Drosophila* with P-element enhancer detectors. *Development*, **114**, 89–98.

Gondim, K.C., Oliveira, P.L. and Masuda, H. (1989) Lipophorin and oogenesis in *Rhodnius prolixus*: Transfer of phospholipids. *J. Insect Physiol.*, **35**, 19–27.

Goode, S., Wright, D. and Mahowald, A.P. (1993) The neurogenic locus *brainiac* cooperates with *Drosophila* EGF receptor to establish the ovarian follicle and to determine its dorsal-ventral polarity. *Development*, **116**, 177– 192.

Goodwin, T.W. (1952) The biochemistry of locust pigmentation. *Biol. Rev.*, **27**, 447–460.

Goss, R.J. (1954) Ovarian development and oogenesis in the book louse, *Liposcelis divergens* Badonnel (Psocoptera, Liposcelidae). *Ann. Entomol. Soc. Am.*, **47**, 190–207.

Gottanka, J. (1992) *Vergleichende Untersuchungen zur Ultrastruktur der Germarien ursprünglicher Insekten*, Dr. Thesis, Institute of Zoology, University of Erlangen, Germany.

Gottanka, J. and Büning, J. (1990) Oocytes develop from interconnected cystocytes in the panoistic ovary of *Nemoura* sp. (Pictet) (Plecoptera, Nemouridae). *Int. J. Insect Morphol. Embryol.*, **19**, 219–225.

Gottanka, J. and Büning, J. (1993) Mayflies (Ephemeroptera), the most 'primitive' winged insects, have telotrophic meroistic ovaries. *Roux Arch. Dev. Biol.*, **203**, 18–27.

Govind, S. and Steward, R. (1991) Dorsoventral polarity in the *Drosophila* embryo: signal transduction and nuclear targeting. *Trends Genet.*, **7**, 119–125.

Granadino, B., Campuzano, S. and Sánchez, L. (1990) The *Drosophila melanogaster fl(2)d* gene is needed for the female-specific splicing of *Sex-lethal* RNA. *EMBO J.*, **8**, 2597–2602.

Granadino, B., San Juán, A.B. and Sánchez, L. (1991) The gene *fl(2)d* is required for various *Sxl*-controlled processes in *Drosophila* females. *Wilhelm Roux Arch. Dev. Biol.*, **200**, 172–176.

Grandi, G. (1988) Oogenesis in *Kalotermes flavicollis* (Fabr.) (Isoptera, Kalotermitidae). II – Prefollicular and follicular cell ultrastructure during oogenesis in female supplementary reproductives. *Boll. Zool.*, **55**, 279–292.

Grandi, G. (1990) Oogenesis in *Kalothermes flavicollis* (Fabr.) (Isoptera, Kalotermitidae) III. Choriogenesis and corpus luteum formation in female supplementary reproductives. *Boll. Zool.*, **57**, 97–107.

Grandi, G., Barbieri, R. and Colombo, G. (1988) Oogenesis in *Kalotermes flavicollis* (Fabr.) (Isoptera, Kalotermitidae). I. Differentiation and maturation of oocytes in female supplementary reproductives. *Boll. Zool.*, **55**, 103–122.

Grassi, B. (1887) Anatomia comparata di tisanuri e considerazioni generali sull organizzazione degli Insetti. *Atti Mem. Accad. Naz. Lincei*, **4**, 543–606.

Graziosi, C., Sakai, R.K., Romans, P., Miller, L.H. and Wellems, T.E. (1990) Method for *in situ* hybridization to polytene chromosomes from ovarian nurse cells of *Anopheles gambiae* Diptera Culicidae. *J. Med. Entomol.*, **27**, 905–912.

Grau, V. and Gutzeit, H.O. (1990) Asymmetrically distributed ecdysteroid-related antigens in follicles and young embryos of *Drosophila melanogaster. Wilhelm Roux Arch. Dev. Biol.*, **198**, 295–302.

Green, M.M. (1953) The *beadex* locus in *Drosophila melanogaster*, on the nature of the mutants Bx^r and Bx^1. *Genetics*, **38**, 91–105.

Gregorio, E.A., Secco, V.N.D.P., Toledo, L.A. and Lello, E. (1990a) Ultrastructure of the ovary of *Dermatobia hominis* (Diptera, Cuterebridae). I. Development during the 3rd larval instar. *Mem. Inst. Oswaldo Cruz. Rio de Janeiro*, **85**, 305–313.

Gregorio, E.A., Secco, V.N.D.P., Toledo, L.A. and Lello, E. (1990b) Ultrastructure of the ovary of *Dermatobia hominis* (Diptera, Cuterebridae). II. Origin of the tunica propria in ovarioles. *Mem. Inst. Oswaldo Cruz. Rio de Janeiro*, **85**, 315–320.

Grell, K.G. (1942) Der Genitalapparat von *Panorpa communis* L. *Zool. Jb. Anat. Ontog.*, **67**, 513–585.

Grenningloh, G., Rehm, E.J. and Goodman, C.S. (1991) Genetic analysis of growth cone guidance in *Drosophila*: Fasciclin II functions as a neural recognition molecule. *Cell*, **67**, 45–57.

Griffin-Shea, R., Thireos, G. and Kafatos, F.C. (1982) Organization of a cluster of four chorion genes in *Drosophila* and its relationship to developmental expression and amplification. *Dev. Biol.*, **91**, 325–336.

Griffith, C.M. and Lai-Fook, J. (1986) Corpus luteum formation and ovulation in the butterfly *Calpodes. Tissue Cell*, **18**, 783–792.

Grillot, J.P. and Raabe, M. (1989) Regulation of last reproduction steps in tsetse flies (*Glossina*, Diptera). *Invert. Reprod. Dev.*, **16**, 119–130.

Grimm, O. (1871) Beiträge zur Lehre von der Fortpflanzung und Entwickelung der Arthropoden. *Mem. Acad. Imp. Sci. St.- Petersbourg*, **17**, 1–20.

Grond, C.J., Rutten, R.G.J. and Hennig, W. (1984) Ultrastructure of the Y chromosomal lampbrush loops in primary spermatocytes of *Drosophila hydei. Chromosoma (Berl.)*, **89**, 85–95.

Grond, C.J., Siegmund, I. and Hennig, W. (1983) Visualization of a lampbrush loop-forming fertility gene in *Drosophila hydei. Chromosoma (Berl.)*, **88**, 50–56.

Gross, J. (1901) Untersuchungen über das Ovarium der Hemipteren, zugleich ein Beitrag zur Amitosenfrage. *Z. Wiss. Zool.*, **69**, 139–201.

Gross, J. (1903) Untersuchungen über die Histologie des Insectenovariums. *Zool. Jb. Abt. Morph.*, **18**, 71–186.

Gross, J. (1906) Untersuchungen über die Ovarien von Mallophagen und Pediculiden. *Zool. Jb. Abt. Anat.*, **22**, 347–386.

Grossniklaus, U., Bellen, H.J., Wilson, C. and Gehring, W.J. (1989) P-element mediated enhancer detection applied to the study of oogenesis in *Drosophila. Development*, **107**, 189–200.

Grünberg, K. (1903) Untersuchungen über die Keim-und Nährzellen in den Hoden und Ovarien der Lepidopteren. *Z. Wiss. Zool.*, **74**, 327–395.

Gruzova, M.N. (1962) The caryosphere formation in the oogenesis of *Parnorpa* (Mecoptera). *Citologiya (Moscow)*, **4**, 150–159.

Gruzova, M.N. (1966) Studies of the role of the nucleus in RNA and protein metabolism of the *Chrysopa perla* (Insecta) oocytes. *Citologiya (Moscow)*, **8**, 713–718.

Gruzova, M.N. (1967) Autoradiographical studies of protein metabolism in the nuclei of oocytes of *Panorpa communis* (Insecta, Mecoptera). *Citologiya (Moscow)*, **9**, 90–92.

Gruzova, M.N. (1979a) Nuclear structure in telotrophic ovarioles of the darkling beetles. I. Trophocytes of *Blaps lethifera* (Tenebrionidae, Polyphaga) (Light and electron microscope data). *Ontogenez*, **10**, 13–23.

Gruzova, M.N. (1979b) Nuclear structures in telotrophic ovarioles of the nocturnal ground beetles (Tenebrionidae, Polyphaga). III. The nucleus of oocytes. Electron microscope data. *Ontogenez*, **10**, 332–339.

Gruzova, M.N. (1989) Oogenesis of the raptorial gall midge *Aphidoletes aphidimyza* Rond. II. Ultrastructure of the trophic chamber. *Citologiya (Moscow)* **31**, 29–33.

Gruzova, M.N. and Batalova, F.M. (1979) Nuclear structures in telotrophic ovarioles of the nocturnal ground beetles (Tenebrionidae, Polyphaga). II.The nucleus of oocytes of *Blabs lethifera* and *Gnaptor spinimanus*. Light microsocope data. *Ontogenez*, **10**, 323–331.

Gruzova, M.N. and Batalova, F.M. (1990) Chromosomes E and S in the oogenesis of the gall midge, in *Nuclear Structure and Function, Proc. 11th Nuclear Workshop*, Plenum, New York, pp. 125–129.

Gruzova, M.N., Batalova, F.M. and Gukina, Y.I. (1989) Oogenesis of the raptorial gall midge *Aphidoletes aphidimyza* Rond. IV. A study of nucleic acid synthesis in ovarioles by light and electron microscopic autoradiography. *Citologiya (Moscow)*, **31**, 739–744.

Gruzova, M.N., Batalova, F.M. and Kozlova-Ermolayeva, L.V. (1987) Oogenesis of the raptorial gall midge *Aphidoletes aphidimyza*. *Citologiya (Moscow)*, **29**, 251–259.

Gruzova, M.N., Sehnal, F. and Gukina, Y.I. (1990) Peculiarities of polytrophic ovariole organization of the wax moth *Galleria melonella* (Lepidoptera). *Citologiya (Moscow)*, **32**, 677–680.

Gruzova, M.N., Zaichikova, Z.P. and Sokolov, I.I. (1972) Functional organization of the nucleus in the oogenesis of *Chrysopa perla* L. *Chromosoma (Berl.)*, **37**, 353–386.

Gu, X., Bei, Y. and Gao, C. (1992) Studies on the ontogeny of *Elenchinus japonicus* Esaki and Hashimoto (Strepsiptera): Egg production and embryo development. Proceedings XIX International Congress of Entomology, Beijing, China, 28 June–4 July 1992, ICE Council, Abstracts, p. 81.

Guddat, U., Bakken, A.H. and Pieler, T. (1990) Protein-mediated nuclear export of RNA, 5S rRNA containing small RNPs in *Xenopus* oocytes. *Cell*, **60**, 619–628.

Günther, J. (1971) Entwicklungsfähigkeit, Geschlechtsverhältnis und Fertilität von *Pimpla turionellae* L. (Hymenoptera, Ichneumonidae) nach Röntgenbestrahlung oder Abschnürung des Eihinterpols. *Zool. Jb. Anat.*, **88**, 1–46.

Günthert, T. (1910) Die Eibildung der Dytisciden. *Zool. Jb. Anat. Ontog.*, **30**, 301–372.

Gustavson, J.F. (1950) The origin and evolution of the genitalia of the Insecta. *Microentomology*, **15**, 35–67.

Gutzeit, H.O. (1985) Oosome formation during *in vitro* oogenesis in *Bradysia tritici* (syn. *Sciara ocellaris*). *Wilhelm Roux Arch. Dev. Biol.*, **194**, 404–410.

Gutzeit, H.O. (1986a) The role of microtubules in the differentiation of ovarian follicles during vitellogenesis. *Wilhelm Roux Arch. Dev. Biol.*, **195**, 173–181.

Gutzeit, H.O. (1986b) Transport of molecules and organelles in meroistic ovarioles of insects. *Differentiation*, **31**, 155–165.

Gutzeit, H.O. (1986c) The role of microfilaments in cytoplasmic streaming in *Drosophila* follicles. *J. Cell Sci.*, **80**, 159–169.

Gutzeit, H.O. (1990a) The microfilament pattern in the somatic follicle cells of mid-vitellogenic ovarian follicles of *Drosophila*. *Eur. J. Cell Biol.*, **53**, 349–356.

Gutzeit, H.O. (1990b) Die Entwicklung der Eizelle bei Insekten. *Biol. Zeit*, **20**, 33–41.

Gutzeit, H.O., Eberhardt, W. and Gratwohl, E. (1991) Laminin and basement membrane-associated microfilaments in wild-type and mutant *Drosophila* ovarian follicles. *J. Cell Sci.,* **100**, 781–788.
Gutzeit, H.O. and Haas-Assenbaum, A. (1991) The somatic envelopes around the germ-line cells of polytrophic insect follicles: Structural and functional aspects. *Tissue Cell,* **23**, 853–865.
Gutzeit, H.O. and Huebner, E. (1986) Comparison of microfilament pattern in nurse cells of different insects with polytrophic and telotrophic ovarioles. *J. Embryol. Exp. Morphol.,* **93**, 291–301.
Gutzeit, H.O. and Koppa, R. (1982) Time-lapse film analysis of cytoplasmic streaming during late oogenesis of *Drosophila. J. Embryol. Exp. Morphol.,* **67**, 101–111.
Gutzeit, H.O., Zissler, D. and Fleig, R. (1993) Oogenesis in the honeybee *Apis mellifera*: cytological observations on the formation and differentiation of previtellogenic ovarian follicles. *Wilhelm Roux Arch. Dev. Biol.,* **202,** 181–191.
Gutzeit, H.O., von Seydlitz-Kurzbach, E. and Neuschroer, R. (1993) How *Drosophila* (Diptera: Drosophilidae) follicles become spatially organized and obtain their ovoid shape. *Int. J. Insect Morphol. Embryol.,* **22**, 335–347.
Hackman, R.H. (1974) Chemistry of the insect cuticle, in *The Physiology of the Insecta,* 2nd edn, (ed. M. Rockstein), Academic Press, Vol. 6, pp. 215–270.
Hackstein, J.H.P. (1987) Spermatogenesis in *Drosophila,* in *Spermatogenesis Genetic Aspects,* (ed. W. Hennig), Springer, Berlin, Vol. 15, pp. 63–116.
Hackstein, J., Hennig, W. and Steinmann-Zwicky, M. (1987) Autosomal control of lampbrush-loop formation during spermatogenesis in *Drosophila hydei* by a gene also affecting somatic sex determination. *Wilhelm Roux Arch. Dev. Biol.,* **196**, 119–123.
Hackstein, J.H.P., Beck, H., Hochstenbach R. *et al.* (1990) Spermatogenesis in *Drosophila hydei,* a genetic survey. *Wilhelm Roux Arch. Dev. Biol.,* **199**, 251–280.
Hagan, H.R. (1951) *Embryology of the Viviparous Insects.* The Ronald Press Company, New York.
Hagan, H. (1954a) The reproductive system of the army ant queen. *Eciton* (Eciton) Part I. General anatomy. *Am. Mus. Novitates,* **1663**, 1–12.
Hagan, H. (1954b) The reproductive system of the army ant queen. *Eciton* (Eciton) Part II. Histology. *Am. Mus. Novitates,* **1664**, 1–17.
Hagan, H. (1954c) The reproductive system of the army ant queen. *Eciton* (Eciton) Part III. The oocyte cycle. *Am. Mus. Novitates,* **1665**, 1–20.
Haget, A. (1970) Premières données infrastructurales sur la surface et le périplasme de l'oeuf du Doriphore. *Leptinotarsa decemlineata* Say. *C. R. Acad. Sci. Paris,* **271D**, 1303–1306.
Haget, A. (1972) Characteristiques ultrastructurales du pôle postérieur et de l'oosome, dans l'oeuf jeune du Coléoptère *Leptinotarsa decemlineata* Say. *C. R. Acad. Sci. Paris,* **275D**, 2737–2740.
Halffter, G. and Lopez, Y. (1977) Development of the ovary and mating behaviour in *Phanaeus. Ann. Entomol. Soc. Am.,* **70**, 203–213.
Halkka, L. (1980) Accumulation of gene products in the previtellogenic oocytes of the dragonfly *Cordulia aenaea*. PhD thesis, Helsinki, Finland.
Halkka, L. and Halkka, O. (1968) RNA and protein in nucleolar structures of dragonfly oocytes. *Science,* **162**, 803–805.
Halkka, L. and Halkka, O. (1975) Accumulation of gene products in the oocytes of the dragonfly *Cordulia aenea* L. I. The nematosomes. *J. Cell. Sci.,* **19**, 103–115.
Hamon, C. and Folliot, R. (1969) Ultrastructure des cordons trophiques de l'ovaire de divers Homoptères Auchènorhynches. *C. R. Hebd. Séanc. Acad. Sci., Paris,* **268**, 577–580.
Happ, G.M. and Happ, C.M. (1975) Fine structure of the spermatheca of the mealworm beetle (*Tenebrio molitor* L.). *Cell Tissue Res.,* **162**, 253–269.

Hardenberg, J.D.F. (1929) Beiträge zur Kenntnis der Pupiparen. *Zool. Jahrb. Anat.*, **50**, 497–570.

Hardie, J. (1984) A hormonal basis for the photoperiod control of polymorphism in aphids. *CIBA Fnd. Symp.*, **104**, 240–258.

Hardy, R.W., Tokuyasu, K.T., Lindsley, D.L. and Garavito, M. (1979) The germinal proliferation center in the testis of *Drosophila melanogaster*. *J. Ultrastr. Res.*, **69**, 180–190.

Harris, S.E. and Forrest, H.S. (1967) RNA and DNA synthesis in developing eggs of the milkweed bug, *Oncopeltus fasciatus* (Dallas). *Science*, **156**, 1613–1615.

Hashimoto, C., Hudson, K.L. and Anderson, K.V. (1988) The *Toll* gene of *Drosophila*, required for dorsal-ventral embryonic polarity, appears to encode a transmembrane protein. *Cell*, **52**, 269–279.

Hashimoto, C., Gerttula, S. and Anderson, K.V. (1991) Plasma membrane localization of the *Toll* protein in the syncytial *Drosophila* embryo: Importance of transmembrane signaling for dorsal-ventral pattern formation. *Development*, **111**, 1021–1028.

Hasper, M. (1911) Zur Entwicklung der Geschlechtsorgane von *Chironomus*. *Zool. Jb. Anat.*, **31**, 543–612.

Hatakeyama, M., Sawa, M. and Oishi, K. (1990) Ovarian development and vitellogenesis in the sawfly, *Athalia rosae ruficornis* Jakovlev (Hymenoptera, Tenthredinidae). *Inv. Reprod. Dev.*, **17**, 237–245.

Hautscheck, E. (1962) Die Cytologie der Pädogenese und der Geschlechtsbestimmung einer heterogonen Gallmücke. *Chromosoma*, **13**, 163–182.

Hay, B., Ackerman, L., Barbel, S. *et al.* (1988a) Identification of a component of *Drosophila* polar granules. *Development*, **103**, 625–640.

Hay, B., Jan L.Y. and Jan, Y.N. (1988b) A protein component of *Drosophila* polar granules is encoded by *vasa* and has extensive sequence similarity to ATP-dependent helicases. *Cell*, **55**, 577–587.

Hay, B., Jan, L.Y. and Jan, Y.N. (1990) Localization of *vasa*, a component of *Drosophila* polar granules, in maternal-effect mutants that alter embryonic anteroposterior polarity. *Development*, **109**, 425–433.

Hazelrigg, T., Watkins, W.S., Marcey, D. *et al.* (1990) The *exuperantia* gene is required for *Drosophila* spermatogenesis as well as for anteroposterior polarity of the developing oocyte, and encodes overlapping sex-specific transcripts. *Genetics*, **126**, 607–617.

Heck, M. and Spradling, A.C. (1990) Multiple replication origins are used during *Drosophila* chorion gene amplification. *J. Cell Biol.*, **110**, 903–914.

Hedge, J. and Stephenson, E.C. (1993) Distribution of swallow protein in egg chambers and embryos of *Drosophila melanogaster*. *Development*, **119**, 457–470.

Hegner, R.W. (1908) Effects of removing germ-cell determinants from the eggs of some chrysomelid beetles. *Biol. Bull. Mar. Biol. Lab., Woods Hole*, **16**, 19–26.

Hegner, R.W. (1911) Experiments with chrysomelid beetles. III. Effect of killing parts of the eggs of *Leptinotarsa decemlineata*. *Biol. Bull. Mar. Biol. Lab., Woods Hole*, **20**, 237–251.

Hegner, R.W. (1914) Studies on germ cells. I. The history of the germ cells in insects with special reference to the 'Keimbahn' determinants. *J. Morphol.*, **25**, 375–509.

Hegner, R.W. (1915) Studies on germ cells. IV. Protoplasmic differentiation in the oocytes of certain Hymenoptera. *J. Morphol.*, **26**, 495–561.

Hegner, R.W. and Russel, C.P. (1916) Differential mitosis in the germ cell cycle of *Dineutes nigrior*. *Proc. Natl. Acad. Sci. USA*, **2**, 356–367.

Heino, T.I. (1989) Polytene chromosomes from ovarian pseudonurse cells of the *Drosophila melanogaster otu* mutant I. Photographic map of chromosome 3. *Chromosoma (Berl.)*, **97**, 363–373.

Heming, B.S. (1979) Origin and fate of germ cells in male and female embryos of *Haplothrips verbasci* (Osborn). *J. Morphol.*, **160**, 323–343.
Hennig, W. (1967) Untersuchungen zur Struktur und Funkton des Lampenbürsten-Y-Chromosoms in der Spermatogenese von *Drosophila. Chromosoma*, **22**, 294–357.
Hennig, W. (1987) The Y chromosomal lampbrush loops of *Drosophila*, in *Results and Problems in Cell Differentiation*, (eds W. Hennig and U. Scheer), Vol. 14, pp. 133–146.
Hennig, W. (1969) *Die Stammesgeschichte der Insekten*, Waldemar Kramer, Frankfurt.
Hennig, W. (1981) *Insect Phylogeny*. (Translator and editor A.C. Pont with revisionary notes by D. Schlee), Wiley and Sons, New York.
Hennig, W. (1982) *Phylogenetische Systematik*, Paul Parey, Berlin.
Henschen, W. (1928) Über die Entwicklung der Geschlechtsdrüsen von *Habrobracon juglandis* Ash. *Z. Morphol. Ökol. Tiere*, **13**, 144–178.
Herter, K. (1965) Vergleichende Beobachtungen und Betrachtungen über die Fortpflanzungsbiologie der Ohrwürmer. *Z. Naturforsch.*, **20b**, 365–375.
Hess, O. and Meyer, G.F. (1963) Chromosomal differentiation of the lampbrush type formed by the Y chromosome in *Drosophila hydei* and *D. neohydei*. *J. Cell Biol.*, **16**, 527–539.
Heymons, R. (1891) Die Entwicklung der weiblichen Geschlechtsorgane von *Phyllodromia* (*Blatta*) *germanica* L. *Z. Wiss Zool.*, **53**, 434–536.
Heymons, R. (1895) *Die Embryonalentwicklung von Dermapteren und Orthopteren*, Gustav Fischer, Jena.
Heymons, R. (1897) Entwicklungsgeschichtliche Untersuchungen an *Lepisma saccarina* L. *Z. Wiss. Zool.*, **62**, 583–632.
Heymons, R. (1930) Über die Morphologie des weiblichen Geschlechtsapparats der Gattung *Scarabaeus* L. *Z. Morphol. Ökol. Tiere*, **18**, 536–574.
Higgins, M.J., Walker, V.K., Holden, J.J. and White, B.N. (1984) Isolation of two *Drosophila melanogaster* genes abundantly expressed in the ovary during vitelline membrane synthesis. *Dev. Biol.*, **105**, 155–165.
Hilfiker, A. and Nöthiger, R. (1991) The temperature-sensitive mutation vir^{ts} *(virilizer)* identifies a new gene involved in sex determination of *Drosophila*. *Wilhelm Roux Arch. Dev. Biol.*, **200**, 240–248.
Hinton, (1958) The phylogeny of the panorpid orders. *Ann. Rev. Entomol.*, **3**, 181–206.
Hirschler, J. (1909) Die Embryonalentwicklung von *Donacia crassipes* L. *Z. Wiss. Zool.*, **92**, 627–744.
Hirschler, J. (1912) Embryologische Studien an Aphiden. *Z. Wiss. Zool.*, **100**, 393–446.
Hirschler, J. (1942) Organisation und Genese des Ei-Nährzellen-Verbandes im Ovarium von *Macrothylacia rubi* L. Lepidoptera. *Biol. Zbl.*, **62**, 555–577.
Hirschler, J. (1945) Gesetzmässigkeiten in den Ei-Nährzellverbänden. *Zool. Jb. Abt. Physiol.*, **60**, 141–236.
Holm, P.B. and Rasmussen, S.W. (1981) Chromosome pairing, crossing over, chiasma formation and disjunction as revealed by three dimensional reconstructions, in *International Cell Biology 1980–1981*, (ed. G. Schweiger), Springer Verlag, Berlin, pp. 194–204.
Honda, B.M. and Roeder, R.G. (1980) Association of a 5S gene transcription factor with 5S RNA and altered levels of the factor during cell differentiation. *Cell*, **22**, 119–126.
Hopkins, C.R. (1964) The histochemistry and fine structure of the accessory nuclei in *Bombus terrestris*. *Q. J. Microsc. Sci.*, **105**, 475–480.
Hourcade, D., Dressler, D. and Wolfson, J. (1973) The amplification of ribosomal RNA genes involves a rolling circle intermediate. *Proc. Natl. Acad. Sci. USA*, **70**, 2926–2930.
Huebner, E. (1980) Spermathecal ultrastructure of the insect *Rhodnius prolixus* Stal. *J. Morphol.*, **166**, 1–25.

Huebner, E. (1981a) Oocyte–follicle cell interaction during normal oogenesis and atresia in an insect. *J. Ultrastruct. Res.*, **74**, 95–104.

Huebner, E. (1981b) Nurse cell–oocyte interaction in the telotrophic ovarioles of an insect, *Rhodnius prolixus*. *Tissue Cell*, **13**, 105–125.

Huebner, E. (1984a) The ultrastructure and development of the telotrophic ovary, in *Insect Ultrastructure*, (eds R.C. King and H. Akai), Plenum Press, New York, Vol. 2, pp. 3–48.

Huebner, E. (1984b) Developmental cell interactions in female insect reproductive organs, in *Advances in Invertebrate Reproduction*, (eds W. Engels, W.H. Clark Jr., A. Fischer *et al.*), Elsevier, Amsterdam, Vol. 3, pp. 97–105.

Huebner, E. and Anderson, E. (1970) The effect of vinblastine sulfate on the microtubular organization of the ovary of *Rhodnius prolixus*. *J. Cell Biol.*, **46**, 191–198.

Huebner, E. and Anderson, E. (1972a) A cytological study of the ovary of *Rhodnius prolixus*. I. The ontogeny of the follicular epithelium. *J. Morphol.*, **136**, 459–494.

Huebner, E. and Anderson, E. (1972b) A cytological study of the ovary of *Rhodnius prolixus*. II. Oocyte differentiation. *J. Morphol.*, **137**, 385–416.

Huebner, E. and Anderson, E. (1972c) A cytological study of the ovary of *Rhodnius prolixus*. III. Cytoarchitecture and development of the trophic chamber. *J. Morphol.*, **138**, 1–40.

Huebner, E. and Diehl-Jones, W. (1993) Nurse cell–oocyte interaction in the telotrophic ovary. *Int. J. Insect Morphol. Embryol.*, **22**, 369–387.

Huebner, E. and Gutzeit, H. (1986) Nurse cell–oocyte interaction, a new F-actin mesh associated with the microtubule-rich core of an insect ovariole. *Tissue Cell*, **18**, 753–764.

Huebner, E. and Injeyan, H.S. (1980) Patency of the follicular epithelium in *Rhodnius prolixus*, a re-examination of the hormone response and technique refinement. *Can. J. Zool.*, **58**, 1617–1625.

Huebner, E. and Injeyan, H. (1981) Follicular modulation during oocyte development in an insect: Formation and modification of septate and gap junctions. *Dev. Biol.*, **83**, 101–113.

Huebner, E., Tobe, S.S. and Davey, K.G. (1975) Structural and functional dynamics of oogenesis in *Glossina austeni*: General features, previtellogenesis and nurse cells. *Tissue Cell*, **7**, 297–317.

Hughes, M. and Berry, S.J. (1970) The synthesis and secretion of ribosomes by nurse cells of *Antheraea polyphemus*. *Dev. Biol.*, **23**, 651–664.

Hughes-Schrader, S. (1925) Cytology of hermaphroditism in *Iceria purchasi* (Coccidae). *Z. Zellforsch.*, **2**, 264–292.

Hughes-Schrader, S. (1927) Origin and differentiation of the male and female germ cells in the hermaphrodite of *Icerya purchasi* (Coccidae). *Z. Zellforsch.*, **6**, 509–540.

Huxley, T.H. (1858) On the agamic reproduction and morphology of *Aphis*. *Trans. Linn. Soc. London*, **22**, 193–236.

Hwang, M.S. and Choi, W.C. (1989) Protein patterns of ovary and hemolymph in *Gerris paludum*. *Korean J. Zool.*, **32**, 337–350.

Hyams, J.S. and Stebbings, H. (1977) The distribution and function of microtubules in nutritive tubes. *Tissue Cell*, **9**, 539–547.

Hyams, J.S. and Stebbings, H. (1979a) The formation and breakdown of nutritive tubes. Massive microtubular organelles associated with cytoplasmic transport. *J. Ultrastr. Res.*, **68**, 46–57.

Hyams, J.S. and Stebbings, H. (1979b) The mechanism of microtubule associated cytoplasmic transport. Isolation and preliminary characterization of a microtubule transport system. *Cell Tissue Res.*, **196**, 103–116.

Illmensee, K. and Mahowald, A.P. (1974) Transplantation of posterior polar plasm in *Drosophila*. Induction of germ cells at the anterior pole of the egg, *Proc. Natl. Acad. Sci. USA*, **71**, 1016–1020.

Illmensee, K., Mahowald, A.P. and Loomis, M.R. (1976) The ontogeny of germ plasm during oogenesis in *Drosophila*. *Dev. Biol.*, **49**, 40–65.

Ingham, P.W. and Nakano, Y. (1990) Cell patterning and segment polarity genes in *Drosophila*. *Dev. Growth Differ.*, **32**, 563–574.

Ivanova-Kasas, O.M. (1965) Trophic connections between the maternal organism and the embryo in paedogenetic Diptera (Cecidomyidae). *Acta Biol. Acad. Sci. Hungaricae*, **16**, 1–24.

Jablonska, A., Szklarzewicz, T., Larink, O. and Bilinski, S. (1993) Structure of ovaries in two collembolans, *Allacma fusca* and *Arrhopalites coecus* (Hexapoda, Entognatha). *Folia Histochem. Cytobiol.*, **31**, 87–92.

Jäckle, H., Gaul, U., Nauber, U. *et al.* (1989) Musterbildung bei *Drosophila*. *Naturwiss.*, **76**, 512–517.

Jacob, J. and Sirlin, J.L. (1959) Cell function in the ovary of *Drosophila*. I. DNA classes in nurse cell nuclei as determed by autoradiography. *Chromosoma (Berl.)*, **10**, 210–228.

Jacobs-Lorena, M. (1980) Dosage of 5S and ribosomal genes during oogenesis of *Drosophila melanogaster*. *Dev. Biol.*, **80**, 134–145.

Jaffe, L.F. and Overall, R. (1985) Patterns of ionic current through *Drosophila* follicles and eggs. *Dev. Biol.*, **108**, 102–119.

Jaffe, L.F. and Woodruff, R.I. (1979) Large electrical currents traverse growing *Cecropia* follicles. *Proc. Natl. Acad. Sci. USA*, **76**, 1328–1332.

Jaglarz, M. (1988) Oogenesis in the common tiger beetle *Cicindela campestris* L. (Coleoptera). I. The structure of the ovary and organization of the egg chamber. *Acta Biol. Crac., Ser. Zool.*, **30**, 99–107.

Jaglarz, M. (1992) Peculiarities of the organization of egg chambers in carabid ground beetles and their phylogenetic implications. *Tissue Cell*, **24**, 397–409.

Jamieson, B.G.M. (1987) *The Ultrastructure and Phylogeny of Insect Spermatozoa.* Cambridge University Press, Cambridge.

Jawale, S.M. and Ranade, D.R. (1990) Morphology of the ovaries of *Sphaerodema (Diplonychus) rusticum* (Heteroptera, Belostomatidae). *J. Morphol.*, **205**, 183–192.

Jaworska, H. and Lima-de-Faria, A. (1973) Amplification of ribosomal DNA in *Acheta*. VI. Ultrastructure of two types of nucleolar components associated with ribosomal DNA. *Hereditas*, **74**, 169–186.

Jentsch, J. (1978) Bienengift – Zusammensetzung und Wirkung, *Biol. Zeit*, **8**, 75–81.

Johnson, E. and Berry, S.J. (1977) Cellular differentiation in a highly specialized insect secretory organ. *Differentiation*, **8**, 39–52.

Johnson, J.H. and King, R.C. (1972) Studies on *fes*, a mutation affecting cystocyte cytokinesis, in *Drosophila melanogaster*. *Biol. Bull. (Woods Hole)*, **134**, 525–547.

Jongens, T.A., Hay, B., Jan, L.Y. and Jan, Y.N. (1992) The *germ cell-less* gene product: a posteriorly localized component necessary for germ cell development in *Drosophila*. *Cell*, **70**, 569–584.

Jung, E. (1966) Untersuchungen am Ei des Speisebohnenkäfers *Bruchidius obtectus* Say. (Coleoptera). I. Mitteilung: Entwicklungsgeschichtliche Ergebnisse zu Kennzeichnung des Eitypus. *Z. Morph. Ökol. Tiere*, **56**, 444–480.

Junquera, P. (1983a) Polar plasm and pole cell formation in an insect egg developing with or without follicular epithelium, an ultrastructural study. *J. Exp. Zool.*, **227**, 441–452.

Junquera, P. (1983b) The role of the follicular epithelium in growing eggs of a dipteran insect during late oogenesis and cleavage. *J. Morphol.*, **178**, 303–312.

Jura, C. (1964a) Cytological and experimental observations on the origin and fate of the pole cells in *Drosophila virilis* Sturt. Part I. Cytological analysis. *Acta Biol. Crac., Ser. Zool.*, **7**, 59–73.

Jura, C. (1964b) Cytological and experimental observations on the origin and fate of the pole cells in *Drosophila virilis* Sturt. II. Experimental analysis. *Acta Biol. Crac., Ser. Zool.*, **7**, 89–103.

Jura, C. (1972) Development of apterygote insects, in *Developmental Systems: Insects*, (eds S.J. Counce and C.H. Addington), Academic Press, London, Vol. 1, pp. 49–94.

Jura, C. (1975) Ovaries structure in *Acerentomon dispar* Stach (Protura). *Acta Biol. Crac., Ser. Zool.*, **18**, 55–65.

Kahle, W. (1908) Die Paedogenesis der Cecidomyiden. *Zoologica, Stuttgart* **21**(55), 1–80.

Kafatos, F.C., Regier, J.C., Mazur, G.D. *et al.* (1977) The egg shell of insects: Differentiation-specific proteins and the control of their synthesis and accumulation during development, in *Results and Problems in Cell Differentiation*, (ed W. Beermann), Springer Verlag, Berlin, **8**, 45–145.

Kalfayan, L.J., Levine, J., Orr-Weaver, T.L. *et al.* (1985) Localization of sequences regulating *Drosophila* chorion gene amplification and expression. *Cold Spring Harbor Symp. Quant. Biol.*, **50**, 527–537.

Kalthoff, K. (1979) Analysis of a morphogenetic determinant in an insect embryo (*Smittia* sp., Chironomidae, Diptera), in *Determinant of Spatial Organization, 37th Symp. Soc. Dev. Biol.*, (eds S. Subtelny and I.R. Konigsberg), Academic Press, New York, pp. 97–126.

Kambysellis, M.P. (1993) Ultrastructural diversity in the egg chorion of Hawaiian *Drosophila* and *Scaptomyza*: Ecological and phylogenetic considerations. *Int. J. Insect Morphol. Embryol.*, **22**, 417–446.

Kappler, C., Golzené, F., Lagueux, M. *et al.* (1986) Role of the follicle cells and the oocyte in ecdysone biosynthesis and esterification in vitellogenic females of *Locusta migratoria*. *Int. J. Invertebr. Reprod. Dev.*, **9**, 17–34.

Kathirithamby, J., Carcupino, M. and Mazzini, M. (1990) Ovarian structure in the order Strepsiptera. *Frustula Entomologica*, **13**, 1–8.

Kaulenas, M.S. (1991) Insect accessory reproductive structures, in *Zoophysiology*, (eds S.D. Bradshaw, W. Burggren, H.C. Heller *et al.*), Vol. 31, Springer Verlag, Berlin.

Kawaguchi, E. (1928) Zytologische Untersuchungen am Seidenspinner und seinen Verwandten. I. Gametogenese von *Bombyx mori* L. und *Bombyx mandarina* M. und ihrer Bastarde. *Z. Zellforsch.*, **7**, 519–552.

Kay, M.A. and Jacobs-Lorena, M. (1985) Selective translational regulation of ribosomal protein gene expression during early development of *Drosophila melanogaster*. *Mol. Cell Biol.*, **35**, 3583–3592.

Kay, M.A. and Jacobs-Lorena, M. (1987) Developmental genetics of ribosome synthesis in *Drosophila*. *Trends Genet.*, **3**, 347–351.

Kern, P. (1912) Über die Fortpflanzung und Eibildung bei einigen Caraben. *Zool. Anz.*, **40**, 345–351.

Kessel, R.G. and Beams, H.W. (1969) Annulate lamellae and 'yolk nuclei' in oocytes of the dragonfly, *Libellula pulchella*. *J. Cell Biol.*, **42**, 185–201.

Kessel, R.G., Tung, N.H., Beams, H.W. and Lin, J.J-C. (1986) Is the nuclear envelope a 'generator' of membrane? Developmental sequences in cytomembrane elaboration. *Cell Tissue Res.*, **245**, 61–68.

Kidder, G.M. (1976) The ribosomal RNA cistrons in clam gametes. *Dev. Biol.*, **49**, 132–142.

Kindås-Mügge, I., Lane, C.D. and Kreil, G. (1974) Insect protein synthesis in frog

cells: The translation of honey bee promellittin messenger RNA in *Xenopus* oocytes. *J. Mol. Biol.*, **87**, 451–462.
King, J.S. and Mortimer, R.K. (1990) A polymerization model of chiasma interference and corresponding computer simulation. *Genetics*, **126**, 1127–1138.
King, P.E. and Fordy, M.R. (1970) The formation of 'accessory nuclei' in the developing oöcytes of the parasitoid hymenopterans *Ophion luteus* (L.) and *Apalanteles glomeratus* (L.). *Z. Zellforsch.*, **109**, 158–170.
King, P.E. and Richards, J. (1968) Accessory nuclei and annulate lamellae in hymenopteran oocytes. *Nature*, **218**, 488.
King, R.C. (1964) Further information concerning envelopes surrounding dipteran eggs. *Q. J. Microsc. Sci.*, **105**, 209–211.
King, R.C. (1969a) The hereditary ovarian tumors of *Drosophila melanogaster*. *Natl. Cancer Inst. Monograph*, **31**, 323–345.
King, R.C. (1969b) Control of oocyte formation by *female sterile (fes) Drosophila melanogaster*. *Natl. Cancer Inst. Monograph*, **31**, 347–349.
King, R.C. (1970) *Ovarian Development in Drosophila melanogaster*. Academic Press, New York.
King, R.C. (1979) Aberrant fusomes in the ovarian cystocytes of the *fs(1)231* mutant of *Drosophila melanogaster* Meigen. *Int. J. Insect Morphol. Embryol.*, **8**, 297–309.
King, R.C. and Aggarwal, S.K. (1965) Oogenesis in *Hyalophora cecropia*. *Growth*, **29**, 17–83.
King, R.C., Aggarwal, S.K. and Aggarwal, U. (1968) The development of the female *Drosophila* reproductive system. *J. Morphol.*, **124**, 143–166.
King, R.C. and Akai, H. (1971a) Spermatogenesis in *Bombyx mori*. I. The canal system joining sister spermatocytes. *J. Morphol.*, **134**, 47–55.
King, R.C. and Akai, H. (1971b) Spermatogenesis in *Bombyx mori*. II. The ultrastructure of synapsed bivalents. *J. Morphol.*, **134**, 181–194.
King, R.C., Bahns, M., Horowitz, R. and Larramendi, P. (1978) A mutation that affects female and male germ cells differentially in *Drosophila melanogaster* Meigen (Diptera, Drosophilidae). *Int. J. Insect Morphol. Embryol.*, **7**, 359–375.
King, R.C. and Büning, J. (1985) The origin and functioning of insect oocytes and nurse cells, in *Comprehensive Insect Physiology, Biochemistry and Pharmacology*, (eds G.A. Kerkut and L.I. Gilbert), Pergamon Press, Oxford, Vol. I, pp. 37–82.
King, R.C. and Burnett, R.G. (1959) An autoradiographic study of uptake of tritiated glycine, thymidine, and uridine by fruit fly ovaries. *Science*, **126**, 562.
King, R.C., Cassidy, J.D. and Rousset, A. (1982) The formation of clones of interconnected cells during gametogenesis in insects, in *Insect Ultrastructure*, (eds R.C. King and H. Akai), **1**, Plenum Press, New York, London, pp. 3–31.
King, R.C. and Koch, E.A. (1963) Studies on the ovarian follicle cells of *Drosophila*. *Q. J. Microsc. Sci.*, **104**, 297–320.
King, R.C., Mohler, D., Riley S.F. *et al.* (1986) Complementation between alleles at the *ovarian tumor* locus of *Drosophila melanogaster*. *Dev. Genet.*, **7**, 1–20.
King, R.C. and Riley, S.F. (1982) Ovarian pathologies generated by various alleles of the *otu* locus in *Drosophila melanogaster*. *Dev. Genet.*, **3**, 69–89.
King, R.C., Riley, S.F., Cassidy, J.D. *et al.* (1981) Giant polytene chromosomes from ovaries of a *Drosophila* mutant. *Science*, **212**, 441–443.
King, R.C., Rubinson, A.C. and Smith, R.F. (1956) Oogenesis in adult *Drosophila melanogaster*. *Growth*, **21**, 281–285.
King, R.C. and Storto, P.D. (1988) The role of the *otu* gene in *Drosophila* oogenesis. *BioEssays*, **8**, 18–24.
King, R.C. and Teasley, M. (1980) Insect oogenesis: Some generalities and their bearing on the ovarian development of fleas, in *Fleas*, (eds R. Traub and H. Starcke), A.A. Balkema, Rotterdam pp. 337–340.

Kinzelbach, R.K. (1971) Morphologische Befunde an Fächerflüglern und ihre phylogenetische Bedeutung (Strepsiptera, Insecta). *Zoologica*, **41**(119), 1–256.

Kisiel, E. (1987) Ultrastructure of oogonia and germarium of adult females of *Allacma fusca* (Collembola), in *Recent Advances in Insect Embryology in Japan and Poland*, (eds H. Ando and C. Jura), ISEBU, Tsukuba.

Klag, J. (1977) Light and electron microscopic study of embryonic development of gonads in *Thermobia domestica* (Packard) (Thysanura) with special reference to intercellular contacts. *Acta Biol. Cracov. Ser. Zool.*, **20**, 229–236.

Klag, J. (1978) Oogenesis in *Acerentomon gallicum* Jonescu (Protura). An ultrastructural analysis of the early previtellogenic stages. *Cell Tissue Res.*, **189**, 365–374.

Klag, J. (1982) Germ line of *Tetrodontophora bielanensis* (Insecta, Collembola). Ultrastructural study on the origin of primordial germ cells. *J. Embryol. Exp. Morphol.*, **72**, 183–195.

Klag, J. and Bilinski, S. (1984) Ovaries of Protura are meroistic: ultrastructural studies. *Cytobios*, **39**, 183–189.

Kleckner, N., Padmore, R. and Bishop, D.K. (1991) Meiotic chromosome metabolism: one view. In: *The Cell Cycle*, Cold Spring Harbor Symp. on Quart. Biol. LVI, Cold Spring Harbor Press, pp. 729–743.

Kleine-Schonnefeld, H. and Engels, W. (1981) Symmetrical pattern of follicle arrangement in the ovary of *Musca domestica*. *Zoomorph.*, **98**, 185–190.

Kloc, M. (1976) Extrachromosomal DNA and RNA synthesis in oocytes of *Creophilus maxillosus* (Staphylinidae, Coleoptera, Polyphpaga). *Experientia*, **32**, 375–377.

Kloc, M. and Matuszewski, B. (1977) Extrachromosomal DNA and the origin of oocytes in the telotrophic-meroistic ovary of *Creophilus maxillosus* (L.) (Staphylinidae, Coleoptera-Polyphaga). *Wilhelm Roux Arch. Dev. Biol.*, **183**, 351–368.

Knaben, N. (1931) Spermatogenese bei *Tischeria angusticolella* Dup. *Z. Zellforsch.*, **13**, 290–323.

Knaben, N. (1934) Oogenese bei *Tischeria angusticolella* Dup. *Z. Zellforsch.*, **21**, 604–625.

Knibiehler, B., Mirre, C., Cecchini, J.P. and Le Parco, Y. (1987) Haemocytes accumulate collagen transcripts during *Drosophila* metamorphosis. *Wilhelm Roux Arch. Dev. Biol.*, **196**, 243–247.

Knibiehler, B., Mirre, C. and Le Parco, Y. (1990) Collagen type IV of *Drosophila* is stockpiled in the growing oocyte and differentially located during early stages of embryogenesis. *Cell Diff. Dev.*, **30**, 147–157.

Knipple, D.C., Seifert, E., Rosenberg, U.B. *et al.* (1985) Spatial and temporal patterns of Krüppel gene expression in early *Drosophila* embryos. *Nature*, **317**, 40–44.

Kobayashi, S. and Okada, M. (1993) Restoration of pole-cell forming ability to u.v.-irradiated *Drosophila* embryos by injection of mitochondrial lrRNA. *Development*, **107**, 733–742.

Kobayashi, S., Kitamura, T., Sasaki, H. and Okada, M. (1993) Two types of pole cells are present in the *Drosophila* embryo, one with and one without splicing activity for the third P-element intron. *Development*, **117**, 885–893.

Koch, E.A. and King, R.C. (1964) Studies of the *fes* mutant of *Drosophila melanogaster*. *Growth*, **28**, 325–369.

Koch, E.A. and King, R.C. (1966) The origin and early differentiation of the egg chamber of *Drosophila melanogaster*. *J. Morphol.*, **119**, 283–304.

Koch, E.A. and King, R.C. (1969) Further studies on the ring canal system of the ovarian cystocytes of *Drosophila melanogaster*. *Z. Zellforsch.*, **102**, 129–152.

Koch, E.A., Smith, P.A. and King, R.C. (1967) The division and differentiation of *Drosophila* cystocytes. *J. Morphol.* **121**, 55–70.

Koch, M. (1929) Die postembryonale Entwicklung der weiblichen Genitaldrüsen und ihrer Ausführgänge von *Psychoda alternata* Say. *Z. Morphol. Ökol. Tiere,* **14**, 1–35.

Komma, D.J., Horne, A.S. and Endow, S.A. (1991) Separation of meiotic and mitotic effects of *claret non-disjunctional* on chromosome segregation in *Drosophila. EMBO J.,* **10**, 419–424.

Korschelt, E. (1886) Über die Entstehung und Bedeutung der verschiedenen Zellenelemente des Insektenovariums. *Z. Wiss. Zool.,* **43**, 537–720.

Kozhanova, N.I., Gruzova, M.N. and Khokhlov, G.N. (1976) Effects of allatectomy and juvenoid on the oogenesis of *Eurygaster integriceps* Put. *Citologiya,* **18**, 824–833.

Kozhanova, N.I. and Pasichnik, M.I. (1979) Differentiation of oocytes and nurse cells in telotrophic ovarioles of the beetle *Coccinella septempunctata. Citologiya,* **21**, 1145–1149.

Krause, G. (1939) Die Eitypen der Insekten. *Biol. Zentralbl.,* **59**, 495–536.

Kriger, F.L. and Davey, K.G. (1982) Ovarian motility in mated *Rhodnius prolixus* requires an intact cerebral neurosecretory system. *Gen. Comp. Endocrinol.,* **48**, 130–134.

Kristensen, N.P. (1975) The phylogey of hexapod 'orders'. A critical review of recent accounts. *Z. Zool. Syst. Evol.,* **13**, 1–44.

Kristensen, N.P. (1981) Phylogeny of insect orders. *Ann. Rev. Entomol.,* **26**, 135–157.

Krzysztofowicz, A. (1971) Histochemical and autoradiographic analysis of RNA synthesis in trophic cells of the female gonad in *Tetrodontophora bielanensis* (Waga) (Collembola). *Acta Biol. Crac., Ser. Zool.,* **14**, 299–305.

Krzysztofowicz, A. (1975) Histochemical and ultrastructural analysis of blocking nurse cells in the female gonad of *Tetrodontophora bielanensis* (Waga) (Collembola). *Acta Biol. Crac., Ser. Zool.,* **18**, 45–53.

Krzysztofowicz, A. (1977) Les études comparatives sur la morphologie des ovaires chez les Collemboles. *Rev. Écol. Biol. Sol.,* **14**, 81–90.

Krzysztofowicz, A. and Kisiel, E. (1989) Further studies on the morphogenesis of the first and second egg envelopes of *Tetrodontophora bielanensis* (Waga) (Collembola), in *3rd Int. Seminar on Apterygota,* (ed. R. Dallai), University of Siena, Siena.

Ksiazkiewicz, M. (1980) Ultrastructure of the trophic chamber and nutritive cord of *Aspidiotus hederae* (Homoptera, Coccoidea). *Cell Tissue Res.,* **213**, 149–157.

Ksiazkiewicz-Kapralska, M. (1985) Organisation of the trophic chamber of homopteran insects, Cercopidae, Cicadomorpha. *Cytobios,* **42**, 133–145.

Ksiazkiewicz-Kapralska, M. (1989a) Inner sheath cells in the larval ovary of *Aphrophora alni* (Insecta, Homoptera). *Zool. Jb. Anat.,* **118**, 1–7.

Ksiazkiewicz-Kapralska, M. (1989b) Organization of the fifth instar larval ovariole of *Aphrophora alni* (Insecta, Homoptera). *Zool. Jb. Anat.,* **118**, 473–481.

Ksiazkiewicz-Kapralska, M. (1991) Organization of the trophic chamber of homopteran insects. 2. Membracidae, Cicadomorpha. *Cytobios,* **66**, 113–119.

Kubrakiewicz, J. (1993) Organization of germ cell clusters in *Osmylus fulcicephalus* (Neuroptera). *Zeszyty Nauk UJ Prace Zool.,* in press, in Polish.

Kukalová-Peck, J. (1987) New carboniferous Diplura, Monura, and Thysanura, the hexapod ground plan, and the role of thoracic side lobes in the origin of wings (Insecta). *Can. J. Zool.,* **65**, 2327–2345.

Kulakosky, P.C. and Telfer, W.H. (1990) Lipophorin as a yolk precursor in *Hyalophora cecropia*: Uptake kinetics and competition with vitellogenin. *Arch. Insect Biochem. Physiol.,* **14**, 269–285.

Kunitskaya, N.T. (1960) On the reproductive organs in female fleas and determination of their physiological age. *Med. Parazit. (Moskva),* **29**, 688–701.

Kunkel, J.G. (1986) Dorsoventral currents are associated with vitellogenesis in cockroach ovarioles, in *Ionic Currents in Development*, (ed. R. Nuccitelli), A.R. Liss, New York, pp. 165–172.

Kunkel, J.G., Koenig, R., Kindle, H. and Lanzrein, B. (1986) Ion flux during vitellogenesis and patterning in insect oocytes, in *Advances in Invertebrate Reproduction 4*, (eds M. Porchet, J.C. Andries and A. Dhainot), Elsevier, Amsterdam, pp. 101–108.

Kunkel, J.G. and Nordin, J.H. (1985) Yolk proteins, in *Comprehensive Insect Physiology, Biochemistry and Pharmacology*, (eds G.A. Kerkut and L.I. Gilbert), Pergamon Press, Oxford, Vol. I, pp. 37–82.

Kunkel, J.G. and Pan, M.L. (1976) Selectivity of yolk protein uptake – comparison of vitellogenins of two insects. *J. Insect Physiol.*, **22**, 809–818.

Kunz, W. (1967a) Funktionsstrukturen im Oocytenkern von *Locusta migratoria*. *Chromosoma (Berl.)*, **20**, 332–370.

Kunz, W. (1967b) Lampenbürstenchromosomen und multiple Nucleolen bei Orthopteren. *Chromosoma (Berl.)*, **21**, 446–462.

Kunz, W. (1969) Die Entstehung multipler Oocytennukleolen aus akzessorischen DNS-Körpern bei *Gryllus domesticus*. *Chromosoma (Berl.)*, **26**, 41–75.

Kunz, W. (1983) How genes are amplified in germ cells and somatic cells, in *Advances in Invertebrate Reproduction 3*, (eds W. Engels, W.H. Clark, A. Fischer *et al.*), Elsevier, Amsterdam, pp. 65–71.

Kunz, W., Trepte, H.H. and Bier, K. (1970) On the function of the germ line chromosomes in the oogenesis of *Wachtliella persicariae* (Cecidomycidae). *Chromosoma (Berl.)*, **30**, 180–192.

Kurihara, M. and Matsuzaki, M. (1989) Ovarian structure and oogenesis in the rice water weevil *Lissorphoptrus oryzophilus* (Coleoptera, Curculionidae). *Jpn. J. Ent.*, **57**, 666–680.

Labhart, P. and Reeder, R.H. (1984) Enhancer-like properties of the 60/81 bp elements in the ribosomal gene spacer of *Xenopus laevis*. *Cell*, **37**, 285–289.

Labhart, P. and Reeder, R.H. (1986) Characterization of three sites of RNA 3′ end formation in the *Xenopus* ribosomal gene spacer. *Cell*, **54**, 431–443.

Landois, L. (1866) Anatomie des Hundeflohs (*Pulex canis* Dugès) mit Berücksichtigung verwandter Arten und Geschlechter. *Nova Acta Acad. Leop. Carol.*, **33**, 1–66.

Landois, L. (1869) Anatomie der Bettwanze (*Cimex lectularius*) mit Berücksichtigung verwandter Hemipterengeschlechter. *Z. wiss. Zool.*, **19**, 206–233.

Langley, P.A. (1977) Physiology of tsetse flies (*Glossina* spp.) (Diptera, Glossinidae), a review. *Bull. Ent. Res.*, **67**, 523–574.

Langley, P.A., Bursell, E., Kabayo, J. *et al.* (1981) Haemolymph lipid transport from fat body to uterine gland in pregnant females of *Glossina mortisans*. *Insect Biochem.*, **11**, 225–231.

Langley, P.A. and Pimley, R.W. (1974) Utilization of U-^{14}C aminoacids or U-^{14}C protein by adult *Glossina mortisans* during *in utero* development of larva. *J. Insect Physiol.*, **20**, 2157–2170.

Lanot, R., Sellam, M.O., Bucher, B. and Thiebold, J.J. (1990) Ecdysone-controlled meiotic reinitiation in oocytes of *Locusta migratoria* involves a decrease in cAMP levels. *Insect Biochem.*, **20**, 639–644.

Lanot, R., Thiebold, J., Costet-Corio, M.F., Benveniste, P. and Hoffmann, J.A. (1988) Further experimental evidence for the involvement of ecdysone in the control of meiotic reinitiation in oocytes of *Locusta migratoria* (Insecta, Orthoptera). *Dev. Biol.*, **126**, 212–214.

Lanot, R., Thiebold, J., Lagueux, M., Golzené, F. and Hoffmann, J.A. (1987) Involvement of ecdysone in the control of meiotic reinitiation in oocytes of *Locusta migratoria* (Insecta, Orthoptera). *Dev. Biol.*, **121**, 174–181.

Larink, O. and Bilinski, S. (1989) Fine structure of the egg envelopes of one proturan and two collembolan genera (Apterygota). *Int. J. Insect Morphol. Embryol.*, **18**, 39–45.

Lasko, P.F. and Ashburner, M. (1990) Posterior localization of *vasa* protein correlates with, but is not sufficient for, pole cell development. *Genes Dev.*, **4**, 905–921.

Lass, M. (1905) Beiträge zur Kenntnis des histologisch-anatomischen Baues des weiblichen Hundeflohes (*Pulex canis* Dugès s. *Pulex serraticeps* Taschenberg). *Z. Wiss. Zool.*, **79**, 73–131.

Lauge, G. (1985) Sex determination: Genetic and epigenetic factors, in *Comprehensive Insect Physiology, Biochemistry and Pharmacology*, (eds G.A. Kerkut and L.I. Gilbert), Pergamon Press, Oxford, Vol. I, pp. 295–318.

Laurie, D.A. (1980) Inter-individual variation in chiasma frequency and chiasma distribution. A study of *Corthippus brunneus* and man. Ph.D. thesis, University of Birmingham, England.

Lawrence, P.A. (1988) Background to *bicoid*. *Cell*, **54**, 1–2.

Lawrence, P.O. (1988) Intraspecific compensation among first instars of the parasitic wasp *Biosteres longicautatus*. *Oecologia (Berl.)*, **74**, 607–611.

Lawrence, P.O. and Akin, D. (1990) Virus-like particles from the poison glands of the parasitic wasp *Biosteres longicaudatus* (Hymenoptera, Braconidae). *Can. J. Zool.*, **68**, 539–546.

Leather, S.R. and Wellings, P.W. (1981) Ovariole number and fecundity in aphids. *Entomol. Exp. Appl.*, **30**, 128–133.

Leclerck-Smekens, M. (1978) Reproduction d'*Euproctis chrysorrhea* L. (Lépidoptère Lymantridae). II. Aspects cytologiques et rôle de la cellule apicale (cellule de Verson) des testicules. *Acad. R. Belg. Cl. Sci.*, **64**, 318–322.

Lehmann, R. (1992) Germ-plasm formation and germ-cell determination in *Drosophila*. *Curr. Opin. Genet. Dev.*, **2**, 543–549.

Lehmann, R. and Nüsslein-Volhard, C. (1991) The maternal gene *nanos* has a central role in posterior pattern formation in the *Drosophila* embryo. *Development*, **112**, 679–691.

Leopold, R.A. (1981) Accessory reproductive gland involvement with the sperm–egg interaction in muscoid flies, in *Advances in Invertebrate Reproduction*, (eds W.H. Clark Jr. and T.S. Adams), Elsevier, New York, Vol. 2, pp. 253–270.

Leopold, R.A. and Degrullier, M.E. (1973) Sperm penetration of house fly eggs, evidence for involvement of a female accessory secretion. *Science*, **181**, 555–557.

Leopold, R.A., Meola, S. and Degrugillier, M.E. (1978) The egg fertilization site within the house fly, *Musca domestica* (L.) (Diptera, Muscidae). *Int. J. Insect Morphol. Embryol.*, **7**, 111–120.

Leydig, F. (1850) Einige Bemerkungen über die Entwicklung der Blattläuse. *Zeitschr. Wiss. Zool.*, **2**, 62–66.

Leydig, F. (1867) Der Eierstock und die Samentasche der Insecten. Zugleich ein Beitrag zur Lehre von der Befruchtung. *Nova Acta Acad. Leop.-Carol.*, **33**, 1–88.

Lifschytz, E. (1987) The developmental program of spermiogenesis in *Drosophila*: A genetic analysis. *Int. Rev. Cytol.*, **109**, 211–257.

Lima-de-Faria, A. (1974) The three-dimensional organization of the amplified ribosomal DNA complex in *Acheta*. *Proc. Natl. Acad. Sci. USA*, **71**, 4778–4782.

Lima-de-Faria, A. and Moses, M.J. (1966) Ultrastructure and cytochemistry of metabolic DNA in *Tipula*. *J. Cell Biol.*, **30**, 177–192.

Lima-de-Faria, A., Nilsson, B., Cave, D., Puga, A. and Jaworska, H.J. (1968) Tritium labelling and cytochemistry of extra DNA in *Acheta*. *Chromosoma (Berl.)*, **25**, 1–20.

Lin, H. and Spradling, A.C. (1993) Germline stem cell division and egg chamber development in transplanted *Drosophila* germaria. *Dev. Biol.*, **159**, 140–152.

Lindsley, D.L. and Tokuyashu, K.T. (1980) Spermatogenesis, in *The Genetics and Biology of Drosophila*, (eds M. Ashburner and T.R.F. Wright), Academic Press, London, Vol. 2d, pp. 225–294.

Lococo, D. and Huebner, E. (1980a) The ultrastructure of the female accessory gland, the cement gland, in the insect *Rhodnius prolixus*. *Tissue Cell*, **12**, 557–580.

Lococo, D. and Huebner, E. (1980b) The development of the female accessory gland in the insect *Rhodnius prolixus*. *Tissue Cell*, **12**, 795–813.

Lohs-Shardin, M. (1982) Dicephalic – a *Drosophila* mutant affecting polarity in follicle organization and embryonic patterning. *Wilhelm Roux Arch. Dev. Biol.*, **191**, 28–36.

Lubbock, J. (1859) On the ova and pseudova of insects. *Phil. Trans. R. Soc. London*, **149**, 341–369.

Lucchesi, J.C. and Manning, J.E. (1987) Gene dosage compensation in *Drosophila melanogaster*. *Adv. Genet.*, **24**, 371–429.

Lunstrum, G.P., Bächinger, H.P., Fessler, L.I. *et al.* (1988) *Drosophila* basement membrane procollagen IVI. Protein characterization and distribution. *J. Biol. Chem.*, **263**, 18318–18327.

Lusis, O. (1963) The histology and histochemistry of development and resorption in the terminal oocytes of the desert locust, *Schistocerca gregarina*. *Q. J. Microsc. Sci.*, **104**, 57–68.

Lutz, D.A. and Huebner, E. (1980) Development and cellular differentiation of an insect telotrophic ovary (*Rhodnius prolixus*). *Tissue Cell*, **12**, 773–794.

Lutz, D.A. and Huebner, E. (1981) Development of nurse cell–oocyte interactions in the insect telotrophic ovary (*Rhodnius prolixus*). *Tissue Cell*, **13**, 321–335.

Ma, W.K. and Ramaswamy, S.B. (1987) Histological changes during ovarian maturation in the tarnished plant bug, *Lygus lineolaris* (Palisot de Beauvois) (Hemiptera-Miridae). *Int. J. Insect Morphol. Embryol.*, **16**, 309–322.

Macdonald, P.M. and Struhl, G. (1988) *Cis*-acting sequences responsible for anterior localisation of *bicoid* mRNA in *Drosophila* embryos. *Nature*, **336**, 595–598.

MacDonald, P.M., Kerr, K., Smith, J.L. and Leask, A. (1993) RNA regulatory element BLE1 directs the early steps of *bicoid* mRNA localization. *Development*, **118**, 1233–1243.

Macgregor, H.C. and Stebbings, H. (1970) A massive system of microtubules associated with cytoplasmic movement in telotrophic ovarioles. *J. Cell. Sci.*, **6**, 431–449.

Macgregor, H.C. (1980) Recent developments in the study of lampbrush chromosomes. *Heredity*, **44**, 3–35.

Machida, J. (1926) The development of the ovary in the silkworm *(Bombyx mori)*. *J. Coll. Agric.*, **7**, 293–351.

Madhavan, M.M. (1973) The dual origin of the nurse chamber in the ovarioles of the gall midge, *Heteropeza pygmaea*. *Wilhelm Roux Arch.*, **173**, 164–168.

Magakyan, Y.A., Makaryan, S.R., Petrossyan, A.V. *et al.* (1976) The oogenesis of the Ararat cochineal (*Porphyrophora hamelli*, Brandt). I. Histomorphological characteristics of the formation of follicular apparatus and the development of oocytes. *Cytologia*, **18**, 932–936.

Mahowald, P.A. (1962) Fine structure of pole cells and polar granules in *Drosophila melanogaster*. *J. Exp. Zool.*, **151**, 201–215.

Mahowald, P.A. (1971a) Origin and continuity of polar granules, in *Origin and Continuity of Cell Organelles*, (eds J. Reinert and H. Ursprung), Springer Verlag, Berlin, pp. 158–169.

Mahowald, P.A. (1971b) Polar granules in *Drosophila*. IV. Cytochemical studies showing loss of RNA from polar granules during early stages of embryogenesis. *J. Exp. Zool.*, **175**, 345–352.

Mahowald, P.A. (1975) Ultrastructural changes in the germ plasm during the life cycle of *Miastor* (Cecidomycidae). *Wilhelm Roux Arch.*, **175**, 223–240.

Mahowald, A.P., Caulton, J.H., Edwards, M.K. and Floyd, A.D. (1979) Loss of centrioles and polyploidization in follicle cells of *Drosophila melanogaster. Exp. Cell Res.*, **118**, 404–410.

Mahowald, A.P. and Kambysellis, M.P. (1980) Oogenesis, in *Biology of Drosophila*, (eds M. Ashburner and T.R.F. Wright), Academic Press, London, **2d**, pp. 141–223.

Mahowald, A.P. and Stoiber, D. (1974) The origin of the nurse chamber in ovaries of *Miastor* (Diptera, Cecidomyidae). *Wilhelm Roux Arch. Dev. Biol.*, **176**, 159–166.

Mahowald, A.P. and Strassheim, J.M. (1970) Intercellular migration of centrioles in the germarium of *Drosophila melanogaster*. An electron microscopic study. *J. Cell Biol.*, **45**, 306–320.

Mahowald, A.P. and Tiefert, M. (1970) Fine structural changes in the *Drosophila* oocyte nucleus during a short period of RNA synthesis. An autoradiographic and ultrastructural study of RNA synthesis in the oocyte nucleus of *Drosophila. Wilhelm Roux Arch.*, **165**, 8–25.

Malpighi, H. (1669) *Dissertatio epistolica de bombyce.* London.

Mandelbaum, I. (1980) Intercellular bridges and the fusome in the germ cells of the *Cecropia* moth. *J. Morphol.*, **166**, 37–50.

Manseau, L.Y. and Schüpbach, T. (1989) The egg came first, of course! *Trends Genet.*, **5**, 400–405.

Margaritis, L.H. (1980) Chorionogenesis in *Drosophila melanogaster*, a model system for the study of cellular differentiation. Monograph, Athens University.

Margaritis, L.H. (1985) Structure and physiology of the egg shell, in *Comprehensive Insect Physiology, Biochemistry and Pharmacology*, (eds G.A. Kerkut and L.I. Gilbert), Pergamon Press, Oxford, Vol. I, pp. 37–82.

Mariani, M. and Bruno-Smiraglia, C. (1957) Presenza di cromosomi politenici nelle cellule vitellogene degli ovari di femmine adulte di *Anopheles. Boll. Soc. Ent. Ital.*, **87**, 23–25.

Mariol, M.C., Preat, T. and Limbourg-Bouchon, B. (1987) Molecular cloning of *fused*, a gene required for normal segmentation in the *Drosophila* embryo. *Mol. Cell Biol.*, **7**, 3244–3251.

Marsh, J.L. and Wieschaus, E. (1978) Is sex determination in germline and soma controlled by separate genetic mechanisms? *Nature*, **272**, 249–251.

Marshall, W.S. (1907a) Contributions towards the embryology and anatomy of *Polistes pallipes* II. The early history of the cellular elements of the ovary. *Z. Wiss. Zool.*, **86**, 173–213.

Marshall, W.S. (1907b) The early history of the cellular elements of the ovary of a Phryganid, *Platiphylax designatus* Walk. *Z. Wiss. Zool.*, **86**, 214–237.

Marshall, W.S. (1908) A study of the follicular epithelium from the ovary of the walking-stick, *Diapheromera femorata. Arch. Zellforsch.*, **3**, 627–643.

Martinez-Arias, A. (1985) The development of *fused* embryos of *Drosophila melanogaster. J. Embryol. Exp. Morphol.*, **87**, 99–114.

Martoja, R. (1977) Les organes génitaux femmelles, in *Traite de Zoologie*, (ed P.P. Grasse), Masson, Paris, Vol. VIII-5-A, pp. 1–123.

Masner, P. (1968) The inductors of differentiation of prefollicular tissue and the follicular epithelium in ovarioles of *Pyrrhocoris apterus* (Heteroptera). *J. Embryol. Exp. Morphol.*, **20**, 1–13.

Matsuda, R. (1976) *Morphology and Evolution of the Insect Abdomen*, Pergamon Press, Oxford.

Matsuzaki, M. (1971) Electron microscopic studies on the oogenesis of dragonfly and cricket with special reference to the panoistic ovaries. *Dev. Growth Differ.*, **13**, 379–398.

Matsuzaki, M. (1972) Oogenesis in adult net-spinning caddisfly, *Parasthenopsyche sauteri* (Trichoptera, Stenopsychidae), as revealed by electron microscopic observation. *Sci. Rep. Fukushima Univ.*, **22**, 27– 40.

Matsuzaki, M. (1973) Oogenesis in the springtail, *Tomocerus minutus* Tullberg (Collembola, Tomoceridae). *Int. J. Insect Morphol. Embryol.*, **2**, 335–349.

Matsuzaki, M. (1975) Ultrastructural changes in developing oocytes, nurse cells, and follicular cells during oogenesis in the telotrophic ovarioles of *Bothrogonia japonica* Ishihara (Homoptera, Tettigellidae). *Kontyu*, **43**, 75–90.

Matsuzaki, M. (1978) Developing ovarian follicles during early oogenesis in the adult green lacewing, *Chrysopa septempunctata* Wesmael (Neuroptera, Chrysopidae). *Annot. Zool. Jpn.*, **51**, 222–235.

Matsuzaki, M. (1987) Oogenesis in *Ascalaphus ramburi* and *Protidricerus japonicus* (Ascalaphidae; Neuroptera). *Proc. Arthropod. Embryol. Soc. Jpn.*, **23**, 23.

Matsuzaki, M. and Ando, H. (1977) Ovarian structures of the adult alderfly, *Sialis mitsuhashi* Okamoto (Megaloptera, Sialidae). *Int. J. Insect Morphol. Embryol.*, **6**, 17–29.

Matsuzaki, M., Ando, H. and Neumann-Visscher, S. (1979) Fine structure of oocyte and follicular cells during oogenesis in *Galloisiana nipponensis* (Caudell and King) (Grylloblattodea, Grylloblattidae). *Int. J. Insect Morphol. Embryol.*, **8**, 257–263.

Matsuzaki, M. and Enomoto, T. (1990) Oogenesis in a spongilla fly, *Sisyla nikkoana* Navas (Neuroptera, Sisylidae). *Proc. Arthropod. Embryol. Soc. Jpn*, **25**, 9–11.

Matsuzaki, M., Sasaki, A. and Komuro, M. (1985) Panoistic ovarioles of the dobsonfly *Protohermes grandis* (Megaloptera, Corydalidae), in *Recent Advances in Insect Embryology in Japan*, (eds H. Ando and K. Miya), ISEBU, Tsukuba pp. 13–23.

Matsuzaki, M. and Satoh, M. (1986) Preliminary report on the early oogenesis in *Eumantispa harmandi* (Neuroptera, Mantispidae). *Proc. Arthropod. Embryol. Soc. Jpn*, **22**, 1–3.

Mattox, W. and Baker, B.S. (1991) Autoregulation of the splicing of transcripts from the *transformer-2* gene of *Drosophila*. *Genes Dev.*, **5**, 786–796.

Mattox, W., Palmer, M.J. and Baker, B.S. (1990) Alternative splicing of the sex determination gene *transformer-2* is sex-specific in the germ line but not in the soma. *Genes Dev.*, **4**, 789–805.

Matuszewski, B. (1968) Regulation of growth of nurse nuclei in the development of egg follicles in Cecidomyiidae (Diptera). *Chromosoma (Berl.)*, **25**, 429–469.

Matuszewski, B., Ciechomski, K., Nurkowska and Kloc, M. (1985) The linear clusters of oogonial cells in the development of telotrophic ovarioles in polyphage coleoptera. *Wilhelm Roux Arch. Dev. Biol.*, **194**, 462–469.

Matuszewski, B. and Hoser, P. (1975) Gene amplification and its effect on the structure and function of the oocyte nucleus in the whirligig beetle *Gyrinus natator* (Gyrinidae, Coleoptera, Adephaga). *Experientia*, **31**, 431–432.

Matuszewski, B., Hoser, P., Hoser, G. and Michalak, M. (1977) Organization of the oocyte nucleus in silphid beetles (Silphidae, Coleoptera-Polyphaga). *Experientia*, **33**, 883–885.

Matuszewski, B. and Kloc, M. (1976) Gene amplification in oocytes of the rove beetle *Creophilus maxillosus* (Staphylinidae, Coleoptera, Polyphaga). *Experientia*, **32**, 34–36.

Mays, U. (1969) Der Stofftransport von den Nährzellen zur Oocyte in einem telotrophen Insektenovar und seine strukturellen Grundlagen; Untersuchungen an der Feuerwanze, *Pyrrhocoris apterus* L. Thesis (Dr.), University of Münster, Germany.

Mays, U. (1972) Stofftransport im Ovar von *Pyrrhocoris apterus* L. Autoradiographische Untersuchungen zum Stofftransport von den Nährzellen zur Oocyte der Feuerwanze, *Pyrrhocoris apterus* L.(Heteroptera). *Z. Zellforsch.*, **123**, 395–410.

Maziarski, S. (1913) Sur la persistance des résidus fusoriaux fendant les nombreuses générations cellulaires au cours de l'ovogénèse de *Vespa vulgaris* L. *Arch. Zellforsch.*, **10**, 507–532.

Mazur, G.D., Regier, J.C. and Kafatos, F.C. (1982) Order and defects in the silkmoth chorion, a biological analogue of a cholesteric liquid crystal, in *Insect Ultrastructure*, (eds R.C. King and H. Akai), Plenum Press, New York, Vol. 1, pp. 150–185.

Mazzini, M. and Gaino, E. (1988) Oogenesis of the mayfly *Habrophlebia eldae*: Synthesis of vitelline and chorionic envelopes. *Gamete Res.*, **21**, 439–450.

Mazzini, M. and Giorgi, F. (1985) The follicle cell–oocyte interaction in ovarian follicles of the stick insect *Bacillus rossius* (Rossi) (Insecta, Phasmatodea). *J. Morphol.*, **185**, 37–49.

Mazzini, M., Carpucino, M. and Fausto, A.M. (1993) Egg chorion architecture in stick insect (Phasmatodea). *Int. J. Insect Morphol. Embryol.*, **22**, 391–415.

McGill, C. (1905) The behavior of the nucleoli during oogenesis of the dragonfly with special reference to synapsis. *Zool. Jb. Abt. Morph.*, **23**, 207–226.

McKearin, D.M. and Spradling, A.C. (1990) *bag-of-marbles*, a *Drosophila* gene required to initiate both male and female gametogenesis. *Genes Dev.*, **4**, 2242–2251.

McKnight, S.L. and Miller, O.L. (1976) Ultrastructural patterns of RNA synthesis during early embryogenesis in *Drosophila melanogaster*. *Cell*, **8**, 305–319.

Meng, C. (1968) Strukturwandel und histochemische Befunde insbesondere am Oosom während der Oogenese und nach Ablage des Eies von *Pimpla turionellae* L. (Hymenoptera, Ichneumonidae). *Wilhelm Roux Arch. Dev. Biol.*, **161**, 162–208.

Meng, C. (1970) Autoradiographische Untersuchungen am Oosom in der Oocyte von *Pimpla turionellae* L. (Hymenoptera). *Wilhelm Roux Arch.*, **165**, 35–52.

Menon, M. (1969) Structure of the apical cells of the testis of the tenebrionid beetles, *Tenebrio molitor* and *Zophobas rugipes*. *J. Morphol.*, **127**, 409–430.

Mercer, E.H. and Brunet, P.C.J. (1959) The electron microscopy of the left colleterial gland of the cockroach. *J. Biophys. Biochem. Cytol.*, **5**, 257–261.

Meola, S.M., Mollenhauer, H.H. and Thompson, J.M. (1977) Cytoplasmic bridges within the follicular epithelium of the ovarioles of two Diptera, *Aedes aegypti* and *Stomoxys calcitrans*. *J. Morphol.*, **153**, 81–86.

Metschnikoff, E. (1866) Embryologische Studien an Insecten. *Z. Wiss. Zool.*, **16**, 389–500.

Mével-Ninio, M., Terracol, R. and Kafatos F.C. (1991) The *ovo* gene of *Drosophila* encodes a zinc finger protein required for female germ line development. *EMBO J.*, **10**, 2259–2266.

Meyer, G.F. (1961) Interzelluläre Brücken (Fusome) im Hoden und im Ei-Nährzellverband von *Drosophila melanogaster*. *Z. Zellforsch.*, **54**, 238–251.

Meyer, G.F. (1963) Die Funktionsstrukturen des Y-Chromosoms in den Spermatocytenkernen von *Drosophila hydei, D. neohydei, D. repleta,* und einigen anderen *Drosophila*-Arten. *Chromosoma*, **14**, 207–255.

Meyer, G.F., Hess, O. and Beermann, W. (1961) Phasenspezifische Funktionsstrukturen in den Spermatocytenkernen von *Drosophila melanogaster* und ihre Abhängigkeit vom Y-Chromosom. *Chromosoma*, **12**, 676–716.

Meyer, G.F., Sokoloff, S., Wolf, B.E. and Brand, B. (1979) Accessory nuclei (nuclear membrane balloons) in the oocytes of the dipteran *Phryne*. *Chromosoma (Berl.)*, **75**, 89–99.

Miller, O.L. and Beatty, B.R. (1969) Visualization of nucleolar genes. *Science*, **164**, 955–957.

Mindrinos, M.N., Scherer, L.J., Garcini, F.J. *et al.* (1986) Isolation and chromosomal location of putative vitelline membrane genes in *Drosophila melanogaster*. *EMBO J.*, **4**, 147–153.

Miya, K. (1959) Studies on the embryonic development of the gonad in the silkworm, *Bombyx mori* L. Part II. Formation of the gonad. *J. Fac. Agric. Iwate Univ.*, **4**, 126–151.

Miya, K. (1975) Ultrastructural changes of embryonic cells during organogenesis in the silkworm, *Bombyx mori*. I. The gonad. *J. Fac. Agric., Iwate Univ.*, **12**, 329–364.

Miya, K. (1978) Electron microscope studies on the early embryonic development of the silkworm, *Bombyx mori*. I. Architecture of the newly laid egg and the changes by sperm entry. *J. Fac. Agric., Iwate Univ.*, **14**, 11–34.

Miya, K., Kurihara, M. and Tanimura, J. (1969) Electron microscope studies in the oogenesis of the silkworm, *Bombyx mori*. I. (Fine structure of oocyte and nurse cells in the early developmental stages). *J. Fac. Agric., Iwate Univ.*, **9**, 221–237.

Moens, P.B. (1968) The structure and function of the synaptinemal complex in *Lilium longiflorum* sporocytes. *Chromosoma (Berl.)*, **23**, 418–452.

Moens, P.B. (1978) Ultrastructural studies of chiasma distribution. *Ann. Rev. Genet.*, **12**, 433–450.

Moens, P.B. and Hugenholtz, A.D. (1975) The arrangement of germ cells in the rat seminiferous tubule: An electron-micrososocope study. *J. Cell Sci.*, **19**, 487–507.

Montell, D.J., Keshishian, H. and Spradling, A.C. (1991) Laser ablation studies of the role of the *Drosophila* oocyte nucleus in pattern formation. *Science*, **254**, 290–293.

Moses, M.J. (1958) The relation between the axial complex of meiotic prophase chromosomes and chromosome pairing in a salamander (*Plethodon cinereus*). *J. Biophys. Biochem. Cytol.*, **4**, 633–638.

Moses, M.J. (1968) Synaptinemal complex. *Ann. Rev. Genet.* **2**, 363–412.

Mougey, E.B., O'Reilly, M., Osheim, Y. *et al.* (1993) The terminal balls characteristic of eukaryotic rRNA transcription units in chromatin spreads are rRNA processing complexes. *Genes Dev.*, **7**, 1609–1619.

Muckenthaler, F.A. (1964) Autoradiographic study of nucleic acid synthesis during spermatogenesis in the grasshopper *Melanoplus differentialis*. *Exp. Cell Res.*, **35**, 531–597.

Muhlach, W.L. and Schwalm, F.E. (1977) Utilization of uridine in developing ovarioles of the kelp fly *Coelopa frigida* (Diptera). *J. Insect Physiol.*, **23**, 931–937.

Müller, J. (1825) Über die Entwicklung der Eier im Eierstock bei den Gespenstheuschrecken und eine neu entdeckte Verbindung des Rückengefässes mit den Eierstöcken bei den Insecten. *Nova Acta Acad. Leop.-Carol.*, **12**, 555–672.

Muller, H.J. (1916) The mechanism of crossing over. IV. *Am. Nat.*, **50**, 421–434.

Mulligan, P.K., Mohler, J.D. and Kalfayan, L.J. (1988) Molecular localization and developmental expression of the *otu* locus of *Drosophila melanogaster*. *Mol. Cell. Biol.*, **8**, 1481–1488.

Münz, A. (1988) Heterologous gap junctions between oocytes and follicle cells of an insect, *Dysdercus intermedius*, and their role as ion current pathways. *Cell Tissue Res.*, **252**, 147–155.

Münz, A. and Dittmann, F. (1987) Voltage gradients and microtubules both involved in intercellular protein and mitochondria transport in the telotrophic ovariole of *Dysdercus intermedius*. *Wilhelm Roux Arch. Dev. Biol.*, **196**, 391–396.

Nagoshi, R.N. and Baker, B.S. (1990) Regulation of sex-specific RNA splicing at the *Drosophila doublesex* gene. *cis*-acting mutations in exon sequences alter sex-specific RNA splicing patterns. *Genes Dev.*, **4**, 89–97.

Netzel, H. (1965) Über die Übereinstimmung von Polarität und Bilateralsymmetrie in Embryo, Ei, Oocyte und Oocytenkern bei *Gryllus domesticus* L. *Roux Arch. Entwicklungsmech.*, **156**, 88–95.

Niki, Y. (1986) Germline autonomous sterility of the P-M dysgenic hybrids and their applications to germ line transfers in *Drosophila melanogaster*. *Dev. Biol.*, **113**, 255–258.

Noirot, C. and Quennedey, A. (1974) Fine structure of insect epidermal glands. *Ann. Rev. Entomol.*, **19**, 61–80.

Nöthiger, R., Jonglez, M., Leuthold, M., Meier-Gerschwiler, P. and Weber, T. (1989) Sex determination in the germ line of *Drosophila* depends on genetic signals and inductive somatic factors. *Development*, **107**, 505–518.

Nüsslein-Volhard, C. (1991) Determination of the embryonic axes of *Drosophila*. *Dev. Suppl.*, **1**, 1–10.

Nüsslein-Volhard, C., Frohnhöfer, H.G. and Lehmann, R. (1987) Determination of antero-posterior polarity in *Drosophila*. *Science*, **238**, 1675–1681.

Nüsslein-Volhard, C. and Wieschaus, E. (1980) Mutations affecting segment number and polarity in *Drosophila*. *Nature*, **287**, 795–801.

O'Kane, C. and Gehring, W.J. (1987) Detection *in situ* of genomic regulatory elements in *Drosophila*. *Proc. Natl. Acad. Sci. USA*, **84**, 9123–9127.

Oliver, B., Perrimon, N. and Mahowald, A.P. (1988) Genetic evidence that the *sans fille* locus is involved in *Drosophila* sex determination. *Genetics*, **120**, 159–171.

Orlando, E. and Crema, R. (1968) Growth of follicular cells in aphids. A cytophotometric and autoradiographic study. *Experientia*, **24**, 1038–1039.

Orr-Weaver, T.L. (1991) *Drosophila* chorion genes, cracking the eggshell's secrets. *BioEssays*, **13**, 97–105.

Osheim, Y.N. and Miller, O.L. (1983) Novel amplification and transcriptional activity of chorion genes in *Drosophila melanogaster* follicle cells. *Cell*, **33**, 543–553.

Osheim, Y.N., Miller, O.L. and Beyer, A.L. (1985) RNP particles at splice junction sequences on *Drosophila* chorion transcripts. *Cell*, **43**, 143–151.

Osheim, Y.N., Miller, O.L. and Beyer, A.L. (1988) Visualization of *Drosophila melanogaster* chorion genes undergoing amplification. *Mol. Cell. Biol.*, **8**, 2811–2821.

Osman, S.E. (1978) Die Wirkung der Sekrete der weiblichen Genitalanhangsdrüsen von *Pimpla turionella* L. (Hym.) auf die Hämocyten und die Einkapselungsreaktion von Wirtspuppen. *Z. Parasitenk.*, **57**, 89–100.

Paglia, L.M., Berry, S.J. and Kastern, W.H. (1976) Messenger RNA synthesis, transport, and storage in silkworm ovarian follicles. *Dev. Biol.*, **51**, 173–181.

Painter, T.S. and Reindorp, E.C. (1939) Endomitosis in the nurse cells of the ovary of *Drosophila melanogaster*. *Chromosoma*, **1**, 276–283.

Pal, S.G. and Gosh, D. (1982) Electron microscopic study of the spermatheca of *Gesonula punctifrons* Acrididae Orthoptera. *Proc. Indian Acad. Sci. (Anim. Sci.)*, **91**, 99–112.

Palevody, C. (1971) L'ovogenése chez les Collemboles. Structure et évolution de l'ovaire. *C. R. Acad. Sci. Paris*, **272**, 3165–3168.

Palevody, C. (1972) Présence de noyaux accessoires dans l'ovaocyte du Collembole *Folsomia candida* Willem (Insecte, Aptérygote).*C. R. Acad. Sci. Paris*, **274**, 3258–3261.

Palevody, C. (1973) Différentiation du noyaux de l'ovocyte au cours de la prophase méiotique chez les Collemboles (Insectes, Aptérygotes). Etude ultrastructurale. *C. R. Acad. Sci. Paris*, **277**, 2201–2204.

Palevody, C. (1976) L'ovogenése chez les Collemboles Isotomides. Cytologie et Approche Physiologique, Ph.D. thesis 697, Univ. Toulouse.

Palmén, J.A. (1884) *Über paarige Ausführungsgänge der Geschlechtsorgane bei Insekten. Eine morphologische Untersuchung*, Thesis, Helsingfors, Norway.

Panelius, S. (1968) Germ line and oogenesis during paedogenetic reproduction in *Heteropeza pygmaea* Winnertz (Diptera, Cecidomyiidae). *Chromosoma*, **23**, 333–345.

Parks, S., Wakimoto, B.T. and Spradling, A.C. (1986) Replication and expression of an X-linked cluster of *Drosophila* chorion genes. *Dev. Biol.*, **117**, 294–305.

Paulcke, W. (1900) Ueber die Differenzirung der Zellelemente im Ovarium der Bienenkönigin (*Apis mellifica*). *Zool Jb. Morphol.*, **14**, 177–202.

Pauli, D. and Mahowald, A.P. (1990) Germ-line sex determination in *Drosophila melanogaster. Trends Genet.*, **6**, 259–264.

Patel, N.H., Snow, P.M. and Goodman, C.S. (1987) Characterization and cloning of fasciclin III, a glycoprotein expressed in a subset of neurons and axon pathways in *Drosophila. Cell*, **48**, 975–988.

Pauli, D., Oliver, B. and Mahowald, A.P. (1993) The role of the *ovarian tumor* locus in *Drosophila melanogaster* germ line sex determination. *Development*. **119**, 123–134.

Peifer, M., Orsulic, S., Sweeton, D. and Wieschaus, E. (1993) A role for the *Drosophila* segment polarity gene *armadillo* in cell adhesion and cytoskeletal integrity during oogenesis. *Development*, **118**, 1191–1207.

Pelham, H.R.B. and Brown, D.D. (1980) A specific transcription factor that can bind either the 5S RNA gene or 5S RNA. *Proc. Natl. Acad. Sci. USA*, **77**, 4170–4174.

Pehani, H. (1925) Die Geschlechtszellen der Phasmiden. Zugleich ein Beitrag zur Fortpflanzungsbiologie der Phasmiden. *Biol. Zentralbl.*, **44**, 167–238.

Perrimon, N. and Mahowald, A.P. (1987) Multiple functions of segment polarity genes in *Drosophila. Dev. Biol.*, **193**, 587–600.

Perkowska, E., Macgregor, H.C. and Birnstiel, M.L. (1968) Gene amplification in the oocyte nucleus of mutant and wild-type *Xenopus laevis. Nature*, **217**, 649–650.

Picard, B. and Wegnez, M. (1979) Isolation of a 7S particle from *Xenopus laevis* oocytes. A 5S RNA–protein compex. *Proc. Natl. Acad. Sci. USA*, **76**, 241–245.

Pieler, T. and Erdmann, V.A. (1983) Isolation and characterization of a 7S RNP particle from mature *Xenopus laevis* oocytes. *FEBS Lett.*, **157**, 283–287.

Pijnacker, L.P. (1966) The development of follicle cells in the ovarioles of *Carausius morosus* Br. (Orthoptera, Phasmidae). *Experientia*, **22**, 158.

Pijnacker, L.P. (1967) Oogenesis in the parthenogenetic stick insect *Sipyloidea sipylus* Westwood (Orthoptera, Phasmidae). *Genetica*, **38**, 504–515.

Pijnacker, L.P. and Godeke, J. (1984) Development of ovarian follicle cells of the stick insect, *Carausius morosus* Br. (Phasmatodea), in relation to their function. *Int. J. Insect Morphol. Embryol.*, **13**, 21–28.

Pluot, D. (1979) Evolution regressive des ovarioles chez les coleopteres Scarabaeinae. *Ann. Soc. Ent. Fr.*, **15**, 575–588.

Pollak, S.B. and Telfer, W.H. (1969) RNA in cecropia moth ovaries: Sites of synthesis, transport and storage. *J. Exp. Zool.*, **170**, 1–24.

Postlethwait, J.H. and Giorgi, F. (1985) Vitellogenesis in insects, in *Developmental Biology. A Comprehensive Synthesis*, (ed. L.W. Browder), Plenum Press, New York, **1**, 85–126.

Préat, T., Thérond, P., Lamour-Isnard, C. *et al.* (1990) A putative serine/threonine protein kinase encoded by the segment polarity *fused* gene of *Drosophila. Nature*, **347**, 87–89.

Pritsch, M. and Büning, J. (1989) Germ cell cluster in the panoistic ovary of Thysanoptera (Insecta). *Zoomorph.*, **108**, 309–313.

Procunier, D. and Dunn, R.J. (1978) Genetic and molecular organization of the 5S locus and mutants in *Drosophila melanogaster*. *Cell*, **15**, 1087–1093.

Prost, E., Deryckere, F., Roos, C. *et al.* (1988) Role of the oocyte nucleus in determination of the dorsoventral polarity of *Drosophila* as revealed by molecular analysis of the *K10* gene. *Genes Dev.*, **2**, 891–900.

Purkinje, J.E. (1825) *Symbolae ovi avium historiam ante incubationem*, Breslau.

Quattropani, S.L. and Anderson, E. (1969) The origin and structure of the secondary coat of the egg of *Drosophila melanogaster*. *Z. Zellforsch.*, **95**, 495–510.

Quennedey, A. (1971) Les glandes exocrines des termites I. Etude histochimique et ultrastructurale de la glande sternale de *Kalotermes flavicollis* Fab. (Isoptera, Kalotermitidae). *Z. Zellforsch.*, **121**, 27–47.

Quian, S., Hongo, S. and Jacobs-Lorena, M. (1988) Antisense ribosomal protein gene expression specifically disrupts oogenesis in *Drosophila melanogaster*. *Proc. Natl. Acad. Sci. USA*, **85**, 9601–9605.

Quian, S., Hongo, S. and Jacobs-Lorena, M. (1989) A heat shock promoter-driven antisense ribosomal protein gene disrupts oogenesis in *Drosophila*, in *Stress-Induced Proteins*, (eds M.L. Pardue, J.R. Feramisco and S. Lindquist), Alan R. Liss, New York, pp. 137–176.

Raabe, M. (1986) Insect reproduction, regulation of successive steps, in *Advances in Insect Physiology*, (eds P.D. Evans and V.B. Wigglesworth), Academic Press, London, **19**, 29–154.

Raff, J.W., Whitfield, W.G.F. and Glover, D.M. (1990) Two distinct mechanisms localise cyclin B transcripts in syncytial *Drosophila* embryos. *Development*, **110**, 1249–1261.

Raikhel, A.S. and Dhadialla, T.S. (1992) Accumulation of yolk proteins in insect oocytes. *Annu. Rev. Entomol.*, **37**, 217–251.

Ramadan, A.A., Shalaby A.M., Abdel-Ghaffar, H. and Abdel-Sabour, S. (1986a) Histological studies of developmental stages during oogenesis in *Culex pipiens molestus* (Forscål) I. Fourth larval instar and pupal stage. *Folia Morphologica*, **34**, 109–113.

Ramadan, A.A., Shalaby A.M., Abdel-Ghaffar, H. and Abdel-Sabour, S. (1986b) Histological studies of developmental stages during oogenesis in *Culex pipiens molestus* (Forscål) II. The adult stage. *Folia Morphologica*, **34**, 153–168.

Ramamurty, P.S. (1963) Über die Herkunft der Ribonukleinsäure in den wachsenden Eizellen der Skorpionsfliege *Panorpa communis* (Insecta, Mecoptera). *Naturwiss.*, **50**, 383–384.

Ramamurty, P.S. (1964a) Intercellular bridges in the ovarian follicles of *Panorpa communis* L. (Mecoptera-Insecta). *Curr. Sci.*, **16**, 493–495.

Ramamurty, P.S. (1964b) On the contribution of the follicle epithelium to the deposition of yolk in the oocyte of *Panorpa communis*. *Exp. Cell Res.*, **33**, 601–605.

Ramamurty, P.S. (1967) The cytoplasmic inclusions of the trophocytes in the ovary of the scorpion fly *Panorpa communis* L. (Mecoptera, Panorpidae). *Proc. R. Ent. Soc. Lond. (A)*, **42**, 87–92.

Ramamurty, P.S. (1968) Origin and distribution of glycogen during vitellogenesis of the scorpion fly *Panorpa communis*. *J. Insect Physiol.*, **14**, 1325–1330.

Ramamurty, P.S. and Engels, W. (1977) Occurrence of intercellular bridges between follicle epithelial cells in the ovary of *Apis mellifica* queens. *J. Cell Sci.*, **24**, 195–201.

Rankin, M.A. and Jäckle, H. (1980) Hormonal control of vitellogenin synthesis in *Oncopeltus fasciatus*. *J. Insect Physiol.*, **26**, 671–684.

Rasch, E.M., King, R.C. and Rasch, R.W. (1984) Cytophotometric studies on cells from the ovaries of *otu* mutants of *Drosophila melanogaster*. *Histochemistry*, **81**, 105–110.

Rasmussen, S.W. (1975) Synaptonemal polycomplexes in *Drosophila melanogaster*. *Chromosoma (Berl.)*, **49**, 321–331.

Rasmussen, S.W. (1976) The meiotic prophase in *Bombyx mori* females analyzed by three dimensional reconstructions of synaptonemal complexes. *Chromosoma (Berl.)*, **54**, 245–293.

Rasmussen, S.W. (1977a) The transformation of the synaptonemal complex into the 'elimination chromatin' in *Bombyx mori* oocytes. *Chromosoma (Berl.)*, **60**, 205–221.

Rasmussen, S.W. (1977b) Chromosome pairing in triploid females of *Bombyx mori* analyzed by three dimensional reconstructions of synaptonemal complexes. *Carlsberg Res. Commun.*, **42**, 163–197.

Rasmussen, S.W. and Holm, P.B. (1980) Mechanics of meiosis. *Hereditas*, **97**, 187–225.

Rasmussen, S.W. and Holm, P.B. (1982) The meiotic prophase in *Bombyx mori*, in *Insect Ultrastructure*, (eds R.C. King and H. Akai), Plenum Press, New York, Vol. 1, pp. 61–85.

Ray, A. and Ramamurty, P.S. (1979) Sources of RNA supply to the oocytes in *Crynodes peregrinus* Fuessly (Coleoptera, Chrysomelidae). *Int. J. Insect Morphol. Embryol.*, **8**, 113–122.

Reddy, J.T. and Locke, M. (1990) The size limited penetration of gold particles through insect basal laminae. *J. Insect Physiol.*, **36**, 397–407.

Redfern, C.P.F. (1981) Homologous banding pattern in the polytene chromosomes from the larval salivary glands and ovarian nurse cells of *Anopheles stephensi* Liston (Culicidae). *Chromosoma (Berl.)*, **83**, 221–240.

Reeder, R.H. and Roan, J.G. (1985) The mechanism of nucleolar dominance in *Xenopus* hybrids. *Cell*, **38**, 39–44.

Regier, J.C. and Kafatos, F.C. (1985) Molecular aspects of chorion formation, in *Comprehensive Insect Physiology, Biochemistry and Pharmacology*, (eds G.A. Kerkut and L.I. Gilbert), Pergamon Press, Oxford, Vol. I, pp. 113–151.

Reitberger, A. (1940) Die Cytologie des paedogenetischen Entwicklungszyklus der Gallmücke *Oligarces paradoxus* Mein. *Chromosoma (Berl.)*, **1**, 391–473.

Renkawitz-Pohl, R. and Kunz, W. (1975) Underreplication of satellite DNAs in polyploid tissue of *Drosophila virilis*. *Chromosoma (Berl.)*, **49**, 375–382.

Ribbert, D. (1979) Chromosomes and puffing in experimentally induced polytene chromosomes of *Calliphora erythrocephala*. *Chromosoma (Berl.)*, **74**, 269–298.

Ribbert, D. and Bier, K. (1969) Multiple nucleoli and enhanced nucleolar activity in the nurse cells of the insect ovary. *Chromosoma (Berl.)*, **27**, 178–197.

Ribbert, D. and Weber, F. (1970) Homologous pairing and synaptinemal complexes in the nurse-cell nuclei of Carabid ovaries (Ins. Coleoptera). *Experientia*, **26**, 800–801.

Richard-Mercier, N. (1977) Organogenesis of sterile ovaries and testes after cauterization of pole cells of Colorado beetle embryos (*Leptinotarsa decemlineata* Say). *Wilhelm Roux Arch.*, **183**, 171–176.

Richard-Mercier, N., Bettane-Casanova, M., Fellous, M. and Charniaux-Cotton, H. (1986) Possible relationship between H-Y antigen and female gonadogenesis in the Colorado potato beetle, *Leptinotarsa decemlineata* (Coleoptera, Chrysomelidae). *Int. J. Invert. Reprod. Dev.*, **10**, 243–252.

Richard-Mercier, N., Thomas-Orillard, M. and Fellous, M. (1988) Abnormal ovarian morphogenesis in *Drosophila melanogaster* after injection of embryos with conditioned media from the Daudi cell line. *Wilhelm Roux Arch. Dev. Biol.*, **197**, 370–374.

Richards, J.G. (1968) The structure and formation of the egg membranes in *Nasonia vitripennis* (Walker) (Hymenoptera, Pteromalidae). *J. Micros.*, **89**, 43–53.

Ries, E. (1932) Die Prozesse der Eibildung und des Eiwachstums bei Pediculiden und Mallophagen. *Z. Zellforsch.*, **16**, 314–388.

Ries, E. and van Weel, P.B. (1934) Die Eibildung der Kleiderlaus, untersucht an lebenden, vital gefärbten und fixierten Präparaten. *Z. Zellforsch. Mikr. Anat.*, **20**, 565–618.

Riley, R.C. and Forgash, A.J. (1967) *Drosophila melanogaster* egg shell adhesive. *J. Insect Physiol.*, **13**, 509–518.

Ritossa, F.M. and Spiegelman, S. (1965) Localization of DNA complementary to ribosomal RNA in the nucleolus organiser region of *Drosophila melanogaster*. *Proc. Natl. Acad. Sci. USA*, **53**, 737–745.

Ritzki, R.M. and Ritzki, T.M. (1984) Selective destruction of a host blood cell type by a parasitoid wasp. *Proc. Natl. Acad. Sci. USA,* **81**, 6154–6158.

Roberts, J.M., Buck, L.B. and Axel, R. (1983) A structure for amplified DNA. *Cell,* **33**, 53–63.

Robertson, J.G. (1961) Ovariole number in Coleoptera. *Can. J. Zool.,* **39**, 345–358.

Rochaix, J.D., Bird, A. and Bakken, A. (1974) Ribosomal RNA gene amplification by rolling circles. *J. Mol. Biol.,* **87**, 473–487.

Röhrkasten, A. and Ferenz, H.J. (1986a) Properties of the vitellogenin receptor of isolated locust oocyte membranes. *Int. J. Invert Reprod. Dev.,* **10**, 133–142.

Röhrkasten, A. and Ferenz, H.J. (1986b) Solubilization of the locust vitellogenin receptor. *Biochem. Biophys. Acta,* **860**, 577–582.

Röhrkasten, A., Ferenz, H.J., Buschmann-Gebgardt, B. and Hafer, J. (1989) Isolation of the vitellogenin-binding protein from locust ovaries. *Arch. Insect Biochem. Physiol.,* **10**, 141–149.

Rosciszewska, E. (1976) Oogenesis in *Tetrodontophora bielanensis* (Waga) (Collembola). Histochemical and autoradiographical analysis of proteins. *Acta Biol. Crac., Ser. Zool.,* **19**, 41–54.

Roth, S., Stein, D. and Nüsslein-Volhard, C. (1989) A gradient of nuclear localization of the dorsal protein determines dorsoventral pattern in the *Drosophila* embryo. *Cell,* **59**, 1189–1202.

Roth, T.F. (1966) Changes in the synaptinemal complex during meiotic prophase in mosquito oocytes. *Protoplasma,* **61**, 346–386.

Roth, T.F. and Porter, K.R. (1964) Yolk protein uptake in the oocyte of the mosquito *Aedes aegypti* L. *J. Cell Biol.,* **20**, 313–332.

Rothe, M., Nauber, U. and Jäckle, H. (1989) Three hormone receptor-like *Drosophila* genes encode an identical DNA-binding finger. *EMBO J.,* **8**, 3087–3094.

Rothschild, M. (1976) Notes on fleas (Part II): The internal organs; Can they throw any light on relationships within the order? *Proc. Br. Ent. Nat. Hist. Soc.,* **9**, 97–110.

Rottmar, B. (1966) Über Züchtung, Diapause und postembryonale Entwicklung von *Panorpa communis* L. *Zool. Jb. Anat.,* **83**, 497–570.

Rousset, A. (1977) Formation et développement d'hétérochromatine extrachromosomienne dans le noyau des cellules germinales femelles de *Chrysopa perla* (L.) (Neuroptera). *C.R. Acad. Sci. Paris, D.,* **285**, 65–68.

Rousset, A. (1978a) La formation des cystes dans l'ovariole de *Chrysopa perla* (L.)(Neuroptera, Chrysopidae). *Int. J. Insect Morphol. Embryol.,* **7**, 45–57.

Rousset, A. (1978b) La differenciation des cellules germinales dans l'ovariole de *Chrysopa perla* (L.) (Neuroptera, Chrysopidae). *Int. J. Insect Morphol. Embryol.,* **7**, 59–71.

Rubenstein, E.C. (1979) The role of an ephithelial occlusion zone in the termination of vitellogenesis in *Hyalophora cecropia* ovarian follicles. *Dev. Biol.,* **71**, 115–127.

Ruby, J.R., Dyer, R.F. and Skalko, R.S. (1969) The occurrence of intercellular bridges during oogenesis in the mouse. *J. Morphol.,* **127**, 307–338.

Ruohola, H., Bremer, K.A., Baker, D. *et al.* (1991) Role of neurogenic genes in establishment of follicle cell fate and oocyte polarity during oogenesis in *Drosophila. Cell,* **66**, 433–449.

Ruohola-Baker, H., Grell, E., Chou, T. *et al.* (1993) Spacially localized *rhomboid* is required for establishment of the dorsal-ventral axis in *Drosophila* oogenesis. *Cell,* **73**, 953–965.

Rushlow, C.A., Han, K., Manley, J.L. and Levine, M. (1989) The graded distribution of the dorsal morphogen is initiated by selective nuclear transport in *Drosophila. Cell,* **59**, 1165–1177.

Sachtleben, H. (1918) Über die Entwicklung der Geschlechtsorgane von *Chironomus* mit besonderer Berücksichtigung der keimbahnbegleitenden Substanzen. Thesis (Dr.), University München, Germany.

Sahai, Y.N. (1975) Histochemical and cytological studies on the oocyte development of *Sphaerodema rusticum* Fabr. *Zool Jb. Anat.*, **94**, 200–217.

Sahai, Y.N. (1989) Cytochemical studies of oocyte development in *Sarcophaga lineaticollis* Macq. (Muscidae, Diptera). *Folia Morphol.*, **37**, 257–263.

Saifuddin, U.T., Baker, R.H. and Sakai, R.K. (1978) The chromosomes of *Anopheles culicifacies. Mosq. News*, **38**, 233–240.

Sakai, S. (1984) Integrated taxonomy on Dermaptera. I: Numerical taxonomic information on male genitalia. *17th International Congress of Entomology, 1984, Hamburg.* ICE Council, abstract.

Saling, T. (1907) Zur Kenntnis der Entwicklung der Keimdrüsen von *Tenebrio molitor* L.. *Z. Wiss. Zool.*, **86**, 238–303.

Salz, H.K., Maine, E.M., Keyes, L.N. *et al.* (1989) The *Drosophila* female-specific sex-determination gene, *Sex-lethal*, has stage-, tissue-, and sex-specific RNAs suggesting multiple modes of regulation. *Genes Dev.*, **3**, 708–719.

Sander, K. (1976) Specification of the basic body pattern in insect embryogenesis. *Adv. Insect Physiol.*, **12**, 125–238.

Sander, K., Gutzeit, H.O. and Jäckle, H. (1985) Insect embryogenesis, morphology, physiology, genetical and molecular aspects, in *Comprehensive Insect Physiology, Biochemistry and Pharmacology*, (eds G.A. Kerkut and L.I. Gilbert), Pergamon Press, Oxford, Vol. I, pp. 321–385.

Sato, H. (1932) Die postembryonale Differenzierung der Gonaden von *Lymantria dispar. Z. Zellforsch. Mikr. Anat.*, **16**, 63–87.

Schäfer, M. and Kunz, W. (1985) rDNA in *Locusta migratoria* is very variable: two introns and extensive restriction site polymorphisms in the spacer. *Nucl. Acids Res.*, **13**, 1251–1266.

Schäfer, M. and Kunz, W. (1987) Ribosomal gene amplification does not occur in the oocytes of *Locusta migratoria. Dev. Biol.*, **120**, 418–424.

Scheer, U. (1981) Identification of a novel class of tandemly repeated genes transcribed on lampbrush chromosomes of *Pleurodeles waltlii. J. Cell Biol.*, **88**, 599–603.

Scheer, U., Trendelenburg, M.F. and Franke, W.W. (1976a) Regulation of transcription of genes of ribosomal RNA during amphibian oogenesis. A biochemical and morphological study. *J. Cell Biol.*, **69**, 465–489.

Scheer, U., Franke, W.W., Trendelenburg, M.F. and Spring, H. (1976b) Classification of loops of lampbrush chromosomes according to the arrangement of transcriptional complexes. *J. Cell Sci.*, **22**, 503–519.

Scheer, U., Trendelenburg, M.F., Krohne, G. and Franke, W.W. (1977) Lengths and patterns of transcriptional units in the amplified nucleoli of oocytes of *Xenopus laevis. Chromosoma (Berl.)*, **60**, 147–167.

Scheer, U. and Zentgraf, H. (1978) Nucleosomal and supranucleosomal organization of transcriptionally inactive rDNA circles in *Dytiscus* oocytes. *Chromosoma (Berl.)*, **69**, 243–254.

Schleiden, M.J. (1838) Beiträge zur Phytogenesis. *Arch. Anat. Physiol. Wiss. Med.*, **5**, 137–176.

Schlein, Y. (1980) Morphological similarities between the skeletal structures of Siphonaptera and Mecoptera, in *Fleas*, (eds R. Traub and H. Starcke), A.A. Balkema, Rotterdam, pp. 359–367.

Schlottman, L.L. and Bonhag, P.F. (1956) Histology of the ovary of the adult mealworm *Tenebrio molitor* L. (Coleoptera, Tenebrionidae). *Univ. Calif. Publ. Entomol.*, **11**, 351–394.

Schmidt, O. and Schuchmann-Feddersen, I. (1989) Role of virus-like particles in parasitoid–host interaction of insects, in *Subcellular Biochemistry*, (ed J.R. Harris), Plenum, New York, Vol. 15, pp. 91–119.

Schneider, D.S., Hudson, K.L., Lin, T.Y. and Anderson, K.V. (1991) Dominant and recessive mutations define functional domains of *Toll*, a transmembrane protein required for dorsal-ventral polarity in the *Drosophila* embryo. *Genes Dev.*, **5**, 797–807.

Schnetter, W. (1965) Experimente zur Analyse der morphogenetischen Funktion der Ooplasmabestandteile in der Embryonalentwicklung des Kartoffelkäfers (*Leptinotarsa decemlineata* Say). *Wilhelm Roux Arch.*, **155**, 637– 692.

Schoenemund, E. (1924) Zur Biologie und Morphologie einiger *Perla*-Arten. *Zool. Jb. Anat.*, **34**, 1–54.

Schreiner, B. (1977a) Vitellogenesis in the milkweed bug, *Oncopeltus fasciatus* Dallas (Hemiptera). A light and electron microscopic investigation. *J. Morphol.*, **151**, 35–80.

Schreiner, B. (1977b) The effect of the hormone(s) from the corpus allatum complex on the ovarian tissue of *Oncopeltus fasciatus*. A light and electron microscopic investigation. *J. Morphol.*, **151**, 81–110.

Schuh, R.T. (1986) The influence of cladistics on heteropteran classification. *Ann. Rev. Entomol.*, **31**, 67–93.

Schüpbach, P.M. and Camenzind, R. (1983) Germ cell lineage and follicle formation in paedogenetic development of *Mycophila speyeri* Barnes (Diptera, Cecidomyiidae). *Int. J. Insect Morphol. Embryol.*, **12**, 211–223.

Schüpbach, P.M. and Went, D.F. (1983) Cell fusions during formation of the oocyte-nurse chamber complex in the ovary of the dipteran insect *Mycophila speyeri*. *Wilhelm Roux Arch. Dev. Biol.*, **192**, 228–233.

Schüpbach, T. (1982) Autosomal mutations that interfere with sex determination in somatic cells of *Drosophila* have no direct effect on the germline. *Dev. Biol.*, **89**, 117–127.

Schüpbach, T. (1985) Normal female germ cell differentiation requires the female X chromosome to autosome ratio and expression of *Sex-lethal* in *Drosophila melanogaster*. *Genetics*, **109**, 529–548.

Schüpbach, T. (1987) Germ line and soma cooperate during oogenesis to establish the dorsoventral pattern of egg shell and embryo in *Drosophila melanogaster*. *Cell*, **49**, 699–707.

Schüpbach, T. and Wieschaus, E. (1986) Maternal-effect mutations altering the anterior-posterior pattern of the *Drosophila* embryo. *Wilhelm Roux Arch. Dev. Biol.*, **195**, 302–317.

Schüpbach, T., Clifford, R.J., Manseau, L.J. and Price, J.V. (1991) Dorsoventral signaling processes in *Drosophila* oogenesis, in *Cell–cell Interactions in Early Development*, (ed. J. Gerhart), Wiley–Liss, New York, pp. 163–174.

Schwann, T. (1839) *Mikroskopische Untersuchungen über die Übereinstimmung in der Structur und dem Wachstum der Thiere und Planzen*, Berlin.

Schwemmler, (1979) *Mechanismen der Zellevolution*, Walter de Gruyter, Berlin.

Seidel, F. (1924) Die Geschlechtsorgane in der embryonalen Entwicklung von *Pyrrhocoris apterus* L. *Z. Morphol. Ökol. Tiere*, **1**, 429–506.

Sikora, A. (1975) Utrastructure of nurse cells in the female gonad of *Tetrodontophora bielanensis* (Waga) (Collembola). *Acta Biol. Crac., Ser. Zool.*, **18**, 79–84.

Silvestri, F. (1943) Studi sugli 'Strepsiptera' (Insecta) III. Descrizione e biologia di 6 specie italiane di Mengenilla. *Boll. Lab. Zool. Gen. Agric., Fac. Agric. Portici*, **32**, 197–282.

Sinha, P., Mishra, A. and Lakhotia, S.C. (1987) Chromosomal organization of *Drosophila* tumours. I. Polytene chromosome organization and DNA synthesis in ovarian pseudonurse cells in *otu* mutants of *D. melanogaster*. *Chromosoma*, **95**, 108–116.

Smith, P.A. and King, R.C. (1968) Genetic control of synaptonemal complexes in *Drosophila melanogaster*. *Genetics*, **60**, 335–351.

Smith, J.L., Wilson, J.E. and Mahowald, P.M. (1992) Overexpression of *oscar* directs ectopic activation of *nanos* and presumptive pole cell formation in *Drosophila* embryos. *Cell*, **70**, 849–859.

Snodgrass, R.E. (1933) Morphology of the insect abdomen, II. The genital ducts and the ovipositor. *Smithonian Misc. Coll.*, **89**, (8)1–148.

Snodgrass, R.E. (1935) *The Principles of Insect Morphology*, McGraw-Hill, London.

Snodgrass, R.E. (1956) *Anatomy of the Honey Bee*, Cornell Univ. Press, Ithaca, New York.

Soldan, T. (1979) The structure and development of the female internal reproductive system in six European species of Ephemeroptera. *Acta Entomol. Bohemoslovacia*, **76**, 353–365.

Soldan, T. (1981) Comparative anatomy of the internal reproductive organs of mayflies (Ephemeroptera). *Studie CSAV*, **6**, 5–118.

Soltani-Mazouni, N. and Bordereau, C. (1987) Changes in the cuticle, ovaries and colleterial glands during the pseudergate and neotenic molt in *Kalotermes flavicollis* (Fabr.) (Isoptera, Kalotermitidae). *Int. J. Insect Morphol. Embryol.*, **16**, 221–235.

Sonnenblick, B.P. (1950) The early embryology of *Drosophila melanogaster*, in *Biology of Drosophila*, (ed. M. Demerec), Wiley, New York, pp. 62–167.

Spradling, A. (1987) Gene amplification in dipteran chromosomes, in *Results and Problems in Cell Differentiation*, (eds W. Hennig and U. Scheer), Springer Verlag, Berlin, pp. 199–212.

Spradling, A.C. and Mahowald, A.P. (1980) Amplification of genes for chorion proteins during oogenesis in *Drosophila melanogaster*. *Proc. Natl. Acad. Sci. USA*, **77**, 1096–1100.

Spradling, A.C. and Orr-Weaver, T. (1987) Regulation of DNA replication during *Drosophila* development. *Ann. Rev. Genet.*, **21**, 373–403.

Sprenger, F., Stevens, L.M. and Nüsslein-Volhard, C. (1989) The *Drosophila* gene *torso* encodes a putative receptor tyrosine kinase. *Nature*, **338**, 478–483.

Sreng, L. and Quennedey, A. (1976) Role of a temporary ciliary structure in the morphogenesis of insect glands. *J. Ultrastr. Res.*, **56**, 78–95.

Staurengo da Cunha, M.A. (1979) Ovarian development in *Scaptotrigona postica* Latr. 1807 (Hym., Apidae) II. A quantitative study. *Ins. Sociaux (Paris)*, **26**, 196–203.

Staurengo da Cunha, M.A. (1988) Ultrastructure of the oocytes of *Scaptotrigona postica* workers (Hymenoptera, Apidae). *Naturalia* **13**, 103–115.

Stay, B. and Coop, A. (1974) Developmental stages and chemical composition in embryos of the cockroach *Diploptera punctata*, with observations on the effect of diet. *J. Insect Physiol.*, **19**, 147–171.

Stay, B. and Roth, L.M. (1962) The colleterial glands of cockroaches. *Ann. Ent. Soc. Am.*, **55**, 124–130.

Stearns, S.C. (1987) *The Evolution of Sex and its Consequences. Experientia Suppl.*, **55**, Birkhäuser Verlag, Basel.

Stebbings, H. (1981) Observations on cytoplasmic transport along ovarian nutritive tubes of polyphagous coleopterans. *Cell Tissue Res.*, **220**, 153–161.

Stebbings, H. (1988) Translocation along microtubules in insect ovaries. *J. Cytol. Subcell. Biol.*, **10**, 1–14.

Stebbings, H. and Bennett, C.E. (1976) The effect of colchicine on the sleeve element of microtubules. *Exp. Cell Res.*, **100**, 419–423.

Stebbings, H. and Hunt, C. (1983) Microtubule polarity in the nutritive tubes of insect ovarioles. *Cell Tissue Res.*, **233**, 133–141.

Stebbings, H. and Hunt, C. (1985) Binding of axonemal dynein to microtubules comprising the cytoplasmic transport system in insect ovarioles. *Cell. Biol. Int. Rep.*, **9**, 245–252.

Stebbings, H. and Hunt, C. (1987) The translocation of mitochondria along insect ovarian microtubules from isolated nutritive tubes, a simple re-activated model. *J. Cell Sci.*, **88**, 641–648.

Stebbings, H. and Sharma, K.K. (1989) 'Corkscrewing', as evidenced for force generation within a detergent-extracted microtubule translocation system from insect ovarioles. *J. Cell Sci.*, **92**, 21–27.

Stebbings, H., Sharma, K. and Hunt, C. (1985) Protein turnover in the cytoplasmic transport system within an insect ovary – a clue to the mechanism of microtubule-associated transport. *FEBS Lett.*, **193**, 22–26.

Stebbings, H., Sharma, K.K. and Hunt, C. (1986) Microtubule-associated proteins in the ovaries of hemipteran insects and their association with microtubule transport system linking nutritive cells and oocytes. *Eur. J. Cell Biol.*, **42**, 135–139.

Stebbings, H. and van Buren, J. (1981) Stability of microtubules from insect ovarioles. *Cell Biol. Int. Rep.*, **5**, 705–710.

Stein, F. (1847) *Vergleichende Anatomie und Physiologie der Insecten. Erste Monographie. Die weiblichen Geschlechtsorgane der Käfer.* Dunker und Humblot, Berlin.

Stein, D. and Nüsslein-Volhard, C. (1992) Multiple extracellular activities in *Drosophila* egg perivitelline fluid are required for establishment of embryonic dorsal-ventral polarity. *Cell*, **68**, 429–440.

Steiner, P. (1937) Ein Beitrag zur Fortpflanzungsbiologie und Morphologie des Genitalapparates von *Boreus hiemalis* L. *Z. Morph. Ökol. Tiere*, **32**, 276–288.

Steinauer, W.R. and Kalfayan, L.J. (1992) A specific ovarian *tumor* protein isoform is required for efficient differentiation of germ cells in *Drosophila* oogenesis. *Genes Dev.*, **6**, 233–243.

Steinhauer, W.R., Walsh, R.C. and Kalfayan, R.J. (1989) Sequence and structure of the *Drosophila melanogaster* ovarian tumor gene and generation of an antibody specific for the ovarian tumor protein. *Mol. Cell Biol.*, **9**, 5726–5732.

Steinmann-Zwicky, M. (1988) Sex determination in *Drosophila*, the X-chromosomal gene *liz* is required for *Sxl* activity. *EMBO J.*, **7**, 3889–3898.

Steinmann-Zwicky, M., Amrein, H. and Nöthiger, R. (1990) Genetic control of sex determination in *Drosophila. Adv. Genet.*, **27**, 189–237.

Steinmann-Zwicky, M., Schmid, H. and Nöthiger, R. (1989) Cell-autonomous and inductive signals can determine the sex of the germ line of *Drosophila* by regulating the gene *Sxl*. *Cell*, **57**, 157–166.

Stephenson, E.C. and Pokrywka, N.J. (1993) Localization of *bicoid* message during *Drosophila* oogenesis. *Curr. Topics Dev. Biol.*, **26**, 23–34.

Stephenson, E.C., Chao, Y.C. and Fackenthal, J.D. (1988) Molecular analysis of the *swallow* gene of *Drosophila melanogaster. Genes Dev.*, **2**, 1655–1665.

Stevens, N.M. (1905) A study of germ cells of *Aphis rosae* and *Aphis oenotherae*. *J. Exp. Zool.*, **2**, 313–334.

Stevens, L.M., Frohnhöfer, H.G., Klingler, M. and Nüsslein-Volhard, C. (1990) Localized requirement for *torsolike* expression in follicle cells for the development of terminal anlagen of the *Drosophila* embryo. *Nature*, **346**, 660–663.

Steward, R. (1989) Relocalization of the dorsal protein from the cytoplasm to the nucleus correlates with its function. *Cell*, **59**, 1179–1188.

Steward, R. and Govind, S. (1993) Dorsal-ventral polarity in the *Drosophila* embryo. *Curr. Opinion Gen. Dev.*, **3** , 556–561.

St Johnston, D. and Nüsslein-Volhard, C. (1992) The origin of pattern and polarity in the *Drosophila* embryo. *Cell*, **68**, 201–219.

St Johnston, D., Driever, W., Berleth, T., Richstein, S. and Nüsslein-Volhard, C.

(1989) Multiple steps in the localization of *bicoid* RNA to the anterior pole of the *Drosophila* oocyte. *Development 1989 Suppl.*, 13–19.

St Johnston, D., Beuchle, D. and Nüsslein-Volhard, C. (1991) *staufen*, a gene required to localize maternal RNAs in the *Drosophila* egg. *Cell*, **66**, 51–63.

St Johnston, D., Brown, N.H., Gall, J.G. and Jantsch, M. (1992) A conserved double-stranded RNA-binding domain. *Proc. Natl. Acad. Sci. USA*, **89**, 10979–10983.

Storto, P.D. and King, R.C. (1987) Fertile heteroallelic combinations of mutant alleles of the *otu* locus of *Drosophila melanogaster*. *Wilhelm Roux Arch. Dev. Biol.*, **196**, 210–221.

Storto, P.D. and King, R.C. (1988) Multiplicity of functions for the *otu* gene products during *Drosophila* oogenesis. *Dev. Genet.*, **9**, 91–120.

Storto, P.D. and King, R.C. (1989) The role of polyfusomes in generating branched chains of cystocytes during *Drosophila* oogenesis. *Dev. Gen.*, **10**, 70–86.

Struhl, G. (1989) Differing strategies for organizing anterior and posterior body pattern in *Drosophila* embryos. *Nature*, **338**, 741–744.

Stuhlmann, F. (1886) Die Reifung des Arthropodeneies nach Beobachtungen an Insekten, Spinnen, Myriapoden und *Peripatus*. *Ber. Naturforsch. Ges. Freiburg*, **1**, 101–228.

Stys, P. and Bilinski, S. (1990) Ovariole types and the phylogeny of hexapods. *Biol. Rev.*, **65**, 401–429.

Stys, P. and Kerzhner, I. (1975) The rank and nomenclature of higher taxa in recent Heteroptera. *Acta Entomol. Bohemoslov.*, **72**, 64–79.

Sugawara, T. (1981) Fine structure of the stretch receptor in the bursa copulatrix of the butterfly, *Pieris rapae crucivora*. *Cell Tissue Res.*, **217**, 23–36.

Sun, Y.A. and Wyman, R.J. (1987) Lack of an oocyte to nurse cell voltage difference in *Drosophila*. *Neuroscience*, **13**, 1139.

Sun, Y.A. and Wyman, R.J. (1989) The *Drosophila* egg chamber: external ionic currents and the hypothesis of electrophoretic transport. *Biol. Bull.*, **176**, 79– 85.

Sun, Y.A. and Wyman, R.J. (1993) Reevaluation of electrophoresis in *Drosophila* egg chamber. *Dev. Biol.*, **155**, 206–215.

Sun, S.C., Lindström, I., Boman H.G., Faye, I. and Schmidt, O. (1990) Hemolin: An insect-immune protein belonging to the immunglobulin superfamily. *Science*, **250**, 1729–1732.

Suter, B., Romberg, L.M. and Steward, R. (1989) *Bicaudal-D*, a *Drosophila* gene involved in developmental asymmetry, localized transcript accumulation in ovaries and sequence similarity to myosin heavy chain tail domains. *Genes Dev.*, **3**, 1957–1968.

Suzzoni, J.P. (1972) Ultrastructure de la glande de la spermatheque chez *Phosphuga atrata* L. (Coleoptera Silphidae). *Z. Zellforsch.*, **128**, 426–437.

Swammerdam, J. (1737) *Bybel der Natuure*. Leyden.

Szabad, J. and Nöthiger, R. (1992) Gynandromorphs of *Drosophila* suggest one common primordium for the somatic cells of the female and male gonads in the region of abdominal segments 4 and 5. *Development*, **115**, 527–533.

Szklarzewicz, T. (1987) Oogenesis of *Nicoletia phytophila* (Zygentoma, Nicoletiidae). Preliminary studies, in *Recent Advances in Insect Embryology in Japan and Poland*, (eds H. Ando and C. Jura), Arthropod. Soc. Jpn., Tsukuba, pp. 69–76.

Szklarzewicz, T. (1989) Ultrastructural studies on the vitellogenesis of *Nicoletia phytophila* (Zygentoma, Nicoletiidae). *Zool. Jb. Anat.*, **118**, 483–491.

Szöllösi, A. (1982) Relationships between germ and somatic cells in the testes of locusts and moths, in *Insect Ultrastructure*, (eds R.C. King and H. Akai), Plenum Press, New York, Vol. 1, pp. 32–60.

Szöllösi, A. and Marcaillou, C. (1980) Gap junctions between germ and somatic cells in the testis of the moth, *Anagasta kuehniella* (Insecta, Lepidoptera). *Cell Tissue Res.*, **213**, 137–147.

Taddei, C., Chicca, M., Maurizii, M.G. and Scali, V. (1992) The germarium of panoistic ovarioles of *Bacillus rossius* (Insecta Phasmatodea): larval differentiation. *Invert. Reprod. Dev.*, **21**, 47–56.

Taddei, C., Maurizii, M.G., Chicca, M. and Scali, V. (1993) The germarium of panoistic ovarioles of *Bacillus rossius* (Insecta Phasmatodea): Structure and function during imaginal life. *Invert. Reprod. Dev.*, **23**, 203–210.

Tamate, H.B., Patel, R.C., Riedl, A.E. and Jacobs-Lorena, M. (1990) Overproduction and translational regulation of rp49 ribosomal protein mRNA in transgenic *Drosophila* carrying extra copies of the gene. *Mol. Gen. Genet.*, **221**, 171–175.

Tannreuther, G.W. (1907) History of the germ cells and early embryology of certain aphids. *Zool. Jb. Anat. Ontog. Tiere*, **24**, 609–642.

Tartof, K.D. (1973) Regulation of ribosomal RNA gene multiplicity in *Drosophila melanogaster*. *Genetics*, **73**, 57–71.

Tartof, K.D. and Perry, R.P. (1970) The 5S RNA genes of *Drosophila melanogaster*. *J. Mol. Biol.*, **51**, 171–183.

Telfer, W.H. (1965) The mechanism and control of yolk formation. *Ann. Rev. Entomol.*, **10**, 161–184.

Telfer, W.H. (1975) Development and physiology of the oocyte–nurse cell syncytium. *Adv. Insect Physiol.*, **11**, 223–319.

Telfer, W.H. (1979) Sulfate and glucosamine labelling of the intercellular matrix in vitellogenic follicles of a moth. *Wilhelm Roux Arch. Dev. Biol.*, **185**, 347–362.

Telfer, W.H., Huebner, E. and Smith, D.S. (1982) The cell biology of vitellogenic follicles in *Hyalophora* and *Rhodnius*, in *Insect Ultrastructure*, (eds R.C. King and H. Akai), Plenum Press, New York, Vol. 1, pp. 118–149.

Telfer, W.H. and Pan, M. (1988) Adsorptive endocytosis of vitellogenin, lipophorin, and microvitellin during yolk formation in *Hyalophora*. *Arch. Insect Biochem. Physiol.*, **9**, 339–355.

Telfer, W.H., Woodruff, R.I. and Huebner, E. (1981) Electrical polarity and cellular differentiation in meroistic ovaries. *Am. Zool.*, **21**, 675–686.

Theurkauf, W.E. and Hawley, R.S. (1992) Meiotic spindle assembly in *Drosophila* females: behavior of nonexchange chromosomes and the effects of mutations in the *nod* kinesin-like protein. *J. Cell Biol.*, **116**, 1167– 1180.

Theurkauf, W.E., Smiley, S., Wong, M.L. and Alberts, B.M. (1992) Reorganization of the cytoskeleton during *Drosophila* oogenesis: implications for axis specification and intercellular transport. *Development*, **115**, 923–936.

Theurkauf, W.E., Alberts, B.M., Jan, Y.N. and Jongens, T.A. (1993) A central role for microtubules in the differentiation of *Drosophila* oocytes. *Development*, **118**, 1169–1180.

Thiébaud, C.H. (1979) Quantitative determination of amplified rDNA and its distribution during oogenesis in *Xenopus laevis*. *Chromosoma (Berl.)*, **73**, 37–44.

Tichomirow, A. (1898) Zur Anatomie des Insectenhodens. *Zool. Anz.*, **21**, 623–630.

Tobe, S.S., Davey, K.G. and Huebner, E. (1973) Nutrient transfer during reproductive cycle in *Glossina austeni* Newst: Histology and histochemistry of the milk gland, fat body, and oenocytes. *Tissue Cell*, **5**, 633–650.

Tobe, S.S. and Davey, K.G. (1974) Autoradiographic study of protein synthesis in abdominal tissues of *Glossina austeni*. *Tissue Cell*, **6**, 255–268.

Tobe, S.S. and Langley, P. (1978) Reproductive physiology of *Glossina*. *Ann. Rev. Entomol.*, **23**, 283–307.

Tokuyashu, K.T. (1974a) Dynamics of spermiogenesis in *Drosophila melanogaster*. III. Relation between axoneme and mitochondrial derivatives. *Exp. Cell Res.*, **84**, 239–250.

Tokuyashu, K.T. (1974b) Dynamics of spermiogenesis in *Drosophila melanogaster*. IV. Nuclear transformation. *J. Ultrastruct. Res.*, **48**, 284–303.

Tokuyashu, K.T. (1974c) Spoke heads in sperm tail of *Drosophila melanogaster*. *J. Cell Biol.*,**63**, 334–337.

Tokuyashu, K.T. (1975a) Dynamics of spermiogenesis in *Drosophila melanogaster*. V. Head–tail alignment. *J. Ultrastruct. Res.*, **50**, 117–129.

Tokuyashu, K.T. (1975b) Dynamics of spermiogenesis in *Drosophila melanogaster*. VI. Significance of 'onion' nebenkern formation. *J. Ultrastruct. Res.*, **53**, 93–112.

Torres, M. and Sánchez, L. (1989) The *scute* (T4) gene acts as a numerator element of the X: A signal that determines the state of activity of *Sex-lethal* in *Drosophila*. *EMBO J.*, **8**, 3079–3086.

Toth, L. (1933) Über die frühembryonale Entwicklung der viviparen Aphiden. *Z. Morphol. Ökol. Tiere*, **27**, 692–731.

Toyama, N.K. (1894) Preliminary note on the spermatogenesis of the silkworm. *Zool. Anz.*, **17**(438), 20–24.

Tremblay, E. and Caltagirone, L.E. (1973) Fate of polar bodies in insects. *Ann. Rev. Entomol.*, **18**, 421–444.

Trendelenburg, M.F. (1974) Morphology of ribosomal RNA cistrons in oocytes of the water beetle, *Dytiscus marginalis* L. *Chromosoma (Berl.)*, **48**, 119–135.

Trendelenburg, M.F., Scheer, U., Zentgraf, H. and Franke, W.W. (1976) Heterogeneity of spacer lengths in circles of amplified ribosomal DNA of two insect species, *Dytiscus marginalis* and *Acheta domesticus*. *J. Mol. Biol.*, **108**, 453–470.

Trendelenburg, M.F. and McKinnel, R.G. (1979) Transcriptionally active and inactive regions of nucleolar chromatin in amplified nucleoli of fully grown oocytes of hibernating frogs, *Rana pipiens* (Amphibia, Anura). A quantitative electron microscopic study. *Differentiation*, **15**, 73–95.

Trendelenburg, M.F., Allen, R.D., Gundlach, H. *et al.* (1986) Recent improvements in microscopy towards analysis of transcriptionally active genes and translocation of RNP-complexes, in *Nucleocytoplasmic Transport*, (eds R. Peters and M. Trendelenburg), Springer Verlag, Heidelberg, pp. 95–112.

Trendelenburg, M.F., Meissner, B., Tröster, H., Berger, S. and Spring, H. (1988) Direct visualisation of intranuclear lampbrusch chromosome gene domains using videomicroscopy. *Cell Biol. Int. Rep.*, **12**, 737–783.

Trepte, H.H. (1981) Rate of follicle growth, duration of oogenesis stages and inter-individual synchrony of egg development in the 1st, 2nd and 3rd egg cycles of the house fly, *Musca domestica*. *Zool. Jb. Anat.*, **105**, 362–370.

Tröster, H., Spring, H., Meissner, B. *et al.* (1985) Structural organization of an active, chromosomal nucleolar organizer region (NOR) identified by light microsocopy and subsequent TEM and STEM electron microscopy. *Chromosoma (Berl.)*, **91**, 151–163.

Tröster, H., Edström, J., Trendelenburg, M.F. and Hofmann, A. (1990) Structural organization of *Acheta* rDNA. Evidence for differential amplification of soma and germ line-specific rDNA sequences. *J. Mol. Biol.*, **216**, 533–543.

Truckenbrodt, W. (1966) Ovariolenverlängerung und Eiablage bei der Termite *Kalotermes flavicollis* Fabr. *Z. Morph. Ökol. Tiere*, **57**, 274–294.

Truckenbrodt, W. (1970) Über die Oogenese der Termite *Kalotermes flavicollis* Fabr. I. Zur Feinstruktur des Terminalfilaments, Germariums und der Prophaseregion. *Z. Zellforsch.*, **108**, 339–356.

Truckenbrodt, W. and Amelung, U. (1986) Postembryonic development of the queen's ovary of *Odontotermes stercorivorus* (Isoptera, Termitidae). *Ent. Gen.*, **11**, 237–248.

Ullerich, F.H. (1981) Autonomie der Ovar-Anlagen für das Genprodukt eines prädeterminierenden Geschlechtsrealisators bei der monogenen Fliege *Chrysomya rufifacies* (Diptera, Calliphoridae). *Ent. Gen.*, **7**, 113–119.

Ullman, J.S., Lima-de-Faria, A., Jaworska, H. and Bryngelsson, T. (1973) Amplification of ribosomal DNA in *Acheta* V. Hybridization of RNA complementary to ribosomal DNA with pachythene chromosomes. *Hereditas,* **74**, 13–24.

Ullmann, S.L. (1973) Oogenesis in *Tenebrio molitor*: Histological and autoradiographical observations on pupal and adult ovaries. *J. Embryol. Exp. Morphol.*, **30**, 179–217.

Ulrich, H. (1963) Vergleichend-histologische und zyklische Untersuchungen an den weiblichen Geschlechtsorganen und den innersekretorischen Drüsen adulter Hippobosciden (Diptera Pupipara). *Dtsch. Entomol. Zeitschr.*, **10**, 28–71.

Underwood, E.M., Caulton, J.H., Allis, C.D. and Mahowald, A.P. (1980) Developmental fate of pole cells in *Drosophila melanogaster. Dev. Biol.*, **77**, 303–314.

Urbani, E. (1970) A survey of some aspects of oogenesis in *Dytiscus, Cybister* and *Hygrobia* (Col., Dytiscidae). *Acta Embryol. Exp.*, **1**, 281–297.

Valdimarsson, G. and Huebner, E. (1989) The development of microtubular arrays in the germ tissue of an insect telotrophic ovary. *Tissue Cell,* **21**, 123–138.

Vale, R. (1987) Intracellular transport using microtubule-based motors. *Ann. Rev. Cell Biol.*, **3**, 347–378.

Vallee, R.B., Wall, J.S., Paschal, B.M. and Shpetner, H.S. (1988) Microtubule-associated protein 1C from brain is a two-headed cytosolic dynein. *Nature,* **332**, 561–563.

Vanderberg, J.P. (1963) Synthesis and transfer of DNA, RNA, and protein during vitellogenesis in *Rhodnius prolixus* (Hemiptera). *Biol. Bull. Mar. Biol. Lab., Woods Hole,* **125**, 556–575.

Van der Meer, J.M. (1978) Region specific cell differentiation during early insect development. Ph. D. Thesis, Catholic University of Nijmegen, The Netherlands.

Van der Meer, J.M., Kemmner, W. and Miyamoto, D.M. (1982) Mitotic waves and embryonic pattern formation. No correlation in *Callosobruchus* (Coleoptera). *Wilhelm Roux Arch. Dev. Biol.*, **191**, 355–365.

Verachtert, B., Ameluncks, M. and De Loof, A. (1989) Potassium and chloride dependence of the membrane potential of vitellogenic follicles of *Sarcophaga bullata. J. Insect Physiol.*, **35**, 143–148.

Verhein, A. (1921) Die Eibildung der Musciden. *Zool. Jb. Anat.*, **42**, 149–212.

Verson, E. (1889) Zur Spermatogenesis. *Zool. Anz.*, **12**, 100–103.

Verson, E. (1894) Zur Spermatogenesis bei der Seidenraupe. *Z. Wiss. Zool.*, **58**, 303–313.

Von Baer, C.E. (1827) *De ovi mammalium et hominis genesi. Epist. Acad. Caesar. Petropolit.* Leibzig.

Von Baehr, W.B. (1909) Die Oogenese bei einigen viviparen Aphididen und die Spermatogenese von *Aphis saliceti*, mit besonderer Berücksichtigung der Chromatinverhältnisse. *Arch. Zellforsch.*, **3**, 269–333.

Von Kraft, A. (1960a) Entwicklungsgeschichtliche und histochemische Untersuchungen zur Oogenese von *Tachycines* (Insecta, Saltatoria). I. Mitteilung, Der weibliche Geschlechtsapparat, insbesondere die Ovarentwicklung. *Zool. Jb. Anat. Ontog. Tiere,* **78**, 457–484.

Von Kraft, A. (1960b) Entwicklungsgeschichtliche und histochemische Untersuchungen zur Oogenese von *Tachycines* (Insecta, Saltatoria). II. Mitteilung, Die Entwicklung der Oocyten. *Zool. Jb. Anat. Ontog. Tiere,* **78**, 485–558.

Wagner, R. (1836) *Prodomus Historiae Generationis,* Leibzig.

Wakimoto, B.T., Kalfayan, L.J. and Spradling, A.C. (1986) Developmentally regulated expression of *Drosophila* chorion genes introduced at diverse chromosomal positions. *J. Mol. Biol.*, **187**, 33–45.

Warn, R.M., Gutzeit, H.O., Smith L. and Warn, A. (1985) F-actin rings are associated with the ring canals of the *Drosophila* egg chamber. *Exp. Cell Res.*, **157**, 355–363.

Weber, H. (1933) *Lehrbuch der Entomologie.* Fischer Verlag, Jena.

Weber, H. (1934) Die postembryonale Entwicklung der Aleurodinen. Ein Beitrag der Metamorphosen der Insekten. *Z. Morphol. Ökol. Tiere*, **23**, 575–753.

Weber, H. (1935) Der Bau der Imago der Aleurodinen. *Zoologica*, **33** (89), 1–71.

Weber, H. and Weidner, H. (1974) *Grundriss der Insektenkunde*, 5.Aufl., Fischer Verlag, Stuttgart.

Wedlich, D., Dreyer, C. and Hausen, P. (1985) Occurrence of a species-specific nuclear antigen in the germ line of *Xenopus* and its expression from paternal genes in hybrid frogs. *Dev. Biol.*, **108**, 220–234.

Weglarska, B. (1961) Oogenesis in the ovoviviparous scale insect *Quadraspidiotus ostreaeformis* (Curt.) (Homoptera, Coccidae, Aspidiotini). Part I. Morphological and cytological investigations. *Zool. Polon.*, **11**, 267–294.

Weglarska, B. (1962) Oogenesis in the ovoviviparous scale insect *Quadraspidiotus ostreaeformis* (Curt.) (Homoptera, Coccidae, Aspidiotini). Part II. Histochemical investigations. *Zool. Polon.*, **12**, 41–68.

Weinstein, A. (1936) The theory of multiple strand crossing over. *Genetics*, **21**, 155–199.

Welch, R.C. (1993) Ovariole development in Staphylinidae (Coleoptera). *Invert. Reprod. Dev.*, **23**, 225–234.

Wellauer, P.K., Dawid, I.B., Brown, D.D. and Reeder, R.H. (1976a) The molecular basis for length heterogeneity in ribosomal DNA from *Xenopus laevis*. *J. Mol. Biol.*, **105**, 461–486.

Wellauer, P.K., Reeder, R.H., Dawid, I.B. and Brown, D.D. (1976b) The arrangement of length heterogeneity in repeating units of amplified and chromosomal ribosomal DNA from *Xenopus laevis*. *J. Mol. Biol.*, **105**, 487–505.

Weismann, A. (1885) *Die Kontinuität des Keimplasmas als Grundlage der Theorie einer Vererbung*. Jena.

Weismann, A. (1892) *Das Keimplasma-Eine Theorie der Vererbung*. Jena.

Went, D.F. (1982) Egg activation and parthenogenetic reproduction in insects. *Biol. Rev.*, **57**, 319–344.

Went, D.F., Fux, T. and Camenzind, R. (1978) Movement pattern and ultrastucture of pulsating oocyte nuclei of the paedogenetic gall midge, *Heteropeza pygmaea* Winnertz (Diptera, Cecidomyiidae). *Int. J. Insect Morphol. Embryol.*, **7**, 301–314.

Went, D.F. and Junquera, P. (1981) Embryonic development of insect eggs formed without follicular epithelium. *Dev. Biol.*, **86**, 100–110.

Werner, F.C. (1970) *Die Benennung der Organismen und Organe nach Grösse, Form, Farbe und anderen Merkmalen*, VEB Max Niemeyer, Halle (Saale).

Weyda, F. (1989) Female reproductive system of the first-instar nymphs of *Machilis helleri* (Ver.) (Thysanura, Machilidae) with special reference to the segmental arrangement of ovarioles in arthropods. *Int. J. Insect Morphol. Embryol.*, **18**, 85–96.

Wharton, R.P. and Struhl, G. (1989) Structure of the *Drosophila* bicaudal D protein and its role in localizing the posterior determinant *nanos*. *Cell*, **59**, 881–892.

Wharton, R.P. and Struhl, G. (1991) RNA regulatory elements mediate control of *Drosophila* body pattern by the posterior morphogen *nanos*. *Cell*, **67**, 955–967.

Whitehouse, H.L.K. (1970) *Towards an Understanding of the Mechanisms of Heredity*, 2nd edn, Edward Arnold, London.

Wick, J.R. and Bonhang, P.F. (1955) Postembryonic development of the ovaries of *Oncopeltus fasciatus* (Dallas). *J. Morphol.*, **96**, 31–60.

Wieschaus, E. and Szabad, J. (1979) The development and function of the germ line in *Drosophila melanogaster*: A cell lineage study. *Dev. Biol.*, **68**, 29–46.

Wigglesworth, V.B. (1973) Haemocytes and basement membrane formation in *Rhodnius*. *J. Insect Physiol.*, **19**, 831–844.

Wightman, J.A. (1973) Ovariole microstructure and vitellogenesis in *Lygocoris pabulinus* (L.) and other mirids (Hemiptera, Miridae). *J. Entomol.*, **48**, 103–115.

Wilkes, A. (1965) Sperm transfer and utilization by the arrhenotokous wasp *Dahlbominus fuscipennis* (Zett.) (Hymenoptera, Eulophidae). *Can. Ent.*, **97**, 647–657.

Willimzik, E. (1930) Über den Bau der Ovarien verschiedener coprophager Lamellicornier und ihre Beziehung zur Brutpflege. *Z. Morphol. Ökol. Tiere*, **18**, 669–700.

Willison, K. and Ashworth, A. (1987) Mammalian spermatogenic gene expression. *Trends Genet.*, **3**, 351–355.

Willmann, R. (1987) The phylogenetic system of the Mecoptera. *Syst. Entomol.*, **12**, 519–524.

Wilson, J.E. and Macdonald, P.M. (1993) Formation of germ cells in *Drosophila*. *Curr. Opinion Gen. Dev.*, **3**, 562–565.

Wimber, D.E. and Steffensen, D.M. (1970) Localization of 5S RNA genes on *Drosophila* chromosomes by RNA–DNA hybridization. *Science*, **170**, 639–641.

Wimber, D.E. and Wimber, D.R. (1977) Sites of the 5S ribosomal genes in *Drosophila*. I. The multiple clusters in the *virilis* group. *Genetics*, **86**, 133–148.

Winter, H. (1975) Charakteristika mütterlicher RNP-Partikel aus dem Ei der Baumwollwanze, *Dysdercus intermedius* Dist. (Heteroptera, Pyrrhoc.). *Verh. Dtsch. Zool. Ges.*, **1974**, 201–204.

Winter, H., Wiemann-Weiss, D. and Duspiva, F. (1977) Endogene Synthese kurzlebiger Messenger-RNS in der Oocyte von *Dysdercus intermedius* Dist. nach Abschluss der Vitellogenese. *Wilhelm Roux Arch. Dev. Biol.*, **182**, 39–58.

Wirth, U. (1984) Die Struktur der Metazoen-Spermien und ihre Bedeutung für die Phylogenetik. *Verh. Naturwiss. Ver. Hamburg*, **27**, 295–362.

Witlaczil, E. (1884) Entwicklungsgeschichte der Aphiden. *Z. Wiss. Zool.*, **40**, 559–696.

Witlaczil, E. (1885) Die Anatomie der Psylliden. *Z. Wiss. Zool.*, **42**, 569–638.

Wolf, B.E. and Sokoloff, S. (1984) The genesis of nuclear membrane balloons (NMBs) ('Accessory nuclei') in the oocyte of *Phryne fenestralis* and *Phryne cincta* (Diptera), in *Advances in Invertebrate Reproduction 3*, (eds W. Engels, W.H. Clark, A. Fischer *et al.*), Elsevier, Amsterdam, p. 652.

Wolf, R. (1978) The cytaster, a colchicine-sensitive migration organelle of cleavage nuclei in an insect egg. *Dev. Biol.*, **62**, 464–472.

Wolf, R. (1985) Migration and division of cleavage nuclei in the gall midge, *Wachtliella persicariae*. *Wilhelm Roux Arch. Dev. Biol.*, **194**, 257–270.

Wolfner, M.F. (1988) Sex-specific gene expression in somatic tissues of *Drosophila melanogaster* .*Trends Genet.*, **4**, 333–337.

Woodruff, R.I. (1979) Electrotonic junctions in *Cecropia* moth ovaries. *Dev. Biol.*, **69**, 281–295.

Woodruff, R.I. (1989) Charge-dependent molecular movement through intercellular bridges in *Drosophila* follicles. *Biol. Bull.*, **176**, 71–78.

Woodruff, R.I. and Anderson, K.L. (1984) Nutritive cord connection and dye-coupling of the follicular epithelium to the growing oocytes in the telotrophic ovarioles in *Oncopeltus fasciatus*, the milkweed bug. *Wilhelm Roux Arch. Dev. Biol.*, **193**, 158–163.

Woodruff, R.I., Kulp, J.H. and LaGaccia, E.D. (1988) Electrically mediated protein movement in *Drosophila* follicles. *Wilhelm Roux Arch. Dev. Biol.*, **197**, 231–238.

Woodruff, R.I., Münz, A. and Telfer, W.H. (1992) Steady-state potentials in ovarian follicles of a moth, *Hyalophora cecropia*. *J. Insect Physiol.*, **38**, 49–60.

Woodruff, R.I. and Telfer, W.H. (1973) Polarized intercellular bridges in ovarian cells of the *Cecropia* moth. *J. Cell Biol.*, **58**, 172–188.

Woodruff, R.I. and Telfer, W.H. (1974) Electrical properties of ovarian cells linked by intercellular bridges. *Ann. N.Y. Acad. Sci.*, **238**, 408–419.

Woodruff, R.I. and Telfer, W.H. (1980) Electrophoresis of proteins in intercellular bridges. *Nature,* **286**, 84–86.

Woodruff, R.I. and Telfer, W.H. (1990) Activation of a new physiological state at the onset of vitellogenesis in *Hyalophora* follicles. *Dev. Biol.,* **138**, 410–420.

Wülker, W. and Götz, P. (1968) Die Verwendung der Imaginalscheiben zur Bestimmung des Entwicklungszustandes von *Chironomus*-Larven (Dipt.). *Z. Morph. Tiere,* **62**, 363–388.

Wülker, W. and Winter, G. (1970) Untersuchungen über die Ultrastruktur der Gonaden von *Chironomus* (Dipt). I. Normalentwicklung der Ovarien im 4. Larvenstadium. *Z. Zellforsch.,* **106**, 348–370.

Wyatt, G.R. (1991) Gene regulation in insect reproduction. *Invertebr. Reprod. Dev.,* **20**, 1–35.

Xu, T., Caron, L.A., Fehon, R.G. and Artavanis-Tsakonas, S. (1992) The involvement of the *Notch* locus in *Drosophila* oogenesis. *Development,* **115**, 913–922.

Yamauchi, H. and Yoshitake, N. (1982) Origin and differentiation of the oocyte–nurse cell complex in the germarium of the earwig, *Anisolabis maritima* Borelli (Dermaptera, Labiduridae). *Int. J. Insect Morphol. Embryol.,* **11**, 293–305.

Yue, L. and Spradling, A.C. (1993) *hu-li tai shao* a gene required for ring canal formation during *Drosophila* oogenesis, encodes a homolog of adducin. *Genes Dev.,* **6**, 2443–2454.

Zaffagnini, F. (1969) Vitellogenesi e differenziamento ovocitario negli artropodi. *Boll. Zool.,* **36**, 263–289.

Zalokar, M. (1965) Études de la formation de l'acide ribonucleique et des proteines chez les insectes. *Rev. Suisse Zool.,* **72**, 241–261.

Zick, K. (1911) Beiträge zur Kenntnis der postembryonalen Entwicklungsgeschichte der Genitalorgane bei Lepidopteren. *Z. Wiss. Zool.,* **98**, 430–477.

Zimowska, G., Silhacek, D.L., Shaaya, E. and Shirk, P. (1991) Immuno-fluorescent analysis of follicular growth and development in whole ovaries of the indianmeal moth. *J. Morphol.,* **209**, 215–228.

Zinn, K., McAllister, L. and Goodman, C.S. (1988) Sequence analysis and neuronal expression of fasciclin I in grasshopper and *Drosophila. Cell,* **53**, 577–587.

Zinsmeister, P.P. (1973) RNA and protein synthesis in the earwig ovary. *J. Insect Physiol.,* **19**, 1765–1770.

Zinsmeister, P.P. and Davenport, R. (1971) An autoradiographic and cytochemical study of cellular interactions during oogenesis in the milkweed bug, *Oncopeltus fasciatus. Exp. Cell Res.,* **67**, 273–278.

Zissler, D. and Sander, K. (1973) The cytoplasmic architecture of the egg cell of *Smittia* spec. (Diptera, Chironomidae) I. Anterior and posterior pole regions. *Wilhelm Roux Arch. Dev. Biol.,* **172**, 175–186.

Zissler, D. and Sander, K. (1977) The cytoplasmic architecture of the egg cell of *Smittia* spec. (Diptera, Chironomidae). II. Periplasm and yolk endoplasm. *Wilhelm Roux Arch. Dev. Biol.,* **183**, 233–248.

Zissler, D. and Sander, K. (1982) The cytoplasmic architecture of the insect cell, in *Insect Ultrastructure,* (eds R.C. King and H. Akai), Plenum Press, New York, Vol. 1, pp. 189–221.

Zhu, J., Indrasith, L.S. and Yamashita, O. (1986) Characterization of vitellin, egg-specific protein, and 30 kDa protein from *Bombyx* eggs, and their fates during oogenesis and embryogenesis. *Biochim. Biophys. Acta,* **882**, 427–436.

Zou, B.X., Stoffolano, J.G., Nordin, J.H. and Yin, C.M. (1988) Subunit composition of vitellin and concentration profiles of vitellogenin and vitellin in *Phormia regina* (Meig.) following a protein meal. *Comp. Biochem. Physiol. B,* **90**, 861–867.

Species index

Subject index